Lecture Notes in Mathematics 1462

D. Mond J. Montaldi (Eds.)

Singularity Theory and its Applications

Warwick 1989, Part I: Geometric Aspects of Singularities

Springer-Verlag
Berlin Heidelberg New York
London Paris Tokyo
Hong Kong Barcelona
Budapest

Editors

David Mond
James Montaldi
Mathematics Institute, University of Warwick
Coventry CV4 7AL, United Kingdom

The figure on the front cover shows two non-isomorphic disentanglements of a projection to $\mathbb{C}^3$ of the cone over the rational normal curve of degree 4 in $\mathbb{P}^4$. For details see the paper, *Disentanglements* by T. de Jong and D. van Straten.

Mathematics Subject Classification (1980): 14B07, 32G11, 58C27, 58A35

ISBN 3-540-53737-6 Springer-Verlag Berlin Heidelberg New York
ISBN 0-387-53737-6 Springer-Verlag New York Berlin Heidelberg

2146/3140-543210 - Printed on acid-free paper

Preface

A year-long symposium on Singularity Theory and its Applications was held at the University of Warwick in the academic year 1988–89. Two workshops were held during the Symposium, the first primarily geometrical and the second concentrating on the applications of Singularity Theory to the study of bifurcations and dynamics. Accordingly, we have produced two volumes of proceedings. One of the notable features of Singularity Theory is the close development of the theory and its applications, and we tried to keep this as part of the philosophy of the Symposium. We believe that we had some success.

It should perhaps be pointed out that not all the papers included in these two volumes were presented at the workshops; these are not Proceedings of the workshops, but of the Symposium as a whole. In fact a considerable amount of the material contained in these pages was developed during the Symposium.

For the record, the Symposium was organized by the four editors of the two volumes: David Mond, James Montaldi, Mark Roberts and Ian Stewart. There were over 100 visitors and 120 seminars. The Symposium was funded by the S.E.R.C., and could not have been such a success without the hard work of Elaine Shiels, to whom we are all very grateful.

Every paper published here is in final form and has been refereed.

David Mond
James Montaldi

University of Warwick,
August 1990

Contents

Symmetric Lagrangian singularities and Gauss maps of theta divisors

Malcolm R. Adams, Clint McCrory, Theodore Shifrin and Robert Varley

Recently three of the authors studied the Thom–Boardman singularities and the local $\mathbb{Z}/2$-stability of the Gauss map of the theta divisor of a smooth algebraic curve of genus three [12]. In this paper we develop a theory of $\mathbb{Z}/2$-symmetric Lagrangian maps appropriate to the study of theta divisors of curves of arbitrary genus, and we apply this theory to the genus three case, obtaining Lagrangian analogues of the results of [12]. We find that the local classification of $\mathbb{Z}/2$-Lagrangian-stable Gauss maps coincides with our previous local classification of $\mathbb{Z}/2$-stable Gauss maps in genus three. The corresponding classifications in higher genus are expected to diverge, as in the nonsymmetric case (cf. [3]).

Let C be a curve of genus g, $J(C)$ its Jacobian variety, and $\Theta \subset J(C)$ the theta divisor. Torelli's theorem states that the curve C is determined by the pair $(J(C), \Theta)$. More precisely, C is determined by the Gauss map γ which assigns to a point of Θ its tangent hyperplane, translated to the origin of $J(C)$. Andreotti proved that the branch locus of γ is the dual hypersurface of the canonical embedding of C, provided C is nonhyperelliptic. Thus the singularities of the Gauss map are directly related to the extrinsic geometry of the canonical embedding, and hence to the intrinsic geometry of the curve C.

Locally Θ can be given as the graph of a function $f\colon \mathbb{C}^{g-1} \to \mathbb{C}$, and the Gauss map $\gamma\colon \Theta \to \mathbb{P}^{g-1*}$ is given locally as the gradient of f. Since the gradient df has a canonical Lagrangian structure, namely the factorization through the Lagrangian submanifold $\mathrm{graph}(df) \subset T^*\mathbb{C}^{g-1}$, γ is locally Lagrangian. However, this local Lagrangian structure depends on the choice of local coordinates used to define f; moreover, the $\mathbb{Z}/2$-symmetry of $T^*\mathbb{C}^{g-1}$ is antisymplectic. To obtain a global symmetric Lagrangian structure, we consider the conormal bundle $\mathcal{C}_\Theta \subset T^*J(C)$ of the theta divisor. The Gauss map γ lifts to the *homogeneous Gauss map* Γ, which is the restriction of the projection to the fiber:

$$\begin{array}{ccc} \mathcal{C}_\Theta & \xrightarrow{\Gamma} & T_0^*J(C) \\ \downarrow & & \downarrow \\ \Theta & \xrightarrow{\gamma} & \mathbb{P}(T_0^*J(C)) \end{array}$$

If we remove the zero-section of $T^*J(C)$, Γ is a *conic Lagrangian map*, with $\mathbb{Z}/2$-symmetry induced by the (-1)-map of $J(C)$. (If Θ is singular, then Γ is defined over the Nash blowup of the Gauss map.) The homogeneous Gauss map is defined in the same way for a complex affine hypersurface $M \subset \mathbb{C}^n$. In contrast to the real case, the Gauss map of a complex affine hypersurface does not seem to have a natural global Lagrangian structure.

For conic Lagrangian submanifolds $\Lambda \subset T^*X$, projection to the *base* X has been studied by several authors (cf. [8], [15]), but projection to the *fiber* has not been previously considered. For example, in the work of Janeczko and Kowalczyk [9] [10], the symmetry of T^*X is also induced from a symmetry of the base X, but Λ is projected to the base.

The homogeneous Gauss map $\Gamma: C_\Theta \to T^*J(C) \to T_0^*J(C)$ is equivariant with respect to commuting actions of $\mathbb{C}^*$ and $\mathbb{Z}/2$, and $\mathbb{Z}/2$ acts trivially on the $\mathbb{C}^*$-orbit space of the target. If $\ell: \mathcal{M} \to \mathcal{E} \to \mathcal{B}$ is any such conic $\mathbb{Z}/2$-Lagrangian map-germ, there are two cases to consider: $\mathbb{Z}/2$ acts either trivially or non-trivially on the $\mathbb{C}^*$-orbit space of the total space $\mathcal{E}$. In the latter case we say that ℓ is *odd*. In both cases we prove that stable conic $\mathbb{Z}/2$-Lagrangian germs are classified by versal *generating families* of functions; if ℓ is odd, its generating family is a family of odd functions. Our method of proof is to pass from a conic Lagrangian map to a Legendrian map by *dehomogenization*, and then to use the work of Zakalyukin on the classification of Legendrian maps (cf. [16], [3]). (The Gauss map of a complex projective hypersurface $M \subset \mathbb{P}^n$ has a natural Legendrian structure; the case $n = 4$ is studied in [11].) Our main result for the homogeneous Gauss map Γ of a theta divisor is the following.

Theorem. For a nonhyperelliptic curve of genus three, Γ is a locally infinitesimally stable conic $\mathbb{Z}/2$-Lagrangian map if and only if the canonical curve $C \subset \mathbb{P}^2$ has no higher flexes. For a hyperelliptic curve of genus three, Γ is a locally infinitesimally stable conic $\mathbb{Z}/2$-Lagrangian map.

In section 1 we show that dehomogenization gives a bijection from isomorphism classes of conic G-Lagrangian germs to isomorphism classes of G-Legendrian germs (1.5), and we construct symmetric Darboux coordinates for an odd conic Lagrangian fibration germ (1.12). In the second section, we show that isomorphism classes of odd conic Lagrangian germs are in one-to-one correspondence with stable isomorphism classes of generating families (2.2), and that infinitesimally stable germs correspond to versal families (2.11). We then derive normal forms for odd versal families with two parameters (2.16). Section three contains the application of our results to genus 3 theta divisors. As in [12], we use the extrinsic geometry of the canonical curve $C \subset \mathbb{P}^2$ to describe the singularities of the homogeneous Gauss map Γ.

All manifolds, maps and group actions are assumed to be complex analytic. (The results in sections 1 and 2 are also valid in the real C^∞ category, with $\mathbb{C}^*$ replaced by $\mathbb{R}^+$, the multiplicative group of positive real numbers.)

The last author expresses his gratitude to the NSF for support under grant #DMS-8803487.

McCrory and Varley wish to thank both the Mathematics Institute of the University of Warwick for its hospitality and support, and also the paticipants in the singularity theory workshops at Warwick for many stimulating conversations. We are especially grateful to J. Montaldi, C. T. C. Wall, and V. M. Zakalyukin for their interest in our work.

1. Conic G-Lagrangian maps

Let G be a finite group. A *conic* G-*manifold* is a manifold together with a proper, free $\mathbb{C}^*$ action and a G action such that the two actions commute. It follows that the orbit map of the $\mathbb{C}^*$ action is a G-equivariant principal $\mathbb{C}^*$-bundle. A *conic* G-*map* is a map of conic G-manifolds which is equivariant with respect to the actions of $\mathbb{C}^*$ and G. A *conic symplectic* G-*manifold* is a conic G-manifold with a (holomorphic) symplectic structure such that the symplectic form Ω is homogeneous and G-invariant. In other words, if κ_t is the action of $t \in \mathbb{C}^*$ then $(\kappa_t)^*\Omega = t\Omega$, and if ν_g is the action of $g \in G$ then $(\nu_g)^*\Omega = \Omega$.

A *conic* G-*Lagrangian map* is a conic G-map $\mathcal{L}: \mathcal{M} \to \mathcal{B}$ of manifolds, together with a factorization $\mathcal{L} = \pi \circ i$,

$$\begin{array}{ccc} & & \mathcal{E} \\ & {}^{i}\nearrow & \downarrow \pi \\ \mathcal{M} & \xrightarrow[\mathcal{L}]{} & \mathcal{B} \end{array}$$

where $\mathcal{E}$ is a conic symplectic G-manifold, $i: \mathcal{M} \to \mathcal{E}$ is a Lagrangian conic G-immersion, and $\pi: \mathcal{E} \to \mathcal{B}$ is a Lagrangian conic G-fibration (cf. [3, ch. 18]). If $\mathcal{E}$ has dimension 2n, then $\mathcal{M}$ and the fibers of π have dimension n.

We will be primarily concerned with two special types of conic G-Lagrangian maps $\mathcal{L}$: either G is the trivial group, or $G = \mathbb{Z}/2$ and the symmetry of $\mathcal{L}$ is odd. An *odd conic Lagrangian fibration* is a conic Lagrangian $\mathbb{Z}/2$-fibration $\mathcal{E} \to \mathcal{B}$ such that

(1) $\mathbb{Z}/2$ acts nontrivially on the $\mathbb{C}^*$-orbit space of $\mathcal{E}$, and

(2) $\mathbb{Z}/2$ acts trivially on the $\mathbb{C}^*$-orbit space of $\mathcal{B}$.

An *odd conic Lagrangian map* is a conic $\mathbb{Z}/2$-Lagrangian map $\mathcal{L} = \pi \circ i: \mathcal{M} \to \mathcal{E} \to \mathcal{B}$ such that $\pi: \mathcal{E} \to \mathcal{B}$ is an odd conic Lagrangian fibration.

(1.1) Example. Let $\iota: \mathbb{C}^n \to \mathbb{C}^n$ be the involution $\iota(x) = -x$, and let M be a smooth hypersurface in $\mathbb{C}^n$ such that $\iota(M) = M$. Let $\mathcal{E} = (T^*\mathbb{C}^n) - \mathcal{Z}$, the cotangent bundle of $\mathbb{C}^n$ minus the zero-section, with standard Darboux coordinates $(x,\xi) = (x_1,\ldots,x_n,\xi_1,\ldots,\xi_n)$ and involution $\tilde{\iota}(x,\xi) = (-x,-\xi)$, and let $i: \mathcal{M} \subset \mathcal{E}$ be the conormal variety of M,

$$\mathcal{M} = \{(x,a) \mid x \in M,\ a \in T_x^*\mathbb{C}^n - \{0\},\ a(T_xM) = 0\}.$$

Let $\pi: \mathcal{E} \to \mathcal{B}$ be the projection to the *fiber* of the cotangent bundle, $\pi: (T^*\mathbb{C}^n) - \mathcal{Z} \to (T_0^*\mathbb{C}^n) - \{0\}$, $\pi(x,\xi) = \xi$. Then the *homogeneous Gauss map* $\Gamma = \pi \circ i: \mathcal{M} \to \mathcal{E} \to \mathcal{B}$ is an odd conic Lagrangian map. The map on $\mathbb{C}^*$-orbits induced by Γ is the *Gauss map*

$$\gamma: M \to \mathbb{P}T_0^*\mathbb{C}^n = G(n-1,\mathbb{C}^n),$$

which is $\mathbb{Z}/2$-invariant (cf. [13, p. 720]).

An *isomorphism* of conic G-Lagrangian maps $\mathcal{L}_1 = \pi_1 \circ i_1$ and $\mathcal{L}_2 = \pi_2 \circ i_2$ is a commutative diagram

$$\begin{array}{ccccc} \mathcal{M}_1 & \to & \mathcal{E}_1 & \to & \mathcal{B}_1 \\ \downarrow & & \downarrow & & \downarrow \\ \mathcal{M}_2 & \to & \mathcal{E}_2 & \to & \mathcal{B}_2 \end{array}$$

where the vertical maps are conic G-isomorphisms, and the isomorphism $\mathcal{E}_1 \to \mathcal{E}_2$ is symplectic.

Given a conic G-manifold $\mathcal{M}$ of dimension n and a conic Lagrangian fibration $\pi: \mathcal{E} \to \mathcal{B}$ with $\mathcal{E}$ of dimension 2n, a topology on the set of conic G-Lagrangian maps $\mathcal{L} = \pi \circ i: \mathcal{M} \to \mathcal{E} \to \mathcal{B}$ is induced from the topology of uniform convergence on compact subsets (or the Whitney topology in the real C^∞ category) on the set of conic G-Lagrangian immersions $i: \mathcal{M} \to \mathcal{E}$.

Let $\mathcal{M}$ be a conic G-manifold, and let $\mathcal{O}$ be a $\mathbb{C}^*$-orbit of $\mathcal{M}$ such that $G\mathcal{O} = \mathcal{O}$. A *conic G-Lagrangian map-germ* $\ell: \mathcal{M} \to \mathcal{E} \to \mathcal{B}$ at $\mathcal{O}$ is an equivalence class of conic G-Lagrangian maps $\ell = \pi \circ i: U \to V \to \mathcal{B}$, such that $U \subset \mathcal{M}$ and $V \subset \mathcal{E}$ are open sets invariant under the actions of $\mathbb{C}^*$ and G, with $\mathcal{O} \subset U$ and $i(\mathcal{O}) \subset V$. Two such maps $\ell_1 = \pi_1 \circ i_1$ and $\ell_2 = \pi_2 \circ i_2$ are equivalent if i_1 and i_2 agree on a neighborhood of $\mathcal{O}$ and π_1 and π_2 agree on a neighborhood of $i_1(\mathcal{O}) = i_2(\mathcal{O})$. Isomorphism of map-germs is defined in the same way as for maps. Given $\mathcal{M}$ and $\mathcal{O}$ as above, and $\pi: \mathcal{E} \to \mathcal{B}$ a conic G-Lagrangian fibration, let $\mathcal{L}(\mathcal{M},\mathcal{O},\pi)$ be the space of conic G-Lagrangian germs $\ell = \pi \circ i: \mathcal{M} \to \mathcal{E} \to \mathcal{B}$ at $\mathcal{O}$.

A conic G-Lagrangian germ $\ell: \mathcal{M} \to \mathcal{E} \to \mathcal{B}$ at $\mathcal{O} \subset \mathcal{M}$ is *stable* if it has a representative $\pi \circ i: U \to V \to \mathcal{B}$ with the following property. For every open set $U' \subset U$ such that $\mathcal{O} \subset U'$ and U' is $\mathbb{C}^*$- and G-invariant, there is a neighborhood N of $i|_{U'}$ in the space of conic G-Lagrangian immersions $U' \to V$ such that if $j \in N$ there exists an orbit $\mathcal{O}' \subset U'$ with $G\mathcal{O}' = \mathcal{O}'$ and ℓ isomorphic to the germ of $\pi \circ j$ at $\mathcal{O}'$.

We will be primarily interested in infinitesimal stability of Lagrangian germs (cf. [5, p. 271]).

Let $i: \mathcal{M} \to \mathcal{E}$ be a conic G-Lagrangian immersion. An *infinitesimal* (first order) *Lagrangian deformation* of i is a $\mathbb{C}^*$- and G-equivariant section u of $i^*T\mathcal{E}$ such that $i^*\mathcal{L}_u\Omega = 0$, i.e., the pullback to $\mathcal{M}$ of the Lie derivative of the symplectic form Ω with respect to (an extension of) u is zero. An *infinitesimal isomorphism* of $\mathcal{M}$ is a $\mathbb{C}^*$- and G-equivariant vector field v on $\mathcal{M}$. An *infinitesimal symplectomorphism* of $\mathcal{E}$ is a $\mathbb{C}^*$- and G-equivariant vector field w on $\mathcal{E}$ such that $\mathcal{L}_w\Omega = 0$. If $\pi: \mathcal{E} \to \mathcal{B}$ is a conic G-Lagrangian fibration, the vector field w on $\mathcal{E}$ is (infinitesimally) *fiber-permuting* if $\pi_*v = 0$ implies $\pi_*[w,v] = 0$. A conic G-Lagrangian germ $\mathcal{L}: \mathcal{M} \to \mathcal{E} \to \mathcal{B}$ at $\mathcal{O} \subset \mathcal{M}$ is *infinitesimally stable* if, for every representative $\pi \circ i: U \to V \to \mathcal{B}$ and every infinitesimal Lagrangian deformation u of i, there exist a $\mathbb{C}^*$- and G-invariant open set $U' \subset U$, an infinitesimal isomorphism v of U' and an infinitesimal fiber-permuting symplectomorphism w of a neighborhood of i(U') in V such that $u = i_*v + w$.

Now we reformulate infinitesimal Lagrangian stability of the germ $\mathcal{L}: \mathcal{M} \to \mathcal{E} \to \mathcal{B}$ in terms of Hamiltonian functions, using the method of Arnold [1, §10]. Given an infinitesimal Lagrangian deformation u of i as above, we define a 1-form τ locally on $\mathcal{M}$ as follows. Let $\tilde{u}$ be a local extension of u to a $\mathbb{C}^*$- and G-equivariant vector field on $\mathcal{E}$. Define $\tilde{\tau}$ on $\mathcal{E}$ by $\tilde{\tau} = \Omega(\tilde{u},\cdot)$, and let $\tau = i^*\tilde{\tau}$. The form τ is independent of the choice of extension $\tilde{u}$, and τ is G-invariant and homogeneous, i.e., $(\kappa_t)^*\tau = t\tau$. The deformation u is Lagrangian if and only if $d\tau = 0$. Two infinitesimal deformations u and u' determine the same 1-form τ if and only if $u - u'$ is tangent to $\mathcal{M}$. We conclude that, up to infinitesimal isomorphisms of $\mathcal{M}$, an infinitesimal Lagrangian deformation of i is the same as a G-equivariant homogeneous closed 1-form τ on $\mathcal{M}$ (cf. [5]).

Let w be an infinitesimal symplectomorphism of $\mathcal{E}$, and let H be a local Hamiltonian function for w, i.e., $H: \mathcal{E} \to \mathbb{C}$ and $dH = \Omega(w,\cdot)$. Then dH is G-invariant and H can be chosen so that it is homogeneous, i.e., $H(\kappa_t e) = tH(e)$. (In fact $H = \Omega(w, t)$ is a homogeneous Hamiltonian for w, where t is the infinitesimal generator of the $\mathbb{C}^*$-action.) The vector field w is fiber-permuting if and only if H is linear with respect to the canonical affine linear structure on each fiber of the Lagrangian fibration $\mathcal{E} \to \mathcal{B}$.

(1.2) <u>Proposition</u>. The conic G-Lagrangian germ $\mathcal{L} = \pi \circ i: \mathcal{M} \to \mathcal{E} \to \mathcal{B}$ is infinitesimally stable if and only if every homogeneous germ $\Phi: \mathcal{M} \to \mathbb{C}$ such that $d\Phi$ is G-invariant can be written as $\Phi = H \circ i$ for some homogeneous germ $H: \mathcal{E} \to \mathbb{C}$ such that dH is G-invariant and H is affine linear on the fibers of $\pi: \mathcal{E} \to \mathcal{B}$.

<u>Proof</u>. Such a germ Φ corresponds to a closed 1-form $\tau = d\Phi$. If u is an infinitesimal deformation of i corresponding to τ, and w is the Hamiltonian vector field of H, then $u = w$ (modulo infinitesimal isomorphisms of $\mathcal{M}$) if and only if $\Phi = H \circ i$. □

Let $\mathcal{L}: \mathcal{M} \to \mathcal{E} \to \mathcal{B}$ be a conic G-Lagrangian map, let $\mathcal{O}$ be a $\mathbb{C}^*$-orbit of $\mathcal{M}$, and let

$G_{\mathcal{O}} = \{g \in G : g\mathcal{O} = \mathcal{O}\}$. Then the germ of $\mathcal{L}$ at $\mathcal{O}$ is a conic $G_{\mathcal{O}}$-Lagrangian germ. We will use the following abbreviated terminology. We say that the conic G-Lagrangian map $\mathcal{L}$ is *locally infinitesimally stable* if the germ of $\mathcal{L}$ at $\mathcal{O}$ is an infinitesimally stable conic $G_{\mathcal{O}}$-Lagrangian germ for all $\mathbb{C}^*$-orbits $\mathcal{O}$ of $\mathcal{M}$.

A G-*Legendrian map* is a G-map $L: M \to B$ of manifolds, together with a factorization $L = \pi \circ i: M \to E \to B$, where E is a contact G-manifold (the action of G on E preserves the contact structure), $i: M \to E$ is a Legendrian G-immersion, and $\pi: E \to B$ is a Legendrian G-fibration (cf. [3, ch. 20]). If E has dimension $2n-1$, then M and the fibers of π have dimension $n-1$. G-Legendrian germs and isomorphism of G-Legendrian maps and germs are defined in the same way as for conic G-Lagrangian maps. Given a G-manifold M, a fixed point x of G, and a G-Legendrian fibration $\pi: E \to B$, let $\mathbf{L}(M,x,\pi)$ be the space of G-Legendrian germs $\ell: M \to E \to B$ at x.

Stability and infinitesimal stability of a G-Legendrian germ are defined in the same way as for conic G-Legendrian germs, replacing the symplectic form Ω with a contact form α. A G-Legendrian germ $\ell: M \to E \to B$ at $x \in M$ is *infinitesimally stable* if, for every representative $\pi \circ i: U \to V \to B$ and every infinitesimal Legendrian deformation u of i, there exist a G-invariant open set $U' \subset U$, an infinitesimal isomorphism v of U' and an infinitesimal fiber-permuting contact transformation w of a neighborhood of $i(U')$ in V such that $u = i_* v + w$.

(1.3) Example. Let $M \subset \mathbb{C}^n$ be as in example (1.1), let $E = \mathbb{P}T^*\mathbb{C}^n = G(n-1, T\mathbb{C}^n)$, with involution induced by the map $x \mapsto -x$ on $\mathbb{C}^n$, and let $i': M \to E$ be defined by $i'(x) = T_x M$. Let B be the total space of the tautological quotient line bundle on $T\mathfrak{J}\mathbb{C}^n = G(n-1, \mathbb{C}^n)$, and define $\pi': E \to B$ as follows. If $x \in \mathbb{C}^n$ and H is a hyperplane of $T_x\mathbb{C}^n$, then $\pi'(H,x)$ is the coset of x in $\mathbb{C}^n/H_0$, where H_0 is the translate of H to 0. Then $\Gamma' = \pi' \circ i': M \to E \to B$ is a $\mathbb{Z}/2$-Legendrian map. The composition of Γ' with the projection $B \to G(n-1, \mathbb{C}^n)$ is the Gauss map γ of M.

Let $\pi: \mathcal{E} \to \mathcal{B}$ be a conic G-Lagrangian fibration, with $\mathcal{E}$ of dimension 2n, and let $\mathcal{O} \subset \mathcal{E}$ be a $\mathbb{C}^*$ orbit with $G\mathcal{O} = \mathcal{O}$. The *dehomogenization* of π at $\mathcal{O}$ is a G-Legendrian fibration germ $\pi': E \to B$ which we now proceed to define. Let E be the orbit space of the $\mathbb{C}^*$ action, and let $e \in E$ correspond to $\mathcal{O}$. There is a unique 1-form β on $\mathcal{E}$ such that,

(1) $d\beta = \Omega$,

(2) β is homogeneous ($(\kappa_t)^*\beta = t\beta$), and

(3) $\beta(v) = 0$ for all vectors v tangent to an orbit of $\mathbb{C}^*$.

Furthermore β is G-invariant. (In fact, $\beta(w) = \Omega(t,w)$, where t is the infinitesimal generator of the $\mathbb{C}^*$ action.) Since the field of hyperplanes $\mathbf{B} = \{\beta = 0\}$ contains the tangent spaces to the orbits of $\mathbb{C}^*$, $\mathbf{B}$ projects to a field of hyperplanes A on the orbit space E. This field of hyperplanes A defines a G-invariant contact structure on E.

The Legendre fibration germ $\pi'\colon E \to B$ is defined as follows. Let $B^{\#}$ be the $\mathbb{C}^*$-orbit space of $\mathcal{B}$. Note that the quotient fibration $\sigma\colon E \to B^{\#}$ is not Legendrian, since the dimension of the fibers of σ is n, not n−1. But if F is a fiber of σ, the field of hyperplanes **A** is transverse to F, and the intersection of **A** with the tangent bundle of F defines an integrable field A_F of hyperplanes on F. (The quotient map $\mathcal{E} \to E$ takes each fibre $\mathcal{F}$ of the Lagrangian fibration π isomorphically onto a fibre F of σ, and the field of hyperplanes A_F corresponds to $\mathbf{B}_{\mathcal{F}}$, the intersection of **B** with the tangent bundle of $\mathcal{F}$. The distribution $\mathbf{B}_{\mathcal{F}}$, satisfies the integrability condition $d\beta \wedge \beta = 0$, since $d\beta = \Omega$ is zero on $\mathcal{F}$.) We define a Legendrian foliation of E with leaves contained in the fibers of σ: the leaves contained in F are the integral manifolds of A_F. On a sufficiently small G-invariant neighborhood U of e in E, these leaves are the fibers of a map $\pi'\colon U \to B$ which represents the desired G-Legendrian fibration germ at e.

Let $\mathcal{M}$ be a conic G-manifold, let $\mathcal{O}$ be a $\mathbb{C}^*$-orbit of $\mathcal{M}$ such that $G\mathcal{O} = \mathcal{O}$, and let $\pi\colon \mathcal{E} \to \mathcal{B}$ be a conic G-Lagrangian fibration with dehomogenization $\pi'\colon E \to B$. Let M be the $\mathbb{C}^*$-orbit space of $\mathcal{M}$, and let $x \in M$ correspond to the orbit $\mathcal{O}$. We define a function

$$T\colon \mathcal{L}(\mathcal{M},\mathcal{O},\pi) \to L(M,x,\pi')$$

as follows. Given a conic G-Lagrangian germ $\ell = \pi \circ i\colon \mathcal{M} \to \mathcal{E} \to \mathcal{B}$, let $T(\ell) = \pi' \circ i'\colon M \to E \to B$, where i′ is the map on $\mathbb{C}^*$-orbits induced by i. We will call $T(\ell)$ the *dehomogenization* of ℓ.

(1.4) <u>Example.</u> Let M be a smooth hypersurface through the origin in $\mathbb{C}^n$, and let $\mathcal{O} = \{a \in T_0^*\mathbb{C}^n - \{0\} \mid a(T_0M) = 0\}$. Let $\Gamma\colon \mathcal{M} \to \mathcal{E} \to \mathcal{B}$ be the conic $\mathbb{Z}/2$-Lagrangian map of example (1.1). The dehomogenization of the germ at $\mathcal{O}$ of Γ is the germ at 0 of the $\mathbb{Z}/2$-Legendrian map Γ' of example (1.3).

(1.5) <u>Proposition</u>. Dehomogenization $T\colon \mathcal{L}(\mathcal{M},\mathcal{O},\pi) \to L(M,x,\pi')$ induces a bijection of isomorphism classes. A conic G-Lagrangian germ is stable (resp. infinitesimally stable) if and only if its dehomogenization is stable (resp. infinitesimally stable).

(1.6) <u>Remark</u>. Presumably stability is equivalent to infinitesimal stability for conic G-Lagrangian maps and for G-Legendrian maps. We have not checked these assertions.

The proof of the proposition relies on a homogenization function. Let $\pi\colon E \to B$ be a G-Legendrian fibration, let e be a point of E, and let α be a 1-form on a neighborhood of e which is a contact form on each tangent space. The α-*homogenization* of π is the conic G-Lagrangian fibration germ $\tilde{\pi}\colon \mathcal{E} \to \mathcal{B}$ defined as follows. Let $\mathcal{E}$ be the symplectization of E, i.e., the set of all contact forms on the contact manifold germ E [2, p.356], with $\mathbb{C}^*$ action given by $\kappa_t(a) = ta$ and G action given by $\nu_g(a) = (\nu_{g^{-1}})^* a$. Let $\mathcal{O} \subset \mathcal{E}$ be the set of contact forms on T_eE. The symplectic form Ω on $\mathcal{E}$ is homogeneous and G-invariant. (Recall that $\Omega = d\beta$, where β is the tautological 1-form on $\mathcal{E}$.) The composition of the given G-Legendrian fibration $\pi: E \to B$ with the projection $\rho: \mathcal{E} \to E$ is a Lagrangian fibration $\pi^{\#}: \mathcal{E} \to B$; if F is a fiber of π then $\rho^{-1}(F)$ is a fiber of $\pi^{\#}$. But $\mathbb{C}^*$ acts trivially on B, and we want $\mathbb{C}^*$ to act freely on $\mathcal{B}$, so we must define the Lagrangian fibration $\tilde{\pi}: \mathcal{E} \to \mathcal{B}$ differently.

The choice of 1-form α on E defining the given contact structure determines an isomorphism $E \times \mathbb{C}^* \to \mathcal{E}$ which sends the pair (x,t) to the contact form $t\alpha_x$ on the tangent space to E at x. Thus we obtain a local coordinate function $t: \mathcal{E} \to \mathbb{C}^*$. Let X_t be the Hamiltonian vector field associated to the function t; that is, $dt(\xi) = -\Omega(X_t, \xi)$ [3, §18.2]. Using X_t we define the fibers of the Lagrangian fibration $\tilde{\pi}: \mathcal{E} \to \mathcal{B}$ as follows. Let $C_x \subset T_x\mathcal{E}$ be the span of the vector field X_t and the (n-1)-plane $(\ker(\pi^{\#})_*) \cap (\ker dt)$. Then C is a Lagrangian, and hence integrable, field of n-planes on $\mathcal{E}$, invariant under $\mathbb{C}^*$ and G. Thus there is a $\mathbb{C}^*$- and G-invariant neighborhood $\mathcal{V}$ of $\mathcal{O}$ such that on $\mathcal{V}$ the integral manifolds of C are the fibers of a map $\tilde{\pi}: \mathcal{V} \to \mathcal{B}$ which represents the desired conic G-Lagrangian fibration germ.

Let M be a G-manifold, let x be a fixed point of G, and let $\pi: E \to B$ be a G-Legendrian fibration with α-homogenization $\tilde{\pi}: \mathcal{E} \to \mathcal{B}$. We define a function

$$S_\alpha: L(M,x,\pi) \to \mathcal{L}(M \times \mathbb{C}^*, \{x\} \times \mathbb{C}^*, \tilde{\pi})$$

as follows. Given a G-Legendrian germ $\ell = \pi \circ i: M \to E \to B$, let $\alpha_{i(x)}$ be the restriction of α to the tangent space of E at i(x). Define $\tilde{i}: M \times \mathbb{C}^* \to \mathcal{E}$ by $\tilde{i}(x,t) = t\alpha_{i(x)}$, and let $S_\alpha(\ell) = \tilde{\pi} \circ \tilde{i}: M \times \mathbb{C}^* \to \mathcal{E} \to \mathcal{B}$. We call $S_\alpha(\ell)$ the *α-homogenization* of ℓ.

Proof of (1.5). The proposition is a consequence of the following properties of the functions T and S_α, the proofs of which are straightforward:

$$T: \mathcal{L}(\mathcal{M}, \mathcal{O}, \pi) \to L(M,x,\pi')$$
$$S_\alpha: L(M,x,\pi') \to \mathcal{L}(M \times \mathbb{C}^*, \{x\} \times \mathbb{C}^*, \widetilde{\pi'}),$$

(1.7)

(a) $\mathcal{l}_1 \cong \mathcal{l}_2 \Rightarrow T(\mathcal{l}_1) \cong T(\mathcal{l}_2)$.
(b) $\ell_1 \cong \ell_2 \Rightarrow S_\alpha(\ell_1) \cong S_\alpha(\ell_2)$.
(c) $TS_\alpha(\ell) = \ell$.
(d) $S_\alpha T(\mathcal{l}) \cong \mathcal{l}$
(e) T and S_α are continuous.

To prove the stability part of Proposition (1.5), one actually uses (1.7)(c) and (d) for germ representatives. If V is a G-invariant open subset of E such that α is defined on E and the corresponding Lagrangian foliation on the preimage $\mathcal{V}$ of V is a fibration, then (c) holds for any representative $\pi\circ i\colon U\to E\to B$ such that $i(U)\subset V$. Similarly, if $\mathcal{V}$ is a $\mathbb{C}^*$- and G-invariant open subset of $\mathcal{E}$ such that the Legendrian foliation defined on its image V in E is a fibration, then (d) holds for any representative $\pi\circ i\colon U\to\mathcal{E}\to\mathcal{B}$ such that $i(U)\subset\mathcal{V}$. The proof of the infinitesimal stability part of (1.5) is easy. □

Proposition (1.5) reduces the classification of stable conic G-Lagrangian germs to the classification of stable G-Legendrian germs. To carry out this classification for odd conic Lagrangian germs in section 2, it will be convenient to have a Darboux coordinate description of a conic Lagrangian fibration germ.

(1.8) Proposition. Let $\pi\colon\mathcal{E}\to\mathcal{B}$ be a conic Lagrangian fibration germ at $\mathcal{O}\subset\mathcal{E}$. Let Ω be the symplectic form of $\mathcal{E}$. There are local coordinates (p,y,q,t) on $\mathcal{E}$ near $\mathcal{O}$ such that $\Omega = dp\wedge d(-tq)+dy\wedge dt$, the action of $\mathbb{C}^*$ is $s\cdot(p,y,q,t) = (p,y,q,st)$, and $\pi(p,y,q,t) = (q,t)$.

Proof. Let $\pi'\colon E^{2n-1}\to B^n$ be the dehomogenization of π at $\mathcal{O}$. Given a 1-form α defining the contact structure near $\pi'(\mathcal{O})$ on E, there are local coordinates $(p,q,z) = (p_1,\dots,p_{n-1},q_1,\dots,q_{n-1},z)$ on E such that $\pi(p,q,z) = (q,z)$, and $\alpha = dz-pdq$ [3, §20]. In these coordinates, the α-homogenization of π' has the following form. The symplectization $\mathcal{E}$ of the contact manifold E has coordinates (p,q,z,t), with $\Omega = d(t\alpha) = dt\wedge dz + d(-tp)\wedge(dq)$, so that $(-tp,t,q,z)$ are Darboux coordinates for $\mathcal{E}$. The action of $\mathbb{C}^*$ is $s\cdot(-tp,t,q,z) = (-stp,st,q,z)$. The composition of π' with the projection $\rho\colon\mathcal{E}\to E$ is the Lagrangian fibration $\pi^{\#}\colon\mathcal{E}\to B$ given by $\pi^{\#}(-tp,t,q,z) = (q,z)$. Thus $(\ker(\pi^{\#})_*)\cap(\ker dt)$ is parallel to the p-coordinate plane. To give coordinates for the Lagrangian fibration $\pi\colon\mathcal{E}\to\mathcal{B}$, we let $y = pq-z$, the Legendre transform of z. Then $\alpha = -(dy-qdp)$, and $\Omega = d(t\alpha) = dy\wedge dt + dp\wedge d(-tq)$, so that $(p,y,-tq,t)$ are Darboux coordinates on $\mathcal{E}$, and $X_t = -\partial/\partial y$. Therefore the distribution C is parallel to the (p,y)-coordinate plane, and the homogenization $\pi\colon\mathcal{E}\to\mathcal{B}$ of $\pi\colon E\to B$ is given by $\pi(p,y,-tq,t) = (-tq,t)$. □

Before adding $\mathbb{Z}/2$ symmetry to the picture, some discussion of the Darboux coordinates $(p,y,-tq,t)$ is in order. To see why the *Legendre transform* enters into homogenization, consider the reverse process of dehomogenization. If the conic Lagrangian fibration $\mathcal{E}\to\mathcal{B}$ is given in coordinates by $(p,y,-tq,t)\mapsto(-tq,t)$, with $s\cdot(p,y,-tq,t) = (p,y,-stq,st)$, then the quotient fibration is $(p,q,y)\mapsto q$. The restriction of the contact form $\alpha = -(dy-qdp)$ to a fiber $F = \{q = q_0\}$ is exact: $\alpha_F = dz$, $z = pq_0-y$. Thus the dehomogenization of $\mathcal{E}\to\mathcal{B}$ is given by $(p,q,z)\mapsto(q,z)$, with $\alpha = dz-pdq$.

It is instructive to work out examples (1.1) and (1.3) using the coordinates $(p,y,-tq,t)\to(-tq,t)$ for the conic Lagrangian bundle $(T^*\mathbb{C}^n)-\mathcal{Z}\to(T_0^*\mathbb{C}^n)-\{0\}$. To avoid confusing p and q, let $(p,q) = (x,\xi)$. Thus $(x_1,\dots,x_{n-1},y)$ are the standard coordinates on $\mathbb{C}^n$, and $(\xi_1,\dots,\xi_{n-1})$ are the coordinates on $T^*\mathbb{C}^{n-1}$ dual to $(x_1,\dots,x_{n-1})$. Suppose that $M\subset\mathbb{C}^n$ is the graph of the function $y = f(x)$. Using $(x_1,\dots,x_{n-1})$ as coordinates on M, the Gauss map γ has a Lagrangian factorization

(1.9) $$\gamma\colon \mathbb{C}^{n-1} \to T^*\mathbb{C}^{n-1} \to (\mathbb{C}^{n-1})^*,$$
$$(x) \mapsto (x, \frac{\partial f}{\partial x}) \mapsto (\frac{\partial f}{\partial x}).$$

The conic Lagrangian map Γ of example (1.1) takes the following form in the coordinates $(x,y,-t\xi,t)$:

(1.10) $$\Gamma\colon (x,t) \mapsto (x, f(x), -t\frac{\partial f}{\partial x}, t) \mapsto (-t\frac{\partial f}{\partial x}, t).$$

The Legendrian map Γ' of example (1.3) takes the following form in the coordinates (x,ξ,z), with $z = x\xi - y$:

(1.11) $$\Gamma'\colon (x) \mapsto (x, \frac{\partial f}{\partial x}, x\frac{\partial f}{\partial x} - f(x)) \mapsto (\frac{\partial f}{\partial x}, x\frac{\partial f}{\partial x} - f(x)).$$

It can be shown that the Lagrangian structure (1.9) on the Gauss map γ depends (even up to isomorphism) on the choice of coordinates, whereas the conic Lagrangian structure (1.10) on Γ and the Legendrian structure (1.11) on Γ' are independent of coordinates, by definition. The Lagrangian structure (1.9) on γ is obtained from the conic Lagrangian structure (1.10) on Γ by *reduction*. More precisely, the symplectic manifold $T^*\mathbb{C}^{n-1}$ with symplectic form $\omega = dx \wedge d\xi$ is obtained from $T^*\mathbb{C}^n - \mathcal{Z}$ by reduction, or symplectic section and projection (cf. [15, p. 11], [3, p. 289]). For $T^*\mathbb{C}^{n-1}$ is the orbit space of the flow of the Hamiltonian field $X_t = -\partial/\partial y$ on the hypersurface $\{t = 1\}$. This reduction depends on the choice of the Hamiltonian function t, that is, on the choice of contact form α on $\mathbb{P}T^*\mathbb{C}^n = (T^*\mathbb{C}^n - \mathcal{Z})/\mathbb{C}^*$.

Next we consider $\mathbb{Z}/2$-symmetry. Let ι be the generator of $\mathbb{Z}/2$; if ν is a $\mathbb{Z}/2$ action we abbreviate ν_ι to ι.

(1.12) Proposition. Let $\pi\colon \mathcal{E} \to \mathcal{B}$ be an odd conic Lagrangian fibration germ at the $\mathbb{C}^*$-orbit $\mathcal{O} \subset \mathcal{E}$. There are local coordinates (p,y,q,t) as in (1.8) such that the action of $\mathbb{Z}/2$ is $\iota(p,y,q,t) = (-p,-y,q,-t)$.

Proof. First we show there is an *equivariant* 1-form α which defines the canonical contact structure on the $\mathbb{C}^*$-orbit space of $\mathcal{E}$. By hypothesis, $\mathcal{O}$ is a $\mathbb{C}^*$-orbit of $\mathcal{E}$ with $\iota\mathcal{O} = \mathcal{O}$. Since $\iota(t\cdot x) = t\cdot\iota(x)$ for $t \in \mathbb{C}^*$, there is a character $\chi\colon \mathbb{Z}/2 \to \mathbb{C}^*$ such that $\iota(x) = \chi(\iota)\cdot x$ for all $x \in \mathcal{O}$. Let $\varepsilon = \chi(\iota) = \pm 1$. Consider the orbit map $\rho\colon \mathcal{E} \to E$, and let $e = \rho(\mathcal{O})$. Define a function ψ from $\mathcal{O}$ to the set of contact forms on T_eE as follows. Let β be the homogeneous ι-invariant 1-form on $\mathcal{E}$ such that $d\beta = \Omega$ and $\beta(u) = 0$ for u tangent to a $\mathbb{C}^*$-orbit. If $x \in \mathcal{O}$, $v \in T_x\mathcal{E}$, and $w = \rho_* v$, then $\langle\psi(x), w\rangle = \beta_x(v)$. The function ψ is well-defined, and ψ is a bijection equivariant with respect to the actions of $\mathbb{C}^*$ and $\mathbb{Z}/2$. Therefore if a is a contact 1-form on T_eE, then $\iota^* a = \varepsilon a$. Let η be a 1-form on E defining the contact structure, and set $\alpha = \frac{1}{2}(\eta + \varepsilon\iota^*\eta)$. Then $\alpha_e = a \neq 0$, so α also defines the contact structure in a neighborhood of e, and $\iota^*\alpha = \varepsilon\alpha$.

Now let $\pi'\colon E \to B$ be the dehomogenization of $\pi\colon \mathcal{E} \to \mathcal{B}$. By averaging, we define local coordinates (p,q,z) in which ι acts *linearly* and such that $\alpha = dz - p dq$. Namely, let (λ,μ,ζ) be local coordinates on E centered at e such that $\alpha = d\zeta - \lambda d\mu$ and $\pi'(\lambda,\mu,\zeta) = (\mu,\zeta)$. Identify a neighborhood of e with the tangent space T_eE using the coordinates (λ,μ,ζ), and let ι_* be the linear involution of T_eE induced by ι. Then set

$$(p,q,z) = \tfrac{1}{2}\big((\lambda,\mu,\zeta) + \iota_*(\iota(\lambda,\mu,\zeta))\big).$$

Finally we use the fact that the conic Lagrangian fibration germ $\mathcal{E} \to \mathcal{B}$ is *odd*. Since ι acts linearly in the coordinates (p,q,z), and $\iota^*\alpha = \varepsilon\alpha$ with $\alpha = dz - p dq$, the hypersurface $V = \{z = 0\}$ of E is ι-invariant, and ι acts on the coordinate z by multiplication by ε. Consider the symplectic form $\omega = dp \wedge dq$ on V. We have $\iota^*\omega = \varepsilon\omega$, since $\omega = d(\alpha|_V)$. If $\varepsilon = +1$, the action of ι on V is *symplectic*. On the other hand, if we introduce Darboux coordinates $(p,y,-tq,t) \to (-tq,t)$ on $\mathcal{E} \to \mathcal{B}$ as in (1.8), then q gives coordinates for the $\mathbb{C}^*$-orbit space of $\mathcal{B}$, so ι acts trivially on q. Thus ι must act trivially on V, which implies that ι acts trivially on E, the $\mathbb{C}^*$-orbit space of $\mathcal{E}$, contrary to hypothesis. Therefore $\varepsilon = -1$, and $\iota^*\omega = -\omega$, i.e., the action of ι on V is *antisymplectic*. Since ι acts trivially on q, by a linear symplectic change of coordinates on (V,ω) we obtain Darboux coordinates (p,q) such that $\iota(p,q) = (-p,q)$. Thus we have coordinates $(p,y,-tq,t) \mapsto (-tq,t)$ as in (1.8) for the germ $\mathcal{E} \to \mathcal{B}$, with $\mathbb{Z}/2$ action $\iota(p,y,-tq,t) = (-p,-y,tq,-t)$. □

2. Generating families

In this section we carry out the classification of infinitesimally stable conic G-Lagrangian germs $\mathcal{L}$ in case either G is trivial or $G = \mathbb{Z}/2$ and $\mathcal{L}$ is odd. For simplicity we will state and prove results in the odd case only. Statements (2.1) through (2.12) below are also true for G trivial; the proofs are simplifications of those given for the odd case.

Let ν be an action of the finite group G on the germ $(\mathbb{C}^n,0)$, and let ν_* be the action induced on $T\mathbb{C}^n$ by ν. Let $H \subset T_0\mathbb{C}^n$ be a hyperplane such that $\nu_*H = H$. The projection $\pi\colon (\mathbb{P}T^*\mathbb{C}^n,H) \to (\mathbb{C}^n,0)$ is a G-Legendrian fibration germ. The action induced by ν_* on the quotient $T_0\mathbb{C}^n/H$ is multiplication by a character χ_ν. In the following discussion, we fix ν and H, and we let $\chi = \chi_\nu$.

A (G,ν)-*family* is a pair $(\mathcal{F},\eta)$, where $\mathcal{F}\colon (\mathbb{C}^{k+n},0) \to \mathbb{C}$ is a germ and η is a G-action on $(\mathbb{C}^{k+n},0)$, such that

(1) The projection $\rho\colon (\mathbb{C}^{k+n},0) \to (\mathbb{C}^n,0)$, $\rho(x,\lambda) = \lambda$, is a G-map, and

(2) $\mathcal{F}(\eta_g(x,\lambda)) = \chi(g)\mathcal{F}(x,\lambda)$ for all $(x,\lambda) \in \mathbb{C}^{k+n}$.

We say that $(\mathcal{F},\eta)$ is a (G,ν)-*Morse family* if in addition

(3) $\mathcal{F}$ is ρ-*regular* , i.e., $d\mathcal{F}_0 \neq 0$, $d\mathcal{F}_0(\ker\ d\rho) \neq 0$, and $[d\mathcal{F}]$ is transverse to $\mathbb{P}N_\rho = \{[\tau] \in \mathbb{P}T^*\mathbb{C}^{k+n} \mid \tau(\ker d\rho) = 0\}$.

Let $\ell = \pi' \circ i'$: $(\mathbb{C}^{n-1},0) \to (\mathbb{P}T^*\mathbb{C}^n,H) \to (\mathbb{C}^n,0)$ be a G-Legendrian germ. Let $\tilde{\rho}$: $\mathbb{P}N_\rho \to \mathbb{P}T^*\mathbb{C}^n$ be the canonical map. It follows from (3) that

$$(\Lambda'_{\mathcal{F}},H_{\mathcal{F}})= \left(\tilde{\rho}(\mathrm{Image}[d\mathcal{F}] \cap \mathbb{P}N_\rho),\ \tilde{\rho}([d\mathcal{F}_0])\right)$$

is a G-invariant Legendrian submanifold germ of $\mathbb{P}T^*\mathbb{C}^n$. The (G,ν)-Morse family $(\mathcal{F},\eta)$ is a *generating family* of ℓ if $\Lambda'_{\mathcal{F}} = i'(\mathbb{C}^{n-1})$. The generating family $\mathcal{F}$ determines the map ℓ up to G-equivariant coordinate changes in $(\mathbb{C}^{n-1},0)$.

Two (G,ν)-families $(\mathcal{F}_1,\eta_1)$, $(\mathcal{F}_2,\eta_2)$, $\mathcal{F}_1,\mathcal{F}_2$: $(\mathbb{C}^{k+n},0) \to \mathbb{C}$, are *isomorphic* (or G-V-*equivalent*) if there exist germs Φ: $(\mathbb{C}^{k+n},0) \to (\mathbb{C}^{k+n},0)$ and u: $(\mathbb{C}^{k+n},0) \to \mathbb{C}$ with the following properties: Φ is a G-equivariant isomorphism which permutes the fibers of ρ (i.e., $\Phi(x,\lambda) = (h(x,\lambda),\varphi(\lambda))$), u is a G-invariant function with $u(0) \neq 0$, and

$$\mathcal{F}_2(x,\lambda) = u(x,\lambda)\mathcal{F}_1(\Phi(x,\lambda)).$$

If $\mathcal{F}_1$ is isomorphic to $\mathcal{F}_2$, then there is a G-equivariant fiber-permuting contact transformation germ Ψ: $(\mathbb{P}T^*\mathbb{C}^n, H_{\mathcal{F}_1}) \to (\mathbb{P}T^*\mathbb{C}^n, H_{\mathcal{F}_2})$ such that $\Psi(\Lambda'_{\mathcal{F}_1}) = \Lambda'_{\mathcal{F}_2}$.

A *suspension* of a (G,ν)-family $(\mathcal{F},\eta)$, $\mathcal{F}$: $(\mathbb{C}^{k+n},0) \to \mathbb{C}$, is a (G,ν)-family $(\mathcal{G},\xi)$, $\mathcal{G}$: $(\mathbb{C}^{k+\ell+n},0) \to \mathbb{C}$, such that $\xi = \eta \oplus \sigma$ for some G-action σ on $(\mathbb{C}^\ell,0)$ and $\mathcal{G} = \mathcal{F} + Q$ for some nondegenerate quadratic form Q on $\mathbb{C}^\ell$ with $Q(\sigma_g(y)) = \chi(g)Q(y)$ (e.g. $Q(y_1,y_2) = y_1y_2$, $G = \mathbb{Z}/2$ and $\iota(y_1,y_2) = (-y_1,y_2)$). If $(\mathcal{G},\xi)$ is a suspension of $(\mathcal{F},\eta)$, then $\Lambda'_{\mathcal{G}} = \Lambda'_{\mathcal{F}}$. Two (G,ν)-families are *stably isomorphic* if they have isomorphic suspensions. If two G-Legendrian germs ℓ_1 and ℓ_2 have stably isomorphic generating families, then ℓ_1 and ℓ_2 are isomorphic.

Let L: $\mathbb{C}^n \to T^*\mathbb{C}^n \to \mathbb{C}^n$ be a conic G-Lagrangian germ. We define a *Legendrian generating family* of L to be a generating family of the corresponding G-Legendrian germ ℓ (the dehomogenization of L). When L is an odd conic Lagrangian germ, a Legendrian generating family of L is a (G,ν)-family with $G = \mathbb{Z}/2$ and ν reflection in a hyperplane.

(2.1) Theorem. Every odd conic Lagrangian germ has a Legendrian generating family. Two odd Morse families $\mathcal{F}_1$ and $\mathcal{F}_2$ give isomorphic odd conic Lagrangian germs if and only if $\mathcal{F}_1$ and $\mathcal{F}_2$ are stably isomorphic.

Proof. (For trivial G, see [16, §2] or [3, §20.7].) Let $\mathcal{L}: \mathcal{M} \to \mathcal{E} \to \mathcal{B}$ be an odd conic Lagrangian germ with dehomogenization $\ell = \pi' \circ i': M \to E \to B$. Let ι be the nontrivial element of $\mathbb{Z}/2$, and let α define the contact structure on E. By the proof of Proposition (1.12), there are local coordinates (p,q,z) on E such that $\alpha = dz - pdq$, $\iota(p,q,z) = (-p,q,-z)$, and $\pi(p,q,z) = (q,z)$. Applying the construction of [3, p. 321] to the Legendrian germ $\Lambda' = i'(M)$ using the coordinates (p,q,z), we obtain a Legendrian generating family

$$\mathcal{F}(x,\lambda) = S(q_{I'},x) + \langle x,q_{J'}\rangle - z,$$
$$x = p_{J'},\ \lambda = (q,z),$$
$$I' \cup J' = \{1,\dots,n-1\},\ I' \cap J' = \emptyset.$$

Moreover, the projection of Λ' to the hyperplane $V = \{z = 0\}$ is a $\mathbb{Z}/2$-Lagrangian germ with generating family

$$F(x,q) = S(q_{I'},x) + \langle x,q_{J'}\rangle,$$

and the action of ι on V is antisymplectic. The proof of the classification theorem of [3, p. 324] carries over to the present case. (The crucial part of the proof is [3, §19.5].) □

The generating family F which occurs in the preceding proof is the key to the classification of odd conic Lagrangian germs. The family of hypersurfaces $\mathcal{F} = 0$ is the *graph* of the family of functions F. We formalize this relationship as follows.

An m-*parameter* G-*family* is a triple (F,ξ,χ), where $F: (\mathbb{C}^{k+m},0) \to \mathbb{C}$ is a germ, ξ is a G-action on $\mathbb{C}^k$ and χ is a character of G such that

$$F(\xi_g(x),q) = \chi(g)F(x,q).$$

(N.B. the action of G on $\mathbb{C}^m$ is *trivial.*) We also say that (F,ξ,χ) is an m-*parameter* G-*deformation* of the germ $f: (\mathbb{C}^k,0) \to \mathbb{C}$, $f(x) = F(x,0)$. An *odd* family is a $\mathbb{Z}/2$-family (F,ξ,χ) such that χ is nontrivial. (N.B. We do not assume that the nontrivial element $\iota \in \mathbb{Z}/2$ acts by $\iota(x) = -x$.) Let $\sigma: (\mathbb{C}^{k+m},0) \to (\mathbb{C}^m,0)$ be the projection, and let $N_\sigma = \{\tau \in T^*\mathbb{C}^{k+m} \mid \tau(\ker d\sigma) = 0\}$. Then F is a G-*Morse family* if dF is transverse to N_σ. Let $\mathcal{L}: \mathbb{C}^n \to T^*\mathbb{C}^n \to \mathbb{C}^n$ be a conic G-Lagrangian germ. The m-parameter G-Morse family F is a *reduced generating family* of $\mathcal{L}$ if $m = n-1$ and the following (G,ν)-Morse family $(\mathcal{F},\eta)$ is a Legendrian generating family of $\mathcal{L}$:

$$\mathcal{F}(x,q,z) = F(x,q) - z,$$
$$\eta_g(x,q,z) = (\xi_g(x),q,\chi(g)z).$$

The G-V-equivalence and suspension of G-families are defined just as for (G,ν)-families. Two G-families F_1 and F_2 are G-R-*equivalent* if there exists a G-equivariant isomorphism germ

$\Phi: (\mathbb{C}^{k+m},0) \to (\mathbb{C}^{k+m},0)$ such that Φ permutes the fibers of σ and $F_2(x,q) = F_1(\Phi(x,q))$. The G-families F_1 and F_2 are G-*graph-equivalent* if the families $\mathcal{F}_i(x,q,z) = F_i(x,q) - z$, $i = 1,2$, are G-V-equivalent. The proof of (2.1) yields the following.

(2.2) Corollary. Every odd conic Lagrangian germ has a reduced generating family. Two odd Morse families F_1 and F_2 give isomorphic odd conic Lagrangian germs if and only F_1 and F_2 are stably $\mathbb{Z}/2$-graph-equivalent. □

(2.3) Example. For the homogeneous Gauss map of a smooth odd hypersurface $M \subset \mathbb{C}^n$ (1.1), given locally as the graph of an odd function $f: \mathbb{C}^{n-1} \to \mathbb{C}$, a reduced generating family is $F(x,\xi) = -f(x) + \xi x$ (cf. (1.9)), and a Legendrian generating family is $\mathcal{F}(x,\xi,z) = -f(x) + \xi x - z$ (cf. (1.11)).

(2.4) Remark. Example (2.3) generalizes to the following method for computing a reduced generating family for an odd conic Lagrangian submanifold $\Lambda \subset T^*\mathbb{C}^n$, with corresponding Legendrian submanifold $\Lambda' \subset \mathbb{P}T^*\mathbb{C}^n$. Let $(x_1,\dots,x_{n-1},y)$ be the standard coordinates on $\mathbb{C}^n$, and let (x,ξ,z) be Darboux coordinates on $\mathbb{P}T^*\mathbb{C}^n$ as in (1.11). Suppose that (x_J,ξ_I) form a coordinate system for Λ'. Write $y = y(x_J,\xi_I)$ on Λ', and also $x_I = x_I(x_J,\xi_I)$. Then $z = \xi x - y = F(x_J,\xi)$ is a reduced generating family for Λ'.

To classify odd families up to graph-equivalence, we will use the following results. (For G trivial, we use the corresponding assertions with R-equivalence replaced by R^+-equivalence [3, p. 304]).

(2.5) Proposition. Let F_1 and F_2 be odd families. If F_1 and F_2 are $\mathbb{Z}/2$-R-equivalent, they are $\mathbb{Z}/2$-graph-equivalent. If F_1 and F_2 are $\mathbb{Z}/2$-graph equivalent, they are $\mathbb{Z}/2$-V-equivalent. □

We will also use the following two consequences of the parametrized Morse lemma for odd families (cf. [14], [3, p.306]).

(2.6) Proposition. Two odd m-parameter families $F_1: (\mathbb{C}^{k_1+m},0) \to \mathbb{C}$, $F_2: (\mathbb{C}^{k_2+m},0) \to \mathbb{C}$ with $k_1 = k_2$ are $\mathbb{Z}/2$-R- (resp. $\mathbb{Z}/2$-V-) equivalent if and only if they are stably $\mathbb{Z}/2$-R- (resp. $\mathbb{Z}/2$-V-) equivalent. □

(2.7) Proposition. Every odd Morse family is $\mathbb{Z}/2$-R-equivalent to an odd family $F: (\mathbb{C}^{k+m},0) \to \mathbb{C}$ such that

$$F(x,q) = S(b,v) + \langle a,v \rangle + Q(w), \quad \frac{\partial^2 S}{\partial v^2}(0,0) = 0,$$

with $x = (v_1,\dots v_i,w_1,\dots w_{k-i})$, $q = (a_1,\dots,a_i,b_1,\dots,b_{m-i})$, $\iota x = (-v,\iota w)$, $S(b,-v) = -S(b,v)$,

$Q(\iota w) = -Q(w)$, and Q a nondegenerate quadratic form. □

Now we describe three closely related notions of infinitesimal versality for an odd family. Let $\mathbb{C}\{x\} = \mathbb{C}\{x_1,\ldots,x_k\}$ be the ring of germs of functions $f: (\mathbb{C}^k,0) \to \mathbb{C}$. Given an action ι of $\mathbb{Z}/2$ on $\mathbb{C}^k$, we define an action of $\mathbb{Z}/2$ on $\mathbb{C}\{x\}$ by $(\iota f)(x) = -f(\iota x)$. Thus $\iota f = f$ if and only if f is *odd*: $f(\iota x) = -f(x)$. Given an odd germ f, let

$$<\frac{\partial f}{\partial x}> = <\frac{\partial f}{\partial x_1},\ldots,\frac{\partial f}{\partial x_k}>$$

be the ideal generated by the partials of f. Since f is odd, this ideal is invariant under the action of $\mathbb{Z}/2$. For a $\mathbb{Z}/2$-module M, let $M^+ = M^{\mathbb{Z}/2} = \{a \in M \mid \iota a = a\}$. We define

$$T_f^+ = \left(\mathbb{C}\{x\}/<\frac{\partial f}{\partial x}>\right)^+,$$

the set of (first order) *infinitesimal deformations* of f. Similarly, we let

$$T_{Vf}^+ = \left(\mathbb{C}\{x\}/<f,\frac{\partial f}{\partial x}>\right)^+,$$

the set of *infinitesimal V-deformations* of f. T_f^+ and T_{Vf}^+ are modules over the ring of *even* germs $f \in \mathbb{C}\{x\}$ $(f(\iota x) = f(x))$.

Let $F: (\mathbb{C}^{k+m},0) \to \mathbb{C}$ be an m-parameter odd deformation of $f: (\mathbb{C}^k,0) \to \mathbb{C}$. We say that the deformation F is *infinitesimally $\mathbb{Z}/2$-versal* if the images in T_f^+ of the odd functions

$$f,\ \dot{F}_i = \frac{\partial F}{\partial q_i}\Big|_{q=0} \quad (i = 1,\ldots,m),$$

generate T_f^+ as a complex vector space. The deformation F is *infinitesimally $\mathbb{Z}/2$-R-versal* (resp. *infinitesimally $\mathbb{Z}/2$-V-versal*) if the images in T_f^+ of the functions $\dot{F}_i$, $(i = 1,\ldots,m)$, generate T_f^+ (resp. T_{Vf}^+) as a complex vector space.

(2.8) Remark. It follows from the definitions that infinitesimal $\mathbb{Z}/2$-R-versality ⇒ infinitesimal $\mathbb{Z}/2$-versality ⇒ infinitesimal $\mathbb{Z}/2$-V-versality.

(2.9) Remark. The corresponding notions of infinitesimal versality for G trivial are obtained by ignoring the group action and adding the constant function 1 to the set of generators. Thus F is *infinitesimally versal* if the images of the germs 1, f, $\dot{F}_i$ generate $T_f = \mathbb{C}\{x\}/<\frac{\partial f}{\partial x}>$ as a complex vector space.

Let $F: (\mathbb{C}^{k+m},0) \to \mathbb{C}$, $F': (\mathbb{C}^{k+m'},0) \to \mathbb{C}$ be odd deformations of the same germ f. If there is a germ $\varphi: (\mathbb{C}^{k+m'},0) \to (\mathbb{C}^{k+m},0)$ such that $F'(x,q') = F(x,\varphi(q'))$, we say the deformation F' is *induced* from F. The odd deformation F of f is $\mathbb{Z}/2$-R-*versal* if every odd deformation F' of f is $\mathbb{Z}/2$-R-equivalent to an odd deformation induced from F. The proof of the equivalence of R-versality and infinitesimal R-versality given in [3, §8.3] adapts to give the following result for odd families.

(2.10) Proposition. If F is an odd family, then F is $\mathbb{Z}/2$-R-versal if and only if F is infinitesimally $\mathbb{Z}/2$-R-versal. □

(2.11) Theorem. Let F be an odd Morse family which is a reduced generating family of the odd conic Lagrangian germ $\mathcal{L}$. Then F is infinitesimally $\mathbb{Z}/2$-versal if and only if $\mathcal{L}$ is infinitesimally stable.

(2.12) Corollary. Isomorphism classes of infinitesimally stable odd conic Lagrangian germs are in one-to-one correspondence with stable $\mathbb{Z}/2$-graph-equivalence classes of infinitesimally $\mathbb{Z}/2$-versal odd families. □

Proof of (2.11): We adapt the method of [1] to the symmetric homogeneous case. First we work out Arnold's stability condition $\Phi = H\circ i$ (1.2) in coordinates. Let $\mathcal{L} = \pi\circ i: \mathcal{M} \to \mathcal{E} \to \mathcal{B}$ be an odd conic Lagrangian germ. Suppose that (P,Q) are Darboux coordinates on $\mathcal{E}$ such that the Lagrangian bundle $\mathcal{E} \to \mathcal{B}$ is given by $(P,Q) \mapsto Q$. Since the Hamiltonian H is linear on fibers, $H(P,Q) = A(Q)P + B(Q)$. (Here $A(Q)P = \sum A_i(Q)P_i$.) We assume that $(P,Q) = (p,y,-qt,t)$ are the coordinates constructed in (1.12). Since H is homogeneous, so are A and B. Since dH is $\mathbb{Z}/2$-invariant, we can assume H is odd; by dehomogenizing we see that $B = 0$. Thus $H(P,Q) = A(Q)P$, where A is odd and homogeneous.

Now suppose that the $\mathbb{Z}/2$-invariant conic Lagrangian submanifold $\Lambda = i(\mathcal{M}) \subset \mathcal{E}$ is given by the generating function $\mathcal{S}(Q_I,P_J)$, i.e., Λ is defined by Hamilton's equations

(2.13)
$$P_I = \frac{\partial \mathcal{S}}{\partial Q_I}, \quad Q_J = -\frac{\partial \mathcal{S}}{\partial P_J}, \quad I \cup J = \{1,\dots,n\}, \quad I \cap J = \emptyset.$$

Then (1.2) says that if $\Phi(Q_I,P_J)$ is an odd homogeneous function, there exists an odd homogeneous map $A(Q)$ such that

(2.14)
$$\Phi(Q_I,P_J) = A\left(Q_I,-\frac{\partial \mathcal{S}}{\partial P_J}\right)\cdot\left(\frac{\partial \mathcal{S}}{\partial Q_I},P_J\right).$$

Let $\ell = \pi'\circ i': M \to E \to B$ be the dehomogenization of $\mathcal{L}$. Let $(P,Q) = (p,y,-qt,t)$ be the coordinates constructed in (1.12). Then the Legendrian submanifold $\Lambda' = i'(M)$ is defined by the equations

$$p_{I'} = \frac{\partial S}{\partial q_{I'}}, \quad q_{J'} = \frac{\partial S}{\partial p_{J'}}, \quad z = S(q_{I'},p_{J'}) + \langle p_{J'},q_{J'}\rangle,$$

with $I = I' \cup \{n\}$, $J = J'$, and $\mathcal{S}(Q_I,P_J) = -\iota S(q_{I'},p_{J'})$. Thus the odd family $F(x,q) = S(q_{I'},x) + \langle x,q_{J'}\rangle$ is a reduced generating family of $\mathcal{L}$. By dehomogenizing the condition (2.14), we see that $\mathcal{L}$ is infinitesimally stable if and only if, for every odd function $\varphi(q_{I'},x)$, there exist $a_1(q)$ and $a_2(q)$ such that

$$\varphi(q_{I'},x) = a_1(q_{I'},-\frac{\partial S}{\partial x})\cdot(\frac{\partial S}{\partial q_{I'}},x) + a_2(q_{I'},-\frac{\partial S}{\partial x})(q_{I'}\frac{\partial S}{\partial q_{I'}} - S(q_{I'},x)) . \tag{2.15}$$

Setting $q = 0$, and noting that

$$\frac{\partial S}{\partial q_{I'}} = \frac{\partial F}{\partial q_{I'}},\ x = \frac{\partial F}{\partial q_{J'}},\ S(0,x) = F(x,0),$$

(2.15) implies that F is infinitesimally $\mathbb{Z}/2$-versal.

Conversely, if $F(x,q) = S(q_{I'},x) + \langle x,q_{J'}\rangle$ is infinitesimally $\mathbb{Z}/2$-versal then (2.15) holds, by the holomorphic version of Bierstone's equivariant preparation theorem. More precisely, let $M \subset \mathbb{C}\{q_{I'},x\}$ be the set of odd functions (a module over the ring of even functions). Consider the homomorphism $\rho: \mathbb{C}^{n-1} \to \mathbb{C}^{n-1}$ given by $\rho(q_{I'},x) = (q_{I'},q_{J'}) = (q_{I'},\frac{\partial S}{\partial x})$. Let $\mathbf{m}$ be the maximal ideal of $\mathbb{C}\{q_{I'},q_{J'}\}$. The preparation theorem [4, Thm. 5.16, p. 130] gives that the images of $S(q_{I'},x), \frac{\partial S}{\partial q_{I'}}, x$ span the vector space $\mathbf{M}/(\rho^*\mathbf{m})\mathbf{M} = T_f^+$ (i.e., F is infinitesimally $\mathbb{Z}/2$-versal) if and only if $S(q_{I'},x), \frac{\partial S}{\partial q_{I'}}, x$ generate $\mathbf{M}$ as a $\mathbb{C}\{q\}$-module (i.e., condition (2.15)).

Finally, (2.7) says that every odd Morse family is stably $\mathbb{Z}/2$-R-equivalent to a family of the form $F(x,q) = S(q_{I'},x) + \langle x,q_{J'}\rangle$. If the odd families F_1 and F_2 are stably $\mathbb{Z}/2$-R-equivalent and F_1 is infinitesimally versal, then F_2 is also infinitesimally versal. This completes the proof of the theorem. □

(2.16) Theorem. (a) A 2-parameter family is versal if and only if it is stably graph-equivalent to one of the following families:

A_1: $F(x_1,q_1,q_2) = x_1^2$
A_2: $F(x_1,q_1,q_2) = x_1^3 + q_1x_1$
A_3: $F(x_1,q_1,q_2) = x_1^4 + q_2x_1^2 + q_1x_1$

(b) A 2-parameter odd family is $\mathbb{Z}/2$-versal if and only if it is stably $\mathbb{Z}/2$-graph-equivalent to one of the following odd families (where the $\mathbb{Z}/2$-action is given by $\iota(x) = -x$):

$A_2(-)$: $F(x_1,q_1,q_2) = x_1^3 + q_1x_1$
$A_4(-)$: $F(x_1,q_1,q_2) = x_1^5 + q_2x_1^3 + q_1x_1$
$D_4(-)$: $F(x_1,x_2,q_1,q_2) = x_1^3 + x_1x_2^2 + q_1x_1 + q_2x_2$

Proof: We finesse the problem of dealing with nonstandard definitions of versality and equivalence as follows. In the odd case, we classify $\mathbb{Z}/2$-V-versal odd families up to stable $\mathbb{Z}/2$-R-equivalence, and then observe that the normal forms so obtained ($A_2(-)$, $A_4(-)$, $D_4(-)$) are in fact $\mathbb{Z}/2$-R-versal and stably $\mathbb{Z}/2$-V-inequivalent. By (2.5) and (2.8), the resulting classification is the desired one. (In the nonsymmetric case, one classifies families F such that the one-parameter extension $F+c$ is V-versal, up to stable R^+-equivalence. We leave this easier case to the reader.) If F is a $\mathbb{Z}/2$-versal odd family, then F is a Morse family. Thus (2.7) gives that F is stably $\mathbb{Z}/2$-R-equivalent to an odd family of the form $F(x,q) = S(q_I,x) + \langle x,q_J\rangle$, with $S(0,0) = 0$, $(\partial^2 S/\partial x^2)(0,0) = 0$, and $\iota(x) = -x$. By (2.6) we can replace stable $\mathbb{Z}/2$-R-equivalence (resp. stable $\mathbb{Z}/2$-V equivalence) by $\mathbb{Z}/2$-R-equivalence (resp. $\mathbb{Z}/2$-V-equivalence). Thus theorem (2.16)(b) reduces to the following proposition.

For the rest of this section, we assume that the involution ι *acts on* $\mathbb{C}^k$ *by* $\iota(x) = -x$.

(2.17) Proposition. Let $F: (\mathbb{C}^{k+2},0) \to \mathbb{C}$ be a 2-parameter odd deformation of $f: (\mathbb{C}^k,0) \to \mathbb{C}$ such that $f(x) \in \langle x\rangle^3$. Then F is infinitesimally $\mathbb{Z}/2$-V-versal if and only if F is $\mathbb{Z}/2$-R-equivalent to one of the three families (2.16)(b).

(2.18) Lemma. (a) Let $f: (\mathbb{C},0) \to \mathbb{C}$ be a nonzero odd germ. Then $f(x)$ is $\mathbb{Z}/2$-R-equivalent to x^{2r+1}, where $r = \dim_{\mathbb{C}} T^+_{Vf}$.

(b) Let $f: (\mathbb{C}^2,0) \to \mathbb{C}$ be an odd germ such that $f(x,y) \in \langle x,y\rangle^3$ and $\dim_{\mathbb{C}} T^+_{Vf} = 2$. Then $f(x,y)$ is $\mathbb{Z}/2$-R-equivalent to $x^3 + xy^2$.

Proof of (2.17): Since the families (2.16)(b) are $\mathbb{Z}/2$-V-versal, so is any family which is $\mathbb{Z}/2$-R-equivalent to one of them. Conversely, suppose that $F(x,q)$ is an infinitesimally $\mathbb{Z}/2$-V-versal 2-parameter odd deformation of f, with $f(x) \in \langle x\rangle^3$. Since f is odd, $x_1,\dots,x_k$ are linearly independent in T^+_{Vf}. Since F is $\mathbb{Z}/2$-V-versal, $k \le 2$. Therefore, by (2.18) F is $\mathbb{Z}/2$-R-equivalent to a deformation of one of the three functions $f_1 = x_1^3$, $f_2 = x_1^5$, or $f_3 = x_1^3 + x_1x_2^2$. Now consider the deformations F_i of the functions f_i given by (2.16)(b): $F_1(x_1,q_1,q_2) = x_1^3 + q_1x_1$, $F_2(x_1,q_1,q_2) = x_1^5 + q_2x_1^3 + q_1x_1$, $F_3(x_1,x_2,q_1,q_2) = x_1^3 + x_1x_2^2 + q_1x_1 + q_2x_2$. Each of these families is infinitesimally $\mathbb{Z}/2$-R-versal, and hence $\mathbb{Z}/2$-R-versal (2.10). Thus any odd 2-parameter $\mathbb{Z}/2$-R-versal deformation of f_i is $\mathbb{Z}/2$-R-equivalent to F_i, $i = 1,2,3$. But f_i is *homogeneous*, so F is a $\mathbb{Z}/2$-V-versal deformation of f_i implies F is a $\mathbb{Z}/2$-R-versal deformation of f_i. □

Proof of (2.18): Assertion (a) is immediate. (Write $f(x) = u(x)x^n$ with u a unit; then n is odd and $u = v^n$ with v even.) Now we prove (b). Write $f(x,y) = c(x,y) + $ (higher order terms), with $c(x,y)$ a homogeneous cubic. If $\dim_{\mathbb{C}} T^+_{Vf} = 2$, then $c(x,y)$ has three distinct linear factors, say $c = \ell_1\ell_2\ell_3$. It follows that $f = f_1 f_2 f_3$, where $f_i = \ell_i + \ldots$ is analytic (cf. [7, Thm. 19, p. 116, Thm. 19, p. 91]). Furthermore, it is easy to arrange that the germs f_i are odd. Since f_1 and f_2 have linearly independent linear terms, there is a $\mathbb{Z}/2$-equivariant analytic isomorphism germ φ of $(\mathbb{C}^2,0)$ such that $f_1(\varphi(x,y)) = x$ and $f_2(\varphi(x,y)) = y$. Thus $f(x,y)$ is $\mathbb{Z}/2$-R-equivalent to xyh, where h is an odd germ whose linear term is not a scalar multiple of x or y. The germ xyh is $\mathbb{Z}/2$-R-equivalent to $xy(x+y)$, which is $\mathbb{Z}/2$-R-equivalent to $x^3 + xy^2$. □

This completes the proof of Theorem (2.16). □

3. The Gauss map of a genus three theta divisor

Let C be a smooth complex curve of genus three. The Jacobian variety $J(C)$ is a three-dimensional abelian variety; the theta divisor Θ of $J(C)$ is smooth provided C is not hyperelliptic. We define the *Gauss map* of Θ

$$\gamma\colon \Theta \to \mathbb{P}(T_0 J(C))^* = \mathbb{P}^{2*};$$

$\gamma(x)$ is the translate of the tangent plane $T_x\Theta$ to the origin of $J(C)$. If C is hyperelliptic, then Θ has a unique singular point x_o corresponding to the unique g^1_2 (linear system of dimension one and degree two) on C; x_o is an ordinary node. Letting $\hat{\Theta}$ denote the Nash blowup of Θ, the Gauss map

$$\hat{\gamma}\colon \hat{\Theta} \to \mathbb{P}(T_0 J(C))^* = \mathbb{P}^{2*}$$

is the extension of the tangent map $\Theta - \{x_o\} \to \mathbb{P}(T_0 J(C))^*$.

The map $x \mapsto -x$ of $J(C)$ takes the theta divisor Θ to itself, inducing an involution $\iota\colon \Theta \to \Theta$ (and $\hat{\iota}\colon \hat{\Theta} \to \hat{\Theta}$ in the hyperelliptic case), and the Gauss map is ι-invariant. We can see this geometrically as follows. Consider first the case that C is not hyperelliptic. Then, by Riemann's theorem, Θ is isomorphic to the symmetric square $C^{(2)}$ of C; if $\omega_1, \omega_2, \omega_3$ are a basis for $H^0(C,\Omega^1)$ and Λ is the period lattice of C (cf. [6, p. 334]), the isomorphism is given by the Abel-Jacobi map

(3.1)
$$\mu\colon C^{(2)} \to J(C) = H^0(C,\Omega^1)^*/H_1(C,\mathbb{Z}) = \mathbb{C}^3/\Lambda,$$

$$\mu(P+Q) = \left(\int_{P_o}^{P} \omega_j + \int_{P_o}^{Q} \omega_j \right)_{j=1,2,3}, \quad \text{mod } \Lambda.$$

The canonical embedding of C as a smooth plane quartic is given by

$$(3.2)\qquad \begin{array}{c}\phi\colon C \to \mathbb{P}^2 = \mathbb{P}(H^0(C,\Omega^1)^*)\\ \phi(P) = [(\omega_1(P),\omega_2(P),\omega_3(P))] = [\vec{\omega}(P)].\end{array}$$

Thus, when $P \neq Q$, $T_{P+Q}(\Theta) = \mu_* T_{P+Q}C^{(2)} = [\vec{\omega}(P)\wedge\vec{\omega}(Q)] \in G(2, H^0(C,\Omega^1)^*) \cong \mathbb{P}(\Lambda^2 H^0(C,\Omega^1)^*)$. Now, identifying C with its canonical model, we write

$$(3.3)\qquad \gamma(P+Q) = P\wedge Q.$$

When $P = Q$ the secant line is replaced as usual by the tangent line, *viz.*, $\gamma(2P) = [\phi(P)\wedge\phi'(P)] = P\wedge P'$. The fact that γ is ι-invariant can be seen explicitly as follows: if a line cuts C in P+Q+R+S, then by Abel's theorem $\iota(P+Q) = R+S$, and $\gamma(P+Q) = [P\wedge Q] = [R\wedge S] = \gamma(R+S)$.

In the hyperelliptic case, the Abel-Jacobi map $C^{(2)} \to \Theta$ is the Nash blowup $\hat{\Theta}$ of Θ. The canonical map $\phi\colon C \to \mathbb{P}^2$ is a two-to-one mapping of C to a plane conic; let $E \subset C^{(2)}$ be the collection of the fibers of ϕ, i.e., the g^1_2. Then the Gauss map $\hat{\gamma}\colon \hat{\Theta} \cong C^{(2)} \to \mathbb{P}^{2*}$ is given as follows:

$$(3.4)\qquad \hat{\gamma}(P+Q) = \begin{cases} \phi(P)\wedge\phi(Q), & P+Q \notin E\\ \phi(P)\wedge\phi'(Q), & P+Q \in E,\ P \text{ not a Weierstrass point}\\ \phi(P)\wedge\phi''(Q), & P=Q,\ \text{a Weierstrass point} \end{cases}$$

The involution $\hat{\iota}\colon \hat{\Theta} \to \hat{\Theta}$ is induced from the hyperelliptic involution on C sending P to $\bar{P}$, where $P + \bar{P}$ is a fiber of ϕ.

The Gauss map γ (resp. $\hat{\gamma}$) lifts to the *homogeneous Gauss map* Γ (resp. $\hat{\Gamma}$), which is a conic $\mathbb{Z}/2$-Lagrangian map, as we now explain. Let $\mathcal{M} \to \Theta$ (resp. $\widehat{\mathcal{M}} \to \hat{\Theta}$) be the holomorphic line bundle on Θ (resp. $\hat{\Theta}$) defined by pulling back the tautological bundle $\mathcal{O}_{\mathbb{P}^{2*}}(1)$ by the Gauss map γ (resp. $\hat{\gamma}$). Let $\mathcal{Z}$ and Z be the zero-sections of $\mathcal{M}$ and $T^*J(C)$ respectively; then we have the natural inclusion map

$$(3.5)\qquad i\colon \mathcal{M} - \mathcal{Z} \to T^*J(C) - Z \cong J(C) \times (T_0^*J(C) - \{0\}).$$

Projecting to the fiber, we obtain a conic $\mathbb{Z}/2$-Lagrangian map Γ (resp. $\hat{\Gamma}$) inducing the Gauss map γ (resp. $\hat{\gamma}$) on $\mathbb{C}^*$ orbits:

$$\begin{array}{ccc} & & T^*J(C)-Z \\ & {}^{i}\nearrow & \downarrow \\ & \Gamma & \\ \mathcal{M}-Z & \rightarrow & T_0^*J(C)-\{0\} \\ \downarrow & & \downarrow \\ & \gamma & \\ \Theta & \rightarrow & \mathbb{P}(T_0J(C))^* \end{array}$$

Note that we identify $\mathbb{P}(T_0J(C))^* = G(2,T_0J(C))$ with $\mathbb{P}(T_0^*J(C))=G(1,T_0^*J(C))$. The $\mathbb{Z}/2$ action on Θ lifts to $\mathcal{M}$ as the (-1)-map, and so Γ is a conic $\mathbb{Z}/2$-map. On the other hand, Γ is Lagrangian, because the image of i is the conormal variety $\mathcal{C}_\Theta$ of Θ: If $\Theta \subset J(C)$ is given locally by $\theta = 0$, then the image of i is locally the Lagrangian submanifold

$$\{(x,\tau) \in J(C) \times T_0^*J(C) \mid \theta(x) = 0,\ \tau = \lambda\, d\theta(x) \text{ for some } \lambda \in \mathbb{C}^*\}$$

of $T^*J(C) - Z$.

We now state our main result.

(3.6) Theorem. (a) If C is a nonhyperelliptic curve of genus three, then the homogeneous Gauss map Γ is a locally infinitesimally stable conic $\mathbb{Z}/2$-Lagrangian map if and only if C has only normal Weierstrass points. (b) If C is a hyperelliptic curve of genus three, then the homogeneous Gauss map $\hat{\Gamma}$ is a locally infinitesimally stable conic $\mathbb{Z}/2$-Lagrangian map.

Proof of (a). We must show that, for each fiber $\mathcal{O}$ of the $\mathbb{C}^*$-bundle $\mathcal{C}_\Theta$, the germ of Γ at $\mathcal{O}$ is an infinitesimally stable conic $(\mathbb{Z}/2)_{\mathcal{O}}$–Lagrangian germ. We compute the local form of the map Γ at each fiber $\mathcal{O}$. We refer the reader to [12, Theorem 2.8] for a detailed geometric description of the singular loci of the Gauss map, but here is a summary of the results. For convenience, we assume $P \neq Q$. We abuse notation by using the point $P+Q \in C^{(2)} \equiv \Theta$ to refer to the corresponding fiber of the $\mathbb{C}^*$-bundle $\mathcal{M} \equiv \mathcal{C}_\Theta$ over Θ.

(i) $P+Q$ is an A_1 singularity of Γ if $\overline{PQ}$ is not tangent to C at P nor at Q;
(ii) $P+Q$ is an A_2 singularity of Γ if $\overline{PQ}$ is simply tangent to C at P, but not tangent at Q;
(iii) $P+Q$ is an A_3 singularity of Γ if P is an ordinary flex point of C;
(iv) $P+Q$ is a $D_4(-)$ singularity of Γ if $\overline{PQ}$ is bitangent to C at P and Q;
(v) $2P$ is a $D_6(-)$ singularity of Γ if P is a higher flex point of C (and Γ is therefore not stable at such a point).

(3.7) Remark. The Arnold symbols A_k, D_k are used differently here and in the classification theorem of [12; 2.24, 1.6]. There these symbols refer to the singularity types of the *fibers* of the Gauss map γ. Here they refer to the singularity type of the generating family of the Lagrangian map Γ.

(3.8) Lagrangian stability criterion. Suppose $f: \mathbb{C}^2 \to \mathbb{C}$ and $f(0) = \frac{\partial f}{\partial x}(0) = \frac{\partial f}{\partial y}(0) = 0$. It follows from (2.11) and the proof of (2.16) that the germ ℓ at the orbit $\mathcal{O} = \{0\} \times \mathbb{C}^*$ of the conic Lagrangian map

$$\mathbb{C}^2 \times \mathbb{C}^* \to T^*\mathbb{C}^3 - \mathcal{Z} \to T_0^*\mathbb{C}^3 - \{0\}$$

$$(x,y,\lambda) \mapsto (x,y,f(x,y);\ -\lambda\frac{\partial f}{\partial x}, -\lambda\frac{\partial f}{\partial y}, \lambda) \mapsto (-\lambda\frac{\partial f}{\partial x}, -\lambda\frac{\partial f}{\partial y}, \lambda)$$

is Lagrangian-stable if and only if the reduced generating family

$$F(x,y,\xi,\eta) = -f(x,y) + \xi x + \eta y$$

has the property that $F+c$ is V-versal, i.e., $1,x,y$ span $\mathbb{C}\{x,y\}/\langle f, \frac{\partial f}{\partial x}, \frac{\partial f}{\partial y}\rangle$. If $f(\iota(x,y)) = -f(x,y)$, i.e., f is odd, then ℓ is $\mathbb{Z}/2$-Lagrangian-stable if and only if F is $\mathbb{Z}/2$-V-versal, i.e., x,y span $(\mathbb{C}\{x,y\}/\langle f, \frac{\partial f}{\partial x}, \frac{\partial f}{\partial y}\rangle)^+$, the invariant submodule of $\mathbb{C}\{x,y\}/\langle f, \frac{\partial f}{\partial x}, \frac{\partial f}{\partial y}\rangle$.

We consider first cases (i)-(iv) above. Our first task is to set up local coordinates so that (3.8) applies. We are interested in a point $P+Q \in C^{(2)}$, and we suppose $P \neq Q$, so that we may work locally on C^2. If we parametrize C by s at $P = P(0)$ and by t at $Q = Q(0)$, then near P, $\vec{\omega} = P(s)$, and near Q, $\vec{\omega} = Q(t)$, so that the Abel-Jacobi map is of the form

$$\mu(s,t) = \int P(s)ds + \int Q(t)dt,$$

and the map Γ is given in local coordinates by

$$(s,t,\lambda) \mapsto (x,y,\ z=f(x,y); -\lambda\frac{\partial f}{\partial x}, -\lambda\frac{\partial f}{\partial y}, \lambda) =$$
$$(\int P(s)ds + \int Q(t)dt;\ -\lambda P(s)\wedge Q(t), \lambda) \mapsto (-\lambda P(s)\wedge Q(t), \lambda).$$

If we choose linear coordinates on $\mathbb{P}^2$ so that $P = (1,0,0)$ and $Q = (0,1,0)$, then in particular the tangent plane to Θ at $P+Q$ will be the plane $z = 0$. We may further assume that $P(s) = (1,f_1(s),f_2(s))$, $Q(t) = (g_1(t),1,g_2(t))$, where $f_1,f_2,g_1,g_2 \in \langle s,t\rangle$. Then $(\int P(s)ds + \int Q(t)dt) = (s + O(t^2), t + O(s^2), z(s,t))$, so that $\langle\frac{\partial f}{\partial x}, \frac{\partial f}{\partial y}\rangle = \langle\frac{\partial z}{\partial s}, \frac{\partial z}{\partial t}\rangle = \langle f_2(s), g_2(t)\rangle$. Thus the stability criterion may be reformulated as follows. Let $\mathcal{O} = \{0\} \times \mathbb{C}^*$ be the $\mathbb{C}^*$-orbit of the origin 0.

If the stabilizer subgroup $(\mathbb{Z}/2)_{\mathcal{O}} = \{1\}$, then the germ at $\mathcal{O}$ of Γ is stable provided $1, x(s,t), y(s,t)$ span

$$M = \mathbb{C}\{s,t\}/\langle z(s,t), \frac{\partial z}{\partial s}, \frac{\partial z}{\partial t}\rangle = \mathbb{C}\{s,t\}/\langle z(s,t), f_2(s), g_2(t)\rangle.$$

If the stabilizer subgroup $(\mathbb{Z}/2)_{\mathcal{O}} = \mathbb{Z}/2$, then the germ at $\mathcal{O}$ of Γ is stable provided $x(s,t), y(s,t)$ span the invariant submodule

$$M^+ = (\mathbb{C}\{s,t\}/\langle z(s,t), \frac{\partial z}{\partial s}, \frac{\partial z}{\partial t}\rangle)^+ = (\mathbb{C}\{s,t\}/\langle z(s,t), f_2(s), g_2(t)\rangle)^+.$$

(i) $P(s) = (1, s^2+\ldots, s+\ldots)$, $Q(t) = (\alpha t+\ldots, 1, \beta t+\ldots)$, $\beta \neq 0$. Here the quotient module

$$M = \mathbb{C}\{s,t\}/\langle s^2/2 + \beta t^2/2 + \ldots, s+\ldots, \beta t+\ldots\rangle \cong \mathbb{C},$$

which is trivially spanned by $1, x(s,t), y(s,t)$.

(ii) $P(s) = (1, s+\ldots, s^2+\ldots)$, $Q(t) = (\alpha t+\ldots, 1, \beta t+\ldots)$, $\beta \neq 0$. Now,

$$M = \mathbb{C}\{s,t\}/\langle s^3/3 + \beta t^2/2 + \ldots, s^2+\ldots, \beta t+\ldots\rangle \cong \mathbb{C}\{s\}/\langle s^2\rangle,$$

which is spanned by $1, x(s,t)$.

(iii) $P(s) = (1, s+\ldots, s^3+\ldots)$, $Q(t) = (\alpha t+\ldots, 1, \beta t+\ldots)$, $\beta \neq 0$. Here

$$M = \mathbb{C}\{s,t\}/\langle s^4/4 + \beta t^2/2 + \ldots, s^3+\ldots, \beta t+\ldots\rangle \cong \mathbb{C}\{s\}/\langle s^3\rangle.$$

Here $1, x(s,t) = s + \alpha t^2/2 + \ldots$, and $y(s,t) = t + s^2/2 + \ldots$ span M since $s^2 \equiv 2y$ mod the ideal.

(iv) $P(s) = (1, s+\ldots, s^2+\ldots)$, $Q(t) = (\alpha t+\ldots, 1, \beta t^2+\ldots)$, $\alpha \neq 0, \beta \neq 0$.

$$M = \mathbb{C}\{s,t\}/\langle s^3/3 + \beta t^3/3 + \ldots, s^2+\ldots, \beta t^2+\ldots\rangle \cong \mathbb{C}\{s,t\}/\langle s^2, t^2\rangle.$$

Now, $1, x(s,t) = s + \alpha t^2/2 + \ldots$, and $y(s,t) = t + s^2/2 + \ldots$ do not span M, because st is not in their span. However, $x(s,t)$ and $y(s,t)$ do span $M^+ \cong (\mathbb{C}\{s,t\}/\langle s^2, t^2\rangle)^+$, whence our germ is stable, as required.

We now use (3.8) to consider the last case.

(v) Here we must show that if P is a higher flex point of C, then at the point $2P$ the map Γ is not stable. We may take $P(s) = (1, s, s^4+\alpha s^5+\ldots)$, letting $P = (1,0,0)$ as before, and taking the tangent line to C at P to be $(1,0,0)\wedge(0,1,0)$: we need to work in symmetric local coordinates $u = s + t$, $v = st$ at a point of the diagonal of $C^{(2)}$ (cf. [12]). In terms of these variables, we have

$$\mu(u,v) = (u,\ u^2/2 - v,\ u^5/5 + u^3v - uv^2 + \alpha(u^6/6 + u^4v - 3u^2v^2/2 + v^3/3) + \ldots).$$

Then,

$$\begin{aligned} M &= \mathbb{C}\{u,v\}/\langle z(u,v), \tfrac{\partial z}{\partial u}, \tfrac{\partial z}{\partial v}\rangle \\ &= \mathbb{C}\{u,v\}/\langle -v^2 + 3(u^2v - auv^2) + \ldots, -2uv + \alpha v^2 + u^3 - 3au^2v + \ldots\rangle. \end{aligned}$$

Although we don't know that ι acts by -1 on (u,v), we have that ι acts by $(-1)^k$ on m^k/m^{k+1} and by -1 on $1 \in M$, so

$$\dim(M^+) = \sum_{k\ \mathrm{odd}} \dim(m^kM/m^{k+1}M).$$

We now note that u,v are linearly independent in mM/m^2M and that u^3 represents a nonzero element of m^3M/m^4M, whence $\dim(M^+) \geq 3$, and the proof is complete. (The verification that the singularity is of type D_6 is omitted.) □

<u>Proof of</u> (b). Cf. [12, Thm. 2.20] for a geometric description of the various loci; here are the singularity types we must check in the hyperelliptic case. By Riemann-Hurwitz, the canonical map ϕ defined in (3.2) has eight (simple) branch points $P_1, P_2, \ldots, P_8$; these are the eight (hyperelliptic) Weierstrass points of C (cf. [6, p. 273–4]). Recall that $E \subset C^{(2)}$ is the g^1_2 locus.

(i) $P+Q$ is a stable $A_2(-)$ singularity of $\hat{\Gamma}$ whenever $P+Q \in E$ and a stable A_2 singularity of $\hat{\Gamma}$ whenever $P = P_j$, $j = 1,2,\ldots,8$, $Q \neq P_k$ for any k;

(ii) $2P_j$, $j = 1,2,\ldots,8$, is a stable $A_4(-)$ singularity of $\hat{\Gamma}$;

(iii) P_i+P_j, $i \neq j$, is a stable $D_4(-)$ singularity of $\hat{\Gamma}$.

(Note that the involution $\hat{\iota}$ fixes E pointwise and that the points P_i+P_j, $i \neq j$, are isolated fixed points of $\hat{\iota}$, and so it is crucial to take the $\mathbb{Z}/2$ action into consideration. Note in particular that the symmetry forces $2P_j$ to be an $A_4(-)$ singularity, rather than an A_3 singularity.)

Here we include only the proof of (ii); the remaining parts are quite straightforward. Take a local coordinate s on C centered at the Weierstrass point $P = P_j$, and choose homogeneous coordinates on $\mathbb{P}^2$ so that the canonical map $\phi\colon C \to \mathbb{P}^2$ has the form $\phi(s) = (1, s^2, s^4)$. Then the theta divisor is parametrized by the Abel map $\mu\colon C^{(2)} \to J(C)$ with base point $2P$: using symmetric coordinates $u = s + t$, $v = st$, the map to the projectivized conormal variety in $\mathbb{P}T^*J(C)$ is given by

$$\begin{aligned} \hat{\mu}(s,t) &= (s+t, \tfrac{1}{3}(s^3+t^3), \tfrac{1}{5}(s^5+t^5); s^2t^2, s^2+t^2, 1), \\ \hat{\mu}(u,v) &= (u, \tfrac{1}{3}u^3 - uv, \tfrac{1}{5}u^5 + u^3v - uv^2; v^2, u^2 - 2v, 1) \\ &= ((x_1,x_2,y), (\xi_1,\xi_2,1)). \end{aligned}$$

where $x_1 = u$, $x_2 = \frac{1}{3}u^3 - uv$, $y = \frac{1}{5}u^5 + u^3v - uv^2$, $\xi_1 = v^2$, and $\xi_2 = u^2 - 2v$. We observe that x_1 and ξ_2 can serve as local coordinates on the conormal variety, and so, in order to apply Remark (2.4), we must express x_2 and y as functions of x_1 and ξ_2. Since $v = \frac{1}{2}(x_1^2 - \xi_2)$, $x_2 = -\frac{1}{6}x_1^3 + \frac{1}{2}x_1\xi_2$ and $y = \frac{9}{20}x_1^5 - \frac{1}{4}x_1\xi_2^2$. Then the reduced generating family is given by

$$z(x_1;\xi_1,\xi_2) = x_1\xi_1 + x_2\xi_2 - y = x_1\xi_1 - \frac{9}{20}x_1^5 + \frac{3}{4}x_1\xi_2^2 - \frac{1}{6}x_1^3\xi_2.$$

The initial velocities of this family are $\frac{\partial z}{\partial \xi_1}\Big|_{q=0} = x_1$ and $\frac{\partial z}{\partial \xi_2}\Big|_{q=0} = -\frac{1}{6}x_1^3$. Since these two elements span M^+ for the odd function germ $z(x_1,0,0) = -\frac{9}{20}x_1^5$, the family is versal. Further, this generating family is $\mathbb{Z}/2$-V-equivalent to the $A_4(-)$ normal form $F(x_1;\xi_1,\xi_2) = x_1^5 + \xi_2 x_1^3 + \xi_1 x_1$ (2.16)(b). Thus, the point pair 2P is a stable $A_4(-)$ singularity of $\hat{\Gamma}$, as required. □

References

[1] V.I. Arnold, Normal forms for functions near degenerate critical points, the Weyl groups of A_k, D_k, E_k and Lagrangian singularities, *Funct. Anal. Appl.* 6:4 (1972), 254–272.

[2] ———, *Mathematical Methods of Classical Mechanics*, Springer-Verlag, New York 1978.

[3] V.I. Arnold, S.M. Gusein-Zade and A.N. Varchenko, *Singularities of Differentiable Maps*, vol. 1, Birkhäuser, Boston 1985.

[4] E. Bierstone, The structure of orbit spaces and the singularities of equivariant mappings, *Monografiás de Matemática*, IMPA, Rio de Janiero, 1980.

[5] J.J. Duistermaat, Oscillatory integrals, Lagrange immersions and unfolding of singularities, *Comm. Pure Appl. Math.* 28 (1974), 207–281.

[6] P. Griffiths and J. Harris, *Principles of Algebraic Geometry*, Wiley, New York 1978.

[7] R.C. Gunning and H. Rossi, *Analytic functions of several complex variables*, Prentice-Hall, Inc., Englewood Cliffs 1965.

[8] L. Hörmander, Fourier integral operators I, *Acta Math.* 127 (1971), 71–183.

[9] S. Janeczko, On G-versal Lagrangian submanifolds, *Bull. Pol. Acad. Sci. (Math.)* 31 (1983), 183-190.

[10] S. Janeczko and A. Kowalczyk, Classification of generic 3-dimensional Lagrangian singularities with $(\mathbb{Z}_2)^\ell$-symmetries, preprint, December 1988.

[11] C. McCrory, T. Shifrin and R. Varley, The Gauss map of a generic hypersurface in $\mathbb{P}^4$, *J. Diff. Geom.* 30 (1989), 689-759.

[12] ———, The Gauss map of a genus three theta divisor, to appear.

[13] C.T.C. Wall, Geometric properties of generic differentiable manifolds, *Geometry and Topology*, Lecture Notes in Math. 597, Springer-Verlag, New York (1977), 707-774.

[14] A. Weinstein, Singularities of families of functions, *Differentialgeometrie im Grossen* (W. Klingenberg, ed.), Bibliographisches Institut, Mannheim, 1971, 323-330.

[15] ———, *Lectures on Symplectic Manifolds*, C.B.M.S. Regional Conf. Series in Math., no. 29, Am. Math. Soc., Providence 1977.

[16] V.M. Zakalyukin, Lagrangian and Legendrian singularities, *Funct. Anal. Appl.* 10:1 (1976), 23-31.

University of Georgia
Athens, Georgia 30602
U.S.A.

On Infinitesimal Deformations of Minimally Elliptic Singularities

Kurt Behnke*

1 Introduction

In [B-K] H. Knörrer and the author of the present note developed a method for computing the dimension of the space T^1_X of (first order) infinitesimal deformations of rational surface singularities. It turned out that there are purely topological conditions (i.e. bounds on intersection numbers of exceptional curves on the minimal resolution), which imply that a certain lower estimate for dim T^1_X as proved by J. Wahl in [B-K, Appendix] already gives the correct dimension.

One cannot hope for such results for more general normal surface singularities, since already the most elementary analytic invariants like the geometric genus are in general not determined by the topological data. The minimally elliptic singularities, as introduced by Laufer, however have properties which allow to apply the ideas of [B-K]. For example they are Gorenstein, and so the description of the dual space $(T^1_X)^*$ in [B-K] has a particularly simple form. Moreover, given the resolution data, important analytic invariants like the geometric genus p_g and the irregularity q are fixed or can only vary in a very restricted and obvious way.

We shall briefly describe the methods and the results of this paper. Let $(X, 0)$ be a minimally elliptic singularity. Let $\mathcal{O}_X$ be the structure sheaf, and let Ω^1_X be the sheaf of Kähler differentials. For a small Stein neighbourhood X of the singular point 0 denote the smooth part $X \setminus \{0\}$ by X'. The natural epimorphism $\mu : \mathcal{O}^n_X \to \Omega^1_X$ defined by the differentials of a generating set $(f_1, \ldots, f_n)$ of the maximal ideal at the singular point, gives rise to a map

$$\mu : H^0(X', \mathcal{O}^n_{X'}) \to H^0(\mathcal{O}_{X'}, \Omega^1_{X'})$$

with cokernel $(T^1_X)^*$. If $\pi : \tilde{X} \to X$ is the minimal good resolution (i. e. the exceptional set has smooth irreducible components crossing normally, and $\tilde{X}$ is minimal with this property) then $df_1, \ldots, df_n$ also define a map

$$\tilde{\mu} : \mathcal{O}^n_{\tilde{X}} \to I_Z \Omega^1_{\tilde{X}} \langle \log E \rangle.$$

Here $\Omega^1_{\tilde{X}} \langle \log E \rangle$ is the locally free sheaf of meromorphic 1-forms on $\tilde{X}$ with at most logarithmic poles along E, and $I_Z \cong \mathcal{O}_Z(-Z)$ is the pullback of the reduced ideal sheaf of the singular point.

*supported by a "Heisenberg-Stipendium", Be 1078/1-1, of the DFG

If $\rho : H^0(\tilde{X}, I_Z\Omega^1_{\tilde{X}}(\log E)) \to H^0(X', \Omega^1_{X'})$ denotes the natural injective restriction map, we have a factorization $\mu = \rho \circ \tilde{\mu}$, where we identify $H^0(\tilde{X}, \mathcal{O}_{\tilde{X}})$ and $H^0(X', \mathcal{O}_{X'})$ by normality of X.

The cokernel of ρ can in many cases easily be computed, but in order to say something about coker $\tilde{\mu}$ one has to make quite strong assumptions. There is one result, Theorem 4.5, which is of a rather general nature: If all the components of the exceptional set E are smooth rational curves of self-intersection numbers less than -2, and at least one component has self-intersection number less than -4, then dim T^1_X is the sum of well known invariants and the Euler characteristic of a naturally defined sheaf $\mathcal{R}$. For a series of singularities called $D(r-2; b_1, \ldots, b_r)$ with dual resolution graph

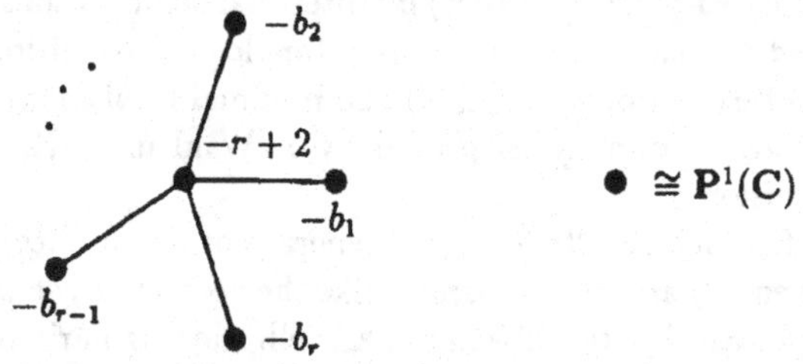

we get dim $T^1_X = (q-1) + d + 3(r-2)$, $d = (\sum_{i=1}^r b_i) - 8$ the degree, once r and the b_i are sufficiently large.

This work started as a computation of dim T^1_X for triangle singularities and their non quasihomogeneous satellites. After the rational double points, the simply elliptic singularities [Pi] and cusp singularities [Be] they are the next interesting case in the hierarchy of Gorenstein surface singularities. The result is that for sufficiently large absolute values of the self-intersection numbers of the exceptional curves on the minimal resolution dim $T^1_X = q + d + r + 1$, where again d is the degree, and r is the number of irreducible components of the exceptional divisor on the minimal resolution. J. Wahl established this formula for all triangle singularities of degree at least 5 using an explicit description of the graded local ring and his method of Jacobian algebras of Gorenstein normal surface singularities [W5, Theorem 4.13].

The non-quasihomogeneous satellites are adjacent to their corresponding triangle singularities, and they deform normally flat into the cusp singularities with the same number of exceptional curves of the same self-intersection numbers. By semicontinuity the dimension result follows. (I am grateful to the referee for pointing this out to me.) We retain part of our original computation as Section 5 though, to serve as an illustration of our method (and to its limitation maybe).

This paper was rewritten while the author stayed at the Mathematics Research Centre at the University of Warwick during the Special Year on Singularity Theory 1988/89. He would like to thank everybody there, and especially the organizers, for their help which made this stay pleasant and fruitful.

2 Schlessinger's description of T^1_X and local duality

In this section we recall from [B-K] the basic facts about the cohomological description of T^1_X of an isolated singularity, and its dual version for the twodimensional Gorenstein case.

Let $(X,0)$ be a normal isolated singularity, and let $i: X \to \mathbf{C}^n$ be an embedding of a Stein representative X as a closed analytic subset of a neighbourhood of the origin of $\mathbf{C}^n$, such that 0 is the only singular point of X. Denote by X' the regular part of X, and let Ω^1_X , $\Omega^1_{\mathbf{C}^n}$ be the sheaves of Kähler differentials on X and $\mathbf{C}^n$ respectively. Finally denote their duals by Θ_X and $\Theta_{\mathbf{C}^n}$. Then we have a natural inclusion map $\Theta_{X'} \to i^*\Theta_{\mathbf{C}^n}|X'$, and Schlessinger's Theorem [Sch] says

Theorem 2.1 *The module T^1_X of first order infinitesimal deformations of $(X,0)$ is naturally isomorphic to the kernel of the map*

$$H^1(X',\Theta_{X'}) \to H^1(X',i^*\Theta_{\mathbf{C}^n}).$$

□

For computational purposes it is often easier to use the dual version of this result. In Section 2 of [B-K] it has been worked out more generally, but for the applications in this paper we restrict ourselves to the case that X is a normal *Gorenstein* surface singularity. Local duality then has a particularly simple form and Corollary 1.2 of [B-K] reads

Proposition 2.2 *For a normal Gorenstein surface singularity X the dual of T^1_X is isomorphic to the cokernel of the restriction map*

$$H^0(X',i^*\Omega^1_{\mathbf{C}^n}) \to H^0(X',\Omega^1_{X'})$$

□

As in Remark 1.3 of [B-K] this result can be given an even more explicit form. Let $f_1,\ldots,f_n$ be the generators of the maximal ideal of $\mathcal{O}_{(X,0)}$ corresponding to the restriction to X of the coordinate functions on the ambient $\mathbf{C}^n$, say. Then $\Omega^1_{\mathbf{C}^n}|X$ is free with basis $df_1,\ldots,df_n$. Let $\mu: \mathcal{O}^n_X \to \Omega^1_X$ be the surjection defined by $\mu(g_1,\ldots,g_n) = g_1df_1 + \cdots + g_ndf_n$. The space $(T^1_X)^*$ is isomorphic to the cokernel of the map

$$H^0(X',\mathcal{O}^n_{X'}) \to H^0(X',\Omega^1_{X'})$$

induced by μ.

3 Factorization of the Map μ

In this section we introduce the basic exact sequences which we shall use to compute $(T^1_X)^*$ of an isolated Gorenstein surface singularity. We keep the notations of the previous sections.

Let X be an isolated Gorenstein surface singularity, let $\pi: \tilde{X} \to X$ be a desingularization of X such that the exceptional divisor E has normal crossings, and such that the pullback of the maximal ideal sheaf m is invertible: $\pi^*(m) \cong \mathcal{O}_{\tilde{X}}(-Z) \cong I_Z$ for an effective divisor Z supported on E.

If $f_1, \ldots, f_n$ represent a set of generators of m, they lift to the resolution, and by normality they generate the algebra of holomorphic functions there. The differentials df_i, $i = 1, \ldots, n$, are sections of the locally free sheaf of twisted logarithmic differentials $I_Z\Omega^1_{\tilde{X}}\langle\log E\rangle \cong \Omega^1_{\tilde{X}}\langle\log E\rangle(-Z)$ (see [B-K, Section 2]). Define a map

$$\tilde{\mu} : \mathcal{O}^n_{\tilde{X}} \to I_Z\Omega^1_{\tilde{X}}\langle\log E\rangle$$

by $\tilde{\mu}(h_1, \ldots, h_n) = h_1 df_1 + \cdots + h_n df_n$ as before.

Lemma 3.1 *Let $\rho : H^0(\tilde{X}, I_Z\Omega^1_{\tilde{X}}\langle\log E\rangle) \to H^0(X', \Omega^1_{X'})$ be the restriction map. Then $(T^1_X)^*$ is isomorphic to the cokernel of the composed map*

$$\rho \circ \tilde{\mu} : H^0(\tilde{X}, \mathcal{O}^n_{\tilde{X}}) \to H^0(\tilde{X}, I_Z\Omega^1_{\tilde{X}}\langle\log E\rangle) \to H^0(X', \Omega^1_{X'}).$$

□

Since $I_Z\Omega^1_{\tilde{X}}\langle\log E\rangle$ is locally free ρ is injective. So the length of the cokernel of the composite map is the sum of the dimensions of the cokernels of each of the factors. Let $E = \cup_{i=1}^r E_i$ be the decomposition of the exceptional set into irreducible components, and let ω_{E_i} be the dualizing sheaf of E_i. The length of the cokernel of the inclusion

$$H^0(\tilde{X}, \Omega^1_{\tilde{X}}) \to H^0(X', \Omega^1_{X'}) \tag{1}$$

is the invariant q, often called the *irregularity* of X. It has been studied quite extensively [W4],[Y].

The long exact cohomology sequence associated with the short exact sequence

$$0 \to I_E\Omega^1_{\tilde{X}}\langle\log E\rangle \to \Omega^1_{\tilde{X}} \to \oplus_{i=1}^r \omega_{E_i} \to 0$$

will be used to compute the dimension of the cokernel of

$$H^0(\tilde{X}, I_E\Omega^1_{\tilde{X}}\langle\log E\rangle) \to H^0(\tilde{X}, \Omega^1_{\tilde{X}}) \tag{2}$$

Finally there is an exact sequence

$$\begin{aligned} 0 &\to H^0(\tilde{X}, I_Z\Omega^1_{\tilde{X}}\langle\log E\rangle) \to H^0(\tilde{X}, I_E\Omega^1_{\tilde{X}}\langle\log E\rangle) \\ &\to H^0(|Z-E|, I_E\Omega^1_{\tilde{X}}\langle\log E\rangle \otimes \mathcal{O}_{Z-E}) \to H^1(\tilde{X}, I_Z\Omega^1_{\tilde{X}}\langle\log E\rangle) \to \ldots \end{aligned}$$

In [W4] invariants g and β are defined by

$$g = \sum_{i=1}^r \dim H^0(E_i, \omega_{E_i}), \quad \beta = \dim\Big(\bigoplus_{i=1}^r H^0(E_i, \omega_{E_i}) / (\mathrm{Im}\ H^0(\tilde{X}, \Omega^1_{\tilde{X}}))\Big).$$

Proposition 3.2 *Assume that $H^1(\tilde{X}, I_Z\Omega^1_{\tilde{X}}\langle\log E\rangle) = 0$. Then coker ρ has dimension $q + g - \beta + \dim H^0(|Z-E|, I_E\Omega^1_{\tilde{X}}\langle\log E\rangle \otimes \mathcal{O}_{Z-E})$.* □

Remark: Of course the last two contributions to this formular cannot be computed in general. But if for example all the E_i are rational curves then $g = \beta = 0$. Also $\beta = 0$ if X is quasihomogeneous [W4, Corollary 1.11].

Certainly the more difficult part of 3.1 is the map $\tilde{\mu}$. As in [B-K] define

$$\mathcal{F} = \mathrm{Im}\,\tilde{\mu} \tag{3}$$
$$\mathcal{C} = \mathrm{Coker}\,\tilde{\mu} \tag{4}$$
$$\mathcal{R} = \mathrm{Ker}\,\tilde{\mu} \tag{5}$$

Then $\mathcal{F} \subset I_Z\Omega^1_{\tilde{X}}\langle\log E\rangle$ is torsion free, $\mathcal{C}$ is concentrated on the exceptional set, and $\mathcal{R}$ is a locally free sheaf of rank $n-2$. From the exact sequences

$$0 \to \mathcal{F} \to I_Z\Omega^1_{\tilde{X}}\langle\log E\rangle \to \mathcal{C} \to 0 \tag{6}$$
$$0 \to \mathcal{R} \to \mathcal{O}^n_{\tilde{X}} \to \mathcal{F} \to 0 \tag{7}$$

the next proposition follows immediately.

Proposition 3.3 *Let $p_g = \dim H^1(\tilde{X}, \mathcal{O}_{\tilde{X}})$ be the geometric genus of X. If*

$$H^1(\tilde{X}, I_Z\Omega^1_{\tilde{X}}\langle\log E\rangle) = 0$$

then the cokernel of $\tilde{\mu}$ has dimension

$$\dim H^0(\tilde{X}, \mathcal{C}) + \dim H^1(\tilde{X}, \mathcal{R}) - np_g$$

□

4 The Case of Minimally Elliptic Singularities

In this Section we collect some facts about minimally elliptic singularities which are needed to apply the previous observations. Minimally elliptic singularities have been introduced in [Lau]. They can be characterized as those normal Gorenstein surface singularities which have geometric genus $p_g = 1$. Except a few cases, which are listed below, the minimal resolution of such a singularity already has an exceptional divior with nonsingular components of genus 0 crossing normally. The exceptions are

1. the simply elliptic singularities, where the exceptional set is a smooth elliptic curve,
2. the triangle (or Dolgachev) singularities with exceptional set of the minimal resolution either a rational curve with one ordinary cusp, or two smooth rational curves having second order contact at their unique point of intersection, or three smooth rational curves meeting transversely in one point,

Let Z be the fundamental cycle of the minimal good resolution, and let $d = -Z^2$ be the degree of the singularity. Then the multiplicity of X is $\max(2, d)$, and the embedding dimension is $\max(3, d)$. If d is at least 2 the pullback of the maximal ideal sheaf m is the invertible sheaf $\mathcal{O}_{\tilde{X}}(-Z)$.

In case the exceptional divisor on the minimal resolution has normal crossings the canonical sheaf $\omega_{\tilde{X}}$ is isomorphic to $\mathcal{O}_{\tilde{X}}(-Z)$ too. In cases (2) and (3) above the canonical sheaf on the *minimal good resolution* is $\omega_{\tilde{X}} \cong \mathcal{O}_{\tilde{X}}(-A)$ where the anticanonical cycle A is an effective divisor supported on E and has $A \leq Z$. At the same time A is the *minimally elliptic cycle* i. e. it is the uniquely determined minimal cycle L with $\chi(L) = 0$.

Theorem 4.1 *Let $\pi : \tilde{X} \to X$ be the minimal good resolution of a minimally elliptic singularity. Let Z be the fundamental cycle, and let $\omega_{\tilde{X}} \cong \mathcal{O}_{\tilde{X}}(-A)$. Assume that $d = -Z^2 \geq 2$. Then*

$$H^1(\tilde{X}, I_Z\Omega^1_{\tilde{X}}\langle \log E\rangle) = 0$$

and if A is not equal to Z then also

$$H^1(\tilde{X}, I_A\Omega^1_{\tilde{X}}\langle \log E\rangle) = 0.$$

Proof: If the minimal resolution has an exceptional divisor with normal crossings then $A = Z$, and the result means $H^1(\tilde{X}, \Omega^1_{\tilde{X}}\langle \log E\rangle \otimes \omega_{\tilde{X}}) = 0$. But by Serre duality this is an immediate consequence of Wahl's vanishing theorem $H^1_E(Der_E(\tilde{X}) = 0$ from [W1].

We shall give another argument which covers all the cases. By [W4, 1.6.1], exterior differentiation gives an exact sequence of sheaves

$$0 \to j_!\mathbf{C}_{\tilde{X}\setminus E} \to \mathcal{O}_{\tilde{X}}(-Z) \xrightarrow{d} \Omega^1_{\tilde{X}}\langle \log E\rangle(-Z) \xrightarrow{d} \omega_{\tilde{X}}(-Z+E) \to 0$$

Here $j : \tilde{X} \setminus E \to \tilde{X}$ is the inclusion map, $j_!$ is the direct image functor with compact supports, and $\mathbf{C}_{\tilde{X}\setminus E}$ is the constant sheaf with stalk $\mathbf{C}$. If X is contractible $\tilde{X}$ retracts onto E, so $j_!\mathbf{C}_{\tilde{X}\setminus E}$ has no cohomology. Hence

$$H^i(\tilde{X}, \mathcal{O}_{\tilde{X}}(-Z)) = H^i(\tilde{X}, d\mathcal{O}_{\tilde{X}}(-Z))$$

all i. In the long exact sequence

$$\cdots \to H^1(\tilde{X}, \mathcal{O}_{\tilde{X}}(-Z)) \to H^1(\tilde{X}, \Omega^1_{\tilde{X}}\langle \log E\rangle)(-Z) \to H^1(\tilde{X}, \omega_{\tilde{X}}(-Z+E))$$

the first and the third term vanish by [Lau, Lemma 3.1.1] applied to Z and to $Z+(A-E)$. □

The invariants q, g and β are easy to handle in the minimally elliptic case. Let us look at g and β first. Obviously $g \geq \beta$, and $g = 0$ except in case (1). In the simply elliptic case $g = 1$, and from (2) and 4.1 we see immediately that $\beta = 0$ (in fact $\beta = 0$ for all quasihomogeneous surface singularities by [W4, Corollary 1.11],

Theorem 1.9 of [W4] gives an explicit formula for q in terms of other invariants. Analyzing all possible cases we get the following values for q:

- If X is quasihomogeneous then either $q = 0$ (X simply elliptic) or $q = 1$ (all other cases).
- If X is not quasihomogeneous then $q = 0$ always.

For the computation of $H^0(\tilde{X}, \mathcal{C})$ and of $H^1(\tilde{X}, \mathcal{R})$ it is essential to know how global holomorphic functions restrict to the irreducible components of the exceptional divisor. The following result of Laufer [Lau, Lemma 3.12] provides all the information we need.

Proposition 4.2 *Let $\pi : \tilde{X} \to X$ be the minimal good resolution of a minimally elliptic singularity. Let Y be a positive cycle on the exceptional divisor E such that $YE_i \leq 0$ for all irreducible components E_i of E. Let Z be the fundamental cycle, and A the anticanonical cycle, $A = \sum_{i=1}^{r} r_i E_i$. Let E_1 be an irreducible component of E.*

1. *If E_1 is a rational curve, $E_i Y = 0$ for $i \neq 1$, and $r_1 = 1$ then the image of the restriction map*

$$H^0(\tilde{X}, \mathcal{O}_{\tilde{X}}(-Y)) \to H^0(\tilde{X}, \mathcal{O}_{\tilde{X}}(-Y))/H^0(\tilde{X}, \mathcal{O}_{\tilde{X}}(-Y - E_1)) \cong H^0(E_1, \mathcal{O}_{E_1}(-Y))$$

   is a subspace S of codimension 1 in $H^0(E_1, \mathcal{O}_{E_1}(-Y))$. If $\dim S \geq 2$ the linear system S has no basepoints. If $\dim S = 1$ there is one basepoint $x \in E_1 \setminus (E_2 \cup \ldots \cup E_r)$.

2. *If E_1 is not as in (1) then the restriction map is surjective.*

The next Lemma computes $\mathcal{C}$ in many cases.

Lemma 4.3 *Assume that $ZE \leq -2$.*

1. *If $ZE_1 < 0$ then for $p \in E_1 \setminus (E_2 \cup \ldots \cup E_r)$ the stalk $\mathcal{C}_p = 0$.*

2. *Let E_1,E_2 be components of E which intersect transversely in a point $p \in \tilde{X}$, and assume that $ZE_i < 0$, $i = 1, 2$. Then near p the sheaf $\mathcal{C}$ is concentrated in p with a stalk of length one there.*

Proof: By the assumption $ZE \leq -2$ either the restriction map in Proposition 4.3 is onto or it has an image without basepoints. Hence for a point p as in (1) we have global holomorphic functions f and g such that f is a local equation for Z near p, and g has a divisor $(g) = Z + \Delta$, where Δ is a piece of a smooth curve through p, meeting E transversely. The differentials df and dg generate the stalk $I_Z \Omega^1_{\tilde{X}} \langle \log E \rangle_p$.

The proof of (2) is exactly the same as in [B-K, Lemma 3.3]. □

We assume now that all the components of the exceptional set of the minimal good resolution are rational (i. e. that X is not simply elliptic). Then over the components E_i the restriction of the locally free sheaf $\mathcal{R}$ splits into a direct sum of line bundles on E_i. We want in some cases to compute the degrees of these line bundles. By restriction of the exact sequences (6) and (7) to a component E_1, say, we get

$$0 \to Tor_1^{\mathcal{O}_{\tilde{X}}}(\mathcal{C}, \mathcal{O}_{E_1}) \to \mathcal{F} \otimes \mathcal{O}_{E_1} \to I_Z \Omega^1_{\tilde{X}} \langle \log E \rangle \otimes \mathcal{O}_{E_1} \to \mathcal{C} \otimes \mathcal{O}_{E_1} \to 0 \qquad (8)$$

and

$$0 \to \mathcal{R} \otimes \mathcal{O}_{E_1} \to \mathcal{O}^n_{E_1} \to \mathcal{F} \otimes \mathcal{O}_{E_1} \to 0. \qquad (9)$$

We use (8) to compute sections of $\mathcal{F}\otimes\mathcal{O}_{E_1}(l)$, l a positive integer, and we try to check that

$$H^0(E_1,\mathcal{O}_{E_1}(l)^n)\to H^0(E_1,\mathcal{F}\otimes\mathcal{O}_{E_1}(l))$$

is surjective. Since E_1 is a rational curve, surjectivity would imply that $H^1(E_1,\mathcal{R}\otimes\mathcal{O}_{E_1}(l))=0$, giving a lower bound of $-(l+1)$ for the degrees of the line bundle summands of $\mathcal{R}\otimes\mathcal{O}_{E_1}$. Computing with the twisted sheaves $\mathcal{O}_{E_1}(l)$ means that we allow meromorphic linear combinations $g_1df_1+\cdots+g_ndf_n$, where the coefficients have poles of order at most l along a fixed curve E_i meeting E_1.

Let $\mathcal{F}_1=\mathcal{F}\otimes\mathcal{O}_{E_1}$, and $\tilde{\mathcal{F}}_1$ the image of $\mathcal{F}_1$ in $I_Z\Omega^1_{\tilde{X}}\langle\log E\rangle\otimes\mathcal{O}_{E_1}$. If the restriction of $\mathcal{C}$ to $E_1\setminus(E_2\cup\ldots\cup E_r)$ is zero then $\tilde{\mathcal{F}}_1$ is precisely the torsion free part of $\mathcal{F}_1$. The differential df of a global holomorphic function f gives a nonzero element in $H^0(E_1,\tilde{\mathcal{F}}_1)$ if and only if f vanishes to minimal order along E_1, i. e. if the vanishing order of f along E_1 equals the multiplicity of E_1 in the fundamental cycle Z ([B-K, Lemma 4.6] or check directly in local coordinates).

If D is an effective divisor supported on E such that $E_1\not\subset|D|$, and $\mathcal{C}$ is locally free as an $\mathcal{O}_D$-module, then $Tor_1^{\mathcal{O}_{E_1}}(\mathcal{C},\mathcal{O}_{E_1})=0$ and $\mathcal{F}_1$ has no torsion.

If on the other hand $\mathcal{C}$ is a skyscraper sheaf, concentrated on points of intersection of E_1 with other E_i then every such point contributes a skyscraper of length one to the torsion of $\mathcal{F}_1$. Let r_i be the multiplicity of E_i in Z. Then the differential df generates the torsion of $\mathcal{F}_1$ in $E_1\cap E_2=\{p\}$ if and only if f vanishes to order r_1+1 along E_1 and to order r_2 along E_2 [B-K, Lemma 4.4].

Remark: Of course this discussion does not cover all the cases, but it is sufficient for the applications we have in mind.

Proposition 4.4 *Assume that either $\mathcal{C}$ is a skyscraper sheaf with stalks of length one concentrated in the singular locus of the exceptional set E, or that there exists a divisor D supported on E such that $\mathcal{C}$ is an invertible $\mathcal{O}_D$-module. Let $E_1\cong\mathbf{P}^1(\mathbf{C})$ be a component of E with $ZE_1<0$.*

1. *If the restriction map $H^0(\tilde{X},\mathcal{O}_{\tilde{X}}(-Z))\to H^0(E_1,\mathcal{O}_{E_1}(-Z))$ is onto then $\mathcal{R}\otimes\mathcal{O}_{E_1}$ is a direct sum of line bundles of degree at least -2.*

2. *If the restriction map is not surjective, then $\mathcal{R}\otimes\mathcal{O}_{E_1}$ has one summand of degree -4 or two summands of degree -3, and the other summands have degree at least -2.*

Proof: We have to show that $H^0(E_1,\mathcal{O}_{E_1}(1))^n\to H^0(E_1,\mathcal{F}_1(1))$ is surjective. First consider the case where $\mathcal{F}_1$ has torsion. Let $E_2,\ldots,E_t$ be the components of E which meet E_1. Then Proposition 4.2 applies to $Y_i=Z+E_1+E_i$, $i=1,\ldots,t$, so we get holomorphic functions f_i such that f_i generates the torsion part of $\mathcal{F}_1$ everywhere along E_1, except in $E_1\cap E_i$. Together these f_i generate the torsion subsheaf of $\mathcal{F}_1$.

From the exact sequence (8) one gets $-2ZE_1$ and $-2ZE_1+2$ for the dimensions of $H^0(E_1,\tilde{\mathcal{F}}_1)$ respectively: we have

$$H^1(E_1,\tilde{\mathcal{F}}_1)=H^1(E_1,\tilde{\mathcal{F}}_1(1))=0$$

$$I_Z\Omega^1_{\tilde{X}}\langle\log E\rangle \otimes \mathcal{O}_{E_1} = (\omega_{E_1}(t_1) \oplus \mathcal{O}_{E_1})(-ZE_1)$$

and $H^0(E_1, \mathcal{C} \otimes \mathcal{O}_{E_1})$ has length t_1, the number of curves which meet E_1. On the other hand $H^0(E_1, \mathcal{O}_{E_1}(-Z)$ has dimension $d_1 = -ZE_1 + 1$. Let E_2 be a curve which meets E_1. In case (1) there are holomorphic functions $f_1, \ldots, f_{d_1} \in H^0(\tilde{X}, \mathcal{O}_{\tilde{X}}(-Z))$ with $f_i = v^{r_1}u^{r_2+i-1} +$ higher order terms, $i = 1, \ldots, d_1$, where $E_1 = \{v = 0\}$ and $E_2 = \{u = 0\}$ locally. If we differentiate the f_i and restrict to E_1, by a little computation in local coordinates we see that $df_1, (1/u)df_1, \ldots, df_{d_1}, (1/u)df_{d_1}$ are linearly independent elements of $H^0(E_1, I_Z\Omega^1_{\tilde{X}}\langle\log E\rangle \otimes \mathcal{O}_{E_1}(1)$, so they span $H^0(E_1, \tilde{\mathcal{F}}_1)$.

In case (2) we only have $d_1 - 1$ functions at our disposal. Their linear combinations with coefficients $\alpha + \beta/u$ span a subspace of codimension 2 in $H^0(E_1, \tilde{\mathcal{F}}_1(1)$, hence $H^1(E_1, \mathcal{R} \otimes \mathcal{O}_{E_1}(1))$ is twodimensional. □

Remark: One can even get more precise information. If k of the functions $f_1, \ldots, f_n$ defining the map μ go to 0 in $H^0(E_1, \mathcal{F}_1)$ then $\mathcal{R} \otimes \mathcal{O}_{E_1}$ has a trivial summand of rank k. In case (1) of 4.4 the cokernel of $H^0(E_1, \mathcal{O}^n_{E_1}) \to H^0(E_1, \mathcal{F}_1)$ has length $d_1 - 2$, this gives the number of -2 summands. By (8) and (9) the Euler characteristic of $\mathcal{R} \otimes \mathcal{O}_{E_1}$ equals $n + 2ZE_1 - t_1$ if $\mathcal{C}$ is a skyscraper sheaf near E_1 , and $n + 2ZE_1$ in the second case of 4.4 (1). Hence we have in these two cases

$$\mathcal{R} \otimes \mathcal{O}_{E_1} = \mathcal{O}(-2)^{(d_1-2)} \oplus \mathcal{O}_{E_1}(-1)^{t_1} \oplus \mathcal{O}_{E_1}^{(n-d_1-t_1)} \quad (10)$$

$$\mathcal{R} \otimes \mathcal{O}_{E_1} = \mathcal{O}(-2)^{(d_1-2)} \oplus \mathcal{O}_{E_1}^{(n-d_1)} \quad (11)$$

We now come to the main result of this section. Observe that if the exceptional divisor of the minimal resolution has only smooth components crossing normally then by the adjunction formula $-ZE_i = KE_i = -2 + 2g_i - E_iE_i$ where g_i is the genus of E_i. Therefore in the minimally elliptic case only rational curves of self-intersection number -2 have trivial intersection with the fundamental cycle.

Theorem 4.5 *Let X be a minimally elliptic singularity of degree d, and assume that on the minimal resolution all the components of the exceptional divisor are smooth rational curves and have normal crossings. If there are no (-2)-curves, and $E_iE_i \leq -5$ for at least one i, $1 \leq i \leq r$, then*

$$\dim T^1_X = q + r + b - 1 + \dim H^0(|Z - E|, I_E\Omega^1_{\tilde{X}}\langle\log E\rangle) - (d + \chi(\mathcal{R} \otimes \mathcal{O}_Z))$$

where b is the number of cycles of the dual graph (so $b = 1$ for a cusp and $b = 0$ otherwise).

Proof: We know that $g = \beta = 0$, and that $\dim H^0(\tilde{X}, \mathcal{C}) = r + b - 1$. So the only thing to prove is that under the hypotheses of the Theorem $\dim H^1(\tilde{X}, \mathcal{R}) - n = -\chi(\mathcal{R} \otimes \mathcal{O}_Z) - d$. Since for a coherent sheaf $\mathcal{S}$ on $\tilde{X}$ by the comparison theorem

$$H^1(\tilde{X}, \mathcal{S}) = \varprojlim_n H^1(E, \mathcal{S} \otimes \mathcal{O}_{nZ})$$

and in the exact sequences

$$0 \to \mathcal{O}_Z(-nZ) \to \mathcal{O}_{(n+1)Z} \to \mathcal{O}_{nZ} \to 0$$

the sheaves $\mathcal{O}_Z(-nZ)$ are generated by their global sections the vanishing of $H^1(E, \mathcal{S} \otimes \mathcal{O}_Z)$ is equivalent to the vanishing of $H^1(\tilde{X}, \mathcal{S})$. We use this observation to show that the first cohomology of $\mathcal{R}(-Z)$ vanishes.

By Serre duality $H^1(E, \mathcal{R}(-Z) \otimes \mathcal{O}_Z)$ is dual to $H^0(E, \mathcal{R}^* \otimes \mathcal{O}_Z(Z))$ (recall that $\omega_Z \cong \mathcal{O}_Z$). Now $\mathcal{R}^* \otimes \mathcal{O}_Z(Z)$ restricted to a curve E_i is a line bundle of degree at most 0, and of strictly negative degree for at least one i, say $i = 1$. There is a computation sequence

$$Z_1 = E_1, \ldots, Z_{k+1} = Z_k + E_{i_k}, \ldots, Z_t = Z$$

with $Z_k E_{i_k} > 0$ for all k. From the exact sequences

$$\begin{aligned} 0 \to H^0(E_{i_k}, \mathcal{R}^* \otimes \mathcal{O}_{E_{i_k}}(-Z_k)) &\to H^0(|Z_{k+1}|, \mathcal{R}^*(Z) \otimes \mathcal{O}_{Z_{k+1}}) \\ &\to H^0(|Z_k|, \mathcal{R}^*(Z) \otimes \mathcal{O}_{Z_k}) \to \ldots \end{aligned}$$

one gets inductively that $H^0(|Z_k|, \mathcal{R}^* \otimes \mathcal{O}_{Z_k}) = 0$ for $k = 1, \ldots, t$.

To end the proof of Theorem 4.5 we have to show the equality

$$\dim H^1(\tilde{X}, \mathcal{R} \otimes \mathcal{O}_Z)) - n = -\chi(\mathcal{R} \otimes \mathcal{O}_Z) - d.$$

Among the holomorphic functions $f_1, \ldots, f_n$ used to define the map μ with kernel $\mathcal{R}$ one can choose a minimal set of generators, say $f_1, \ldots, f_d$. Let $\mathcal{R}'$ be the kernel of the corresponding map $\mathcal{O}^d_{\tilde{X}} \to I_Z \Omega^1_{\tilde{X}}(\log E)$. Then it is obvious that $\mathcal{R} \cong \mathcal{R}' \oplus \mathcal{O}^{(n-d)}_{\tilde{X}}$. Since $\chi(\mathcal{O}_Z) = 0$, and $\dim H^1(|Z|, \mathcal{O}_Z) = 1$ the assertion reduces to the following

Lemma 4.6 *Let $\mathcal{R}', Z$ be as above. Then $H^0(E, \mathcal{R}' \otimes \mathcal{O}_Z) = 0$.*

Proof: Recall that $H^0(E, \mathcal{O}_Z) = \mathbf{C}$. A nontrivial global section of $\mathcal{R}' \otimes \mathcal{O}_Z$ is a complex linear combination $c_1 df_1 + \cdots + c_d df_d \in H^0(\tilde{X}, I_Z^2 \Omega^1_{\tilde{X}}(\log E))$. First of all $c_1 df_1 + \cdots + c_d df_d$ cannot be zero as a differential form on $\tilde{X}$ because this would contradict the minimality of the set $f_1, \ldots, f_d$ of generators. But for $f = c_1 f_1 + \cdots + c_d f_d$ the differential $df \in H^0(\tilde{X}, I_Z^2 \Omega^1_{\tilde{X}}(\log E))$ implies that already f is an element of $H^0(\tilde{X}, I_Z^2)$. Considered as a holomorphic function on X it is in the square of the maximal ideal at the singular point — again contradicting the minimality of $f_1, \ldots, f_d$. □

We shall apply Theorem 4.5 to a family of quasihomogeneous surface singularities, whose deformation theory has already been investigated by various authors (cf. [W3]). Denote by $D(r-2; b_1, \ldots, b_r)$ a normal surface singularity with dual graph

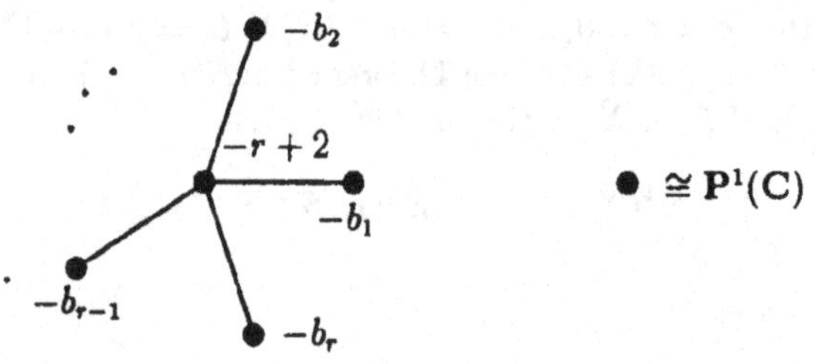

of the minimal resolution. There is an $(r-2)$-parameter family of such singularities. If a $D(r-2; b_1, \ldots, b_r)$ singularity has a $\mathbf{C}^*$-action then the irregularity $q = 1$, and $q = 0$ otherwise.

Assume $r \geq 4$, $b_1, \ldots, b_r \geq 2$, and $1/b_1 + \ldots + 1/b_4 < 2$ for $r = 4$. Then the $D(r-2; b_1, \ldots, b_r)$ singularities are minimally elliptic, of degree $d = b_1 + \ldots + b_r - 8$, and the fundamental cycle on the minimal resolution is $Z = 2E_0 + E_1 + \ldots + E_r$, where E_0 denotes the central curve.

Theorem 4.7 *Let X be a $D(r-2; b_1, \ldots, b_r)$ singularity with $r \geq 5$, and $b_1, \ldots, b_r \geq 3$, and assume that moreover one of the b_i is at least 5, or that $r \geq 7$. Then*

$$\dim T^1_X = (q-1) + d + 3(r-2)$$

□

Remark: It is interesting to compare this to Wahl's lower estimate (see [B-K], Appendix). It asserts that for a minimally elliptic singularity of degree at least 5, if the irreducible components of the exceptional divisor of the minimal resolution are smooth rational curves and intersect normally,

$$\dim T^1_X \geq \dim H^1(\tilde{X}, Der_E(\tilde{X})) + s + Z(Z-E) + d$$

Here s is the number of components of the exceptional set, and $Der_E(\tilde{X})$ is the sheaf of vectorfields on $\tilde{X}$ which are parallel to all components of the exceptional set. In the case of Theorem 4.7 this gives a lower bound of $q + d + r + 1$ for $\dim T^1_X$. So the actual dimension of T^1_X is always strictly bigger, and the difference can be even arbitrarily large.

Proof: $Z - E$ is the central curve E_0, and $I_E\Omega^1_{\tilde{X}}(\log E) \otimes \mathcal{O}_{E_0} \cong \omega_{E_0} \oplus \mathcal{O}_{E_0}(-2)$. So $\dim H^0(|Z-E|, I_E\Omega^1_{\tilde{X}}(\log E) \otimes \mathcal{O}_{Z-E}) = r - 3$. Using (10) the Euler characteristic of $\mathcal{R} \otimes \mathcal{O}_Z$ can be computed to be $(-2)(b_1 + \ldots + b_r) - r + 20$, and by Theorem 4.5 we have

$$\dim T^1_X = q + (r-3) + r + (2d + r - 4) - d = (q-1) + d + 3(r-2).$$

□

We give another example where (4.5) applies: Let X be a cusp singularity with at least 3 irreducible components of the exceptional divisor of the minimal resolution, and such that the hypotheses of Theorem 4.5 apply (i. e. there are no (-2) curves , and at least one component has self-intersection number $E_iE_i \leq -5$). The fundamental cycle is reduced, so it remains to compute the Euler characteristic of $\mathcal{R} \otimes \mathcal{O}_Z$. A little calculation, again using (10), shows that $\chi(\mathcal{R} \otimes \mathcal{O}_Z) = -2d$. So $\dim T^1_X = r + d$, giving back part of the results of [Be].

5 Triangle Singularities

In this Section we illustrate our method for computing T^1_X of minimally elliptic singularities in the case of the triangle singularities $Cu(d)$ and their non-quasihomogeneous satellites, both with a cuspidal exceptional curve of self-intersection $(-d)$ on the minimal resolution, and the dual graph

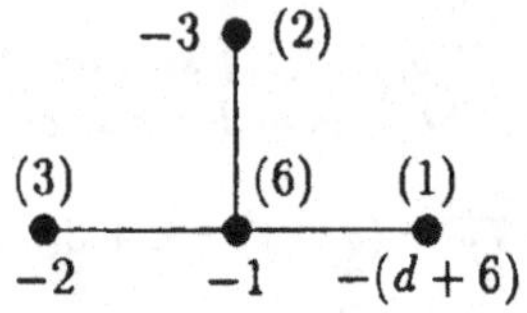

of the minimal good resolution (numbers in brackets denote the multiplicities of the fundamental cycle). This example provides a good opportunity to show that in principle the technique can be applied even though on the minimal good resolution there are curves of low self-intersection number, and to see its limitations.

For each of these graphs there exist exactly two different analytic structures, one is quasihomogeneous, the other is not. In the quasihomogeneous case $q = 1$, and $q = 0$ otherwise.

Let us fix the following notation: $E_0, \dots, E_3$ are the components of the exceptional divisor of the minimal good resolution, E_0 is the central curve, and $E_1^2 = -2$, $E_2^2 = -3$, and $E_3^2 = -(d+6)$. The anticanonical divisor is $A = 2E_0 + E_1 + E_2 + E_3$.

Proposition 5.1 *For a triangle singularity the inclusion ρ (cf. Section 3) has a cokernel of length $q+1$.*

Proof : Since $g = \beta = 0$ it remains to compute the cokernel of the inclusion

$$H^0(\tilde{X}, I_Z\Omega^1_{\tilde{X}}(\log E)) \to H^0(\tilde{X}, I_E\Omega^1_{\tilde{X}}(\log E))$$

We have $E \le A \le Z$, and we proceed in two steps. The inclusion $H^0(\tilde{X}, I_A\Omega^1_{\tilde{X}}(\log E)) \to H^0(\tilde{X}, I_E\Omega^1_{\tilde{X}}(\log E))$ is an isomorphism, because $I_E/I_A = \mathcal{O}_{E_0}(-E)$, and

$$H^0(E_0, I_E\Omega^1_{\tilde{X}}(\log E) \otimes \mathcal{O}_{E_0}) \cong H^0(E_0, \omega_{E_0}(1) \oplus \mathcal{O}_{E_0}(-2)) = 0$$

In the exact sequence

$$0 \to I_Z\Omega^1_{\tilde{X}}(\log E) \to I_A\Omega^1_{\tilde{X}}(\log E) \to I_A\Omega^1_{\tilde{X}}(\log E) \otimes \mathcal{O}_{Z-A} \to 0$$

we have to compute the section of the sheaf on the right. $|Z - A|$ is an exceptional curve of the first kind, so by construction Z meets every component of $|Z - A|$ trivially, and

$$I_A\Omega^1_{\tilde{X}}(\log E) \otimes \mathcal{O}_{Z-A} \cong \Omega^1_{\tilde{X}}(\log E) \otimes \mathcal{O}_{Z-A}(Z - A)$$

There are natural inclusions

$$\begin{aligned} &H^0(E_0, \Omega^1_{\tilde{X}}(\log E) \otimes \mathcal{O}_{E_0}(E_0)) \quad \to \\ &H^0(|Z - A|, \Omega^1_{\tilde{X}}(\log E) \otimes \mathcal{O}_{Z-A}(Z - A)) \quad \to \quad H^1_E(\tilde{X}, \Omega^1_{\tilde{X}}(\log E)) \end{aligned}$$

(cf. [W1, Proposition 2.2]). The space on the left hand side is clearly one dimensional, and for the cohomology group on the right observe that $\Omega^1_{\tilde{X}}\langle\log E\rangle \cong \Omega^1_{\tilde{X}}\langle\log E\rangle^* \otimes \omega_{\tilde{X}}(E)$ so that by Serre duality $H^1_E(\Omega^1_{\tilde{X}}\langle\log E\rangle)$ is dual to $H^1(\tilde{X}, I_E\Omega^1_{\tilde{X}}\langle\log E\rangle)$ which has been computed by Wahl in [W4, 1.9.1] to have dimension one too. □

We shall restrict to a situation where the sheaf C is concentrated on the exceptional curves which arise by resolving the exceptional divisor on the minimal resolution. The next Lemma is elementary, but useful.

Lemma 5.2 *Let $(S,0) \subset (\mathbf{C}^2,0)$ be an isolated plane curve singularity. Let $\sigma : Y \to \mathbf{C}^2$ be a resolution such that Y is smooth, and $\sigma^{-1}(S)$ is a divisor with normal crossings. Then the subsheaf of Ω^1_Y, generated by the differentials $D(\sigma^* f)$, $f \in I_S$, can already be generated by the three sections $d(\sigma^* g)$, $d(\sigma^*(xg))$, and $d(\sigma^*(yg))$, where g is a generator of the principal ideal I_S and x and y are local coordinates of $\mathbf{C}^2$ at the singular point of S.* □

Lemma 5.3 *For X of type $Cu(d)$ let $D = 2E_0 + 2E_1 + E_2$. Then C is an invertible $\mathcal{O}_D$-module, and the Chern class*

$$c_1(C \otimes \mathcal{O}_{E_i}) = \begin{cases} 1 & i = 0 \\ -1 & i = 1,2 \end{cases}$$

In particular $\dim H^0(\tilde{X}, C) = 1$.

Proof: Blowing down to the minimal resolution we find three global holomorphic functions g, xg and yg as in Lemma 5.2. Locally around the singular point of the exceptional curve of the minimal resolution we have $g = y^2 - x^3$. Blowing up until we reach the minimal good resolution, we get coordinate patches $U_i \cong \mathbf{C}^2$, $i = 0, \ldots, 3$ with coordinate functions u_i, v_i on U_i, such that

$$U_0 \cap U_2 = \{u_0 \neq 0\}, \text{and } u_0 = \frac{1}{u_2}, \quad v_0 = u_2^3 v_2$$
$$U_1 \cap U_3 = \{u_1 \neq 0\}, \text{and } u_1 = \frac{1}{v_3}, \quad v_1 = v_3^2 u_3$$
$$U_2 \cap U_3 = \{u_2 \neq 0\}, \text{and } u_2 = \frac{1}{u_3}, \quad v_2 = u_3 v_3$$

The open set $U_0 \cup \ldots \cup U_3$ is a neighbourhood of $|D|$, and the proper transform of the exceptional curve of the minimal resolution meets D in $u_2 = u_3 = 1$, $v_2 = v_3 = 0$. After a bit of calculation one finds that

$$\begin{aligned} g &= v_0^2(1 - u_0^3 v_0) = v_1^3(u_1^2 v_1 - 1) = u_2^2 v_2^6(1 - u_2) = u_3^3 v_3^6(u_3 - 1) \\ xg &= u_0 v_0^3(1 - u_0^3 v_0) = v_1^4(u_1^2 v_1 - 1) = u_2^3 v_2^8(1 - u_2) = u_3^4 v_3^8(u_3 - 1) \\ yg &= v_0^3(1 - u_0^3 v_0) = u_1 v_1^5(u_1^2 v_1 - 1) = u_2^3 v_2^9(1 - u_2) = u_3^5 v_3^9(u_3 - 1) \end{aligned}$$

For example in $u_3 = v_3 = 0$ where the curves E_0 and E_1 of self- intersection numbers -1 and -2 respectively meet the stalk of $I_Z\Omega^1_{\tilde{X}}\langle\log E\rangle$ has a basis consisting of $e_1 = u_3^2 v_3^6 du_3$ and $e_2 = u_3^3 v_3^5 dv_3$. Now $dg = (4u_3 - 3)e_1 + (6u_3 - 6)e_2$, $d(xg) = u_3 v_3^2((5u_3 - 4)e_1 +$

$(8u_3 - 8)e_2)$, and one checks immediately that modulo $dg, d(xg)$ the two basis elements are congruent, and that $u_3^2 v_3^2 e_1 = 0$

To compute the Chern classes observe that $\mathcal{C} \otimes \mathcal{O}_{E_i}$ is the quotient of $I_Z\Omega^1_{\tilde{X}}(\log E) \otimes \mathcal{O}_{E_i} = (\omega_{E_i}(t_i) \oplus \mathcal{O}_{E_i})(-ZE_i)$ by the image of $\mathcal{F} \otimes \mathcal{O}_{E_i}$. Here again t_i denotes the number of curves adjacent to E_i. Locally the two generators are identified modulo $\mathcal{F} \otimes \mathcal{O}_{E_i}$. For the central curve $t_i = 3$, and $t_i = 1$ for the others. □

The result of this section is

Proposition 5.4 *Let X be a minimally elliptic singularity of type $Cu(d)$, $d \geq 8$, then* $\dim T^1_X = q + d + 2$.

Proof: Let P be the divisor $5E_0 + 3E_1 + 2E_2 + E_3$.

There is exactly one curve, namely E_3, with $ZE_i < 0$. Hence by Proposition 4.4 (2) $\mathcal{R} \otimes \mathcal{O}_{E_3}$ is a direct sum of line bundles of degree at least -4. Moreover the cokernel of (9) is $(b_3 - 6)$-dimensional, and $\mathcal{R} \otimes \mathcal{O}_{E_3}$ has a trivial summand of rank $n - b_3 + 6$. In a neighbourhood of $E_0 \cup E_1 \cup E_2$ let $\mathcal{L}$ be the nontrivial line bundle summand of $\mathcal{R}$. Since $Tor_1^{\mathcal{O}_{\tilde{X}}}(\mathcal{C}, \mathcal{O}_{E_i}) \cong \mathcal{O}_{E_i}(-D)$ we see from the exact sequence (8) that $\mathcal{F} \otimes \mathcal{O}_{E_i}$ has first Chern class 0 for $i = 0, 2$, and 1 for $i = 1$ (observe that $\mathcal{C}$ has no subsheaves concentrated in points , so $\mathcal{F}$ is locally free). So $c_1(\mathcal{L} \otimes \mathcal{O}_{E_i}) = 0$ for $i = 1, 2$, and $= -1$ for $i = 1$.

Claim: Then $H^1(\tilde{X}, \mathcal{R}(-P)) = 0$, and $H^1(\tilde{X}, \mathcal{R}) \cong H^1(\tilde{X}, \mathcal{R} \otimes \mathcal{O}_P)$ has dimension $n + d$.

For $H^1(\tilde{X}, \mathcal{R}(-P)) = 0$ we need $H^1(|Z|, \mathcal{R}(-P) \otimes \mathcal{O}_Z) = 0$, which by Serre duality is equivalent to $H^0(|Z|, \mathcal{R}^* \otimes \mathcal{O}_Z(Z - A + P)) = 0$. The divisor $Z - A + P = 9E_0 + 5E_1 + 3E_2 + E_3$ has intersection numbers

$$(Z - A + P)E_i = \begin{cases} 0, & i = 0 \\ -1, & i = 1 \\ 0, & i = 2 \\ -(d-3), & i = 3 \end{cases}$$

By $d \geq 8$ the vector bundle $\mathcal{R}^*(Z - A + P)$ restricted to E_i is a direct sum of line bundles of degree at most 0, and strictly less than 0 for $i = 3$. A computation sequence as in the proof of Theorem 4.5, starting with E_3, shows what we want.

For the computation of $H^1(|P|, \mathcal{R} \otimes \mathcal{O}_P)$ observe that $P \leq Z$. Hence as in the proof of (4.5) we compute $\chi(\mathcal{R} \otimes \mathcal{O}_P) - d$, which equals the dimension of $H^1(|P|, \mathcal{R} \otimes \mathcal{O}_P) - n$. An easy calculation yields $\chi(\mathcal{R} \otimes \mathcal{O}_P) = -2d$. Now $P \geq A$, so $H^1(|P|, \mathcal{O}_P)$ has dimension one, and

$$\dim H^1(|P|, \mathcal{R} \otimes \mathcal{O}_P) = -\chi(\mathcal{R} \otimes \mathcal{O}_P) + n - d = n + d.$$

□

References

[Be] K. Behnke. Infinitesimal deformations of cusp singularities. Math. Ann. 265, 407–422,1983.

[B-K] K. Behnke, H. Knörrer. On infinitesimal deformations of rational surface singularities. Compositio Math. 61, 103–127,1987.

[Lau] H. Laufer. On minimally elliptic singularities. Amer. Journal of Math. 99, 1257–1295, 1977.

[Pi] H. Pinkham. Deformations of normal surface singularities with $\mathbf{C}^*$ action. Math. Ann. 232, 65–84, 1978.

[Sch] M. Schlessinger. Rigidity of quotient singularities. Inventiones Math. 14, 17–26,1971.

[W1] J. Wahl. Vanishing theorems for resolutions of singularities. Inventiones Math. 31, 17–41, 1975.

[W2] —. Simultaneous resolution and discriminantal loci. Duke Math. Journal 46, 341–375, 1979.

[W3] —. Derivations of negative weight and non-smoothability of certain singularities. Math. Ann. 258, 383-398, 1982.

[W4] —. A characterization of quasihomogeneous Gorenstein surface singularities. Compositio Math. 55, 3-32, 1984.

[W5] —. The Jacobian Algebra of a quasihomogeneous Gorenstein surface singularity. Duke Math. Journal 55 , 843–871, 1987.

[Y] S. T. Yau. $s^{(n-1)}$-invariant for isolated n-dimensional singularities and its applications to moduli problems. Amer. Journal of Math. 104, 829–841, 1982.

Author's address : Mathematisches Seminar der Universität Hamburg
Bundesstraße 55
D-2000 Hamburg 13 , FRG

C-Régularité et Trivialité Topologique

K. Bekka

Introduction

Dans ce travail on introduit une condition de régularité pour les espaces stratifiés, définie par des fonctions de contrôle sans éclatements. Nous dirons pour le moment que cette condition est plus faible que celle de Whitney (voir [5], [12], [15], [18]), et entraine la structure de Thom-Mather (voir [5], [12], [15]).

A) En théorie des singularités on rencontre souvent des stratifications qui ne vérifient pas les conditions de Whitney, mais qui sont localement topologiquement triviales; on trouve de tels exemples dans les travaux de Lê-Ramanujam[11], Bruce-Giblin[2] et Damon[4].

Exemple: Briançon et Speder[3] ont donné un exemple d'une famille de polynômes quasihomogènes complexes, $f(x,y,z,t) = z^5 + ty^6z + y^7x + x^{15}$, à singularité isolée en $(x,y,z) = (0,0,0)$ et à nombre de Milnor constant, telle que l'hypersurface $f(x,y,z,t) = 0$, stratifiée par $(f^{-1}(0) - \Sigma f, \Sigma f)$ (Σf: ensemble des points singuliers de f), soit topologiquement triviale mais ne vérifie pas les conditions de Whitney. On peut montrer que c'est un espace (C)-régulier; plus généralement, la stratification obtenue à partir d'une famille de polynômes quasihomogènes à singularité isolée et à nombre de Milnor constant est (C)-régulière et donc topologiquement triviale.

B) Hironaka (Oslo 1976) a posé le problème suivant: soit $f: M \to N$ une application lisse telle que $f|_A: (A,\Sigma) \to (B,\Sigma')$ soit une application de Thom ([5], [12], [15]), où (A,Σ) (resp.(B,Σ')) est un espace de Whitney de M (resp.de N). Question: peut-on raffiner les stratifications de A et de B tel que f soit une application de Thom et que l'image réciproque d'un sous-Whitney de B soit un sous-Whitney de A?

La réponse est en général non, comme le montre le contre-exemple suivant. Soit

$$F: \mathbf{R}^2 \longrightarrow \mathbf{R}^2$$

$$(x,y) \longmapsto (xe^{-1/r^2}, ye^{-1/r^2})$$

$(r = \sqrt{x^2+y^2})$. On peut alors stratifier la source $\mathbf{R}^2$ par $(\mathbf{R}^2 - \{0\}, \{0\})$ et le but $\mathbf{R}^2$ par $(\mathbf{R}^2 - \{0\}, \{0\})$; pour cette stratification F est une application de Thom (en effet, F est un difféomorphisme sur chaque strate).

Soit la spirale définie par

$$S = \{(x,y) \in \mathbf{R}^2; x = e^{-\sqrt{t}}\cos t \qquad y = e^{-\sqrt{t}}\sin t; 1 \le t < +\infty\} \cup \{0\};$$

cet espace stratifié n'est pas de Whitney, car $\lim_{t\to 0} \frac{\rho(t)}{\rho'(t)} = \infty$, bien qu'il soit l'image réciproque de la spirale de Whitney

$$S' = \{(x,y) \in \mathbf{R}^2; x = e^{-e^{2\sqrt{t}+t}}\cos t \qquad y = e^{-e^{2\sqrt{t}+t}}\sin t; 1 \le t < +\infty\} \cup \{0\}.$$

Montrons maintenant que pour tout raffinement des stratifications de la source et du but S' est toujours un sous-Whitney.

Il suffit de montrer que dans un voisinage de 0, S coupe transversalement les strates du raffinement de Σ; comme la stratification est localement finie il suffit de le vérifier pour une strate de dimension 1. Soit V une telle strate, comme $(V,0)$ vérifie la condition de Whitney, l'angle entre $\vec{0x}$ et T_xV tend vers 0 lorsque x tend vers 0, $x \in V$. Mais pour $x \in S$ l'angle entre $\vec{0x}$ et T_xS tend vers $\frac{\pi}{2}$ lorsque x tend vers 0. Ceci veut dire que T_xS et T_xV sont transverses d'où la conclusion. On termine en remarquant que F est un difféomorphisme hors de 0.

On notera que la réponse est oui, si on remplace la condition (B) de Whitney par la condition (C).

Ce papier est organisé comme suit:

Au § 1 on rappelle les notions d'espace de Thom-Mather et d'applications de Thom et on donne une condition équivalente à la condition (a_f) de Thom (cette équivalence est démontré différemment par Koike [10]). Au § 2 on introduit la condition (C) et on donne une forme équivalente. On termine ce chapitre en montrant que les espaces (C)-réguliers admettent une structure de Thom-Mather. Au § 3 on discute les théorèmes d'isotopie. Ce travail est extrait de ma thèse, faite sous la direction de D.Trotman (le nom de (C)-régularité m'a été suggéré par C.T.C Wall). Je tiens a remercier l'I.H.E.S de Bures sur Yvette pour son hospitalité durant la rédaction de ce papier.

§ 1) Espace stratifié abstrait et application de Thom

1) - Structure d'espace stratifié abstrait. Dans cette partie on donne quelques définitions dont l'essentiel se trouve dans [12].

Soit A un espace topologique, séparé, localement compact, paracompact, et à base dénombrable de voisinages. Soient $X \subset A$ un ensemble localement fermé, T_X un voisinage ouvert de X dans A, $\pi_X : T_X \longrightarrow X$ une rétraction continue $(\pi_X|_X = Id_X)$ et $\rho_X : T_X \longrightarrow \mathbf{R}_+$ une application continue telle que: $\rho_X^{-1}(0) = X$. Soient $\varepsilon, \delta : X \longrightarrow \mathbf{R}_+$ des applications continues. On note $\varepsilon < \delta$ si $\varepsilon(x) < \delta(x)$ pour tout $x \in X$. On définit:

$$X \times (\varepsilon, \delta) = \{(x,t) \in X \times \mathbf{R}_+ ; \varepsilon(x) < t < \delta(x)\};$$

de la même façon on définit $X \times [\varepsilon, \delta], X \times (\varepsilon, \delta]$ et $X \times [\varepsilon, \delta)$;

$$T_X^\varepsilon = \{a \in T_X ; \rho_X(a) < \varepsilon(\pi_X(a))\}$$
$$S_X^\varepsilon = \{a \in T_X ; \rho_X(a) < \varepsilon(\pi_X(a))\}$$

On dira que les triplets (T_X, π_X, ρ_X) et (T'_X, π'_X, ρ'_X) sont équivalents si

$$(T_X, \pi_X, \rho_X) = (T'_X, \pi'_X, \rho'_X)$$

dans un voisinage de X, i.e. s'il existe un voisinage U de X dans A tel que

$$T_X \cap U = T'_X \cap U \quad , \quad \pi_X|_{T_X \cap U} = \pi'_X|_{T_X \cap U}$$

et

$$\rho_X|_{T_X \cap U} = \rho'_X|_{T_X \cap U}.$$

1.1) - Définition. *On appelle tube de X dans A une classe d'équivalence pour cette relation, qu'on note (T_X, π_X, ρ_X).*

Remarques: Quitte à restreindre T_X on peut avoir:

(1) - Si $X \subset U \subset A$ avec U ouvert dans A, alors il existe $\varepsilon > 0$ telle que $T_X^\varepsilon \subset U$.

(2) - $(\pi_X^\varepsilon, \rho_X^\varepsilon) : T_X^\varepsilon \longrightarrow X \times [0, \varepsilon)$ est une application propre pour un certain ε positif.

(3) - Etant donné un compact $K \subset X$ et un voisinage U de K dans X, il existe $\varepsilon \in \mathbf{R}_+^*$ et un voisinage V de K dans X tels que $\pi_X^{-1}(V) \cap T_X^\varepsilon \subset U$.

Remarque: Un voisinage tubulaire est un tube.

1.2) - Définition.

Une structure d'espace stratifié abstrait sur A est la donnée d'une partition localement finie Σ de A en sous-ensembles localement fermés, connexes et un système de tubes

$$T = \{(T_X, \pi_X, \rho_X), X \in \Sigma\},$$

tels que

1) - $X, Y \in \Sigma$ et $X \cap \overline{Y} \neq \emptyset$ implique que $X \subset \overline{Y}$. On note cela $X < Y$.

2) - Chaque strate (i.e. $X \in \Sigma$) est une variété de classe C^1 (sans bord), telle que la structure différentielle soit compatible avec la topologie induite.

3) - Pour tout $X \in \Sigma$ il existe ε_X tel que pour tout $Y \in \Sigma$, $Y \neq X$, et $T_X \cap Y \neq \emptyset$, on a $X < Y$ et $(\pi_X, \rho_X) : T_X^{\varepsilon_X} \cap Y \longrightarrow X \times (0, \varepsilon_X)$ une submersion de classe C^1 (en particulier $dim(X) < dim(Y)$).

(4) - Si $X < Y$ on a pour tout $a \in T_X^{\varepsilon_X} \cap T_Y^{\varepsilon_Y}$ et $\pi_Y(a) \in T_X$

$$a) \pi_X(\pi_Y(a)) = \pi_X(a) \quad \text{qu'on note } (CR\pi)$$
$$b) \rho_X(\pi_Y(a)) = \rho_X(a) \quad \text{qu'on note } (CR\rho)$$

Puisque A est normal on peut supposer de plus que:

(5) -soit $X, Y \in \Sigma$ alors $T_X^{\varepsilon_X} \cap T_Y^{\varepsilon_Y} \neq \emptyset$ si et seulement si $X \leq Y$ ou $X > Y$.

On note (A, Σ, T) l'espace stratifié (A, Σ) muni de cette structure. On aura besoin par la suite de la définition suivante:

On appelle profondeur de $X \in \Sigma$ dans (A, Σ, T) l'entier

$$\text{depth}_\Sigma(X) = Sup\{n; \text{il existe une chaîne de strates}, X = X_0 < X_1 < \cdots < X_n\}.$$

La profondeur de la stratification (A, Σ) est

$$\text{depth}(\Sigma) = Sup\{depth_\Sigma(X); X \in \Sigma\}$$

On remarquera que

$$dim(A) = Sup\{dim(X), X \in \Sigma\} < +\infty \quad \text{entraine que} \quad depth(\Sigma) < +\infty.$$

Si le système de tubes de (A, Σ) vérifie toutes les conditions de la définition d'une structure d'espace stratifié abstrait sauf (peut-être) $(CR\rho)$, on dira que (A, Σ, T) est un espace stratifié abstrait faible.

Soit $A_0 \subset A$ un sous-ensemble localement fermé. On note

$$\Sigma|_{A_0} = \{X \cap A_0 / X \in \Sigma\}$$

et

$$T_A = \{(T_{X\cap A_0}, \pi_{X\cap A_0}, \rho_{X\cap A_0})\}_{X\in\Sigma}.$$

Il n'est pas toujours possible de munir A_0 d'une structure d'espace stratifié abstrait. Des exemples simples de sous-ensembles localement fermés pour lesquels on peut le faire sont les sous-ensembles réunion de strates et les ouverts de A.

Soit $f : (A,\Sigma,T) \longrightarrow (B,\Sigma',T')$ une application continue entre deux espaces stratifiés. Soient X une strate de (A,Σ) et X' une strate de (B,Σ') telles que $f(X) \subset X'$. Si $\{T_X, \pi_X, \rho_X\}$ et $\{T'_X$, et $\pi'_X, \rho'_X\}$ sont les tubes de X et X', on définit la condition (CRf) comme suit:

Il existe δ_X telle que $f(T_X^{\delta_X}) \subset T_{X'}$ et $f \circ \pi_X(a) = \pi_{X'} \circ f(a)$ pour $a \in T_X^{\delta_X}$.

On termine par les deux définitions suivantes:

Bon tube: On appelle bon tube de X dans (A,Σ) un tube vérifiant:

$(\pi_X, \rho_X) : T_X \cap Y \longrightarrow X \times \mathbf{R}$ est une submersion de classe C^1 pour tout $Y \in Et(X)$, où $Et(X) =$ étoile de$X = \{Y \in \Sigma / X < Y\}$.

Ouvert de type (σ) Un ouvert U de A est de type σ, si pour toute strate X dans (A,Σ) on a l'une des deux conditions:

i) $X \cap U \neq \emptyset$

ii) il existe un voisinage V de X dans A tel que $V \cap U = \emptyset$.

Remarque: La réunion d'ouverts de type (σ) est un ouvert de type (σ).

2) Applications de Thom. Dans ce sous-chapitre on se place dans un contexte plongé, toutes les variétés sont de classe C^r où $r \geq 1$ et toutes les rétractions sont des submersions.

2.1) - Soit $f : M \longrightarrow N$ une application de classe C^1 et soient $A \subset M$, $B \subset N$ des ensembles tels que $f(A) \subseteq B$. Une stratification de $f : A \longrightarrow B$ est une paire (Σ, Σ') avec Σ (resp. Σ') une stratification de A (resp. B) qui vérifie les conditions suivantes:

i) f envoie toute strate de Σ dans une strate de Σ', i.e. pour tout $X \in \Sigma$, il existe $X' \in \Sigma'$ tel que $f(X) \subset X'$

ii) Soit X une strate de Σ, qui est envoyée par f dans X' une strate de Σ', alors $f_{|X} : X \longrightarrow X'$ est une submersion.

Une telle application est dite stratifiée. Dans le cas analytique on a le résultat suivant (dû à Hardt [14] et Hironaka [15]):

Soient $f : M \longrightarrow N$ une application analytique, $A \subseteq M$ et $B \subseteq N$ des ensembles sous-analytiques tels que $f : A \longrightarrow B$ soit une application sous-analytique. Alors il existe des stratifications de Whitney Σ de A, Σ' de B telles que f soit une application stratifiée.

2.2) - La condition de Thom (a_f)

Soit $f : (A,\Sigma) \longrightarrow (B,\Sigma')$ une application stratifiée, (X,Y) un couple de strates de (A,Σ) et x un point de $\overline{Y} \cap X$. On dit que le triplet (X,Y,x) vérifie la condition (a_f) si, pour toute suite $\{y_i\}$ de points de Y qui converge vers x, telle que la suite $\{\ker d_{y_i}(f|_Y)\}$ converge dans la grassmannienne appropriée vers τ, alors $\ker d_x(f|_X) \subset \tau$.

Le couple de strates (Y,X) vérifie la condition (a_f) s'il la vérifie en tout point $x \in X$.

Définition. *Soient $f: M \longrightarrow N$ une application de classe C^1, $A \subseteq M$, $B \subseteq N$ des sous-ensembles avec $f(A) \subseteq B$ et (Σ, Σ') une stratification de $f|_A$. On dit que $f:(A,\Sigma) \longrightarrow (B,\Sigma')$ est une application de Thom si la condition (a_f) est vérifiée pour tout couple de strates de Σ.*

Voici un exemple où la condition de régularité de Thom a un sens géométrique simple.

Soit $f: \mathbf{R}^2 \longrightarrow \mathbf{R}$ définie par $f(x,y) = xy$. On définit une stratification de $\mathbf{R}^2$ par $\{0\}$, $\{(x,0), x \neq 0\}$, $\{(0,y), y \neq 0\}$ et $\{(x,y); x \neq 0, y \neq 0\}$ et la stratification de $\mathbf{R}$ par $\{0\}$ et $\mathbf{R} - \{0\}$.

On prend une suite $\{\mathbf{y}_i\}$ dans le premier quadrant qui converge vers un point de l'axe des $\mathbf{x}$. La condition (a_f) dit que les tangentes aux fibres $f^{-1}(f(\mathbf{y}_i)) = \{(x,y), xy = f(\mathbf{y}_i)\}$ convergent vers l'axe des x. Dans l'exemple précédent on a une stratification naturelle qui verifie la condition (a_f).

Ce n'est pas toujours le cas. En effet l'application $f: \mathbf{R}^2 \longrightarrow \mathbf{R}^2$ définie par $f(x,y) = (x, xy)$ n'admet pas de stratification de Thom (pour plus de details voir [5]).

Cet exemple montre que l'on doit imposer des conditions supplémentaires pour que les applications polynomiales soient des applications de Thom. On montre (voir Gibson [5], Mather [9]) que si f est polynomiale et la restriction de f à l'ensemble de ses points critiques $\Sigma(f)$ est à fibre finie, alors f est une application de Thom.

Pour étudier certaines propriétés des applications de Thom on a besoin de remplacer la condition (a_f) par une condition équivalente; pour cela on a besoin des définitions suivantes:

2.3) - Soient V_1, V_2, V_1', V_2' des espaces vectoriels réels et f_1, f_2, f_1', f_2' des applications linéaires tels que le diagramme

$$\begin{array}{ccc} V_2 & \xrightarrow{f_1} & V_1 \\ \downarrow f_2 & & \downarrow f_2' \\ V_2' & \xrightarrow{f_1'} & V_1' \end{array}$$

soit commutatif. Le carré est dit <u>régulier</u> (ou cartésien) si pour tout $v_1 \in V_1$ et $v_2' \in V_2'$ tels que si $f_2'(v_1) = f_1'(v_2')$; il existe $v_2 \in V_2$ tel que $f_1(v_2) = v_1$ et $f_2(v_2) = v_2'$.

2.4) - Remarque: Soit un diagramme commutatif

$$\begin{array}{ccc} V_2 & \xrightarrow{f_1} & V_1 \\ \downarrow g_2 & & \downarrow g_1 \\ V_2' & \xrightarrow{f_1'} & V_1' \\ \downarrow g_2' & & \downarrow g_2' \\ V_2'' & \xrightarrow{f_1''} & V_1'' \end{array}$$

d'espaces vectoriels et d'applications linéaires. Si le carré supérieur et le carré inférieur sont réguliers alors on voit facilement que le grand carré est régulier.

2.5) - Soit un carré commutatif

$$\begin{array}{ccc} M_2 & \xrightarrow{g_1} & M_1 \\ \downarrow f_2 & & \downarrow f_1 \\ N_2 & \xrightarrow{g_2} & N_1 \end{array}$$

de variétés différentiables et d'applications différentiables.

On dira que ce carré est régulier si pour tout $m \in M_2$ le carré commutatif

$$\begin{array}{ccc} T_m M_2 & \xrightarrow{d_m g_1} & T_{g_1(m)} M_1 \\ \downarrow d_m f_2 & & \downarrow d_{g_2(m)} f_1 \\ T_{f_2(m)} N_2 & \xrightarrow{d_{f_2(m)} g_2} & T_{f_1 \circ g_1(m)} N_1 \end{array}$$

est un carré régulier d'espaces vectoriels et d'applications linéaires.

On va maintenant voir le lien entre les diagrammes réguliers et les applications de Thom.

Pour cela, prenant une application stratifiée $f : (A, \Sigma) \longrightarrow (B, \Sigma')$, la proposition suivante montre qu'il y a équivalence entre la condition (a_f) et le fait que certains carrés soient réguliers.

Soient X une strate dans Σ et X' une strate de Σ' telles que $f(X) \subset X'$. Soient π_X une retraction locale en $x_0 \in X$ de classe C^1, $\pi_X : M \longrightarrow X$ et $\pi_{X'}$ une rétraction locale en $f(x_0)$ de classe C^1, $\pi_{X'} : N \longrightarrow X'$.

On dira que ces rétractions vérifient la condition de compatibilité (CRf) si l'on a $f \circ \pi_X = \pi_{X'} \circ f$ dans un voisinage de x_0 dans M.

2.6) - Remarques:

1) Pour toute application stratifiée f (pas nécessairement sans éclatements), on a la propriété suivante:

Pour toute rétraction locale de classe C^1 en $f(x_0)$, $\pi_{X'} : N \longrightarrow X'$, il existe une rétraction locale de classe C^1 en x_0, $\pi_X : M \longrightarrow X$, telle que (CRf) soit satisfaite dans un voisinage de x_0.

En effet, considérant l'application $\pi_{X'} \circ f$ dans un voisinage de x_0, puisque $\pi_{X'} \circ f|_X : X \longrightarrow X'$ est une submersion, il existe une rétraction C^1 en x_0, $\pi_X : M \longrightarrow X$ telle que $\pi_{X'} \circ f \circ \pi_X = \pi_{X'} \circ f$ c.à.d que $f \circ \pi_X = \pi_{X'} \circ f$, donc la condition (CRf) est satisfaite.

2) Dans l'autre sens, ce n'est pas toujours vrai, i.e. pour toute rétraction C^1 en x_0 il existe une rétraction C^1 en $f(x_0)$ telle que (CRf) soit satisfaite.

En effet, considérons l'application $f : \mathbf{R}^3 \longrightarrow \mathbf{R}^2$ définie par $f(x, y, z) = (xy, z)$. On stratifie f à la source par la stratification de $\mathbf{R}^3$ en

$$\begin{aligned} X_1 &= \{(0,0,z) \in \mathbf{R}^3\}, \\ X_2 &= \{(x,0,z) | x \neq 0\}, \\ X_3 &= \{(0,y,z) | y \neq 0\}, \text{et} \\ X_4 &= \mathbf{R}^3 - (X_1 \cup X_2 \cup X_3), \end{aligned}$$

et au but par la stratification de $\mathbf{R}^2$ en

$$\begin{aligned} X_1' &= \{(0,z) \in \mathbf{R}^2\}, \text{ et} \\ X_2' &= \mathbf{R}^2 - X_1'. \end{aligned}$$

Considérons la rétraction sur X_1, $\pi_{X_1} : \mathbf{R}^3 \longrightarrow X_1$ définie par

$$\pi_{X_1}(x, y, z) = (0, 0, y + z).$$

Soient $(0, y_1, z)$ et $(0, y_2, z)$ des point de $\mathbf{R}^3$ avec $y_1 \neq y_2$ alors

$$\pi_{X_1}(0, y_1, z) = (0, 0, y_1 + z) \neq (0, 0, y_2 + z) = \pi_{X_1}(0, y_2, z),$$

par suite

$$f \circ \pi_{X_1}(0, y_1, z) = (0, y_1 + z) \neq (0, y_2 + z) = f \circ \pi_{X_1}(0, y_2, z),$$

mais on a

$$f(0, y_1, z) = f(0, y_2, z),$$

donc il n'existe pas de rétraction locale, $\pi_{X_1'} : \mathbf{R}^2 \longrightarrow X_1'$ telle que (CRf) soit vérifiée. (On remarquera que f est une application de Thom)

2.7) - Proposition.

Soit $f : M \longrightarrow N$ une application de classe C^1 telle que $f : A \longrightarrow B$ soit stratifiée à la source par Σ et au but par Σ'.

Alors les conditions suivantes sont équivalentes:

1) - f est une application de Thom

2) - Pour tout couple de strates (X, X') dans $\Sigma \times \Sigma'$ et de rétractions locales de classe C^1, $\pi_X : M \longrightarrow N$ en $x_0 \in X$ et $\pi_{X'} : N \longrightarrow X'$ en $f(x_0)$ vérifiant (CRf), il existe un voisinage U de x_0 dans M tel que pour tout $y \in Y \cap U$ l'application $d_y(\pi_X|_Y) : \ker d_y(f|_Y) \longrightarrow \ker d_x(f|_X)$ est surjective, avec $x = \pi_X(y)$.

3) - Pour tout couple de strates (X, X') dans $\Sigma \times \Sigma'$ et de rétractions locales de classe C^1 les applications $\pi_X : M \longrightarrow X$ en x_0 et $\pi'_X : N \longrightarrow X'$ en $f(x_0) \in X'$ sont telles que le carré commutatif:

$$\begin{array}{ccc} Y & \xrightarrow{\pi_X|Y} & X \\ \downarrow f|Y & & \downarrow f|X \\ Y' & \xrightarrow{\pi_{X'}|Y'} & X' \end{array}$$

soit régulier en tout point $y \in Y \cap U$.

Preuve :

1) implique 2): Supposons que ce ne soit pas vraie. Comme le carré

$$\begin{array}{ccc} T_yY & \xrightarrow{d_y\pi_X} & T_{\pi_X(y)}X \\ \downarrow d_yf|Y & & \downarrow d_{\pi_X(y)}f|X \\ T_{f(y)}Y' & \xrightarrow{d_{f(y)}\pi_{X'}} & T_{f\circ\pi_X(y)}X' \end{array}$$

est commutatif, il existe une suite $\{y_i\}$ dans Y convergeant vers $x_0 \in X$ telle que

$$d_{y_i}\pi_X : \ker d_{y_i}(f|_Y) \longrightarrow \ker d_{\pi_X(y_i)}(f|_X)$$

ne soit pas surjectif. En passant à une sous-suite on peut supposer que tous les sous-espaces $d_{y_i}(\pi_X)(\ker d_{y_i}(f|_Y))$ ont la même dimension, que cette suite de sous-espaces converge vers $T \subset T_xX$ et que la suite $\{\ker d_{y_i}(f|_Y)\}$ converge vers τ. Comme $\ker d_{\pi_X(y_i)}(f|_X)$ converge vers $\ker d_x(f|_X)$, on a l'inclusion propre

$$d_x\pi_X(\tau) = T \subset \ker d_x(f|_X).$$

Puisque

$$d_x\pi_X(\ker d_x(f|_X)) = \ker d_x(f|_X)$$

et d'après ce qui précède on ne peut pas avoir $\ker d_x(f|_X) \subset \tau$, alors f n'est pas une application de Thom.

2) implique 3): Soit $y \in U \cap Y$, on veut montrer que le carré commutatif

$$\begin{array}{ccc} T_yY & \stackrel{d_y\pi_X|Y}{\longrightarrow} & T_{\pi_X(y)}X \\ \downarrow d_yf|Y & & \downarrow d_{\pi_X(y)}f|X \\ T_{f(y)}Y' & \stackrel{d_{f(y)}\pi_{X'}|Y'}{\longrightarrow} & T_{f\circ\pi_X(y)}X' \end{array}$$

est régulier.

Soient $v_1 \in T_{\pi(y)}X$ et $v_2 \in T_{f(y)}Y'$ tels que

$$d_{\pi(y)}(f|_X)(v_1) = d_{f(y)}(\pi_{X'}|_{Y'})(v_2),$$

comme $d_y(f|_Y)$ est surjective, il existe $v_2' \in T_yY$ tel que $d_y(f|_Y)(v_2') = v_2$ donc

$$d_y(\pi_X) : \ker d_y(f|_Y) + v_2' \longrightarrow \ker d_x(f|_X) + v_1$$

est surjectif, car

$$d_y(\pi_X) : \ker d_y(f|_Y) \longrightarrow \ker d_x(f|_X)$$

l'est. Par suite il existe $\omega \in \ker d_y(f|_Y) + v_2'$ tel que $d_y(\pi_X)(\omega) = v_1$, mais on a aussi $d_y(f|_Y)(\omega) = v_2$, ce qui veut dire que le carré est régulier.

3) implique 1): Supposons que ce ne soit pas vraie. Il existe alors une suite de points $\{y_i\}$ dans Y convergeant vers $x_0 \in X$ telle que la suite de sous-espaces $\{\ker d_{y_i}(f|_Y)\}$ converge vers τ et que $\ker d_x(f|_X)$ ne soit pas inclus dans τ.

On veut montrer qu'il existe une rétraction locale de classe C^1, $\pi_X : M \longrightarrow X$, et un carré commutatif

$$\begin{array}{ccc} Y & \stackrel{\pi_X|Y}{\longrightarrow} & X \\ \downarrow f|Y & & \downarrow f|X \\ Y' & \stackrel{\pi_{X'}|Y'}{\longrightarrow} & X' \end{array}$$

non régulier.

On utilise un théorème de perturbation dû à N.Perkal (voir [16]):

Théorème.

Soit Y une sous-variété de $\mathbf{R}^n$ de classe C^1 et soit $\{y_i\}$ une suite de points de Y convergeant vers x_0.

Alors pour toute suite de bijections linéaires, $\{L_i\}$, de $\mathbf{R}^n$ convergeant vers l'identité, il existe une carte de classe C^1, (ϕ, U), en x_0 dans $\mathbf{R}^n$ telle que

(i) $\phi(y_i) = y_i$ pour i assez large

(ii) $d_{y_i}(\phi) = L_i$ pour i assez large.

On peut supposer que $\ker d_{y_i}(f|_Y) = \tau$ pour i assez grand.

En effet, il suffit de prendre pour L_i la bijection linéaire qui envoie la base de $\ker d_{y_i}(f|_Y)$ dans la base de τ et telle que la suite converge vers $Id_{\mathbf{R}^n}$. On supposera que X est linéaire.

Soit $\pi_{X'} : N \longrightarrow X'$ une rétraction locale de classe C^1. D'après la remarque (2.6) on peut construire une rétraction π_X de sorte que le carré

$$\begin{array}{ccc} Y & \xrightarrow{\pi_X} & X \\ \downarrow f & & \downarrow f \\ Y' & \xrightarrow{\pi_{X'}} & X' \end{array}$$

soit commutatif. De plus on peut choisir π_X tel que $d\pi_X(\tau) = \tau \cap X$, cela est possible car l'espace tangent à la fibre de $\pi_{X'} \circ f$ en y_i contient τ. Cette rétraction convient, en effet comme le carré est commutatif, on a $d\pi_X(\tau) \subset \ker df|X$ donc $\tau \cap X \subset \ker df|X$ et cette inclusion est propre (car $\ker df|X \not\subset \tau$) donc $d\pi : \tau \longrightarrow \ker df|X$ n'est pas surjectif. On en déduit que le carré n'est pas régulier. □

Remarques: Dans la preuve de la proposition on a utilisé implicitement que la condition (a_f) est invariante par difféomorphisme local et que la régularité du carré

$$\begin{array}{ccc} V_1 & \xrightarrow{f_1} & V_2 \\ \downarrow f_3 & & \downarrow f_2 \\ V_4 & \xrightarrow{f_4} & V_3 \end{array}$$

est équivalente à la régularité du carré

$$\begin{array}{ccc} B(V_1) & \xrightarrow{A\circ f\circ B^{-1}} & A(V_2) \\ \downarrow f_3\circ B^{-1} & & \downarrow f_2\circ A^{-1} \\ V_4 & \xrightarrow{f_4} & V_3 \end{array}$$

où A et B sont des bijections linéaires.

2.8) - Soit

$$\begin{array}{ccc} V_1 & \xrightarrow{f_1} & V_2 \\ \downarrow f_3 & & \downarrow f_2 \\ V_4 & \xrightarrow{f_4} & V_3 \end{array}$$

un carré commutatif d'espaces vectoriels et d'applications linéaires.

On remarque alors que la régularité du carré est équivalente au fait que l'application $g = (f_1, f_3)$ de V_1 dans le produit fibré

$$V_1 \times_{V_3} V_4 = \{(v_2, v_4) \in V_2 \times V_4 \ ; \ f_2(v_2) = f_4(v_4)\}$$

est surjective.

Dans la situation de la proposition (2.7) on peut former la variété produit fibré (car les applications sont transverses) et on obtient facilement que la condition (a_f) pour le couple (Y, X) est équivalente au fait que (π_X, f) de Y dans $X \times_{X'} Y$ est une submersion pour toute rétraction π_X. Il résulte de cela que l'application

$$(\pi_X, f) : Et(X) \longrightarrow X \times_{X'} Et(X')$$

est un morphisme stratifié.

En formant le carré commutatif correspondant, un calcul simple montre qu'il est régulier, et on a alors

2.9) - Proposition.
Dans la situation précédente l'application

$$(\pi_X, f) : T_X \longrightarrow X \times N$$

est une application de Thom de $Et(X)$ dans l'espace stratifié $X \times_{X'} Et(X')$, où X' est la strate de (B, Σ') telle que $f(X) \subseteq X'$.

Remarque: La composée de deux applications de Thom n'est pas nécessairement une application de Thom.Il suffit de considérer les applications stratifiées

$$f : (\mathbf{R}^2 - \{(0,t)\}, \{(0,t)\}) \longrightarrow (\mathbf{R}^2 - \{(0,t)\}, \{(0,t)\})$$
$$(x,t) \mapsto (x^2(x^4 + t^2), t)$$

$$g : (\mathbf{R}^2 - \{(0,t)\}, \{(0,t)\}) \longrightarrow (\mathbf{R} - \{0\}, \{0\})$$
$$(x,t) \mapsto x.$$

La composée $g \circ f$ n'est pas une application de Thom (on peut raffiner la stratification pour que $g \circ f$ devient une application de Thom).

§ 2 - Stratification régulière

Soit A un sous-ensemble d'une variété M de classe C^1. On suppose que A est muni d'une partition localement finie, Σ, en variétés de classe C^1 connexes telles que la condition de frontière soit vérifiée, i.e. pour tout couple de strates (X, Y) dans $\Sigma \times \Sigma$ tel que $X \cap \overline{Y} \neq \emptyset$ on a $X \subset \overline{Y}$ (c-à-d que l'adhérence de toute strate soit une réunion de strates).

Un espace muni d'une telle partition est appelé espace stratifié, qu'on note (A, Σ).

Remarque: On n'impose pas de filtrage par la dimension (le bord d'une strate peut avoir la même dimension que la strate).

On définit maintenant sur cet espace stratifié une condition de régularité entre les strates de (A, Σ). On montrera plus loin qu'elle est plus faible que la condition classique de Whitney (B).

1) - Condition (C)

1.1) Définition. *On dit que la stratification vérifie la condition (C) si pour toute strate X de (A, Σ), il existe un voisinage U_X de X dans la variété ambiante M et une application f_X de classe C^1 de U_X dans $\mathbf{R}_+$ telle que*

$$f|_{Et(X) \cap U_X} : Et(X) \cap U_X \longrightarrow (\mathbf{R}_+ - \{0\}, \{0\})$$

soit une application de Thom et $f_X^{-1}(0) = X$.

La proposition suivante donne une caractérisation de la condition (C).

1.2) - Proposition. *Les conditions suivantes sont equivalentes:*

α) *condition (C)*

β) *pour toute rétraction locale de classe C^1, $\pi_X : U \longrightarrow X$ il existe un voisinage T_X de X dans M tel que (T_X, π_X, f_X) soit un <u>bon tube</u> en X de (A, Σ) dans M.*

Preuve : Il suffit d'appliquer la proposition (I 2.7) à cette situation et de remarquer que pour tout rétraction locale π_X le diagramme suivant est commutatif

$$\begin{array}{ccc} Y & \xrightarrow{\pi_X} & X \\ \downarrow f_X & & \downarrow f_X \\ Y' & \xrightarrow{\pi_0} & 0 \end{array}$$

□

On en déduit

1.3) Corollaire.

La condition (C) entraine

γ) *Pour toute strate X de (A, Σ), il existe un voisinage U_X de X dans la variété ambiante M et une application f_X de classe C^1 de U_X dans R tels que:*

1) $f_X^{-1}(0) = X$

2) la restriction de f_X à toute strate de l'étoile de X est une submersion.

3) toute application lisse g d'une variété lisse N dans M transverse à X, est transverse aux variétés de niveau de f_X restreint à $Y \in Et(X)$ dans un voisinage de X.

La réciproque est fausse (contre-exemple semi-algébrique communiqué par D. Trotman). On notera aussi que cette condition n'entraine pas la trivialité topologique.

Remarques:

1- La condition γ) est la condition proposée par R. Thom dans son article "Sur l'homologie des variétés algébriques réelles" (Differential and combinatorial topology, University Press Princeton, 1965, p.255-265).

2- β) entraine que pour toute strate Y dans $Et(X)$, on a $\dim Y \geq \dim X + 1$. Donc une stratification (C)-régulière est un filtrage par la dimension.

3- β) et γ) entrainent: si X et Y sont des strates telles que $X \subseteq \bar{Y}$, alors les variétés de niveau de f_X et f_Y sont transverses dans un voisinage de X. En effet, il suffit de remarquer que les variétés de niveau de f_X sont transverses à Y, donc sont transverses aux variétés de niveau f_Y. Par récurrence on obtient que la famille des variétés de niveau de $\{f_Y\}_{Y \in Et(X)}$ est multitransverse.

4- La condition (C) entraine la condition (a) de Whitney. En effet $T_x X \subseteq \lim \ker d_y f | Y \subset \lim T_y Y$.

5- D. Trotman [16] a montré que la condition (B) de Whitney est équivalente à β) lorsque f_X parcourt l'ensemble des fonctions tubulaires (i.e. $f_X = \sum x_i^2$ dans une carte de X dans M). On en déduit que les espaces de Whitney sont des espaces (C)-réguliers.

6- La condition (C) est invariante par difféomorphisme.

2) - Voisinage tubulaire

2.1) - Définition: Soit M une variété différentiable et X une sous-variété de M. Un voisinage tubulaire de X dans M est un quadruplet $T_X = (E, \pi, \varepsilon, \phi)$ où

1) $\pi : E \longrightarrow X$ est un fibré vectoriel sur X muni d'un produit scalaire sur chaque fibre (i.e. une métrique riemannienne sur E),

2) $\varepsilon : X \longrightarrow \mathbf{R}_+$ est une fonction différentiable, et

3) $\phi : B_\varepsilon(E) \longrightarrow M$ est un difféomorphisme sur un voisinage ouvert U de X dans M tel que:

$$\begin{array}{ccc} & X & \\ {}^{\mu}\swarrow & & \searrow^{i} \\ B_\varepsilon(E) & \xrightarrow{\phi} & M \end{array}$$

où $B_\varepsilon(E) = \{e \in E | \; ||e|| < \varepsilon(\pi(e))\}$, $i : X \longrightarrow M$ est l'injection canonique et $\mu : X \longrightarrow B_\varepsilon(E)$ est la section nulle.

On notera ce voisinage tubulaire par $(|T_X|, \pi_X, \rho_X)$ où

$$|T_X| = \phi(B_\varepsilon),$$

$$\pi_X : |T_X| \longrightarrow X \quad \text{où} \quad \pi_X = \pi \circ \phi^{-1},$$

$$\rho_X : |T_X| \longrightarrow \mathbf{R}_+ \quad \text{où} \quad \rho_x(m) = < \phi^{-1}(m), \phi^{-1}(m) >_\pi .$$

2.2) - Théorème d'extension des voisinages tubulaires (Mather).

Soit $f : M \longrightarrow N$ une application de classe C^1 telle que $f|_X$ soit une submersion où $X \subset M$ est une sous-variété.

Soient X_0 et X_1 deux ouverts de X tels que $\overline{X}_1 \cap X \subset X_0$. Soit $T_0 = \{|X_0|, \pi_0, \rho_0\}$ un voisinage tubulaire de X_0 dans M compatible avec f. Alors il existe un voisinage tubulaire de X dans M compatible avec f, $T_X = \{|T_X|, \pi_X, \rho_X\}$, tel que:

$$T_X|_{X_1} = T_0|_{X_1}.$$

Preuve : On peut, quitte à remplacer M par un ouvert, supposer que f est une submersion sur M.

On se donne une métrique prolongeant celle au dessus de X_0 et possédant la propriété suivante: si $N_x(X)$ désigne le fibré normal à X en x, alors $N_x(X) \subset \ker d_x f$. Soit $E(X) = \cup_{x \in X} N_x(X)$ et ϕ une exponentielle définie sur un voisinage U de la section nulle de $E(X)$ et égale à ϕ_0 au dessus de $\overline{X}_1$.

Soit $\varepsilon : X \longrightarrow \mathbf{R}_+ - \{0\}$ une fonction C^∞ suffisamment petite pour que $B_\varepsilon(E(X)) \subset U$.

Si V désigne $\phi(B_\varepsilon(E(X)))$, on obtient alors un voisinage tubulaire de X

$$T_X = \{E(x), \varepsilon : X \to \mathbf{R}_+, \phi : B_\varepsilon(E(X)) \to V\}$$

qui est compatible avec f. Pour plus de détails voir Mather ([9]). □

Si $f : M \longrightarrow N$ est une application de classe C^1 telle que $f|_X$ soit une submersion, le théorème d'extension des voisinages tubulaires (2.1) permet de trouver un voisinage tubulaire, T_X, de X dans M tel que (CRf) soit vérifiée, de plus si l'espace stratifié (A, Σ) est un espace de Whitney, le tube $T_X = (T_X, \pi_X, \rho_X)$ (dans ce cas c'est même un voisinage tubulaire) est automatiquement bon (voir Mather [9]).

On a même l'équivalence entre le fait que tout voisinage tubulaire est bon et la condition de Whitney (voir à ce sujet Trotman [11] ou Perkal [16]), mais dans le cas où (A,Σ) est seulement un espace (C)-régulier le tube obtenu par le théorème (2.1) n'est pas nécessairement bon.

La proposition suivante montre l'existence d'un bon tube vérifiant (CRf).

2.3) - Proposition.

Soit (A,Σ) un espace (C)-régulier. Soit $f: M \longrightarrow N$ une application de classe C^1 telle que $f|_X$ soit une submersion, où X est une strate de (A,Σ). Soient X_0 et X_1 deux ouverts de X tels que $\overline{X_1} \cap X \subset X_0$. Soit $T_0 = \{|X_0|, \pi_0, \rho_X\}$ un bon tube de X_0 dans M compatible avec f. Alors il existe un bon tube de X dans M compatible avec f, $T_X = \{|T_X|, \pi_X, \rho_X\}$, tel que:

$$T_0|_{X_1} = T_X|_{X_1}.$$

Preuve : D'après le théorème des voisinages tubulaires, on peut étendre le voisinage tubulaire $T_0 = \{|T_0|, \pi_0, ?\}$, à X tel que $\{|T_X^1|, \pi_X^1, ?\}$ soit compatible avec f et $T_X^1|_{X_1} = T_0|_{X_1}$.

D'après la proposition (2.2), on peut étendre le bon tube $T_0 = \{|T_0|, \pi_0, \rho_0\}$ à X tel que $T_X^2 = (|T_X^2|, |\pi_X^2|, \rho_X^2)$ soit un bon tube et $T_X^1|_{X_1} = T_0|_{X_1}$. Il suffit alors de prendre pour T_X le tube defini par $|T_X| = |T_X^1| \cap |T_X^2|$, $\pi_X = \pi_X^1|_{|T_X|}$ et $\rho_X = \rho_X^2|_{|T_X|}$.

Alors $T_X = (|T_X|, |\pi_X|, \rho_X)$ est un bon tube, en effet ρ_X est une application de Thom donc $(\pi_X, \rho_X)|_{Y\cap T_X}$ est une submersion pour tout $Y \in Et(X)$.

T_X est compatible avec f, car

$$f \circ \pi_X(a) = f \circ \pi_X^1(a) = f(a) \quad \text{pour tout } a \in |T_X|$$

et bien sûr $T_X|_{X_1} = T_0|_{X_1}$. □

On va montrer la proposition principale de ce chapitre qui généralise celle de Mather [12] pour le cas des espaces stratifiés de Whitney. On obtiendra comme corollaire le fait qu'un espace (C)-régulier admet une structure d'espace stratifié abstrait. Dans la proposition qui va suivre on introduit une nouvelle condition qu'on note $(CR\rho)^*$ et qui est la condition $(CR\rho)$ seulement pour les strates qui sont envoyées dans une même strate par l'application de Thom f, i.e. si X, Y sont des strates de (A,Σ) telles que les images $f(X)$ et $f(Y)$ soient incluses dans X' strate de (B,Σ'), alors $(CR\rho)$ est satisfaite.

2.4) - Proposition.

Soit (A,Σ) un espace stratifié de (C)-régulier dans M. Soient A_0 un fermé de A, réunion de strates, U_0 et U_1 des ouverts de A (de type (σ)) tels que $\overline{U}_1 \cap A \subset U_0$. Si $f: M \longrightarrow N$ est une application de classe C^1 et (B,Σ',T') un espace stratifié abstrait faible plongé dans N, tels que $f(A) \subseteq B$ et $f|_A : (A,\Sigma) \longrightarrow (B,\Sigma')$ soit une application de Thom et $T^0 = \{(T_X^0, \pi_X^0, \rho_X^0)\}_{X\in\Sigma_{|A_0\cup U_0}}$ soit un système de bons tubes de $(A_0 \cap U_0, \Sigma_{|A_0\cap U_0})$ dans M vérifiant (CRf), $(CR\pi)$ et $(CR\rho)^*$, alors il existe un système de bons tubes de (A,Σ) dans M, $T = \{(T_X, \pi_X, \rho_X)\}_{X\in\Sigma}$, vérifiant (CRf), $(CR\pi)$ et $(CR\rho)^*$ tel que

$$T|_{A_0\cup U_1} = T_0|_{A_0\cup U_1}$$

Preuve : On va procéder par récurrence sur la profondeur de la stratification.

Si $depth(\Sigma) = 0$ alors A est une variété, A_0 est une réunion de composantes connexes de A et par suite la proposition est une conséquence du théorème d'extension des voisinages tubulaires (2.1).

Supposons que la proposition soit vraie pour tout espace stratifié (C)-régulier de profondeur $< \mathrm{depth}(\Sigma)$.

Soit X une strate de (A, Σ) telle que $\mathrm{depth}_\Sigma(X) = \mathrm{depth}(\Sigma)$. Puisque la construction qui va suivre peut-être faite simultanément pour toutes les strates de profondeur maximale, on peut supposer, sans perdre de généralité, que X est la seule strate avec cette propriété. Il est clair que X est fermée dans A.

D'après la proposition (2.2) il existe un bon tube de X dans M tel que (CRf) soit vérifiée et

$$T_X|_{X\cap U_1} = T_X^0|_{X\cap U_1}$$

(ou bien $T_X = T_X^0$ si $X \subset A_0$).

Soit $A_1 = T_X \cap (A - X)$, alors $(A_1, \Sigma_{|_{A_1}})$ est un espace stratifié (C)-régulier.

Soit X' la strate de (B, Σ') telle que $f(X) \subseteq X'$. Si on note A_1' l'ensemble des strates de A_1 qui sont envoyées dans X', alors A_1' est fermé dans A_1, cela parce que si Y est une strate qui est envoyée par f dans X' alors la strate Z vérifiant $X < Z < Y$ est envoyée dans X'.

D'après ce qui précède l'application

$$g = (\pi_X, \rho_X)_{|_{A_1'}} : A_1' \longrightarrow X \times \mathbf{R}$$

est une application de Thom, de plus le système de bons tubes $T^0_{|_{(A_0\cup U_0)\cap A_1'}}$ vérifie $(CR\pi), (CR\rho)^*$ et (CRg). ((CRg) est satisfaite car $(CR\pi_X)$ et $(CR\rho_X)$ le sont).

Comme la profondeur de $(A_1', \Sigma_{|_{A_1'}}) < \mathrm{depth}(\Sigma)$, par hypothèse de récurrence il existe un système de bons tubes de $(A_1', \Sigma_{|_{A_1'}})$ dans T_X, qu'on note $T_{A_1'}$, tel que

$$T_{|_{(A_0\cup U_0)\cap A_1'}} = T^0_{|_{(A_0\cup U_0)\cap A_1'}}$$

Dans ce cas $(CR\rho)^* = (CR\rho)$ car $f(A_1') \subseteq X'$.

La condition (CRg) nous donne les conditions $(CR\pi_X)$ et $(CR\rho_X)$, donc la compatibilité de $T_{A_1'}$ avec T_X, il ne reste qu'à vérifier (CRf) c-à-d $f \circ \pi_Y = \pi_{X'} \circ f$ pour toute strate Y dans A_1'.

$$\begin{aligned} f \circ \pi_Y &= \pi_{X'} \circ f \circ \pi_Y \quad \text{car} \quad f(Y) \subset X' \\ &= f \circ \pi_X \circ \pi_Y \quad \text{car} \quad f \circ \pi_X = \pi_{X'} \circ f \quad (CRf \,\text{pour}\, T_X) \\ &= f \circ (\pi_X \circ \pi_Y) \\ &= f \circ \pi_X \quad \text{car ((CR}\pi\text{) est vérifiée)} \\ &= \pi_{X'} \circ f. \end{aligned}$$

Donc finalement on a un système de bons tubes de A_1',

$$T_{A_1'} = \{(T_Y, \pi_Y, \rho_Y); Y \in \Sigma|_{A_1'}\}$$

vérifiant $(CRf), (CR\pi)$ et $(CR\rho)$.

On va maintenant étendre ce système de tubes à A_1.

D'après la proposition (I.2.9) l'application
$$h = (\pi_X, f) : A_1 \longrightarrow X \times_{X'} Et(X')$$
est une application de Thom. De plus le système de bons tubes $\{T_{A'_1}, T^0_{(A_0 \cup U_0) \cap A_1}\}$ vérifie $(CRh), (CR\pi)$ et $(CR\rho)^*$, et comme la profondeur de $(A_1, \Sigma_{|A_1}) < depth(\Sigma)$, par hypothèse de récurrence il existe un système de bons tubes de $(A_1, \Sigma_{|A_1})$ dans T_X, soit T_{A_1}, tel que $(CRh), (CR\pi)$ et $(CR\rho)^*$ soient vérifiées, et
$$T_{A_1}|_{(A'_0 \cup U_1) \cap A_1} = T^0_1|_{(A'_0 \cup U_1) \cap A_1}$$
où A'_0 est le fermé $A_0 \cup A'_1) \cap A_1$ de A_1 et $T^0_1 = \{T^0, T_{A'_1}\}$.

Donc maintenant on a obtenu un système de bons tubes de $A_1 = T_X \cap (A - X)$ dans T_X vérifiant $(CRf), (CR\pi)$ et $(CR\rho)^*$ $((CRf)$ est donné par $(CRh))$, de plus $(CR\pi_X)$ est vérifiée, i.e. $\pi_X \circ \pi_Y = \pi_X$ pour toute strate Y de A_1, ainsi que $(CR\rho)^*$, i.e. $\rho_X \circ \pi_Y = \rho_X$ pour toute strate Y de A_1 telle que $f(Y) \subseteq X'$.

Il ne reste plus qu'à étendre ce système de bons tubes à l'espace stratifié (A, Σ).

Soit $\varepsilon : X \longrightarrow \mathbf{R}^*_+$ une application lisse, telle que
$$\overline{T^\varepsilon_X} = \text{adhérence dans } M \text{ de } \{a \in T_X | \rho_X(a) < \varepsilon(\pi_X(a))\} \subset T_X.$$

Comme $(A - X, \Sigma - X)$ est un espace stratifié (C)-régulier de profondeur $< depth(\Sigma)$, $f|_{A-X}$ est une application de Thom et $(T^\varepsilon_X \cup U_1) \cap (A - X)$ et $(T_X \cup U_0) \cap (A - X)$ sont des ouverts de $(A - X)$ (de type σ) tels que
$$\overline{(T^\varepsilon_X \cup U_1) \cap (A - X)} \subset (T_X \cup U_0) \cap (A - X),$$
l'hypothèse de récurrence nous permet d'obtenir un système de bons tubes de $A - X$ dans M, T_{A-X}, vérifiant $(CRf), (CR\pi)$ et $(CR\rho)^*$, et tel que
$$T_{A-X}|_{(T^\varepsilon_X \cup U_1 \cup A_0) \cap (A-X)} = (T_{A_1}, T^0)|_{(T^\varepsilon_X \cup U_1 \cup A_0) \cap (A-X)}.$$

On termine la démonstration en remarquant que le système de bons tubes $\{T_X, T_{A-X}\}$ est le système de bons tubes recherché. □

Remarques

a) On a utilisé implicitement dans la preuve de la proposition précédente que si X est une strate de (A, Σ) qui est envoyée par f dans X' alors, puisque f est continue, on a que $f(T_X) \subset T_{X'}$, quitte à restreindre T_X, et que T_X est un ouvert de type (σ) car X est une strate de profondeur maximale.

b) Le fait de prendre des ouverts particuliers est important (type (σ)), car si on prend l'ouvert $U = A - X$, qui n'est pas de type (σ),(voir (1.2)), on ne peut pas en général prolonger le système de bons tubes de $(A - X, \Sigma_{|A-X})$ à (A, Σ).

2.5) - Théorème.

Soit (A, Σ) un espace stratifié de (C)-régulier dans M, il existe un système controlé de bons tubes, $T = \{(T_X, \pi_X, \rho_X)\}_{X \in \Sigma}$, de (A, Σ) dans M.

Preuve : En effet il suffit d'appliquer la proposition (2.4) à la situation suivante:

$f : A \longrightarrow \{\text{point}\}$, $A_0 = U_0 = U_1 =$ l'ensemble vide. □

Remarque On pourrait prendre $\rho_X = f_X$ où les $\{f_X\}_{X \in \Sigma}$ sont les applications de Thom qui définissent la régularité de la stratification.

Corollaire.

Tout espace stratifié (C)-régulier admet une structure d'espace stratifié abstrait dans M (structure d'espace de Thom-Mather).

2.6) - Proposition.

Soit $f : M \longrightarrow N$ une application de classe C^1, et (B, Σ') un espace stratifié (C)-régulier dans N. Si f est transverse à chaque strate X', $X' \in \Sigma'$, alors $f^{-1}(B)$ est stratifié par les composantes connexes de $f^{-1}(X')$, $X' \in \Sigma'$; de plus l'espace stratifié obtenu est (C)-régulier.

Preuve : On factorise f en

$$M \xrightarrow{Id \times f} M \times N \xrightarrow{\pi_2} N$$

définies par

$$(Id \times f)(x) = (x, f(x)) \quad \text{et} \quad \pi_2(x, f(x)) = f(x).$$

On commence par montrer le lemme suivant:

Lemme 1.

La stratification $\Sigma'' = \{M \times X' \quad ; \quad X' \in \Sigma\}$ est une stratification (C)-régulière de $M \times B$.

Preuve : Cette partition vérifie trivialement les conditions de frontière et de finitude locale. Vérifions la condition (C).

Soit l'application $g_{M \times X'}$ définie par $g_{M\times X'}(x, y) = f_{X'}(y)$, où X' est une strate de (B, Σ') et $f_{X'} : U_{X'} \to \mathbf{R}$ est l'application de Thom. Nous allons montrer que c'est une application de Thom. En effet, on a

$$\ker d_{(m,y)}(g|_{M \times Y'}) = T_M \times \ker d_y (f_{X'}|_{Y'}),$$

donc si (m_i, y_i') est une suite de points de $M \times Y'$ convergeant vers $(m, x') \in M \times X'$ et telle que la suite $\{\ker d_{(m_i, y_i)}(g|_{M\times Y'})\}$ converge vers τ, comme $\ker d_{y_i}(f_{X'}|_Y)$ tend vers τ' et $\tau = T_m M \times \tau'$, alors le fait que $f_{X'}$ soit une application de Thom entraine que la condition (a_g) est vérifiée.

On remarque que $g^{-1}_{M\times X'}(0) = M \times X$; ce qui termine la démonstration. □

Le graphe de f est transverse à l'espace stratifié $M \times B$; en effet soient $(m, x') \in graphe(f)$ et $(u, v) \in T_{(m,x')}M \times N$ alors ils existent $u_1 \in T_m M$ et $v_1 \in T_{x'}X'$, où X' est la strate de Σ' contenant x'; tels que $v = d_m f(u_1) + v_1$ alors $(u, v) = (u - u_1, v_1) + (u_1, d_m f(u_1))$ donc $(u, v) \in T_{(m,x')}M \times X' + T_{(m,x')}graphe(f)$.

Comme on le verra plus tard (la proposition du § 3) l'intersection d'espaces (C)-réguliers transverses est un espace (C)-régulier, donc $M \times B \cap graphe(f)$ est un espace (C)-régulier de $M \times N$.

La projection $\Pi : M \times N \longrightarrow M$ est un difféomorphisme de $graphe(f)$ dans M et donc on a que $\Pi(M \times B \cap graphe(f)) = f^{-1}(B)$ est un espace (C)-régulier.

Par contre la condition de frontière n'est pas nécessairement vérifiée. Pour remédier à cela on va considérer la partition de A par les composantes connexes des strates de Σ: on note Σ^C cette partition.

Cette partition de A, i.e. Σ^C, vérifie la condition (C), et il ne reste plus qu'à montrer qu'elle est localement finie et vérifie la condition de frontière. Pour cela on a besoin du lemme suivant:

Lemme 2.

Soit A un sous-ensemble localement fermé de M et Σ une partition de A en sous-variétés. Si Σ est localement finie et vérifie la condition (C), alors (A, Σ^C) est un espace stratifié (C)-régulier.

Preuve : On va montrer cela par récurrence sur la profondeur de la partition Σ, qui est égale à celle de Σ^C.

Si $depth(\Sigma) = 0$ alors A est une variété,et dans ce cas les composantes connexes de A vérifie toutes les conditions.

Supposons que le lemme soit vrai pour toute partition localement finie et vérifiant la condition (C), de profondeur $< depth(\Sigma)$.

Comme nous l'avons remarqué, la partition Σ^C de A vérifie la condition (C) et il ne reste plus qu'a montrer qu'elle est localement finie et vérifie la condition de frontière.

Soit X une strate de Σ telle que $depth_\Sigma(X) = depth(\Sigma)$, on peut supposer sans perdre de généralité que X est la seule strate vérifiant cette propriété. Il est clair que X est fermée dans A. On peut aussi supposer que X est connexe.

Comme Σ est localement finie et vérifie la condition (C), on a d'après la proposition (2.2) un tube (T_X, π_X, ρ_X) de X dans M, tel que T_X ne rencontre qu'un nombre fini de strates et $(\pi_X, \rho_X) : T_X \cap (A - X) \longrightarrow X \times (0, \varepsilon)$ soit une application de Thom.

D'après la proposition (2.4) il existe un système de bons tubes, T_{A_1}, de $A_1 = T_X \cap (A - X)$ vérifiant (CRg) (CRπ) et (CRρ), où $g = (\pi_X, \rho_X)$. Il existe alors un champ de vecteurs controlé ζ tel que $dg(\zeta) = (0, \frac{d}{dt})$ (pour la définition de champ de vecteurs et leurs intégrations voir Mather [12]); ζ induit un groupe à un paramètre $\gamma : J \longrightarrow A_1$, où

$$J = \{(y,t) \in A_1 \times \mathbf{R} \quad ; \quad -\rho_X(y) < t < \varepsilon - \rho_X(y)\}.$$

Comme $\pi_X : T_X \cap A_1 \longrightarrow X$ est une submersion sur chaque strate de $A_1 \cap T_X$, $S_X^{\varepsilon/2} = g^{-1}(X \times \varepsilon/2)$ est un espace stratifié (C)-régulier, stratifié par $\Sigma_{S_X} = \{S : S = S_X^{\varepsilon/2} \cap Y, Y \in \Sigma\}$.

Comme $depth(\Sigma_{S_X}) < depth(\Sigma)$, $\Sigma^C_{S_X}$ est une stratification (C)-régulière de $S_X^{\varepsilon/2}$, de plus g est une submersion propre sur chaque strate de $\Sigma^C_{S_X}$. On en déduit qu'elle est surjective sur chaque strate. En effet, si S est une strate de $S_X^{\varepsilon/2}$, comme $g : S \longrightarrow X \times \frac{\varepsilon}{2}$ est une submersion, alors g est une application ouverte. Mais g est aussi fermée car si $S_1 \subset \overline{S}$ alors $g(S_1) \subset g(S)$ ce qui veut dire que $g(\overline{S}) \cup X \subset g(S)$ et comme g est propre, $\overline{g(S)} \cup X \subset X$ ce qui implique $g(S) = g(\overline{S})$.

Si X est connexe on a $g(S) = X$.

On peut alors considérer le cylindre d'application $Z_g = S_X^{\varepsilon/2} \times [0, \varepsilon) \bigcup X \times \{0\}$ (réunion disjointe) sous la relation d'équivalence $(a, 0) \sim (g(a), 0)$ où $a \in S_X^{\varepsilon/2}$. Alors Z_g est muni d'une structure d'espace stratifié par les strates de type X, $\{S \times \{0\}, S \in \Sigma^C_{S_X}\}$ et $\{S \times (0,1), S \in \Sigma^C_{S_X}\}$. En effet la condition de frontière est vérifiée car X ne peut rencontrer que l'adhérence des strates de type $S \times (0,1)$ et à cause de la définition de Z_g, $X \subset \overline{S \times (0,1)}$.

Cette stratification est localement finie car Σ_{S_X} l'est. □

Revenant au groupe à un paramètre d'homéomorphisme $\gamma : J \longrightarrow A_1$, quitte à restreindre T_X, on obtient un isomorphisme d'espace stratifié

$$h : Z_g \longrightarrow A_1 \cup X$$

défini par

$$h([a,t]) = \begin{cases} \gamma(a, -\frac{\epsilon}{2}t) & 0 \leq t \leq 1 \quad a \in S_X^{\epsilon/2}, \\ \pi_X(a) & t = 1 \quad a \in S_X^{\epsilon/2}, \end{cases}$$

et $h(x) = x$ pour $x \in X$, où $[a,t]$ est la classe d'équivalence de $(a,t) \in S_X^{\epsilon/2} \times (0,1)$ dans Z_g.

Comme h est un homéomorphisme qui préserve la stratification, donc envoie chaque strate de $\Sigma^c_{|A_1 \cup X}$ sur une strate de Z_g et inversement, $\Sigma^c_{|A_1 \cup X}$ est localement finie et vérifie la condition de frontière.

On termine la démonstration en remarquant que $depth(\Sigma_{|A-X}) < depth(\Sigma)$ et par suite, l'espace stratifié $(A - X, \Sigma^c_{|A-X})$ est (C)-régulier, donc (A, Σ^C) l'est aussi. □

Soient (A, Σ) et (B, Σ') deux espaces stratifiés (C)-réguliers. On dit que (A, Σ) est transverse à (B, Σ') si pour tout couple de strates (X, X') dans $\Sigma \times \Sigma'$ on a X transverse à X' dans la variété ambiante.

On notera par

$$\Sigma \cap \Sigma' = \{\text{les composantes connexes de} \quad X \cap X'; \quad X \in \Sigma, X' \in \Sigma'\}$$

la stratification de $A \cap B$, et par

$\Sigma \cup \Sigma' = \{$les composantes connexes de Z; $Z = X \cap X'$
ou $X - X \cap X'$ ou $X' - X \cap X'$, $X \in$ la stratification de $A \cup B$.

La proposition suivante montre que la catégorie des espaces (C)-réguliers est stable par les coupes transverses.

2.7) - Proposition. *Les espaces stratifiés $(A \cap B, \Sigma \cap \Sigma')$ et $(A \cup B, \Sigma \cup \Sigma')$ sont (C)-réguliers.*

Preuve : On va montrer que $A \cap B$ est (C)-régulier.

Soit X (resp. X') une strate de Σ (resp. Σ'), ρ_X (resp. $\rho_{X'}$) la fonction de contrôle. On propose pour $X \cap X'$ la fonction de contrôle $\rho = \rho_X + \rho_{X'}$; on a tout de suite $\rho^{-1}(0) = X \cap X'$, il reste à montrer que ρ est une application de Thom.

Soient $Y > X$ et $Y' > X'$ deux strates de A et B, on sait d'après ce qui précède que $X \cap X' < Y \cap Y'$. Soit $\{y_i\}$ une suite de points de $Y \cap Y'$ qui converge vers $x_0 \in X \cap X'$ alors:

il existe une sous-suite telle que $\ker d_{y_i}\rho_X|Y \longrightarrow \tau$ et $\ker d_{y_i}\rho_{X'}|Y' \longrightarrow \tau'$ lorsque $y_i \to x_0$; comme X est transverse à X' on a que τ est transverse à τ' et que $X \cap X' \subsetneq \tau \cap \tau'$, donc la condition de Thom est vérifiée car $\ker d_y\rho_X \cap \ker d_y\rho_{X'} \subset \ker d_y\rho$. De plus il existe un voisinage U de x_0 dans la variété ambiante tel que pour tout $y \in (Y \cap Y') \cap U$ on ait $\ker d_y\rho_X|Y$ transverse à $\ker d_y\rho_{X'}|Y'$.

Il ne reste plus qu'a montrer que $\rho|_{Y \cap Y'} : Y \cap Y' \longrightarrow R^+$ est une submersion. Pour cela il suffit de trouver un vecteur tangent à $Y \cap Y'$ en tout point $y \in U$ et qui n'annule pas la différentielle de ρ.

Soit $y \in Y \cap Y' \cap U$, il existe $v \in T_yY$ tel que $d_y\rho_X|Y(v) \neq 0$ et comme $\ker d_y\rho_X|Y$ et $\ker d_y\rho_{X'}|Y'$ sont transverses, il existe des vecteurs $v_1 \in \ker d_y\rho_X|Y$

et $v_2 \in \ker d_y \rho_{X'}|Y'$ tels que $v = v_1 + v_2$. Donc $v_2 \in T_y(Y \cap Y')$ et de plus v_2 n'annule pas la différentielle de $\rho|Y \cap Y'$, ainsi v_2 convient.

De la même façon on obtient que $A \cup B$ est (C)-régulier. □

En utilisant les propositions (2.6) et (2.7), on obtient:

2.8) - Proposition.

Soit $f : M \longrightarrow N$ une application lisse entre variétes lisses et soient $A \subset M$ et $B \subset N$ des espaces stratifiés (C)-réguliers. On suppose que f envoie A transversalement à B. Alors $A \cap f^{-1}(B) = (f|A)^{-1}(B)$ est un espace stratifié (C)-régulier.

§ 3 Les théorèmes d'isotopie

Aprés avoir muni un espace (C)-régulier d'une structure de Thom-Mather, on obtient les théorèmes d'isotopie de Thom (pour les démonstrations voir [5], [12], [15] et [17])

Théorème 1. *Soit $f: M \to N$ une application lisse, (A, Σ) un (C)-régulier de M telle que $f|_A$ soit une submersion stratifiée propre, alors pour tout $a \in N$ il existe un isomorphisme local h tel que:*

$$\begin{array}{ccc} A & \xrightarrow{h} & f^{-1}(a) \times \mathbf{R}^n \\ & {\scriptstyle f}\searrow \quad \swarrow{\scriptstyle \pi} & \\ & \mathbf{R}^n & \end{array}$$

soit commutatif. En particulier, les fibres de $f|_A$ sont homéomorphes (de plus chaque fibre est un espace (C)-régulier).

Soit (A, Σ) (resp. (B, Σ')) un espace C-régulier de M (resp. N) et $f: A \to B$ une application de Thom propre (resp. $g: B \to P$ une submersion propre).

Théorème 2. *Sous les hypothèses précédentes, il existe des isomorphismes $h_1: A \to A_0 \times U$ et $h_2: B \to B_0 \times U$ tels que le diagramme suivant soit commutatif*

$$\begin{array}{ccc} A_0 \times U & \xrightarrow{f|_{A_0} \times Id} & B_0 \times U \\ \downarrow h_1^{-1} & & \downarrow h_2^{-1} \\ A & \xrightarrow{f} & B \\ & {\scriptstyle g\circ f}\searrow \quad \swarrow{\scriptstyle g} & \\ & P & \end{array}$$

où $A_0 = (g \circ f)^{-1}(p) \cap A$, $B_0 = g^{-1}(p) \cap B$ et U est un voisinage de p dans P. On dit que f est localement triviale.

Pour démontrer ces théorèmes on utilise habituellement la structure de Thom-Mather. Dans ce cas on pourrait faire autrement. En effet on peut voir directement (voir plus haut) que les variétés de niveau des fonctions de contrôle forment une famille multitransverse, on va alors lisser ces variétés à coins en utilisant les techniques de lissages élaborées par Hirsch [9], de telle manière à obtenir une fonction tapissante tel que le bord de la strate forme une bonne stratification au sens de Lê D.T. [6]. On obtient alors:

1. Si $f: A \subseteq M \to N$ est une submersion stratifiée, alors tout champ de vecteurs sur N peut être relevé en un champ continu tangent aux variétés de niveau des fonctions de contrôle et transverse à toute rétraction locale compatible avec f (car la fonction tapissante est une fonction de Thom).
2. On obtient le premier théorème d'isotopie en integrant ce champ de vecteurs. L'unicité est dû au fait que ce champ ne quitte pas les strates et l'existence au fait que ce champ est continu.

L'existence d'une stratification (C)-régulière pour un sous-analytique est donnée par l'existence d'une stratification de Whitney, mais cette stratification est trop fine (voir l'exemple de l'introduction). On pourrait construire une stratification (C)-régulière directement en utilisant la proposition suivante. Soit E un sous-analytique de R^n, de dimension m, muni d'une stratification sous-analytique.

Proposition. *Si $f: U \to R$ est une application sous-analytique, où U est un voisinage de $\overline{E}$, alors on peut trouver une stratification de U qui vérifie la condition (a_f) de Thom et qui soit compatible avec la stratification de E.*

On note E^i la réunion des strates de dimension $\leq i$. On applique cette proposition à $f(x) = distance(x, E)$ dans une première etape et on obtient une stratification de E verifiant la condition (a) de Whitney. Dans une seconde étape on prend $f(x) = distance(x, E^{m-1})$ et on poursuit le raffinement ainsi de suite jusqu'à $f(x) = distance(x, E^0)$; on obtiendrait ainsi une stratification (C)-régulière avec comme fonctions de contrôle les fonctions distances aux strates. On obtient:

Théorème. *Tout sous-analytique E admet une stratification (C)-régulière.*

Les détails seront publiés ailleurs.

Bibliographie

[1] K. Bekka: Sur les propriétés topologiques et métriques des espaces stratifiés. *Thèse de doctorat* Université Paris-Sud (1988).

[2] J.W. Bruce & P.J. Giblin: A stratification of the space of plane quartic curves. *Proc. London Math. Soc.* **42** (1981), 270–298.

[3] J. Briançon & J.P. Speder: La trivialité topologique n'implique pas les conditions de Whitney. *C. R. Acad. Sc., Paris,* **280** (1975), 365–367.

[4] J.N. Damon: Topological triviality and versality for subgroups of A and K. *Memoirs of the A.M.S.* **39** (1988).

[5] C.G. Gibson, K. Wirthmuller, A.A. du Plessis & E.J.N. Looijenga: Topological stability of smooth mappings. *Lectures notes in Math.* **552**, Springer Verlag (1976).

[6] H.A. Hamm & D.T. Lê: Un théorème de Zariski du type de Lefschetz. *Ann. Sci. Ec. Norm. Sup.* 4e *serie* **6** (1973), 317–366.

[7] R. Hardt: Stratification of real analytic mappings and images. *Invent. Math.* **28** (1975) 193–208.

[8] H. Hironaka: Stratification and flatness.*Real and complex singularities.* Nordic summer school (Oslo 1976) Sijthoff-Noordhoff, Groningen (1977).

[9] M.W. Hirsch & B. Mazur: Smoothings of piecewise linear manifolds. *Annals of Mathematics Studies,* **80**. Princeton University (1974).

[10] S. Koike: On condition (a_f) of stratified mappings *Ann. Inst. Fourier* **33** (1983), 177–184.

[11] D.T. Lê & C.P. Ramanujam: The invariance of Milnor's number implies the invariance of topological type. *Amer. J. Math.* **98** (1976), 67–78.

[12] J. Mather: Notes on topological stability. *University of Harvard* (1970).

[13] W. Pawlucki: Le théorème de Puiseux pour une application sous-analytique. *Bull. Pol. Acad. Sci.* **32** (1984), 555–560.

[14] N. Perkal: On proving the geometric version of Whitney regularity. *J. London Math. Soc.* **29** (1984), 343–351.

[15] R. Thom: Ensembles et morphismes stratifiés. *Bull. A.M.S.* **75** (1969), 240–284.

[16] D.J.A. Trotman: Comparing regularity conditions on stratifications. Singularities, Arcata 1981, *Proc. Sympos. Pure Math.* **40**, *(A.M.S.)* (1983), 575-586.

[17] A. Verona: Stratified mappings - Structures and triangulability. *Lecture notes in Math.* **1102**, Springer Verlag (1983).

[18] H. Whitney: Tangents to an analytic variety. *Ann. of Math.* **81** (1965), 496–549.

Département de Mathématiques
Université Paris-Sud.

Adresse actuelle:

Department of Pure Mathematics
University of Liverpool
P.O. Box 147
Liverpool L69 3BX.

Folding Maps and Focal Sets

J.W. Bruce and T.C. Wilkinson

§1. Introduction

Given a surface X in Euclidean space $\mathbb{R}^3$ one can study its geometry by considering its infinitesimal symmetries (a viewpoint influenced by Klein's Erlangen programme). For example reflectional symmetry in a plane already picks out the principal directions at any point x of X. Infinitesimally the surface is reflectionally more symmetric across the planes containing these directions and the normal at x than any other planes through x (see Figure 1).

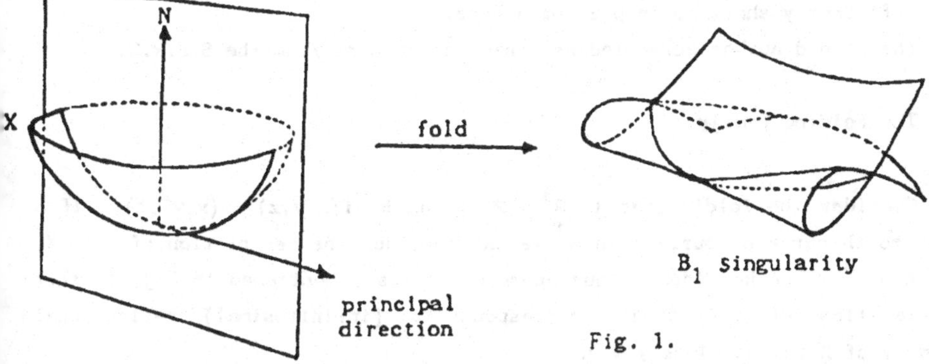

Fig. 1.

We study this reflectional symmetry by means of a family of folding maps, the (essentially) 3-parameter family of mappings obtained by conjugating the fold $(x,y,z) \longmapsto (x,y^2,z)$ by Euclidean motions. Restricting this family of maps to a generic surface $X \subset \mathbb{R}^3$ yields a new family containing the infinitesimally reflective geometry of X. This idea was first discussed in [3] and a list of the singularities arising for a generic surface X was given. In the present work a much more detailed study of the geometry has been undertaken aided by Mond's independent and more useful classification [7]. The most striking result is that the bifurcation set of this family of mappings $X \to \mathbb{R}^3$ is the dual of the union of the focal and symmetry sets of X. These objects arise also as bifurcation sets, for if $D: X \times \mathbb{R}^3 \to \mathbb{R}$ is the family of distance squared functions $D(x,a) = \|x-a\|^2$, then the focal set is the catastrophe set of D, the symmetry set its Maxwell set. So our folding maps shed additional light on this much studied family of distance squared functions. (See [9] and [10].) It also provides a tool for studying the affine geometry of these rather singular objects. (See [4] where other results on the duals of singular sets are given.) We would however contend that the dual viewpoint provided by the family of folding maps F has its own intrinsic value in studying the geometry of a surface in $\mathbb{R}^3$ and is not merely an adjunct to the study via contact with spheres. (After all the focal and symmetry sets are dual to $\underline{F}$'s bifurcation set.) Indeed it has some advantages over that provided by the family D of distance squared functions, since it picks up geometric features of the surface X undetected by D, and blows up the rather subtle geometry at umbilics into more easily understood bite-sized chunks.

In §2 we introduce the family of folding maps, in §3 we describe the associated geometry, and in §4 we consider the blowing up at umbilics.

The full proofs of all our results can be found in [11], although some are sufficiently short to be presented here.

The second author acknowledges financial support from the S.E.R.C..

§2. The Folding Family

Consider the folding map $f: \mathbb{R}^3 \to \mathbb{R}^3$ given by $f(x,y,z) = (x,y^2,z)$. If X is a smooth curve or surface in $\mathbb{R}^3$ we can consider the restriction of f to X, $f|X: X \to \mathbb{R}^3$. The key idea in our approach, already mentioned in [3], is that singularities of $f|X$ should correspond to (infinitesimal) reflectional symmetry of X in the plane $y = 0$.

Of course there is nothing sacrosanct about the plane $y = 0$. If we are interested in all reflectional symmetries of X we should consider the natural family of folding maps parametrised by all planes in $\mathbb{R}^3$. One convenient way of doing this is as follows. Let Euc denote the Euclidean group of motions of $\mathbb{R}^3$. We now define

$$\overline{F}: \mathbb{R}^3 \times \text{Euc} \to \mathbb{R}^3$$

by $\overline{F}(x,A) = A^{-1} \circ f \circ A(x)$. This gives a 6-parameter family of folding maps. If H denotes the subgroup of Euc preserving the region $y \geq 0$ then $\overline{F}$ gives rise to a family

$$F: \mathbb{R}^3 \times \text{Euc}\backslash H \to \mathbb{R}^3.$$

The quotient space Euc\H parametrizes the planes in $\mathbb{R}^3$, and F is the required family of foldings.

Given an embedding $g: X \to \mathbb{R}^3$, that is a smooth curve or surface in $\mathbb{R}^3$, we obtain a family

$$F_g: X \times \mathrm{Euc}\backslash H \to \mathbb{R}^3$$

by restriction. The first result we need is

Theorem 2.1. For a residual set of embeddings $g: X \to \mathbb{R}^3$ the family F_g is a generic family of mappings. (The term generic is defined in the usual way, in terms of transversality to submanifolds of multi-jet spaces.)

Proof: This result was first proved in [11], where, in addition, interpretations of the genericity of F_g in terms of the geometry of X were given. It also follows from a beautiful result of Montaldi [8]. The key fact here is that the fold mapping $f: \mathbb{R}^3 \to \mathbb{R}^3$ is stable. □

Consider next the case when X is a surface. The classification of the map germs which arise was done in [3], but Mond independently classified all $\mathcal{A}$-simple map germs $\mathbb{R}^2, 0 \to \mathbb{R}^3, 0$, in [7], of which the required germs form a subset, and pointed out an important connection with functions on manifolds with boundary. We use his notation below.

Proposition 2.2. For a residual set of embeddings $g: X \to \mathbb{R}^3$ the folding maps $X \to \mathbb{R}^3$ in the family F_g have singularities $\mathcal{A}$-equivalent to one of the following types:

	$\mathcal{A}_e$ codimension	Name	C
$f(x,y) = (x,y,0)$	0	Immersion	0
$f(x,y) = (x,y^2,xy)$	0	Cross-cap	1
$f(x,y) = (x,y^2,x^2y \pm y^{2k+1})$, $1 \leq k \leq 3$	k	$B_k^{\pm}$	2
$f(x,y) = (x,y^2,y^3 \pm x^{k+1}y)$, $2 \leq k \leq 3$	k	$S_k^{\pm}$	k+1
$f(x,y) = (x,y^2,xy^3 \pm x^ky)$, $k=3$	k	$C_k^{\pm}$	k

Moreover these singularities are versally unfolded, by the family F_g.

Proof: It is a straightforward exercise to check that we can write the relevant jet - space as a union of orbits of the indicated kind together with a subset of codimension greater than 5. We now Whitney stratify the jet space with the orbits as strata and apply theorem 2.1. Since the jet extension mapping will be transverse to the indicated orbits (for a generic embedding) it follows that these singularities when they occur, are versally unfolded by the family F_g. □

The $\mathcal{A}_e$- codimension is, of course, the codimension of the set of planes L with f_g^L having a singularity of the given type, while C is the number of cross caps associated to each singularity, an invariant due to Mond. Geometrically it means that given a folding f_g^L possessing such a singularity, one can perturb L to a nearby plane L′ in such a way that folding in L′ we find C cross caps have split off from the singularity. Moreover no perturbation of L splits more than C cross caps from the singular point in question.)

We actually know from Theorem 2.1 that for a generic embedding $g: X \to \mathbb{R}^3$ and an open dense set of planes L in $\mathbb{R}^3$ the map $f_g^L: X \to \mathbb{R}^3$, folding in the plane L, is stable. Since each preimage $(f^L)^{-1}(y)$ contains at most two points, we deduce that the image $f_g^L(X)$ is locally modelled by an immersion, a cross cap or a pair of transversal planes. Moreover the mapping f_g^L can fail to be stable in exactly two ways. First we might have a singularity more degenerate than a cross cap (one of the B_k, S_k, C_k singularities above) or secondly the image might have a point of self tangency. In other words two distinct immersive points $x, x' \in X$ of f_g^L with $f_g^L(x) = f_g^L(x')$ and $\text{im}\, df_g^L(x) = \text{im}\, df_g^L(x')$. The set of planes L for which the fold f_g^L is not smoothly stable is the <u>bifurcation set</u> $\mathcal{B}(F_g)$ of F_g, this is the principal geometric invariant of our family.

It has a natural stratification (see Figure 2) with two open strata. The "local" open stratum consists of planes L with f_g^L having one B_1 and no other non stable singularities, we label this stratum B_1. The "selftangency" open stratum consists of planes L with f_g^L folding two distinct immersive points together, with the corresponding pieces of folded surface tangent at this common point, and having A_1 contact with each other. Again we insist that f_g^L has no other non-stable singularities; this stratum is denoted by A_1^*, and $\mathcal{B}(F_g)$ is the closure of the union of the B_1 and A_1^* strata.

A similar state of affairs holds when we consider the family of distance squared functions on a surface $X \subset \mathbb{R}^3$. The function can fail to be stable locally via a degenerate critical point, or globally via distinct critical points with the same value. The corresponding smooth open strata are labelled A_2 and A_1^2 respectively (a single A_2 singularity or two A_1 singularities having the same critical value) and their closures are respectively the <u>focal set</u> and <u>symmetry set</u> of X. The focal set is the locus of centres of curvature of X, the symmetry set the closure of the locus of centres of spheres bitangent to X.

It turns out that the bifurcation set of the family of folds and that of the family of distance squared functions are intimately related.

Proposition 2.3. The bifurcation set $\mathcal{B}(F_g)$ is the dual of the focal and symmetry sets of X. More precisely the closure of the dual of the A_2 stratum is $\overline{B_1}$ while the closure of the dual of the A_1^2 stratum is $\overline{A_1^{*}}$.

So roughly speaking the "local" part of $\mathcal{B}(F_g)$ is dual to the focal set, the self tangency part is dual to the symmetry set.

Proof: This is fairly straightforward. For example f_g^L has a singularity at $x \in X$ if and only if the plane L contains the normal to X at x. This singularity is worse than a cross cap if and only if L also contains one of the principal directions. If the point p on the focal set corresponding to the <u>other</u> principal direction is smooth it is well known that L is the tangent plane to the focal set at p.

The other cases are dealt with similarly. □

Since the family F_g is an $\mathcal{A}$-versal unfolding of each of its singularities one can deduce local models for this dual set $\mathcal{B}(F_g)$. Ignoring multigerm singularities we have six such models, given in Figure 2. The bifurcation

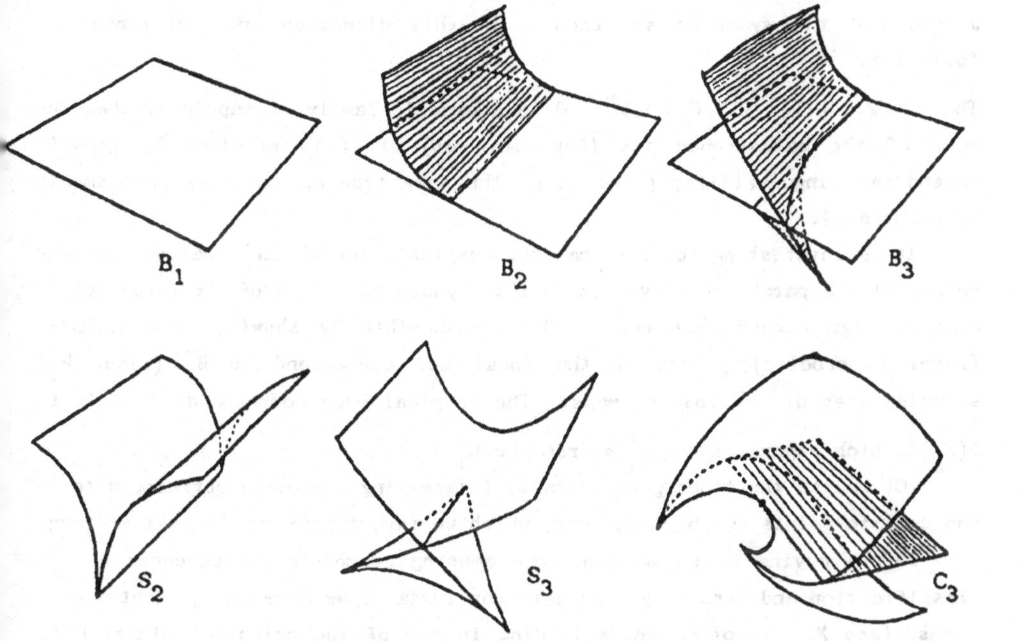

Fig. 2. Bifurcation Sets

sets of the S_1, S_2, S_3 singularities correspond to the discriminants of A_1, A_2, A_3 function germs respectively. Those of the B_k and C_k correspond to the discriminants of the singularities on manifolds with boundary bearing the same labels (cf [1], [7] this is a consequence of Mond's classification). In the pictures that part of the bifurcation set corresponding to self tangencies is shaded. It is fairly easy to deduce what the corresponding geometry on the focal set is:

$S_1 = B_1$	general smooth point of focal set
S_2	parabolic smooth point of focal set
S_3	cusp of gauss at smooth point of focal set
B_2	general cusp point of focal set
B_3	(cusp) point of focal set in closure of parabolic curve on symmetry set
C_3	intersection point of cuspidal edge and parabolic curve on focal set.

The bifurcation sets of multigerms of F_g model interior points of the dual of the symmetry set, and are locally diffeomorphic to the discriminants of A_k singularities. One is led to conclude that the focal set and symmetry set, when separated, have duals like those of any smooth surface. The interesting geometry arises near the cuspidal edge of the focal set where these sets merge, and the symmetry set ends. In this direction one can prove the following.

Theorem 2.4 [11]. If $G:M \times \mathbb{R}^n \to \mathbb{R}$ is a generic family of mappings, then the dual of the catastrophe set (Lagrange set) of G is modelled by generic Legendrian singularities, (i.e. is of the same type as the dual of a smooth hypersurface).

It is interesting to note that the cuspidal edge of the focal set behaves rather like a parabolic curve, in that the gauss curvature of the focal set changes sign across this edge. One proves this by showing that elliptic (resp. hyperbolic) points of the focal set correspond to B_1^+ (resp. B_1^-) singularities of the folding maps. The cuspidal edge corresponds to B_2's in $\mathcal{B}(F_g)$, which separates the B_1^+'s from the B_1^-'s.

Of course the other geometrically interesting phenomena correspond to the umbilic points on the focal set, which we investigate in the next section.

Before moving on we mention some amusing geometric consequences of our classification and versality. Suppose for instance we have an S_3 point x on our surface X. In other words folding in one of the principal planes L (a plane containing the normal and a principal direction) yields an S_3 singularity. If the surface X is generic this singularity is versally unfolded and has a local bifurcation set diffeomorphic to a swallowtail, see Figure 2. Geometrically each point in this space represents a plane, the swallowtail point the given principal plane, and nearby points on the swallowtail surface are principal planes at points near x. Points inside the swallowtail "pocket" represent planes close to L which are normal to X at four points near to x (the four cross caps of the folding). This provides a geometric characterization of S_3 points. The self intersection of the bifurcation set provides planes close to L which are simultaneously principal at two points close to x, another characterization. The same sort of analysis provides interesting geometric interpretations of the other singularities on our list. We hope this also helps convince the reader that the viewpoint provided by the family F_g has its own merits and geometric significance.

§3. Umbilics

At an umbilic every tangent direction is principal and so folding in any plane containing the normal produces a singularity more degenerate than a cross cap. In particular the dual map blows up, and the umbilic point corresponds to a (real) projective line of planes. This merely reflects the fact that to second order a surface at an umbilic is reflectionally symmetric in any "normal plane". Across certain special planes the surface is, infinitesimally, even more symmetric, and folding in these planes produces singularities of type B_2, S_2 (and exceptionally C_3).

The conditions for these singularities are governed conveniently by the cubic part of the immersion. More precisely first note that a change of coordinate frame allows us to write g(X) locally at an umbilic as

$$z = (\kappa/2)(x^2+y^2) + C(x,y) + \text{higher order terms}$$

where C is a cubic binary form, and $1/\kappa$ is the constant sectional radius of curvature. The assertion is that the position of the S_2 and B_2 planes is determined by the cubic C(x,y). Of course this cubic is only determined up to the natural SO(2) action on the variables. Writing the general cubic as $\mathfrak{Re}(\alpha z^3 + \beta \bar{z}^2 z)$, where $\alpha,\beta \in \mathbb{C}$, $z = x + iy$ it is well known that all (bar one) C(x,y)'s are SO(2) equivalent to a form of this type with $\alpha = 1$ (see [10], [12]). So one can think of the space of our cubic forms, indeed umbilics, as a (β-) plane. The folding geometry of the umbilics leads to a partition of this plane, Figure 3, explained below.

Of course we know that the B_2 singularities correspond to ribs - A_3 points of the distance squared functions. So the distinction 3/1 B_2's is the same as elliptic/hyperbolic umbilic. The "B_2 directions" are given by $C(-y,x) = 0$. The S_2 planes however are new - these correspond to parabolic curves on the focal set passing through the umbilic, and the "S_2 directions" are given by the vanishing of the Jacobian of $C(x,y)$.

Curve	Condition
Central hypocycloid $\beta=2e^{2i\theta}+e^{-4i\theta}$	Three B_2's inside, one outside
Circle $\|\beta\|=3$	S_2 is not versally unfolded by F_g
Outer hypocycloid $\beta=-3(2e^{2i\theta}+e^{-4i\theta})$	Three S_2's inside, one outside
Tangents to hypocycloids $\arg\beta=0,\pi/3,2\pi/3$	C_3 direction occurs.

(Note the triangular symmetry of this diagram is no accident, for β, $\bar{\beta}$ and $\beta e^{2\pi i/3}$ yield O(2)-equivalent forms.)

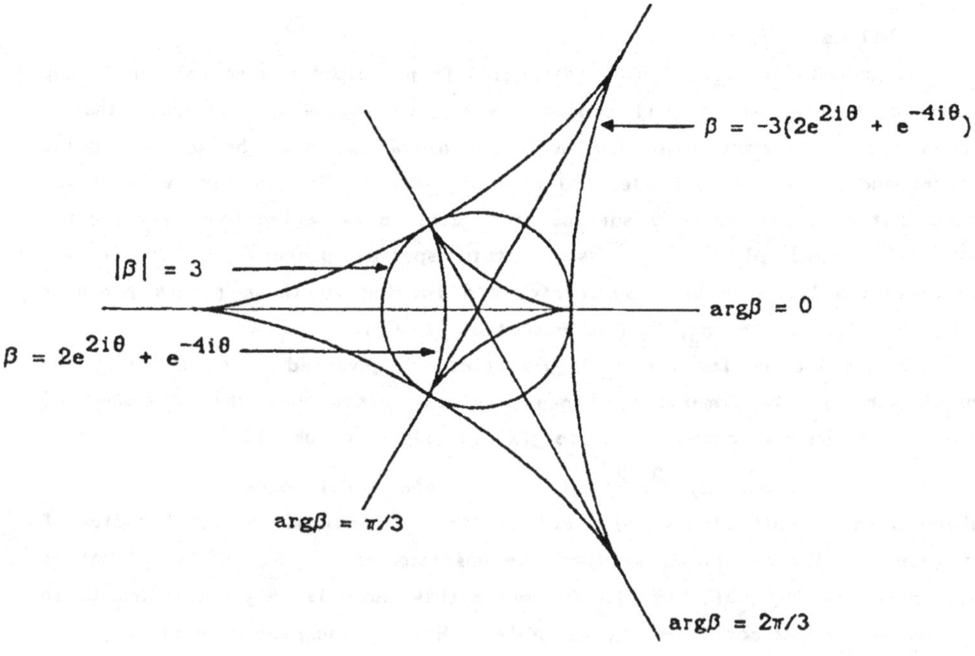

Fig. 3. Space Of Umbilics

We see that elliptic umbilics always have three such parabolic curves, hyperbolic umbilics three or one. (We can deduce from our genericity result that there are no parabolic curves on the symmetry set near the umbilic). The C_3 singularities arise for some folding across an umbilic precisely when the conditions for both an S_2 and a B_2 are satisfied; of course they do not arise on a generic surface. It is interesting to compare the geometry laid bare by the folding maps with that given by the distance squared functions, see especially [9], [10].

The geometry at umbilics is especially rich, with fascinating connections between the singularities of our family, and the configuration of lines of curvature. It is amusing to note that standard pictures of say an elliptic umbilic occurring in books like [2], [6], show no sign of the 3 parabolic curves which must be present in the generic case. Richard Morris has produced computer drawn pictures of these geometrically generic focal sets.

§4. Concluding Remarks

All of the results of §2 hold in a more general setting. We can clearly consider generic submanifolds X of any Euclidean space $\mathbb{R}^n$ and the natural n-parameter family of folding maps $X \to \mathbb{R}^n$. The case X a plane curve has been used, at our suggestion, by Giblin and Tari in their detailed investigation of rotational and reflexional symmetry in the plane [5]. When X is a hypersurface in $\mathbb{R}^n$, $n \geq 4$, the classification of singularities and geometry is very similar to that discussed here, again exhibiting close connections with functions on manifolds with boundary. Indeed one can make a classification of all $\mathcal{A}$-simple singularities which arise from folding submanifolds X^k of dimension k in $\mathbb{R}^n$, for any k and n. This allows one to determine the dimensions (k,n) for which the family of foldings $X^k \to \mathbb{R}^n$ is smoothly stable. One can also prove that if $n \geq 4k - 1$ then a generic submanifold X has no infinitesimal reflectional symmetry. More precisely each member of the family of folds is either an immersion or $\mathcal{A}$-equivalent to

$$(x_1, \ldots, x_k) \longmapsto (x_1, \ldots x_{k-1}, x_k^2, x_1 x_k, \ldots, x_{k-1} x_k, x_k^3, 0, \ldots, 0)$$

in the case when the folding plane contains the normal space at the corresponding point of X. See [11] for details.

References

[1] V.I. ARNOL'D, "Critical points of functions on manifolds with boundary", Russian Math. Surveys, 33 (1978), 99-116.

[2] TH. BROCKER & L. LANDER, Differentiable Germs and Catastrophes, London Math. Soc. Lecture Notes 17, Cambridge University Press, (1975).

[3] J.W. BRUCE, "Projections and reflections of generic surfaces in $\mathbb{R}^3$", Math. Scand., 54 (1984), 262-278.

[4] J.W. BRUCE, "Geometry of singular sets", Math. Proc. Camb. Phil. Soc., 106 (1989), 495-509.

[5] P.J. GIBLIN & F. TARI, "Local reflexional and rotational symmetry in the plane", these proceedings.

[6] M. GOLUBITSKY & V. GUILLEMIN, Stable Mappings and their Singularities, Graduate Texts in Math. 14, Springer-Verlag, Heidelberg, (1973).

[7] D. MOND, "On the classification of germs of maps from $\mathbb{R}^2$ to $\mathbb{R}^3$", Proc. London Math. Soc., (3), 50 (1985), 333-369.

[8] J.A. MONTALDI, "On generic composites of maps", to appear in Bull. Lond. Math. Soc.

[9] I.R. PORTEOUS, "The normal singularities of a submanifold", Jour. Diff. Geom., 5 (1971), 543-564.

[10] I.R. PORTEOUS, "The normal singularities of surfaces in $\mathbb{R}^3$", Proceedings of Symposia in Pure Math., volume 40, (1983), Part 2, 379-393.

[11] T.C. WILKINSON, Ph.D. thesis, University of Newcastle Upon Tyne, (In preparation).

[12] E.C. ZEEMAN, "The umbilic bracelet and the double-cusp catastrophe", Structural stability, the theory of catastrophes and applications in the sciences, Springer Lecture notes in Mathematics, Volume 525 (1976), 328-366.

Department of Mathematics and Statistics
The University
Newcastle Upon Tyne, NE1 7RU,
England.

The Dual Graph for Space Curves

Julio Castellanos[1]

ABSTRACT

In this paper we give an analogue of the dual graph for a space curve. Taking an embedded resolution, we consider the graph whose vertices are the components of the exceptional divisor, with edges joining intersecting components, and appropriate weights which determine the complete geometry of all infinitely near points associated with the curve.

Introduction.

Let K be an algebraically closed field, and let $X = \mathrm{Spec}R_N$, where $R_N = K[[\underline{X}]]$, $\underline{X} = (X_1,\ldots,X_N)$, the X_i are indeterminates over K, and $N \geq 3$. Let $C = \mathrm{Spec}\,\theta$ be an irreducible algebroid space curve, and $i: C \to X$ an embedding of the curve into X. It is well known that by means of successive quadratic transforms of X we obtain a smooth curve C* as transform of C. Let $X \xleftarrow{\pi_1} X_1 \xleftarrow{\pi_2} \cdots X_i \xleftarrow{\pi_{i+1}} X_{i+1} \longleftarrow \cdots \xleftarrow{\pi_m} X_m$ be one of these sequences of quadratic transformations where the centre of π_1 is P_0, the closed point given by the maximal ideal of R_N, C_1 is the strict transform of C, E_1 is the exceptional divisor of π_1, i.e. the projective space $\mathbb{P}^{N-1}$ of all directions through P_0, and $P_1 = E_1 \cap C_1$. The centre of π_{i+1} is $P_i = C_i \cap E_i$, where C_i is the strict transform of C_{i-1} and E_i is the exceptional divisor of π_i. Let us suppose that C_m is a smooth curve and that it is transverse to the exceptional divisor E_m. In the space X_m we have a projective space $\mathbb{P}^{N-1} = E_m$, and some smooth varieties obtained by the sequence of quadratic transformations over the exceptional divisors E_i, the total exceptional divisor of the sequence. In this paper we describe the geometry of these smooth varieties at X_m and we see that they determine the singularity of the curve from the point of view of its resolution.

1. The dual graph.

We can consider the following varieties in X_k:

(1) The exceptional divisor E_k;

(2) The strict transform C_k, with $C_k \cap E_k = P_k$;

(3) The smooth projective varieties $E_1^{(k)},\ldots,E_{k-1}^{(k)}$, where $E_i^{(k)}$, $i < m$ is the strict transform,

[1] Partially supported by D.G.I.C.Y.T. n° PB 88-0344-C-03-01

by the sequence, of the exceptional divisor E_i. $E_i^{(k)}$ is the $(k-i)^{\text{th}}$ quadratic transform of the projective space $\mathbb{P}^{N-1} = E_i$;

(4) Smooth linear subspaces $L_{jk} \subseteq E_k$, for k,j with $0 < k \leq m$, defined as follows: given $j \leq m$, we set $L_j^j = \{p_j\}$ and inductively define $L_{j,k}$ to be the linear subspace of E_k determined by $(T_{p_k}E_k) \cap L_k^j$, where $L_k^j = \left\{p \in T_{p_k}X_k \,\middle|\, \overline{PP}_k \in L_{j,k+1}\right\}$. See [1, Theorem 2.3] and [3, Definition 1 and Lemma 1].

(5) The smooth subvarieties $L_{ji}^{(k)} \subset E_i^{(k)}$, $i < k$, $j \leq m$ where $L_{ji}^{(k)}$ is the strict transform of L_{ji} by the sequence of quadratic transforms.

We are interested in the geometry of this picture, and in particular in knowing the dimensions d_{ij} of $L_{ji}^{(k)}$ and the intersections $E_i^{(k)} \cap E_j^{(k)}$ and $E_i^{(k)} \cap E_k$. Thus associated with the above situation in X_m, i.e. with the varieties E^m, $E_i^{(m)}$, $L_{ji}^{(m)}$, C_m, $0 \leq i \leq j \leq m$ we define a weighted graph as follows:

Definition 1. The graph associated with the curve C, is the following weighted labeled graph formed by:

(1) m vertices, each of $i \leq m$ representing $E_i^{(m)}$,

(2) the edges (i,j) if $E_j^{(n)} \cap E_i^{(m)} \neq \emptyset$ (we set $E_m^{(m)} = E_m$),

(3) a weight for each edge (i,i+1), consisting of an (n−i)-tuple (d_{ji}), $i < j \leq m$, where d_{ji} is the dimension of $L_{ji}^{(m)}$ if it is different from zero.

Example 1. The curve given by parametric equations $\{x_1 = t^8, x_2 = t^{10}, x_3 = t^{15}\}$ has the following dual graph:

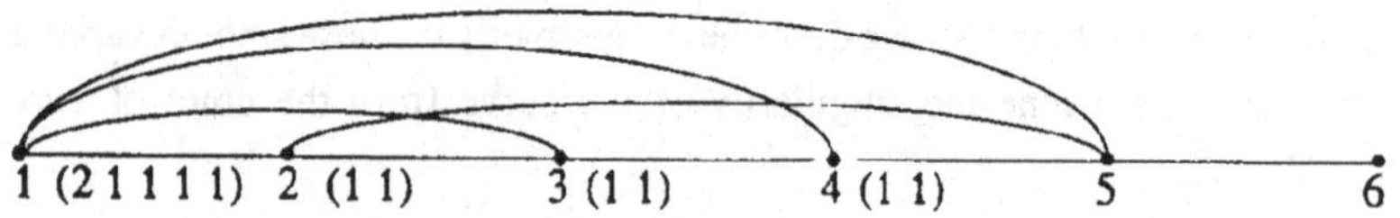

Remark 1. (1) The dual graph does not depend on the embedding of the curve. In fact the isomorphism between the ambient spaces obtained from the embedding, gives us one between the spaces of directions through the base points i.e. between the projective spaces given by the exceptional divisors. Following the process of resolution step by step, we get equality between the dimension of L_{jk} in both embeddings and also the dimensions of $E_j^{(k)} \cap E_k$ in both embeddings.

(2) The dual graph G associated with C determines the dual graph G_k associated with the kth

transform C_k, for all $k < m$. In fact the graph for C_k is obtained from G by taking off the vertices i, edges (i,) for $i < k$ and changing the numbers j, for $k \leq j \leq m$ to $j - k + 1$. The dimensions d_{ij} remain as above.

(3) The graph G determines the geometry at each step k of the resolution i.e. we know the intersections among the $E_i^{(k)}$ and the dimensions of L_{jk} and $L_{ji}^{(k)}$.

In fact suppose $E_j^{(k)} \cap E_i^{(k)} = H \neq \emptyset$; since $\dim E_j^{(k)} = \dim E_i^{(k)} = N - 1$, and $N > 2$, we have $\dim H \geq 1$. Let H' be the strict transform by π_k; if $P_k \in H$, $H' = E_j^{(k+1)} \cap E_i^{(k+1)}$ has the same dimension as H too, and so $E_j^{(k+1)} \cap E_i^{(k+1)} \neq \emptyset$. Then $E_i^{(m)} \cap E_j^{(m)} \neq \emptyset$ if and only if $E_i^{(k)} \cap E_j^{(k)} \neq \emptyset$ for all $i,j \leq k$ and $k \leq m$.

So $E_i^{(k)} \cap E_k \neq \emptyset$ if (i,k) is an edge, and $E_i^{(k)} \cap E_j^{(k)} \neq \emptyset$ for $i < j < k$, if (i,j) belongs to G. Obviously $E_{k-1}^{(k)} \cap E_k \neq \emptyset$.

The smooth subspace $L_{jk} \subset E_k$ such that $L_{jk} \neq \{P_k\}$ has the same dimension as its transform $L_{jk}^{(i)}$ for each i,j, with $k < j \leq m$ and, $k < i \leq m$, so $\dim L_{jk} = \dim L_{jk}^{(m)} - d_{jk}$, and $\dim L_{ji}^{(k)} = \dim L_{ji}^{(m)} = d_{ji}$ for $i < k$. Then the d_{ji} appearing in the graph, determine the dimensions of L_{ji}, $L_{ji}^{(k)}$.

Associated with the resolution of the singularity of an irreducible algebroid space curve C one can consider the following invariants:

(I) The multiplicity sequence $E(C) = \{n, n_1, n_2, \ldots, n_r = 1\}$ with $n = e(C) = e(\theta)$ the multiplicity of E, and $n_i = e(C_i) = e(\theta_i)$ the multiplicity of the ith quadratic transform $D_i = \mathrm{Spec}(\theta_i)$.

(II) The Arf numbers $\{d, d_1, \ldots, d_r\}$ of C and C_i for $i = 1, \ldots, r$. The Arf number of a curve C is the minimum dimension of a linear variety on which the curve C can be generically projected preserving the multiplicity sequence [5].

Theorem 1. The dual graph F of a curve C determines the multiplicity sequence and the Arf numbers of C.

Proof.

(1) The multiplicity sequence can be obtained from the proximity relations among the infinitely near points associated with the curve, following Noether's formula [6]. One can see this classical result for plane curves in Casas's paper [2]. Given $i < k$, P_k is proximate to P_i if $P_k \in E_{i+1}$ $(k = i + 1)$ or $P_k \in E_{i+1}^{(k)}$, i.e. $E_{k+1} \cap E_{i+1}^{(k+1)} \neq \emptyset$ and this means $E_{k+1}^{(m)} \cap E_{i+1}^{(m)} \neq \emptyset$, and can be determined from the graph. Thus the graph determines also the satellitism order in P_k,

the multiplicity of the total exceptional divisor in P_k, i.e. the number of $E_i^{(k)}$ such that $P_k \in E_i^{(k)}$, and the relative satellitism order OS_i^j by considering the sequence starting on P_i.

We obtain a matrix (a_{ij}) $1 \le i,j \le m$ from the graph: $a_{ij} = 1$ if $i \le j$ and P_j is proximate to P_i and zero otherwise. We get the multiplicity sequence $n, n_1, \ldots, n_r = 1$ from the relations among them given by $(n_1, \ldots, n_m)(a_{ij}) = (n, n_1, \ldots, n_{m-1})$.

(2) The Arf number d of a curve C is the maximum of the dimensions of L_0^j, d_0^j for $0 \le j \le m$, for C_i, $d_i = \max\{d_i^j \,/\, i \le j \le m\}$ [1,3]. Let us see that we get d_{ji} from d_i^j; suppose $L_{i+1}^j \subset E_{i+1}$, then $\dim(L_{i+1}^j) = \dim(L_{ji+1})$, and $\dim(L_i^j) = \dim(L_{ji+1}) + 1$. If $L_{i+1}^j \not\subset E_{i+1}$, $L_{i+1}^j \cap E_{i+1} = H$, with $\dim H = \dim(L_{i+1}^j) - 1$ then $\dim(L_{i+1}^j) = \dim(L_{ji+1}) - 1$ and $\dim(L_i^j) = \dim(L_{ji+1})$. So for each $j \le m$, given i, we put

$$d_i^j = \begin{cases} d_{ji} + 1 & \text{if} \quad d_{ji} = d_{ji} + 1 \\ d_{ji} & \text{if} \quad d_{ji} = d_{ji+1} + 1 \end{cases}$$

Now we will consider the dual graph as an abstract object i.e. we define a weighted labelled graph, and we will show that there exists a space curve with precisely this considered graph.

Definition 2. A dual graph G of a space curve is the labelled weighted graph which consists of

(i) m vertices,

(ii) the edges $(i,i+1)$ for $i < m$, and (i,j) for some i,j with $i < j < m$, such that if $k < i - 1$ and (k,i) does not belong to the graph, then (k,j) for $i < j$ are not in the graph.

(iii) a weight for each edge $(i,i+1)$, consisting of an $(m-i)$-tuple (a_i^j) $i < j \le m$, $a_i^j \in \mathbb{N}_+$ verifying

(a) $a_i^j = a_i^{j+1} + \begin{Bmatrix} 1 \\ 0 \end{Bmatrix}$, $a_i^j = a_{i+1}^j - \begin{Bmatrix} 1 \\ 0 \end{Bmatrix}$,

(b) there exists j with $a_i^j = 1$,

(c) let $k_1, \ldots, k_t$ with $1 \le k_i \le m$ be such that for each i $(k_i - 1, k_i + 1)$ belongs to the graph, then $a_{k_i}^j = a_{k_i+1}^j = \ldots = a_{k_{i+1}-1}^j$ for $i = 1, \ldots, t$, and $a_\ell^{k_j} = a_\ell^{k_{j+1}-1}$ for $\ell \le k_j$., and $i = 1, \ldots, t$.

Theorem 2. Given a dual graph G as above there is a space curve C over k such that the dual graph associated to it in definition 1 is G.

Proof.

We find the curve by building a Hamburger-Noether matrix over K with a matrix associated to the sequence of infinitely near points of the curve (see [3]). The embedding dimension of C, i.e. the number of rows of the matrix, will be $N = \max\left\{a_i^j\right\} + 1$, where the a_i^j are defined in (iii). The number of boxes will be $t + 1$, the maximum subindex of the k_i's defined in (iii). The length of each box B_i is $h_{i-1} = k_i - k_{i-1}$, except that the length of b_0 is $h = k_1$. We set $h_{t+1} = \infty$.

We mark the first row in the last box B_t and the second one in B_{t-1}. In the box B_{t-2}, if $a^m_{k_{t-2}-1} = a^m_{k_{t-2}}$ we mark the first one, and if $a^m_{k_{t-2}-1} = a^m_{k_{t-2}} + 1$ the third one. Now for each case mentioned we have two possibilities with $a^m_{k_{t-3}-1} = a^m_{k_{t-3}} + \left\{\begin{matrix}1\\0\end{matrix}\right\}$ and there are four possibilities to mark the 1st, 2nd, 3rd, 4th rows. Recursively we mark one row in each box.

The satellitism order is given in the matrix by $OS_i^j = \#$ {marked rows having zeros from the 1 of the marked rows in column i to column j}. (See [1].) Since we know the satellitism order given by the graph G (proof of theorem 1), we put as entries into the matrix after the marked rows all the necessary zeros to get for the matrix the same satellitism order, putting any element different from zero after the last zero. The remaining entries of the matrix can be any elements of K.

From the above, it is not difficult to see that the curve C associated with the matrix has the graph G as its dual graph.

We notice that the numbers labelling the points of the graph, correspond to "the age" of the components of the exceptional divisor. If we take off these numbers in the graph, the question is how we can get these numbers L_n by looking at the varieties $E_i^{(m)}$?

This should be possible using as a new weight in each point instead of the labelling, the rank of the 1-codimensional Chow group. One has $H^1(\mathbb{P}_{n-1}) = \mathbb{Z}$ and $H^1(E^{(r)}) = \mathbb{Z}^r$ if $E^{(r)}$ is obtained from $\mathbb{P}_{N-1}$ by $r-1$ successive quadratic transformations with centre in points of each $E^{(i)}$, so at the last step one has $H^1(E_m) = \mathbb{Z}^1$, and $H^1(E_i^{(m)}) = \mathbb{Z}^{r+1}$ means that among the transform $E_i^{(k)}$ of E_i, $k = 1,\ldots,m$, there are r of them, $E_i^{(i_j)}$ for $j = 1,\ldots,r$, such that the centre of the blowing up, P_{i_j} belongs to $E_i^{(i_j)}$.

Theorem 3. The dual graph where the labelling has been changed by the ranks of the codimension 1 Chow groups as above determines the age of all components of the total exceptional divisor.

Proof.

The proof is by induction on the number of the transformations, i.e. on the number of points. We consider m points, then there is only one with 1 as rank, and this is the last one, i.e. the point m. If we take off this point and consider the graph formed by the remaining m−1 points the ranks of the groups decrease by 1 if they were connected in the graph with the point m (see proof of theorem 2).

Remark 2. The converse of theorem 2 is also true in the sense that we can get the ranks of the Chow groups of $E_i^{(m)}$ from the invariants of the dual graph considered in the definition 1. In fact rank $(E_i^{(m)}) = \#\{k, i < k < m$ such that k and i are connected by the graph$\} +1$.

2. Equisingular deformations of curves and dual graphs:

The dual graph of a space curve determines the invariants associated with the resolution process of the curve, but it doesn't determine other invariants of the curve, for example, the semigroup of the curve and the equisingularity type of the generic plane projection, as we can see in the following examples:

Example 2. The curves given by the parametric equations $C_1 \equiv \{x_1 = t^8, x_2 = t^{10}, x_3 = t^{15}\}$ (example 1), $C_2 \equiv \{x_1 = t^8 + t^9, x_2 = t^{10} + t^{11}, x_3 = t^{15} + t^{16}\}$ have the same dual graph and the exponents for the equisingularity type of the generic plane projections are $\{8,10,15\}$ for C_1 and $\{8,10,11\}$ for C_2.

Example 3. The curves $C_1 \equiv \{x_1 = t^5, x_2 = t^6, x_3 = t^7\}$ and $C_2 \equiv \{x_1 = t^5, x_2 = t^6, x_3 = t^8\}$ have different semigroup but their dual graphs are the same:

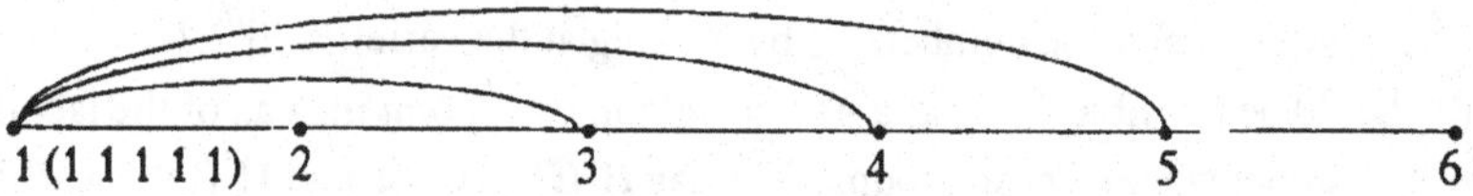

One of the reasons to consider the dual graph as above is that it is constant along a deformation given by a Hamburger-Noether matrix.

In [4] we have studied Hamburger-Noether (H-N) matrices over a ring A, where $A = K[[V_1,...,V_1]]$, with V_i, $1 \le i \le n$ indeterminants over k. We have a deformation associated with a H-N matrix over A, (Z, Z_0, Y, s) where $Y = \text{Spec}A$, $Z = \text{Spec}R$, $R = \dfrac{A[[\underline{X}]]}{(\text{Ker}\,\varphi)}$,

$\varphi : A[[\underline{X}]] \to A[[t]]$ is the homorphism given by the matrix, $Z_0 = \left(\text{Spec}\left(\dfrac{R}{\mathcal{M}_A R}\right)\right)_{\text{red}}$ is the branch given by the H-N matrix making the residue over $\mathcal{M}_A$ of the entries and s is the section given by the ideal $\mathcal{B} = (x_1,...,x_n)R$. (See section 2 of [4]). In the same paper, we show that the above deformation is equisingular, following the treatment of the plane curves case by Zariski and Stuz. In fact we get that the curves Z_0 and Z_g, the generic curve along the section i.e. the curve obtained from the matrix by considering the entries in the field $\overline{k((V_1,\ldots,V_n))}$, have the same multiplicity sequence (proposition 2.4 [4]).

Let us consider the H-N matrix over A such that condition (ii) is changed to:

(ii) If the marked row i_k in the box C_g was marked last time on the box C_s, then all marked rows from C_s to C_g have a zero in the first column of the box C_g.

Proposition 1. The deformation (Z,Z_0,Y,s) associated with an H-N matrix as above has constant dual graph, i.e. the dual graphs of Z_0 and Z_g are the same.

Proof.

The units after the marked rows in the matrix force the numbers of quadratic transformations for Z_g and Z_0 needed to get smooth curves and transversal to exceptional divisor to be the same. So the number of vertices of the two graphs remains invariant. The units force the satellitism orders OS_i^j (see proof of theorem 2) to be equal for both curves, so the intersection $E_i^{(m)} \cap E_j^{(m)}$ is non-empty, for $1 \le i, j \le m$, and for both curves (see the proof of theorem 1).

In an H-N matrix of a curve the dimensions d_i^j = # {different marked rows from column i to column j}. The condition (ii) considered, gives that the possibility of choosing a new marked row is the same for the H-N matrices for Z_g and Z_0, so the d_i^j's are the same for both curves and so the d_{ji} and the weights coincide for them.

Remark 3. A deformation (Z, Z_0, A, s) preserving the dual graph is equisingular in the sense of Zariski, since the multiplicity sequences of Z_0 and Z_g are the same.
The converse is not true in general as we can see in the following example.

Example 4. The deformation of $K[[V]]$ $\{x_1 = t^4 + Vt^5, x_2 = t^6 + Vt^7, x = t^9 + Vt^{10}\}$ given by the matrix $\left(\begin{array}{c|cc|cc} 1 & 0 & 1 & V & 0 \\ 0 & 1 & 0 & 0 & 1 \\ 0 & 0 & 0 & 1 & 0 \end{array}\right)$ is equisingular following Stuz and Zariski [4] but the dual graph Z_0 is

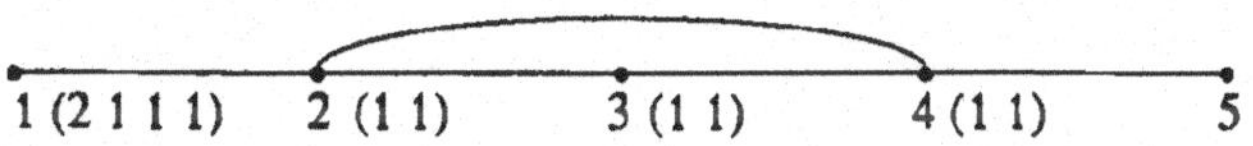

and the graph of Z_g is

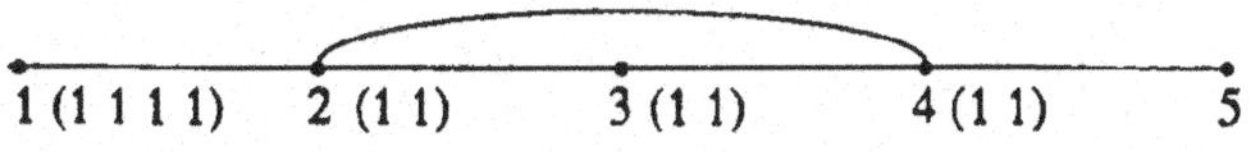

REFERENCES.

[1] Campillo, A. & Castellanos, J.: "On projections of space curves", in: "Algebraic Geometry", Lecture Notes in Math. 961 (Springer, 1982), 22-31.

[2] Casas, E.: Manuscript notes, 1989.

[3] Castellanos, J.: "A relation between the multiplicity sequence and the semigroup of values of an algebroid curve", Journal of Pure and Applied Algebra, vol. 43 (119-127) 1986.

[4] Castellanos, J.: "Hamburger-Noether matrices over rings", Journal of Pure and Applied Algebra, vol. 64, 1990.

[5] Lipman, J.: "Stable ideals and Arf rings", Amer. J. Math. 93 (1971) 649-685.

[6] Van der Waerden, B.L.: "Infinitely near points", Indiana Math. J. vol. 12 (1950) 401-410.

Departamento de Matemática Fundamental,
Universidad de la Laguna,
Tenerife,
Islas Canarias,
Spain.

On the Components and Discriminant of the Versal Base Space of Cyclic Quotient Singularities

JAN ARTHUR CHRISTOPHERSEN[1]

INTRODUCTION

This is meant as a complementary note to Jan Stevens' paper *On the versal deformation of cyclic quotient singularities* in this proceedings, which I rely on for definitions, notation, results and introduction to the problem. My "proof", referred to in Stevens' introduction, of Stevens' theorem on the component structure of this versal base space, contains a mistake. Though salvageable, it hardly seems worthwhile to pursue, given Stevens' ingenious construction. Still, some of the partial results may be of interest.

My attack was much more elementary (an outline is sketched below). It was based on an explicit description of the families over each component in terms of the continued fractions described in our §1 and [**Stevens**,1.3]. Among the benefits of this description are

(1) the discovery (but not the meaning) of a natural finite Galois covering of each component, generalizing the well known covering of the Artin component, and
(2) a complete description of the discriminant of each component.

The families and the covering are described in §2 and the discriminant in §3. There is also a method for finding all the singularities in the nearby fibers in §3, but no closed formula or list is given.

Though probably no published work on this question will ever contain a single piece of obstruction calculus, this type of computation (as e.g. described in [**Laudal**]) was crucial for finding the correct combinatorical tool to handle the problem. After computing the base space for low embedding dimension, it became clear that there was a simple pattern in the way the obstructed monomials were generated.

Recall that a cyclic quotient surface singularity is the same thing as a two dimensional affine toric variety. If $w_i \in \mathbf{Z}^2$ are the weights of the generators of the maximal ideal, there exists multiplicities a_i such that $w_{i-1} + w_{i+1} = a_i w_i$ ([**Stevens**, 2.1]). To find the pattern in the obstructions in terms of the torus action on T^1 and T^2 one was lead to solve equations of the type

$$w_i + w_j = \sum_{h=i+1}^{j-1} \alpha_h (a_h - k_h) w_h$$

where α_h and k_h are positive integers.

The chains representing zero, [**k**],and their $\alpha([\mathbf{k}])$ described in §1 and [**Stevens**, 1.3], are "canonical" solutions. Amazingly it turned out that the number of such solutions was the same as the upper bound on the number of components conjectured by Arndt. (Only later, thanks to David Eisenbud, did we learn that this number is the famous Catalan number.)

[1]partially supported by the Norwegian Research Council for Science and the Humanities

The reason for this is now clear. The ideals $\mathbf{p}_{[\mathbf{k}]}$ of the components (or at least $\mathrm{Gr}_{\mathbf{m}}\,\mathbf{p}_{[\mathbf{k}]}$, the ideal of leading forms), can be defined in terms of these chains (see 2.1.1 and [**Stevens**, 5.1]). A purely combinatorical argument affirms that $\mathrm{Gr}_{\mathbf{m}}(\cap\mathbf{p}_{[\mathbf{k}]})$ is generated by (in general) monomials which may also be completely described in terms of such chains with smaller length. I will not discuss this in the present paper — see instead [**Christophersen**] — but it was the main observation in my "proof".

The idea was to treat the $\mathbf{p}_{[\mathbf{k}]}$ as candidates to minimal prime ideals of $\mathbf{a}$, the ideal of the versal base space as say described in [**Arndt**], and then using combinatorics and the obstruction calculus to prove $\cap\mathbf{p}_{[\mathbf{k}]} \subseteq \sqrt{\mathbf{a}}$. The other inclusion is guaranteed by the construction of flat families over each $\mathbf{p}_{[\mathbf{k}]}$. What makes it seem feasible to do this, is that the statement can be proven for the subspace, T, where all the $s_i = 0$ (see [**Stevens**, 2.2] for notation). Unfortunately, the mistake in my "proof", is essentially that the computations are much more complicated for the whole space S than for T.

As mentioned already we assume the reader is familiar with Stevens' paper. Yet to understand the statements in this paper, all one "really" has to know about these singularities is that they are labeled by a tuple of integers $a_i \geq 2$. Equations are given in 1.2.2.

Notation. Our notation is slightly different from Stevens', (we take the point of view that the indices of cyclic quotient singularities should conform to the notation of chains that represent zero), so here is a dictionary.

We label the cyclic quotient singularities by $X(a_1,\ldots,a_r)$, i.e. $r = e-2$ is the codimension and we shift the indices one to the left. The maximal ideal of $\mathcal{O}_X$ is generated by $z_0, z_1, \ldots, z_r, z_{r+1}$. This has consequences for the dual base of T^1, $\{t_i^{(l)}, s_i\}$, as well (2.1.1 and [**Stevens**,2.2]). Here also all indices must be shifted one to the left. Moreover, we have a reverse order on the upper indices of the $t_i^{(l)}$; our $t_i^{(l)}$ = Stevens' $t_{i+1}^{(a_{i+1}-l)}$.

Acknowledgements. My work on deformations of cyclic quotient singularities is part of my doctoral thesis at the University of Oslo, done under the supervision of Olav Arnfinn Laudal. I would like to take this opportunity to thank him not only for constant help and guidance during this endeavor, but also for introducing me to and teaching me deformation theory. Thanks to both Jürgen Arndt and Jan Stevens for sharing their ideas with me throughout this project.

§1. Chains representing zero

1.1. Some of the following statements may be found in in [**Stevens**,§1], but even if they are not, we omit their proofs. Most of them are exercises in the blowing up and down procedure described below and in [**Stevens**, 1.1].

1.1.1. We recall the definition of chains, or continued fractions, representing zero. Let $k_1,\ldots,k_n$ be a sequence of n non-negative integers. Such a sequence will be called a *n-chain* and denoted $[k_1,\ldots,k_n]$, often shortened to $[\mathbf{k}]$. Define now $n+2$ integers $\alpha_i = \alpha_i([\mathbf{k}])$ by $\alpha_0 = 0$, $\alpha_1 = 1$ and $\alpha_{i+1} = k_i\alpha_i - \alpha_{i-1}$ for $i = 1,\ldots,n$.

We say that $[\mathbf{k}]$ represents zero if $\alpha_i > 0$ for $i = 1,\ldots,n$ and if $\alpha_{n+1} = 0$. This is equivalent to the continued fraction

$$k_1 - \cfrac{1}{k_2 - \cfrac{1}{\cdots - \cfrac{1}{k_n}}}$$

being both well-defined and actually equal to zero. (See also **[Orlik–Wagreich]**.) The set of all such n-chains is denoted K_n.

Blowing up and *blowing down* are methods for passing back and forth between K_n and K_{n-1}. Recall that if $[\mathbf{k}] \in K_n$ and $n \geq 2$ then at least one $k_i = 1$, and $[k_1, \ldots, k_{i-1} - 1, k_{i+1} - 1, \ldots, k_n] \in K_{n-1}$. This is "blowing down at the i'th position". (Modify suitably at the ends.) The inverse operation is blowing up.

We also define *chopping off*; if $\alpha_i = \alpha_j = 1$, $i < j$, then $[k_i - \alpha_{i-1}, k_{i+1}, \ldots, k_{j-1}, k_j - \alpha_{j+1}] \in K_{j-i+1}$. We call this chopping off at i and j. If we write $j = n+1$ we mean chop off only on the left, and vice versa if $i = 0$.

1.1.2. On the set of all n–chains we take the *partial order* $[\mathbf{k}] \geq [\mathbf{k}']$ if $k_i \geq k'_i$ for all i. (The elements in K_n are not related by this order.)

Besides blowing down there is a another way of passing from K_n to K_{n-1}.

LEMMA – DEFINITION. *If* $[\mathbf{k}] \in K_n$, *then there exists a unique* $\Phi_1([\mathbf{k}]) \in K_{n-1}$ *such that* $[k_1, \ldots, k_{n-1}] > \Phi_1([\mathbf{k}])$ *and a unique* $\Phi_2([\mathbf{k}]) \in K_{n-1}$ *such that* $[k_2, \ldots, k_n] > \Phi_2([\mathbf{k}])$.

Also $\sum_{i=1}^{n} k_i - \sum_{i=1}^{n-1} \Phi_1([\mathbf{k}])_i = 1 + k_n$ *and* $\sum_{i=1}^{n} k_i - \sum_{i=1}^{n-1} \Phi_2([\mathbf{k}])_i = 1 + k_1$. □

We have now at our disposal two well defined maps $\Phi_i : K_n \to K_{n-1}$. Define the *norm* of $[\mathbf{k}]$ by $N([\mathbf{k}]) = \sum_{i=1}^{n} k_i$. One checks that for $[\mathbf{k}] \in K_n$, $3n - 4 \geq N([\mathbf{k}]) \geq 2(n-1)$. Define $d([\mathbf{k}]) = N([\mathbf{k}]) - 2(n-1)$, $K_{n,d} = \{[\mathbf{k}] \in K_n : d([\mathbf{k}]) = d\}$ and $c(n,d) = \operatorname{card} K_{n,d}$. (One can also show that $d([\mathbf{k}]) = \operatorname{card}\{i : \alpha_i([\mathbf{k}]) > 1\}$, so it is the l in **[Stevens**,5.3].)

Using e.g. Φ_1 one gets another (less elegant) proof of Theorem 1.3.4 in **[Stevens]**. More precisely one shows that $\Phi_1 : K_{n,d} \to \bigcup_{i=0}^{d} K_{n-1,i}$ is 1-1 and onto. This yields by induction that $c(n,d) = \frac{n-d-1}{n-1}\binom{n+d-2}{n-2}$. Thus $\operatorname{card} K_n = \frac{1}{n}\binom{2(n-1)}{n-1}$, the Catalan number.

1.2. The Φ_i are of fundamental importance for our later description of the family over each component. Here we will first see how they may be used to get (for each $[\mathbf{k}]$) different sets of generators for I_X. This should be compared with the algorithm in **[Stevens, 6.2]**.

1.2.1. We may iterate the process in 1.1.2 to get a unique "pyramid lattice" of chains representing zero from one $[\mathbf{k}] \in K_n$. Think of the given $[\mathbf{k}]$ as the top node in a lattice, where the second to top level consists of $\Phi_1([\mathbf{k}])$ and $\Phi_2([\mathbf{k}])$ with lines attaching it to $[\mathbf{k}]$. Now apply the Φ_i to get the next row using that $\Phi_1\Phi_2 = \Phi_2\Phi_1$, etc.. For example, if $[\mathbf{k}] = [1,4,1,2,2]$ the pyramid lattice looks like this

$$
\begin{array}{ccccc}
 & & [1,4,1,2,2] & & \\
 & [1,3,1,2] & & [3,1,2,2] & \\
[1,2,1] & & [2,1,2] & & [1,2,1] \\
[1,1] \quad [1,1] & & [1,1] & & [1,1] \\
[0] \quad [0] & & [0] & & [0] \quad [0]
\end{array}
$$

Thus, starting at the bottom, at each level m we get $n-m+1$ (not necessarily distinct) elements $[\mathbf{k}^{i,j}] \in K_m$ which we index with a pair i,j, $1 \leq i+1 \leq j-1 \leq n$ and $j-i-1=m$, such that $[k_{i+1},\dots,k_{j-1}] \geq [\mathbf{k}^{i,j}]$. In the above example $[\mathbf{k}^{0,5}] = [1,3,1,2]$, $[\mathbf{k}^{1,5}] = [2,1,2]$, etc..

Denote $\alpha_h^{i,j} = \alpha_h([\mathbf{k}^{i,j}])$. One checks that also $\alpha_{i+h} = \alpha_{i+h}([\mathbf{k}]) \geq \alpha_h^{i,j}$.

There is a more explicit description of the Φ_i. If $l \in \{1,\dots,n-1\}$ is such that $\alpha_l = 1$ and $\alpha_i > 1$ for $n-1 \geq i \geq l+1$, then $\Phi_1([\mathbf{k}]) = [k_1,\dots,k_{l-1},k_l-1,k_{l+1},\dots,k_{n-1}]$. The following formula will be important for us later (2.1.2):

$$\alpha_{h-i}^{i,j} = \alpha_{h-i}^{i,j-1} + \alpha_{h-l^{i,j}}^{l^{i,j},j},$$

where $l^{i,j}$ is the l as above for $[\mathbf{k}^{i,j}]$ and $j > i+2$. Completely symmetric formulas are valid when we consider Φ_2.

1.2.2. Define $K_r(X) := \{[\mathbf{k}] \in K_r : [\mathbf{k}] < [\mathbf{a}]\}$. For each $[\mathbf{k}] \in K_r(X)$ define $p_{i,j}([\mathbf{k}]) \in \mathbf{C}[z_0,\dots,z_{r+1}]$ by

$$p_{i,j}([\mathbf{k}]) = z_{i+1}^{\alpha_1^{i,j}(a_{i+1}-k_1^{i,j})} z_{i+2}^{\alpha_2^{i,j}(a_{i+2}-k_2^{i,j})} \cdots z_{j-1}^{\alpha_m^{i,j}(a_{j-1}-k_m^{i,j})}.$$

LEMMA. *For each $[\mathbf{k}] \in K_r(X)$ the ideal of X is generated by the set $\{g_{i,j} = z_i z_j - p_{i,j}([\mathbf{k}]) : 1 \leq i+1 \leq j-1 \leq r\}$.* □

We leave the proof to the reader. Notice that the standard generating set found by Riemenschneider ([**Stevens**,2.1]) corresponds to $[1,2,\dots,2,1]$.

1.2.3. We must say something about the relations among these generators. Fix one of the above generating sets. Define $I_{i,j}$ to be the ideals generated by $g_{\epsilon,\delta}$ with $i \leq \epsilon$ and $\delta \leq j$. Assume $z_j g_{i,k} - z_i g_{j,k} \in I_{i,k-1}$ when $i < j < k-1$ and $z_j g_{i,k} - z_k g_{i,j} \in I_{i+1,k}$ when $i+1 < j < k$. Then these inclusions define $\frac{1}{3}(r-1)r(r+1)$ relations $\mathbf{r}_{i,k}^j$ and $\mathbf{s}_{i,k}^j$.

LEMMA. *There exists $\mathbf{r}_{i,k}^j$ and $\mathbf{s}_{i,k}^j$ as above and they generate the relations among the $g_{i,j}$.* □

Again we omit the proof. The existence of such relations is proved with parameters in 2.1.2.

§2. The components

2.1. Fix a cyclic quotient singularity $X = X(\mathbf{a})$, and let $\tau = \dim T^1_X = (\sum_{i=1}^r a_i) - 2$. Let $S \subset (\mathbf{C}^\tau, 0)$ be the versal base space. For every $[\mathbf{k}] \in K_r(X)$ we will construct a linear subspace, $V_{[\mathbf{k}]}$, of $\mathbf{C}^\tau$ and a flat family $\mathcal{X}_{[\mathbf{k}]} \to V_{[\mathbf{k}]}$ specializing to X. Unfortunately these families do not glue together. In particular their corresponding first order deformations are expressed in different bases for T^1_X. We exhibit an analytic coordinate change (one for each $[\mathbf{k}]$) that maps $V_{[\mathbf{k}]}$ into S, and then conclude from Arndt's and Stevens' results that their images, $S_{[\mathbf{k}]}$, are exactly the components of S_{red}.

2.1.1. We index the basis for T^{1*}_X (recall the difference from Stevens' notation mentioned in the Introduction) by $t_i^{(l)}$ for $i = 1, 2, \ldots, r$ and $l = 1, 2, \ldots, a_i - 1$ and s_i for $i = 2, \ldots, r-1$. These are the coordinate functions on $\mathbf{C}^\tau$. Whenever necessary we use the convention that $t_i^{(a_i)} = 1$.

For every $[\mathbf{k}] \in K_r(X)$, define

$$\mathbf{p}_{[\mathbf{k}]} := < \{t_i^{(l_i)} : i = 1, \ldots, r, 1 \le l_i \le k_i - 1\} \cup \{s_i : \alpha_i > 1\} > \subset \mathcal{O}_{\mathbf{C}^\tau}.$$

For example $\mathbf{p}_{[1,2,1]} = (t_2^{(1)})$ and $\mathbf{p}_{[2,1,2]} = (t_1^{(1)}, t_3^{(1)}, s_2)$. Let $V_{[\mathbf{k}]}$ be the zero set of this ideal.

Define $Z_i^{(l)}$ for $i = 1, \ldots, r$ and $1 \le l \le a_i$ by

$$Z_i^{(l)} = z_i^{a_i - l} + t_i^{(a_i-1)} z_i^{a_i-l-1} + \cdots + t_i^{(l+1)} z_i + t_i^{(l)}.$$

Notice that on $V_{[\mathbf{k}]}$, $Z_i^{(l)} = z_i^{k_i - l} Z_i^{(k_i)}$ if $k_i \ge l$.

Using the notation of the pyramid lattice in 1.2.1, set

$$y_j = \begin{cases} z_j & \text{if } \ j = 0, 1, r, r+1 \quad \text{or} \quad \alpha_j > 1 \\ z_j + s_j & \text{otherwise}, \end{cases}$$

for $i + 1 \le h \le j - 1$ set

$$X_h^{i,j} = \begin{cases} Z_h^{(k_{h-i}^{i,j})^{\alpha_{h-i}^{i,j}}} & \text{if } \ h = 1, r \quad \text{or} \quad \alpha_h > 1 \\ y_h^{\alpha_{h-1} - \alpha_{h-i-1}^{i,j}} Z_h^{(k_{h-i}^{i,j} + \alpha_{h-1} - \alpha_{h-i-1}^{i,j})} & \text{otherwise} \end{cases}$$

and finally set

$$P_{i,j} = X_{i+1}^{i,j} X_{i+2}^{i,j} \cdots X_{j-1}^{i,j}.$$

2.1.2.

Proposition. *The equations*

$$G_{i,j} = z_i y_j - P_{i,j}$$

for $1 \le i+1 \le j-1 \le r$, define a flat family $\mathcal{X}_{[\mathbf{k}]}$ over $V_{[\mathbf{k}]}$ with special fiber X.

Proof: First of all it is readily seen that for the special fiber we get the equations of 1.2.2. Let $\mathfrak{m}_\tau$ be the maximal ideal in $\mathcal{O}_{\mathbf{C}^\tau}$ and $\mathcal{R} = \mathcal{O}_{\mathbf{C}^\tau}\{z_0, \ldots, z_{r+1}\}$. Define as in

1.2.3 $\mathcal{I}_{i,j}$ to be the ideals generated by $G_{\epsilon,\delta}$ with $i \le \epsilon$ and $\delta \le j$. It is enough to prove that

$$(1) \qquad z_\delta G_{\epsilon,\gamma} - z_\epsilon G_{\delta,\gamma} = z_\epsilon P_{\delta,\gamma} - z_\delta P_{\epsilon,\gamma} \in \mathcal{I}_{\epsilon,\gamma-1} + \mathbf{m}_r\mathcal{R}$$

when $\epsilon < \delta < \gamma - 1$ and

$$(2) \qquad y_\delta G_{\epsilon,\gamma} - y_\gamma G_{\epsilon,\delta} = y_\gamma P_{\epsilon,\delta} - y_\delta P_{\epsilon,\gamma} \in \mathcal{I}_{\epsilon+1,\gamma} + \mathbf{m}_r\mathcal{R}$$

when $\epsilon + 1 < \delta < \gamma$, since by Lemma 1.2.3 this will define relations lifting generating relations for I.

We show that (2) holds. Notice that $\mathcal{I}_{\epsilon,\delta}/\mathbf{m}_r\mathcal{I}_{\epsilon,\delta} = I_{\epsilon,\delta}$, which is a prime ideal. So if $x, y \in \mathcal{R}$ and $xy \in \mathcal{I}_{\epsilon,\delta}$, then one of them, say $x \in \mathcal{I}_{\epsilon,\delta} + \mathbf{m}_r\mathcal{R}$.

It follows that to prove (2) it is enough to show

$$(3) \qquad y_\gamma P_{\epsilon,\gamma-1} - y_{\gamma-1} P_{\epsilon,\gamma} \in \mathcal{I}_{\epsilon+1,\gamma}$$

for all $1 \le \epsilon + 1 \le \gamma - 2 \le r - 1$. Indeed

$$y_{\delta+1} y_{\delta+2} \cdots y_{\gamma-1}(y_\gamma P_{\epsilon,\delta} - y_\delta P_{\epsilon,\gamma}) = \sum_{h=\delta}^{\gamma-1} y_\delta y_{\delta+1} \cdots \hat{y}_h \hat{y}_{h+1} \cdots y_\gamma (y_{h+1} P_{\epsilon,h} - y_h P_{\epsilon,h+1})$$

and $\mathcal{I}_{\epsilon+1,\gamma-1} \subset \mathcal{I}_{\epsilon+1,\gamma}$.

Consider now

$$(4) \qquad y_\gamma P_{\epsilon,\gamma-1} - y_{\gamma-1} P_{\epsilon,\gamma} = y_\gamma X_{\epsilon+1}^{\epsilon,\gamma-1} X_{\epsilon+2}^{\epsilon,\gamma-1} \ldots X_{\gamma-2}^{\epsilon,\gamma-1} - y_{\gamma-1} X_{\epsilon+1}^{\epsilon,\gamma} X_{\epsilon+2}^{\epsilon,\gamma} \ldots X_{\gamma-1}^{\epsilon,\gamma}.$$

Recall from 1.2.1 that $\alpha_{h-\epsilon}^{\epsilon,\gamma} \ge \alpha_{h-\epsilon}^{\epsilon,\gamma-1}$ when $\epsilon \le h \le \gamma - 1$. Also, if $\alpha_h = 1$, then $k_{h-\epsilon}^{\epsilon,\delta} + \alpha_{h-1} - \alpha_{h-\epsilon-1}^{\epsilon,\delta} = \alpha_{h-1} + \alpha_{h-\epsilon+1}^{\epsilon,\delta}$. It follows that when $\alpha_h = 1$,

$$X_h^{\epsilon,\gamma-1} = y_h^{\alpha_{h-\epsilon-1}^{\epsilon,\gamma} - \alpha_{h-\epsilon-1}^{\epsilon,\gamma-1}} z_h^{\alpha_{h-\epsilon+1}^{\epsilon,\gamma} - \alpha_{h-\epsilon+1}^{\epsilon,\gamma-1}} X_h^{\epsilon,\gamma}$$

for $\epsilon + 1 \le h \le \gamma - 2$. If $\alpha_h > 1$, then in general,

$$X_h^{\epsilon,\delta} = Z_h^{(k_{h-\epsilon}^{\epsilon,\delta})^{\alpha_{h-\epsilon}^{\epsilon,\delta}}} = z_h^{\alpha_{h-\epsilon}^{\epsilon,\delta}(k_h - k_{h-\epsilon}^{\epsilon,\delta})} Z_h^{(k_h)^{\alpha_{h-\epsilon}^{\epsilon,\delta}}}.$$

But it follows from 1.2.1 that if $k_h > k_{h-\epsilon}^{\epsilon,\delta}$, then $\alpha_{h-\epsilon}^{\epsilon,\delta} = 1$, so actually if $\alpha_h > 1$, then

$$X_h^{\epsilon,\delta} = z_h^{k_h - k_{h-\epsilon}^{\epsilon,\delta}} Z_h^{(k_h)^{\alpha_{h-\epsilon}^{\epsilon,\delta}}}.$$

Therefore we may factor out $z_h^{k_h - k_{h-\epsilon}^{\epsilon,\gamma}} Z_h^{(k_h)^{\alpha_{h-\epsilon}^{\epsilon,\gamma-1}}}$ from (4) whenever $\alpha_h > 1$ and $\epsilon + 1 \le h \le \gamma - 2$. Thus (4) becomes

$$\Bigl(\prod_{\substack{\alpha_h = 1 \\ \epsilon+1 \le h \le \gamma-2}} y_h^{\alpha_{h-1} - \alpha_{h-\epsilon-1}^{\epsilon,\gamma}} Z_h^{(\alpha_{h-1} + \alpha_{h-\epsilon+1}^{\epsilon,\gamma})} \prod_{\substack{\alpha_h > 1 \\ \epsilon+1 \le h \le \gamma-2}} z_h^{k_h - k_{h-\epsilon}^{\epsilon,\gamma}} Z_h^{(k_h)^{\alpha_{h-\epsilon}^{\epsilon,\gamma-1}}} \Bigr) F$$

where

$$F = y_\gamma \prod_{\substack{\alpha_h=1 \\ \epsilon+1\le h\le\gamma-2}} y_h^{\alpha^{\epsilon,\gamma}_{h-\epsilon-1}-\alpha^{\epsilon,\gamma-1}_{h-\epsilon-1}} z_h^{\alpha^{\epsilon,\gamma}_{h-\epsilon+1}-\alpha^{\epsilon,\gamma-1}_{h-\epsilon+1}} \prod_{\substack{\alpha_h>1 \\ \epsilon+1\le h\le\gamma-2}} z_h^{k^{\epsilon,\gamma}_{h-\epsilon}-k^{\epsilon,\gamma-1}_{h-\epsilon}}$$

$$- y_{\gamma-1} \prod_{\substack{\alpha_h>1 \\ \epsilon+1\le h\le\gamma-2}} Z_h^{(k_h)\alpha^{\epsilon,\gamma}_{h-\epsilon}-\alpha^{\epsilon,\gamma-1}_{h-\epsilon}} X^{\epsilon,\gamma}_{\gamma-1}\,.$$

Now let l be the $l^{\epsilon,\gamma}$ of 1.2.1, i.e. $l \in \{\epsilon+1,\ldots,\gamma-2\}$ is chosen so that $\alpha^{\epsilon,\gamma}_{l-\epsilon} = 1$ and $\alpha^{\epsilon,\gamma}_{h-\epsilon} > 1$ for $l+1 \le h \le \gamma-2$. It follows from the explicit description of Φ_1 that $\alpha_h^{\epsilon,\gamma} = \alpha_h^{\epsilon,\gamma-1}$ if $h \le l$ and $k_h^{\epsilon,\gamma} = k_h^{\epsilon,\gamma-1}$ if $h \ne l$ and $h \ne \gamma-1$. Notice also that if $\alpha^{\epsilon,\gamma}_{h-\epsilon} > 1$, then also $\alpha_h > 1$. Either $\alpha_l = 1$ or $\alpha_l > 1$, but in both cases one can use these remarks to check that

$$F = z_l y_\gamma - y_{\gamma-1} \prod_{l+1\le h\le\gamma-2} Z_h^{(k_h)\alpha^{\epsilon,\gamma}_{h-\epsilon}-\alpha^{\epsilon,\gamma-1}_{h-\epsilon}} X^{\epsilon,\gamma}_{\gamma-1}\,.$$

The formula in 1.2.1 relating the various α says that indeed $F = G_{l,\gamma} \in \mathcal{I}_{\epsilon+1,\gamma}$.
(1) is proven in the exact same manner. □

2.1.3. The families defined above will not fit together in the ambient space to define a flat family over the union of the $V_{[\mathbf{k}]}$. Even the corresponding first order deformation may differ from the standard basis for T^1_X as described in [**Stevens**, 2.2]. To fix this we may perform a coordinate change for each $[\mathbf{k}]$ sending $V_{[\mathbf{k}]}$ into the base space defined by Arndt's equations.

It follows from Arndt's algorithm in [**Arndt**] that it is enough to change $y_i^{\alpha_i-1} Z_i^{(\alpha_i-1)}$ to $y_i Z_i^{(1)}$. One can write up an analytic coordinate change that does this from the following observation. If $f = \sum_{i=0}^n f_i z^i$ is a polynomial in one variable z and s is some parameter, then $f = \sum_{i=0}^n g_i(z+s)^i$ where $g_i = \sum_{j=0}^{n-i}(-1)^j\binom{i+j}{i} f_{i+j} s^j$. We leave the details to the reader.

In any case let u be such an analytic coordinate transformation and set $S_{[\mathbf{k}]} = u(V_{[\mathbf{k}]})$. Then from [**Stevens**,4.1 and 5.1] we get

CLAIM. *The $S_{[\mathbf{k}]}$ for $[\mathbf{k}] \in K_r(X)$ are exactly the irreducible components of the reduced versal base space of X.* □

REMARK: One should of course rather say that there *is* a mini-versal base such that the above holds, but if we canonize the base in [**Arndt**], the statement has meaning.

2.2. Thus when dealing with the components, we may study the nice detailed description of $\mathcal{X} \to V$. It may look messy as written here, but compared to the family over S (see e.g. [**Arndt**]) it is quite transparent. If one writes it down in the form of the "pyramid lattice", it is also quite easy to work with. One immediate observation is that there is a natural covering of every component. It is identical to the covering by simultaneous resolution deformations (see e.g. [**Wahl**]) in the case the component is the Artin component ($[\mathbf{k}] = [1,2,2,\ldots,2,1]$).

2.2.1. Define $\Xi_i := \prod_{\epsilon=1}^{a_i-k_i}(z_i - \xi_{i,\epsilon})$, and think of the $\xi_{i,\epsilon}$ as roots of $Z_i^{(k_i)}$. Consider now the family $\mathcal{Y}_{[\mathbf{k}]} \to U_{[\mathbf{k}]}$ defined by $F_{i,j}$ $(1 \le i+1 \le j-1 \le r)$, where $F_{i,j}$ is gotten from $G_{i,j}$ by exchanging $Z_i^{(k_i)}$ with Ξ_i, and $U_{[\mathbf{k}]}$ is just $\mathbf{C}^{\dim V}$ with coordinate functions $\xi_{i,\epsilon}$ for $i = 1, \ldots, r$ and $\epsilon = 1, \ldots, a_i - k_i$ and the s_i where $\alpha_i = 1$. (Recall that $\dim V = \tau - (r-2) - 2d([\mathbf{k}])$).

Let W be the product of $\mathbf{S}_{a_i-k_i}$, the groups of permutations of $a_i - k_i$ letters, for $i = 1$ to r. Introducing this covering group yields the following base change:

$$\begin{array}{ccc} \mathcal{Y}_{[\mathbf{k}]} & \longrightarrow & \mathcal{X}_{[\mathbf{k}]} \\ \downarrow & & \downarrow \\ U_{[\mathbf{k}]} & \xrightarrow{W} & V_{[\mathbf{k}]} \end{array}$$

where of course W acts by permuting the $\xi_{i,\epsilon}$. Notice also that after an obvious coordinate change $\mathcal{Y}_{[\mathbf{k}]}$ is defined by bi-monomial equations, so it is a $2 + \dim V$ dimensional affine toric variety.

2.2.2. Certainly one hopes for a conceptual explanation of this covering. We conjecture (with Kurt Behnke and Jan Stevens) that at least it is the monodromy cover. Hopefully it has an even more deformation theoretic meaning (like the Artin component) that may generalize to other quotient singularities. In any case we will see in the next section how it relates to the discriminant of $V_{[\mathbf{k}]}$.

§3. The discriminants

3.1. We shall now described $D = D_{[\mathbf{k}]}$, the (reduced) discriminant of $S_{[\mathbf{k}]}$. It is the set of all points $s \in S_{[\mathbf{k}]}$ for which the fiber X_s is singular. There are ways of defining a "good" algebraic (or analytic) structure for this set. One could in our case e.g. consider the ideal of $r \times r$ minors of the Jacobian of $(G_{i,j})$ with respect to the z_i, and then eliminate the z_i. Our case though is no exception to the general rule that such a structure is non-reduced. We will be content with the reduced space that comes about from a more naive set theoretic discussion.

3.1.1. For our purposes we may consider the explicit family $\mathcal{X}_{[\mathbf{k}]} \to V_{[\mathbf{k}]}$ defined in 2.1.1, and we will describe D as subset of $V_{[\mathbf{k}]}$. Recall that $V_{[\mathbf{k}]}$ is defined by $t_i^{(l)} = 0$ if $1 \le l \le k_i - 1$ and $s_i = 0$ if $\alpha_i > 1$.

Let Δ be the discriminant of the covering $U_{[\mathbf{k}]} \to V_{[\mathbf{k}]}$, i.e. Δ is the set of W orbits containing fewer than $|W|$ points. If $p_i = \prod_{\epsilon<\delta}(\xi_{i,\epsilon} - \xi_{i,\delta})^2$, then as variety, Δ is defined by the vanishing of the invariant polynomial $\prod p_i$. Of course Δ is part of D, but far from all of it.

Set $\tilde{t}_i^{(k_i)} = \sum_{j=0}^{a_i-k_i}(-1)^j t_i^{(k_i+j)} s_i^j$ and remember our convention that $t_i^{(a_i)} = 1$. Define

$$D_i^{(k_i)} \text{ by } \tilde{t}_i^{(k_i)} = 0$$
$$D_{i,j}^{(k_i,k_j)} \text{ by } \tilde{t}_i^{(k_i)} = \tilde{t}_j^{(k_j)} = 0 \text{ and } s_h = 0 \text{ for all } h \in [i+1, j-1] \text{ with } \alpha_h = 1.$$

Notice that if $a_i = k_i$, then $\tilde{t}_i^{(k_i)} = 1$, so if $a_i = k_i$ then $D_i^{(k_i)} = \emptyset$ and if $a_i = k_i$ or $a_j = k_j$ then $D_{i,j}^{(k_i,k_j)} = \emptyset$. Finally let $A_{>1} = \{i : \alpha_i > 1\}$ and $A_{=1} = \{i : \alpha_i = 1\}$.

PROPOSITION. *The discriminant $D = D_{[k]}$ of the family $\mathcal{X}_{[k]} \to V_{[k]}$ is*

$$D = \Delta \cup \bigcup_{i \in A_{>1}} D_i^{(k_i)} \cup \bigcup_{i,j \in A_{=1}} D_{i,j}^{(k_i,k_j)}$$

3.1.2. For the proof of the Proposition 3.1.1 we need the following lemma. (We use the notation of §2.)

LEMMA. *At a singular point in a fiber over D, $z_0 = z_{r+1} = 0$ and if $z_h \neq 0$ and $y_h \neq 0$ then $Z_h^{(k_h)} = \partial Z_h^{(k_h)}/\partial z_h = 0$.*

PROOF: We call an $n \times n$ matrix $(a_{i,j})$ *semi-triangular* if there is an $h \in [1, n]$ such that $a_{i,j} = 0$ when $1 \leq j < h$ and $i > j$ and $a_{i,j} = 0$ when $h < j \leq n$ and $j > i$. If $(a_{i,j})$ is semi-triangular then $\det(a_{i,j}) = \prod_{i=1}^n a_{i,i}$.

Given an $h \in [0, r+1]$ consider the $r \times r$ submatrix of the Jacobian of $(G_{i,j})$ with respect to the z_i, with row indices $(0,h), \ldots, (h-2,h), (h-1,h+1), (h,h+2), \ldots, (h,r+1)$ (just omit the row if it doesn't make sense, e.g. $G_{-2,0}$) and column indices $0, \ldots, h-2, h, h+2, \ldots, r+1$. One checks that this is a semi-triangular matrix with diagonal entries z_0 if $h = 0$, z_{r+1} if $h = r+1$ and otherwise both z_h, y_h and $\partial P_{h-1,h+1}/\partial z_h$.

Now consider the $r \times r$ submatrix with rows as above but column indices $0, \ldots, h-1, h+2, \ldots, r+1$. This is also semi-triangular and diagonal entries are z_h, y_h and y_{h+1}. If we exchanged $h-1$ with $h+1$ above, we would get a semi-triangular matrix with diagonal entries z_h, y_h and z_{h-1}.

The claim follows from the vanishing of these minors and that $z_{h-1}y_{h+1} = P_{h-1,h+1}$. □

3.2. We discuss here the nature of the singular points of fibers over D. The type of analysis we give below certainly leads to a complete description of the adjacencies of X. The purpose here though is to help us prove Proposition 3.1.1. To label the types of singularities we find in the fibers, it is enough to give the data $([a], [k], p, q)$. By this tuple we mean the singularity in the fiber over p of the family $\mathcal{X}([a])_{[k]} \to V_{[k]}$ at the point q. In other words, the germ (X_p, q) where X_p is defined by the vanishing of the $G_{i,j}$ of 2.1.2 for the given $[a]$ and $[k]$ and with the values of the $t_i^{(l)}$ and s_i at p.

3.2.1. If $p \in \Delta$, say ξ is a multiple root of $Z_i^{(k_i)}$, it is easy to find a q with $z_i = \xi$, such that (X_p, q) is a singularity. The type of this singularity will vary depending on whether or not ξ is a multiple root of more than one $Z_i^{(k_i)}$ and on if $\xi = 0$ or not. Obviously the most general situation is that $\xi \neq 0$ and has multiplicity 2 in only one $Z_i^{(k_i)}$. This is of course a quadratic hypersurface singularity (the A_1) as it should be.

3.2.2. Assume now that $p \in D \setminus \Delta$. We may conclude from 3.1.2 that at a singular point q, either $z_i = 0$ or $y_i = 0$. Start with a singularity labeled $([a], [k], p, q)$. We want to construct an induction scheme where (under certain conditions) this singularity may be identified with singularities appearing in the fibers of families for cyclic quotients with smaller $[a]$.

The first case we treat is when all $s_i = 0$ at p. Then we may also assume that at q all $z_i = 0$. If $k_i = 1$, we want to describe the effect of $t_i^{(1)}$ being non zero. This corresponds to blowing down $[k]$ at i in the following sense; $Z_i^{(1)}$ becomes a unit in the local ring at

(X_p, q), so $z_i = uz_{i-1}z_{i+1}$ for some unit u. Plugging this into the rest of the equations one sees that our singularity is the same as one with data $([\mathbf{a}'], [\mathbf{k}'], p', 0)$ where

- $[\mathbf{k}']$ is the blow down of $[\mathbf{k}]$ at i
- $[\mathbf{a}']$ is the blow down of $[\mathbf{a}]$ at i
- $p' \in V_{[\mathbf{k}']}(X(\mathbf{a}'))$ is gotten by letting $t'^{(l)}_h = t^{(l)}_h$ for $h < i-1$, $t'^{(l)}_{i-1} = t^{(l-1)}_{i-1}$,$t'^{(l)}_i = t^{(l-1)}_{i+1}$ and $t'^{(l)}_h = t^{(l)}_{h+1}$ for $h > i$.

Of course this doesn't really make sense if $a_{i-1} = 2$ or $a_{i+1} = 2$ since then $a'_{i-1} = 1$ or $a'_{i+1} = 1$. To make sense of this assume e.g. $a_{i-1} = 2$. Then $1 \leq k_{i-1} \leq 2$ and if $k_{i-1} = 1$, r must be 2, $[\mathbf{k}'] = [0]$ and $[\mathbf{a}'] = [1]$. We treat this case separately below, so assume $k_{i-1} = a_{i-1} = 2$. Then $Z^{(k_{i-1})}_{i-1} = 1$ and the equation $G_{i-2,i+1}$ says now that $z_{i-1} = uz_{i-2}z_{i+1}$. Plugging this into the equations we get the same data as if we blew down $([\mathbf{a}'], [\mathbf{k}'], p', 0)$ at $k'_{i-1} = a'_{i-1} = 1$.

The same reasoning works symmetrically if $a_{i+1} = 2$. In other words if after a blow down as above the new $[\mathbf{a}']$ has a 1 in it, we may blow down this 1 and continue blowing down until we have reached either of the following situations labeled by their data $([\mathbf{a}''], [\mathbf{k}''], p'', 0)$;

(i) $[\mathbf{k}''] = [\mathbf{a}''] = [1, 1]$: This is a "smooth singularity", and means that our original X is of type T and the component is the special component (**[Stevens,3.5]**). We conclude that for such data, if $p \in D \setminus \Delta$ then $t^{(1)}_i = 0$.

(ii) $[\mathbf{k}''] = [0]$ and $[\mathbf{a}''] = [1]$: This is another "smooth singularity", and means again that for our original X, $(X_p, 0)$ singular implies $t^{(1)}_i = 0$.

(iii) $[\mathbf{k}''] < [\mathbf{a}'']$ and no $a''_i = 1$: (Note we may have to go down to $r = 1$,$[\mathbf{k}''] = [0]$ before this happens.) Now we have an honest singularity and our convention is that it is this data we mean when an $[\mathbf{a}']$ with a 1 appears via blowing down.

3.2.3. Let now p be any point in $D \setminus \Delta$ and $q \in X_p$ a singular point with either z_i or y_i vanishing at q. We want to reduce this general case to the one above where all $s_i = 0$. More precisely we will show that a given singularity with label $([\mathbf{a}], [\mathbf{k}], p, q)$ is the same as one of type $([\mathbf{a}'], [\mathbf{k}'], p', q')$ where all $s'_i = 0$ at p'.

If some $s_i \neq 0$ then either z_i or y_i will become a unit at q. The effect of this is related to the chopping off procedure described in 1.1.1. Choose i to be the largest index (if there is one) with y_i a unit, i.e. $i = \max\{i \in [2, r-1] : s_i \neq 0 \text{ at } p \text{ and } z_i = 0 \text{ at } q\}$. If this set is empty, set $i = 0$. Choose j to be the smallest index (if there is one) with z_j a unit, i.e. $j = \min\{j \in [2, r-1] : s_j \neq 0 \text{ at } p \text{ and } z_j = -s_j \text{ at } q\}$. If this set is empty, set $j = r+1$. In particular, at q, $z_0 = z_1 = \cdots = z_{j-1} = 0$ and $y_{i+1} = \cdots = y_{r+1} = 0$. On the other hand, from the equations, one checks that $z_0, \ldots, z_{i-2}$ and $y_{j+2}, \ldots, y_{r+1}$ are not needed for generating the maximal ideal at q. It follows that since (X_p, q) is singular, we must have $j > i$. Then $z_h = y_h = 0$ for $i+1 \leq h \leq j-1$, therefore $s_h = 0$ for $i+1 \leq h \leq j-1$.

Note that by definition of $V_{[\mathbf{k}]}$, $\alpha_i = 1$ if $i \neq 0$ and $\alpha_j = 1$ if $j \neq r+1$. By considering the equations one sees that our singularity with data $([\mathbf{a}], [\mathbf{k}], p, q)$ is the same as the one with data $([\mathbf{a}'], [\mathbf{k}'], p', q')$ where the conversion is achieved by setting

- $[\mathbf{k}']$ to be $[\mathbf{k}]$ chopped off at i and j
- $[\mathbf{a}']$ to be $[\mathbf{a}]$ chopped off at i and j
- $q' = 0$ after the coordinate change $z'_0 = z_{i-1}, z'_1 = z_i, z'_2 = z_{i+1} = y_{i+1}, \ldots, z'_{r'-1} = z_{j-1} = y_{j-1}, z'_{r'} = y_j, z'_{r'+1} = y_{j+1}$

– p' from p by $t'^{(l)}_1 = t_i^{(l-\alpha_{i-1})}$, $t'^{(l)}_h = t^{(l)}_{i+h-1}$ when $1 < h < r'$ and $t'^{(l)}_{r'} = \tilde{t}_j^{(l-\alpha_j+1)}$ where $\tilde{t}_j^{(l)}$ are gotten from $Z_j^{(k_j)}$ and the substitution z_j to y_j as in 2.1.3. All $s'_i = 0$.

If $a'_1 = 1$ or $a'_{r'} = 1$, we may resolve this as in 3.2.2.

3.3. We can now prove the proposition, but first we need a lemma telling us what is necessary for a fiber to contain a point where either z_i or y_i vanishes.

3.3.1. For given [a] and [k] consider the $P_{i,j}$ as polynomials in z_i, $t_i^{(l)}$ and s_i modulo the ideal $_{[k]}$. Let $< \mathbf{z} >=< z_0, \ldots, z_{r+1} >$ and $< \mathbf{y} >=< y_0, \ldots, y_{r+1} >$. (Remember, by definition many of the y_i equal z_i modulo $_{[k]}$.)

LEMMA. *We have the following inclusions*

$$P_{i,j} \in \begin{cases} < \mathbf{z} > & \text{if } j \neq r+1 \\ < \mathbf{y} > & \text{if } i \neq 0. \end{cases}$$

□

3.3.2.

PROOF OF PROPOSITION 3.1.1: Let $D' = \Delta \cup \left(\bigcup_{i \in A_{>1}} D_i^{(k_i)}\right) \cup \left(\bigcup_{i,j \in A_{=1}} D_{i,j}^{(k_i,k_j)}\right)$. We show first that $D' \subseteq D$. As stated above $\Delta \subset D$, so assume first that $p \in D_i^{(k_i)}$, $a_i > k_i$ and $\alpha_i > 1$.

I claim that X_p contains the point q given by $z_h = 0$ for $h = 0, \ldots, i$ and $y_h = 0$ for $h = i+1, \ldots r+1$. This follows from Lemma 3.3.1 and the fact that $Z_i^{(k_i)} = z_i Z_i^{(k_i+1)}$ if $t_i^{(k_i)} = 0$. If (X_p, q) then $\mathbf{m}_{(X_p,q)} =< z_0, z_{r+1} >$. We say that z_h or y_h are *redundant*, if they vanish at q but are not needed to generate $\mathbf{m}_{(X_p,q)}$. By considering the equations, z_h is redundant iff there exists $P_{\epsilon,\delta} = u z_h$ where u is locally a unit. We show that this is impossible for z_i.

There are two possibilities; if $(\epsilon, \delta) \neq (0, r+1)$ then it follows from the combinatorics of the lattice of [k] that $t_i^{(k_i)}$ is non zero, so we can at most have $P_{0,r+1} = u z_i$. But this is impossible since $\alpha_i > 1$.

Assume now that $p \in D_{i,j}^{(k_i,k_j)}$. One checks that X_p contains the point q given by $z_h = 0$ for $h = 0, \ldots, j-1$ and $y_h = 0$ for $h = j, \ldots r+1$. A similar argument to the above case shows that if z_i is redundant, then y_j cannot be.

To prove the other inclusion we have only to show that $D \setminus \Delta \subseteq D' \setminus \Delta$. So we may assume the singular point in X_p is one with vanishing z_i or y_i. We reduce immediately to the case where all $s_i = 0$. Indeed, if some $s_i \neq 0$ at the point $p \in D \setminus \Delta$, then by the construction in 3.2.3 and the definition of D, $p \in D'$ iff the p' of 3.2.3 is in the D' defined for [a′] and [k′]. Note that at p' all $s'_i = 0$.

Now we may use induction on r, starting at $r = 2$. When $r = 2$ there is only one component and an easy calculation shows that the discriminant is $\Delta \cup D_{1,2}^{(1,1)}$. For an X with $r > 2$, we may assume that at p at least one $t_h^{(1)} \neq 0$ or else we are through. Then we may "blow down" at h and we find by induction that $p \in D$ iff the p' of 3.2.3 is in the D' defined for [a′] and [k′]. Check then that the identification of the different $t_i^{(l)}$ with the $t'^{(l)}_i$ implies that $p \in D'$. □

REFERENCES

[Arndt] Arndt, J., "Verselle Deformationen zyklischer Quotientensingularitäten," Dissertation, Universität Hamburg, 1988.

[Christophersen] Christophersen, J.A., "Obstructions for rational singularities and deformations of cyclic quotients," Thesis, Universitetet i Oslo, 1990.

[Laudal] Laudal, O.A., *Matric Massey products and formal moduli*, in "Algebra, Algebraic Topology and their Interactions, Stockholm 1983," Lecture Notes in Mathematics 1183, Springer Verlag, Berlin–Heidelberg–New York.

[Orlik–Wagreich] Orlik, P., Wagreich, P., *Algebraic surfaces with k^* - action*, Acta Math. 138 (1977), 43–81.

[Stevens] Stevens, J., *On the versal deformation of cyclic quotient singularities*, in "These proceedings."

[Wahl] Wahl, J.M., *Simultaneous resolution of rational singularities*, Compositio Math. 38 (1979), 43–54.

University of Oslo, Matematisk Institutt, PB 1053 Blindern, 0316 Oslo 3 , Norway

$\mathcal{A}$ - equivalence and the Equivalence of Sections of Images and Discriminants

James Damon*

In [Mo2] and [MM] Mond and Marar obtain a formula relating the $\mathcal{A}_e$- codimension of map germs $f_0: \mathbb{C}^2,0 \longrightarrow \mathbb{C}^3,0$ to the Euler characteristic of the image of a stable perturbation f_t of f_0. This has been proven to hold quite generally for such map germs by de Jong and Pellikaan (unpublished) and by de Jong and van Straten [JS]. One curious aspect of this formula is the presence of the $\mathcal{A}_e$- codimension, which seems to have little relation with the image of f_t. This codimension is related by de Jong and van Straten to the dimension of the space of deformations of $X = \text{Image}(f_0)$ for which the singular set of X deforms flatly. Their arguments depend strongly upon X being a surface singularity in $\mathbb{C}^3$.

In this paper, we derive another relation between $\mathcal{A}$- equivalence and properties of image(f_0). This relation is valid for all dimensions and directly relates the $\mathcal{A}_e$- codimension of f_0 with a codimension of a germ defining Image(f_0) as a section of the image of a stable germ.

$$\begin{array}{ccc} \mathbb{C}^{n'},0 & \xrightarrow{\quad F \quad} & \mathbb{C}^{p'},0 \\ \uparrow & & \uparrow\ g_0 \\ \mathbb{C}^{n},0 & \xrightarrow{\quad f_0 \quad} & \mathbb{C}^{p},0 \end{array} \qquad \text{diagram 1}$$

We recall that by Mather [M-IV], if $f_0: \mathbb{C}^n,0 \longrightarrow \mathbb{C}^p,0$ is a holomorphic germ of finite singularity type (i.e. finite contact codimension) then there is a stable germ $F: \mathbb{C}^{n'},0 \longrightarrow \mathbb{C}^{p'},0$ and a germ of an immersion $g_0: \mathbb{C}^p,0 \longrightarrow \mathbb{C}^{p'},0$ with g_0 transverse to F such that f_0 is obtained as a pull-back in diagram 1 (F is the stable unfolding of f_0 [M-IV]).

The germ g_0 has been used to determine $\mathcal{A}$-determinacy properties of f_0 by Martinet [Ma2] and topological determinacy properties by du Plessis [DP]. However, there was lacking a precise relation between equivalence for the germ g_0 and the $\mathcal{A}$- equivalence of f_0. In this paper we derive such a relation.

Let V = D(F) denote the discriminant of F (which is also Image(F) when $n' < p'$). Given a variety-germ $V,0 \subset \mathbb{C}^{p'},0$ there is a notion of "contact equivalence preserving V" on

*Partially supported by a grant from the National Science Foundation and a Fulbright Fellowship

germs h: $\mathbb{C}^m,0 \longrightarrow \mathbb{C}^{p'},0$, defined by the action of a group $\mathcal{K}_V$ [D2].

The main results here concern the relation between $\mathcal{K}_V$ - equivalence for g_0 and $\mathcal{A}$- equivalence for f_0. They are:

1) g_0 has finite $\mathcal{K}_V$ - codimension if and only if f_0 has finite $\mathcal{A}$- codimension;

2) if we denote the extended tangent spaces to the $\mathcal{A}$- orbit of f_0 and the $\mathcal{K}_V$ - orbit of g_0 by $T\mathcal{A}_e \cdot f_0$ and $T\mathcal{K}_{V,e} \cdot g_0$, with associated normal spaces

$$N\mathcal{A}_e \cdot f_0 = \theta(f_0)/T\mathcal{A}_e \cdot f_0 \quad \text{and} \quad N\mathcal{K}_{V,e} \cdot g_0 = \theta(g_0)/T\mathcal{K}_{V,e} \cdot g_0$$

then these normal spaces are isomorphic as $\mathcal{O}_{\mathbb{C}^p,0}$-modules (theorem 2);

3) taking dimensions in (2) we obtain (theorem 1)

$$\mathcal{A}_e\text{-codimension}(f_0) = \mathcal{K}_{V,e}\text{-codimension}(g_0).$$

4) if we replace the germ f_0 and the stable germ F by multi-germs $f_0: \mathbb{C}^n,S \longrightarrow \mathbb{C}^p,0$ and $F: \mathbb{C}^{n'},S \longrightarrow \mathbb{C}^{p'},0$ with f_0 finitely determined and F stable then 1) - 3) remain valid (see theorem 3; however, to keep notation simple we give the proofs for the case where $|S|=1$ and observe that they work for all finite S).

The third result allows us to place the Mond-Marar formula into a common context with other formulas which relate the algebraic codimension of (nonlinear) sections of varieties to Euler characteristics of their perturbations.

As corollaries of these results and their proofs we obtain: i) sufficient conditions for unfoldings of f_0 to be $\mathcal{A}$- trivial in terms of the corresponding unfoldings of g_0 being $\mathcal{K}_V$-trivial (but it is unknown whether the converse holds); ii) a proof that unfoldings of f_0 are $\mathcal{A}$-versal if and only if the corresponding unfoldings of g_0 are $\mathcal{K}_V$ - versal and iii) a characterization of the versality discriminant as the set of points where g_0 fails to be transverse to V and an explicit method for computing the versality discriminant for unfoldings of hypersurfaces.

The author is especially grateful to the organizers of the special year in bifurcation and singularity theory at the University of Warwick for their generous hospitality and support.

§1 $\mathcal{A}$ and $\mathcal{K}_V$-equivalence

Here we recall several basic properties of $\mathcal{A}$ and $\mathcal{K}_V$-equivalence; while those of $\mathcal{A}$-equivalence are generally well-known, those of $\mathcal{K}_V$-equivalence are less so. The key properties of these groups which we are interested in are: their tangent spaces and infinitesimal conditions for versality, infinitesimal conditions for triviality of unfoldings, and geometric characterizationsoffinitedeterminacy.

All germs which we consider will be holomorphic. The two principal notions of equivalence for map germs are $\mathcal{A}$ and $\mathcal{K}$-equivalence. We denote the space of holomorphic germs $f_0 : \mathbb{C}^s,0 \to \mathbb{C}^t,0$ by $C_{s,t}$ and use local coordinates x for $\mathbb{C}^s$ and y for $\mathbb{C}^t$. With $\mathcal{D}_n$ denoting the group of germs of diffeomorphisms $\varphi : \mathbb{C}^n,0 \to \mathbb{C}^n,0$, the group $\mathcal{A} = \mathcal{D}_s \times \mathcal{D}_t$ acts on $C_{s,t}$ by $(\varphi,\psi)\cdot f_0 = \psi \circ f_0 \circ \varphi^{-1}$. The group $\mathcal{K}$ (contact equivalence) consists of $H \in \mathcal{D}_{s+t}$ such that there is an $h \in \mathcal{D}_s$ so that $H \circ i = i \circ h$ and $\pi \circ H = h \circ \pi$, where $i(x) = (x,0)$ is the inclusion $i : \mathbb{C}^s \hookrightarrow \mathbb{C}^{s+t}$ and $\pi(x,y) = x$ is the projection $\pi : \mathbb{C}^{s+t} \to \mathbb{C}^s$. Then, $\mathcal{K}$ acts on $C_{s,t}$ by

$$(h(x), H\cdot f(x)) = H(x, f(x)),$$

i.e. graph(H·f) = H(graph(f)). Germs are $\mathcal{A}$ or $\mathcal{K}$-equivalent if they lie in common orbits of the group actions.

An unfolding of f_0 is a germ $f : \mathbb{C}^{s+q},0 \to \mathbb{C}^{t+q},0$ of the form $f(x,u) = (\bar{f}(x,u),u)$ with $\bar{f}(x,0) = f_0(x)$ (here u denotes local coordinates for $\mathbb{C}^q,0$). Both $\mathcal{A}$ and $\mathcal{K}$ extend to actions on unfoldings: if $\varphi \in \mathcal{D}_{s+q}$ and $\psi \in \mathcal{D}_{t+q}$ are unfoldings then $(\varphi,\psi)\cdot f = \psi \circ f \circ \varphi^{-1}$ while if $H \in \mathcal{D}_{s+t+q}$ is an unfolding with an unfolding $h \in \mathcal{D}_{s+q}$ so that $H \circ i' = i' \circ h$ and $\pi' \circ H = h \circ \pi'$ (and i' and π' are inclusions and projections for $\mathbb{C}^{s+q}$ and $\mathbb{C}^{s+t+q}$). Then

$$(\bar{h}(x,u), H\cdot f(x,u)) = H(x,\bar{f}(x,u), u).$$

If $(V,0) \subset \mathbb{C}^t,0$ is a germ of a variety then we can define a subgroup of $\mathcal{K}$

$$\mathcal{K}_V = \{H \in \mathcal{K} : H(\mathbb{C}^s \times V) \subseteq \mathbb{C}^s \times V\}$$

and similarly for unfoldings. This yields $\mathcal{K}_V$-equivalence. Just as $\mathcal{K}$-equivalence captures the equivalence of the germs of varieties $f_0^{-1}(0)$, so too $\mathcal{K}_V$-equivalence captures the equivalence of the germs of varieties $f_0^{-1}(V)$.

For $\mathcal{G} = \mathcal{A}, \mathcal{K}$ or $\mathcal{K}_V$, we say that an unfolding f of f_0 is a *$\mathcal{G}$-trivial unfolding* if it is $\mathcal{G}$-equivalent to the trivial unfolding $f_0 \times id_{\mathbb{C}^q}$. It is *$\mathcal{G}$-trivial as a family* if the $\mathcal{G}$-equivalence preserves the origin for all parameter values. If $f_1(x,u,v) = (\bar{f}_1(x,u,v), u,v)$ is an unfolding of f_0 so that $\bar{f}_1(x,u,0) = \bar{f}_1(x,u)$, then f_1 will be said to extend f. An *extension* f_1 of f *is $\mathcal{G}$-trivial* if it is $\mathcal{G}$-equivalent to f × id by an equivalence which is the identity when v = 0. Lastly, an unfolding f is $\mathcal{G}$-versal if for any other unfolding $g : \mathbb{C}^{s+r},0 \to \mathbb{C}^{t+r},0$ of f_0, there is a germ $\lambda : \mathbb{C}^r,0 \to \mathbb{C}^q,0$ such that $\lambda^*f(x,v) = (\bar{f}(x,\lambda(v)), v)$ is $\mathcal{G}$-equivalent to g.

Tangent spaces

For $\mathbb{C}^s,0$ with local coordinates x, we denote the ring of holomorphic germs $\mathcal{O}_{\mathbb{C}^s,0}$ by $\mathcal{C}_x$ with maximal ideal m_x, and similarly with u also denoting local coordinates for $\mathbb{C}^q$, $\mathcal{C}_{x,u}$ denotes $\mathcal{O}_{\mathbb{C}^{s+q},0}$, etc. Also for $f_0 : \mathbb{C}^s,0 \to \mathbb{C}^t,0$, the ring homomorphism $f_0^*: \mathcal{C}_y \to \mathcal{C}_x$, induced by composition, will be understood without being explicitly stated.

The tangent space to $\mathcal{C}_{s,t}$ at f_0 consists of germs of vector fields $\zeta : \mathbb{C}^s,0 \to T\mathbb{C}^t$ such that $\pi \circ \zeta = f_0$ and is denoted by $\theta(f_0) \tilde{\to} \mathcal{C}_x \left\{\frac{\partial}{\partial y_i}\right\}$ (here the R-module generated by $\varphi_1,\ldots,\varphi_k$ is denoted by $R\{\varphi_1,\ldots,\varphi_k\}$ or $R\{\varphi_i\}$ if k is understood). Also, $\theta_s = \theta(\mathrm{id}_{\mathbb{C}^s}) \tilde{\to} \mathcal{C}_x \left\{\frac{\partial}{\partial x_i}\right\}$ and similarly for θ_t. The extended tangent spaces to $\mathcal{A}$ and $\mathcal{K}$ (which allow movement of the source and/or target) are given by

$$T\mathcal{A}_e \cdot f_0 = \mathcal{C}_x \left\{\frac{\partial f_0}{\partial x_i}\right\} + \mathcal{C}_y \left\{\frac{\partial}{\partial y_i}\right\}$$

$$T\mathcal{K}_e \cdot f_0 = \mathcal{C}_x \left\{\frac{\partial f_0}{\partial x_i}\right\} + f_0^* m_y \cdot \mathcal{C}_x \left\{\frac{\partial}{\partial y_i}\right\}.$$

For the tangent space for $\mathcal{K}_V$, we consider the module of vector fields tangent to V. If I(V) denotes the ideal of germs vanishing on V, then we let

$$\theta_V = \{\zeta \in \theta_t : \zeta(I(V)) \subseteq I(V)\}.$$

This is denoted Derlog(V) by Saito [Sa]; however, we use this simpler notation as there is no danger of confusion with other notions. θ_V extends to a sheaf of vector fields tangent to V, Θ_V which is easily seen to be coherent [Sa]. If $\{\eta_i\}_{i=1}^m$ denotes a set of generators for θ_V, then

$$T\mathcal{K}_{V,e} \cdot f_0 = \mathcal{C}_x \left\{\frac{\partial f_0}{\partial x_i}\right\} + \mathcal{C}_x \{\eta_i \circ f_0\}.$$

For $\mathcal{G} = \mathcal{A}, \mathcal{K}$ or $\mathcal{K}_V$, we denote the normal space by

$$N\mathcal{G}_e \cdot f_0 = \theta(f_0)/T\mathcal{G}_e \cdot f_0$$

and the $\mathcal{G}_e$-codimension of f_0 is $\dim_{\mathbb{C}} N\mathcal{G}_e \cdot f_0$.

The versality theorem allows us to relate several different approaches to versality (Martinet [Ma1] for $\mathcal{A}$ and $\mathcal{K}$, [D2] or [D1] for $\mathcal{K}_V$). For any unfolding $f:\mathbb{C}^{s+q},0\to\mathbb{C}^{t+q},0$

we let $\partial_j f = \frac{\partial \bar{f}}{\partial u_j}|_{u=0}$.

Theorem (versality theorem) : *For $\mathcal{G} = \mathcal{A}$, $\mathcal{K}$ or $\mathcal{K}_V$, and an unfolding* f *of* f_0 *the following are equivalent:*

i) f *is* $\mathcal{G}$*-versal*

ii) $T\mathcal{G}_e\cdot f_0 + \langle\partial_1 f,\dots,\partial_q f\rangle = \theta(f_0)$

iii) *any unfolding* f_1 *of* f_0 *which extends* f *is a* $\mathcal{G}$*-trivial extension.*

Note: If $f:\mathbb{C}^{s+q},0\to\mathbb{C}^{t+q},0$ is $\mathcal{G}$-versal then $q \geq \mathcal{G}_e$-codim(f_0), and if they are equal f is said to be $\mathcal{G}$*-miniversal.*

Furthermore, f_0 is infinitesimally stable if $T\mathcal{A}_e\cdot f_0 = \theta(f_0)$; by Mather [M-IV], if f_0 has rank 0 then an unfolding f of f_0 is infinitesimally stable when viewed as a germ of a mapping if and only if

$$T\mathcal{K}_e\cdot f_0 + \langle\partial_1 f,\dots,\partial_q f, \frac{\partial}{\partial y_1},\dots,\frac{\partial}{\partial y_t}\rangle = \theta(f_0).$$

Hence, any f_0 with $\mathcal{K}_e$-codim$(f_0) < \infty$ has an unfolding f which is infinitesimally stable. Then, any unfolding of f is $\mathcal{A}$-equivalent to $f \times \mathrm{id}$.

Examples for $\mathcal{K}_V$-equivalence

Example (1.1):

Let $(V,0) \subset (\mathbb{C}^4,0)$, with coordinates (X,Y,Z,W), be defined by $YW^2 - Z^2 = 0$. Then, V = Whitney umbrella $\times\ \mathbb{C}$ and is parametrized by $F(x,y,u) = (x,y^2,uy,u)$. Consider $g_0:\mathbb{C}^3,0 \to \mathbb{C}^4,0$ defined by $g_0(x,y,z) = (x,y,z,p(x,y))$. It can be shown that θ_V is generated by

$$2Y\frac{\partial}{\partial Y} + Z\frac{\partial}{\partial Z},\quad Z\frac{\partial}{\partial Z} + W\frac{\partial}{\partial W},\quad WY\frac{\partial}{\partial Z} + Z\frac{\partial}{\partial W},\quad 2Z\frac{\partial}{\partial Y} + W^2\frac{\partial}{\partial Z},\quad \frac{\partial}{\partial X}.$$

We denote these by $\{\eta_i\}_{i=1}^5$. Since

$$\mathcal{C}_{x,y,z}\left\{\frac{\partial}{\partial X},\dots,\frac{\partial}{\partial W}\right\} = \mathcal{C}_{x,y,z}\left\{\frac{\partial g_0}{\partial x},\frac{\partial g_0}{\partial y},\frac{\partial g_0}{\partial z},\frac{\partial}{\partial W}\right\},$$

then modulo $\mathcal{C}_{x,y,z}\left\{\frac{\partial g_0}{\partial x}, \frac{\partial g_0}{\partial y}, \frac{\partial g_0}{\partial z}\right\}$, $\frac{\partial}{\partial X}$, $\frac{\partial}{\partial Y}$, and $\frac{\partial}{\partial Z}$ are equal respectively to $-\frac{\partial p}{\partial x}\frac{\partial}{\partial W}$, $-\frac{\partial p}{\partial y}\frac{\partial}{\partial W}$, and 0. Consequently, $\eta_i \circ g_0$ for i = 1 to 5 equal, respectively,

$$-2y\frac{\partial p}{\partial y}\cdot\frac{\partial}{\partial W},\quad p(x,y)\frac{\partial}{\partial W},\quad z\frac{\partial}{\partial W},\quad -2z\frac{\partial p}{\partial y}\cdot\frac{\partial}{\partial W},\quad -\frac{\partial p}{\partial x}\frac{\partial}{\partial W}.$$

Hence,

$$N\mathcal{K}_{V,e}\cdot g_0 = \mathcal{C}_{x,y,z}\left\{\frac{\partial}{\partial X},\dots,\frac{\partial}{\partial W}\right\} / \mathcal{C}_{x,y,z}\left\{\frac{\partial g_0}{\partial x},\dots,\frac{\partial g_0}{\partial z}, \eta_i\circ g_0\right\}$$

$$\cong \mathcal{C}_{x,y,z}\left\{\frac{\partial}{\partial W}\right\} / \left\{y\frac{\partial p}{\partial y}, p, z, \frac{\partial p}{\partial x}\right\}\cdot\frac{\partial}{\partial W}$$

$$(1.2)\qquad \cong \mathcal{C}_{x,y} / \left(y\frac{\partial p}{\partial y}, p, \frac{\partial p}{\partial x}\right).$$

If we pull back F via g_0 to form $f_0(x,y) = (x,y^2,yp(x,y^2))$

$$\begin{array}{ccc} \mathbb{C}^3,0 & \xrightarrow{F} & \mathbb{C}^4,0 \\ \uparrow & & \uparrow g_0 \\ \mathbb{C}^2,0 & \xrightarrow{f_0} & \mathbb{C}^3,0 \end{array}$$

then Mond computes $N\mathcal{A}_e\cdot f_0$ be exactly (1.2) [Mo1].

Example (1.3)

Let $f_0 : \mathbb{C}^n,0 \to \mathbb{C},0$ be a weighted homogeneous germ defining an isolated singularity. Also, let $F : \mathbb{C}^{n+q},0 \to \mathbb{C}^{1+q},0$ be its versal unfolding, with V=discriminant of F. Then, Saito [Sa] gives the following construction for the generators of θ_V. Let $\{\varphi_i\}_{i=1}^q$ be a basis for $N\mathcal{A}_e\cdot f_0$ and let $\varphi_0 = 1$. We may assume up to equivalence that F is given by (

$$\bar{F}(x,u), u) = (f_0(x) + \sum_{i=1}^q u_i\varphi_i, u).$$

$$\bar{F}\cdot\varphi_i = \sum_{j=0}^q a_{ij}(u)\varphi_j \mod \left(\frac{\partial\bar{F}}{\partial x_1},\dots,\frac{\partial\bar{F}}{\partial x_n}\right)$$

Let

$$\eta_i = -y\cdot\frac{\partial}{\partial u_i} + \sum_{j=1}^q a_{ij}\frac{\partial}{\partial u_j} + a_{i0}\frac{\partial}{\partial y} \quad \text{for } i>0$$

and

$$\eta_0 = \frac{1}{d}\cdot\text{Euler vector field} \qquad (d = wt(f_0)).$$

Then, $\{\eta_i\}_{i=0}^{q}$ generate θ_V. Let $g_0 : \mathbb{C} \to \mathbb{C}^{1+q}$ be defined by $g_0(y) = (y, 0)$. Then,

$\eta_i \circ g_0 = -y\frac{\partial}{\partial u_i}$ for $i > 0$ or $= y\frac{\partial}{\partial y}$ for $i = 0$. Thus,

$$\begin{aligned} N\mathcal{K}_{V,e}\cdot g_0 &= \mathcal{C}_y\left\{\frac{\partial}{\partial y}, \frac{\partial}{\partial u_i}\right\} / \mathcal{C}_y\left\{\frac{\partial g_0}{\partial y}, \eta_i \circ g_0\right\} \\ &\cong \mathcal{C}_y / m_y\left\{\frac{\partial}{\partial u_i}\right\} \\ &\cong \bigoplus_{i=1}^{q} \mathbb{C} \qquad (\text{here } q = \tau(f_0) - 1) \end{aligned}$$

Again g_0 pulls back F to give $f_0 : \mathbb{C}^n,0 \to \mathbb{C},0$.

$$N\mathcal{A}_e\cdot f_0 \cong \mathcal{C}_x / \left(\frac{\partial f_0}{\partial x_1}, \dots, \frac{\partial f_0}{\partial x_n}\right) + \langle 1\rangle .$$

Since f_0 is weighted homogeneous, $f_0 \in \left(\frac{\partial f_0}{\partial x_1}, \dots, \frac{\partial f_0}{\partial x_n}\right)$. Thus, as a $\mathcal{C}_y$-module.

$$N\mathcal{A}_e\cdot f_0 \cong \bigoplus_{i=1}^{\mu-1} \mathbb{C} .$$

Since $\mu = \tau$, these $\mathcal{C}_y$-modules are isomorphic.

Infinitesimal Conditions for Triviality

Next, the relations we shall establish between $\mathcal{A}$ and $\mathcal{K}_V$-equivalence are most easily established at the infinitesimal level. For this reason, we recall the infinitesimal conditions for triviality.

Let $f : \mathbb{C}^{s+q}, 0 \to \mathbb{C}^{t+q}, 0$ be an unfolding of f_0 and let $f_1 : \mathbb{C}^{s+q+r}, 0 \to \mathbb{C}^{t+q+r}, 0$ extend f (with local coordinates u for $\mathbb{C}^q$ and v for $\mathbb{C}^r$).

Criterion for $\mathcal{A}$-triviality: f_1 is an $\mathcal{A}$-trivial extension of f if and only if there exist vector fields $\xi_i \in C_{x,u,v}\left\{\frac{\partial}{\partial x_i}\right\}$, $\delta_i \in C_{y,u,v}\left\{\frac{\partial}{\partial y_i}\right\}$ and $\zeta_i \in C_{u,v}\left\{\frac{\partial}{\partial u_i}\right\}$ such that

$$\text{(1.4)} \qquad \frac{\partial \bar{f}_1}{\partial v_i} = -\xi_i(\bar{f}_1) - \zeta_i(\bar{f}_1) + \delta_i \circ f_1 \qquad 1 \leq i \leq r.$$

Also, if $q = 0$ and $f_1 = f_0$, then f_1 is an $\mathcal{A}$-trivial unfolding of f_0 if and only if (1.4) can be solved with $\zeta_i \equiv 0$. Furthermore, in this case f_1 is an $\mathcal{A}$-trivial family if and only if we can choose $\xi_i \in m_x \cdot C_{x,v}\left\{\frac{\partial}{\partial x_i}\right\}$, $\delta_i \in m_y \cdot C_{y,v}\left\{\frac{\partial}{\partial v_i}\right\}$ (and again $\zeta_i \equiv 0$).

This criterion follows from the reduction lemma for $\mathcal{A}$-equivalence in Martinet [Ma1]. The converse follows by differentiating the trivialization with respect to coordinates trivializing the unfoldings in the v_i-directions.

Criterion for $\mathcal{K}_V$-triviality: f_1 is a $\mathcal{K}_V$-trivial extension of f if and only if there are vector fields $\xi_i \in C_{x,u,v}\left\{\frac{\partial}{\partial x_i}\right\}$, $\delta_i \in C_{y,u,v}\{\eta_i\}$ (where $\{\eta_i\}$ generate θ_V) and $\zeta_i \in C_{u,v}\left\{\frac{\partial}{\partial u_i}\right\}$ such that (1.4) is satisfied.

Similarly, if $q = 0$, f_1 is a $\mathcal{K}_V$-trivial unfolding of f_0 if (1.4) can be solved with $\zeta_i \equiv 0$, or a $\mathcal{K}_V$-trivial family if (1.4) can be solved with $\xi_i \in m_x C_{x,v}\left\{\frac{\partial}{\partial x_i}\right\}$. This follows for $\mathcal{K}_V$-equivalence by the corresponding reduction lemmas in [D1] or [D2].

Geometric Criteria for Finite Determinancy

Finite $\mathcal{A}$-determinacy and finite $\mathcal{K}_V$-determinacy each have geometric characterizations. For $\mathcal{G} = \mathcal{A}$ and $\mathcal{K}$, by Mather [M-III], and $\mathcal{K}_V$, by [D2], finite $\mathcal{G}$-determinacy of f_0 is equivalent to finite $\mathcal{G}$-codimension of f_0. Via this, there is the geometric characterization of finite $\mathcal{A}$-determinacy by Gaffney and Mather: f_0 is finitely $\mathcal{A}$-determined if and only if f_0 is infinitesimally stable in a punctured neighbourhood of 0, i.e. there is a representative of f, $f_1 : U \to \mathbb{C}^t$ such that f_1 is infinitesimally stable on $U \backslash \{0\}$.

For finite $\mathcal{K}_V$-determinacy, let $\{\eta_i\}_{i=1}^m$ be a set of generators of θ_V. By coherence they also generate $\Theta_{V,y}$ in a neighbourhood of 0. By $f_0 : \mathbb{C}^s,0 \to \mathbb{C}^t,0$ *being transverse to* $(V,0)$ *at* x_0 we shall mean

$$df_0(x_0)(T\mathbb{C}^s) + \langle \eta_{1(f(x_0))},\dots,\eta_{m(f(x_0))}\rangle = T\mathbb{C}^t_{f(x_0)}.$$

Then, f_0 is finitely $\mathcal{K}_V$-determined if and only if f_0 is transverse to V in a punctured neighbourhood of 0 (although this characterization was stated in [D2] for finite map germs f_0, the proof given there works in general).

$\mathcal{K}_V$-equivalence and suspension

Lastly, we relate $\mathcal{K}_V$-equivalence to $\mathcal{K}_{V'}$-equivalence for $V' = V \times \mathbb{C}^r$. Given $f_0 : \mathbb{C}^s,0 \to \mathbb{C}^t,0$ and $g : \mathbb{C}^t,0 \to \mathbb{C}^p,0$ we let $g_* f_0 = g \circ f_0$. For an unfolding $f : \mathbb{C}^{s+q},0 \to \mathbb{C}^{t+q},0$ of f_0, we define $g_* f(x,u) = (g \circ \bar{f}(x,u), u)$ which is an unfolding of $g_* f_0$. We consider $i : \mathbb{C}^t,0 \to \mathbb{C}^{t+r},0$ with $i(y) = (y, 0)$ and $\pi : \mathbb{C}^{t+r},0 \to \mathbb{C}^t,0$ with $\pi(y,w) = y$. We also note g induces a C_x-module homomorphism $g_*:\theta(f_0) \to \theta(g_* f_0)$, defined by $g_*(\zeta) = dg(\zeta)$.

We say that $V,0 \subset \mathbb{C}^t,0$ and $V_1,0 \subset \mathbb{C}^p,0$ are *g-related* if for a set of generators $\{\eta_i\}_{i=1}^m$ of θ_V there are $\eta'_i \in \theta_{V_1}$ so that $g_*(\eta_i) = \eta'_i \circ g$. For example, $V,0 \subset \mathbb{C}^t,0$ and $V' = V \times \mathbb{C}^r,0 \subset \mathbb{C}^{t+r}$ are both i and π related.

Proposition 1.5: *With the preceding notation, let* f *be an unfolding of* f_0 *and* f_1 *an extension of* f.

i) *Suppose* $V,0 \subset \mathbb{C}^t,0$ *and* $V_1,0 \subset \mathbb{C}^p,0$ *are g-related; if* f *is a* $\mathcal{K}_V$*-trivial unfolding (respectively family) then* $g_* f$ *is a* $\mathcal{K}_{V_1}$*-trivial unfolding (respectively family); also if* f_1 *is a* $\mathcal{K}_V$*-trivial extension of* f *then* $g_* f_1$ *is a* $\mathcal{K}_{V_1}$*-trivial extension of* $g_* f$.

ii) i_* *and* π_* *induce isomorphisms of* C_y *(respectively* $C_{y,w}$*)-modules)*

$$i_* : N\mathcal{K}_{V,e} \cdot f_0 \xrightarrow{\sim} N\mathcal{K}_{V',e}\, i_* f_0 \quad \text{and} \quad \pi_* : N\mathcal{K}_{V',e}\, f'_0 \xrightarrow{\sim} N\mathcal{K}_{V,e}\, \pi_* f'_0$$

iii) f *is* $\mathcal{K}_V$*-versal if and only if* $i_* f$ *is* $\mathcal{K}_{V'}$*-versal,*
f' *is* $\mathcal{K}_{V'}$*-versal if and only if* $\pi_* f$ *is* $\mathcal{K}_V$*-versal.*

Proof: i) By the infinitesimal criterion we may solve

$$\frac{\partial \bar{f}_1}{\partial v_i} = -\ \xi_i(\bar{f}_1) - \zeta_i(\bar{f}_1) + \delta_i \circ f_1 .$$

Applying dg, we obtain

$$(1.6) \qquad \frac{\partial (g \circ \bar{f}_1)}{\partial v_i} = -\xi_i\,(g \circ \bar{f}_1) - \zeta_i\,(g \circ \bar{f}_1) + dg\,(\delta_i \circ f_1)$$

If $\delta_i = \sum h_{ij}\eta_j$ with $h_{ij} \in \mathcal{C}_{x,u,v}$, then

$$\begin{aligned} dg\,(\delta_i) \circ f_1 &= \sum h_{ij}(dg(\eta_i) \circ f_1) = \sum h_{ij} \cdot \eta'_i \circ g \circ f_1 \\ &= \eta^{(i)'} \circ (g \circ \bar{f}) \quad \text{with} \quad \eta^{(i)'} = \sum h_{ij}\eta'_i . \end{aligned}$$

Substituting into (1.6) satisfies the criterion for triviality for $g_* f_1$. The cases of triviality of unfoldings or families are similar.

ii) Suppose V and V_1 are g-related:

If $\zeta \in T\mathcal{K}_{V,e} \cdot f_0$, then $\zeta = \frac{\partial f}{\partial t}\big|_{t=0}$ for f a 1-parameter $\mathcal{K}_V$-trivial unfolding of f_0.

Then, $g_*(\zeta) = \frac{\partial (g \circ \bar{f})}{\partial t}\big|_{t=0} \in T\mathcal{K}_{V_1,e}\, g_* f_0$. Thus, g_* induces a map

$$g_* : N\mathcal{K}_{V,e} \cdot f_0 \to N\mathcal{K}_{V_1,e} \cdot g_* f_0 .$$

It remains to show that this is an isomorphism for $g = i$ and $g = \pi$. However, by naturality $\pi_* \circ i_* = (\pi \circ i)_* = id_* = id$. If we can show π_* is an isomorphism on normal spaces then so is i_*. Explicitly if $f'_0 : \mathbb{C}^s,0 \to \mathbb{C}^{t+r},0$ has components $f'_0 = (f'_{0,1}, f'_{0,2})$ then $\theta_{V'}$ is generated by $\{\eta_i\} \cup \left\{\frac{\partial}{\partial w_j}\right\}$ where $\{\eta_i\}$ are a set of generators for θ_V; hence,

$$\begin{aligned} N\mathcal{K}_{V',e} \cdot f'_0 &= \mathcal{C}_X\left\{\frac{\partial}{\partial y_i}, \frac{\partial}{\partial w_j}\right\} / \left(\mathcal{C}_X\left\{\frac{\partial f'_0}{\partial x_i}\right\} + \mathcal{C}_X\left\{\eta_i \circ f'_0, \frac{\partial}{\partial w_j}\right\}\right) \\ &\xrightarrow{\sim} \mathcal{C}_X\left\{\frac{\partial}{\partial y_i}\right\} / \mathcal{C}_X\left\{\frac{\partial f'_{01}}{\partial x_i}\right\} + \mathcal{C}_X\left\{\eta_i \circ f'_{01}\right\} \\ &\xrightarrow{\sim} N\mathcal{K}_{V,e} \cdot \pi_* f'_0 \end{aligned}$$

and the projection is exactly π_*.

iii) Finally since π_* and i_* commute with $\frac{\partial}{\partial u_i}$, condition ii) of the versality theorem yields the results. □

§2. Relating $\mathcal{A}$ and $\mathcal{K}_V$-equivalence

In this section we deduce relations between $\mathcal{K}_V$-equivalence of unfoldings and families and $\mathcal{A}$-equivalence for the corresponding unfoldings and families induced via pullback. As a consequence we obtain the numerical equality between $\mathcal{A}_e$-codimension and $\mathcal{K}_{V,e}$-codimension described in the introduction.

Because the $\mathcal{A}$-equivalence and $\mathcal{K}_V$-equivalence are for germs which map between different spaces, we slightly change notation from the preceding section. Consider a germ $f_0 : \mathbb{C}^n,0 \to \mathbb{C}^p,0$ which has finite $\mathcal{K}$-codimension. As mentioned in the preceding section, there is an unfolding $F : \mathbb{C}^{n'},0 \to \mathbb{C}^{p'},0$ of f_0 which is stable when viewed as a germ. We shall refer to such an unfolding as a *stable unfolding* of f_0. There is an inclusion $g_0 : \mathbb{C}^p,0 \to \mathbb{C}^{p'},0$ given by $g_0(y) = (y,0)$ and g_0 is transverse to F, and f_0 may be viewed as being obtained by the fiber product, i.e. pull-back of F by g_0.

$$\begin{array}{ccc} \mathbb{C}^{n'},0 & \xrightarrow{F} & \mathbb{C}^{p'},0 \\ \uparrow & & \uparrow g_0 \\ \mathbb{C}^n,0 & \xrightarrow{f_0} & \mathbb{C}^p,0 \end{array}$$

Also, given an unfolding g of g_0 we have an induced unfolding f of f_0 obtained as the fiber product of F and $\bar{g}$. We shall relate the $\mathcal{A}$-equivalence of f_0 and its unfoldings with the $\mathcal{K}_V$-equivalence of g_0 and its unfoldings.

By [M2], we may choose a representative of F, again denoted by $F : U \to W$ such that if $\Sigma(F) = \{x \in U : \text{rk } df(x) < p'\}$ denotes the critical set of F, then

1) $F^{-1}(0) \cap \Sigma(F) = \{0\}$
2) $F | \Sigma(F)$ is finite to one
3) F is stable.

We let $D(F) = F(\Sigma(F))$. If $n \geq p$ this is the discriminant of F, while if $n < p$ it is the image of F. We denote D(F) by V.

Remark 2.1: Any unfolding of F is $\mathcal{A}$-equivalent to $F \times id$. If we were to replace F by $F \times id_{\mathbb{C}^r}$ then $D(F \times id_{\mathbb{C}^r}) = D(F) \times \mathbb{C}^r = V'$, say. By proposition 1.5, $\mathcal{K}_V$-equivalence for g_0 and its unfoldings is equivalent to $\mathcal{K}_{V'}$-equivalence for $i_* g_0 = i \circ g_0 : \mathbb{C}^p,0 \to \mathbb{C}^{p'},0 \hookrightarrow \mathbb{C}^{p'+r},0$. Thus, it does not matter which stable unfolding of f_0 we choose.

A principal reason for the close relation between $\mathcal{A}$ and $\mathcal{K}_V$-equivalence is the characterization of θ_V due to Arnold [A] and Saito [Sa] (see also Bruce [Br] and Terao [T]).

Lemma 2.2: *With the preceding notation,*

$$\theta_V = \{\eta \in \theta_{p'} : \textit{there is a } \xi \in \theta_{n'} \textit{ so that } \xi(F) = \eta \circ F\},$$

that is, the set of liftable vector fields.

Proof: The proofs for $n \geq p$ are given in the above references. The argument for $n < p$ is the same; by Hartogs' theorem η lifts if and only if it lifts off a set of codimension 2 in $\mathbb{C}^{n'}$. As F is stable, the only singular points of codimension 1 occur at double points when $p = n + 1$. Clearly η lifts from the regular points of V. At double points, F is a suspension of the germ $\mathbb{C},0 \perp\!\!\perp \mathbb{C},0 \to \mathbb{C}^2,0$ defined by $x \mapsto x$, $y \mapsto y$ in $\mathbb{C}^2$ with image $x \cdot y = 0$. The vector fields tangent to this set are generated by $x\frac{\partial}{\partial x}$ and $y\frac{\partial}{\partial y}$ and clearly lift. The converse is immediate since $dF(\xi)$ is tangent to V_{reg} so for any $h \in I(V)$, $\xi(h) = 0$ on V_{reg} and hence by continuity on V. □

The first question to resolve is the relation between g_0 being finitely $\mathcal{K}_V$-determined and f_0 being finitely $\mathcal{A}$-determined.

Proposition 2.3 *f_0 is finitely $\mathcal{A}$-determined if and only if g_0 is finitely $\mathcal{K}_V$-determined.*

Proof: For both directions we use the geometric criterion from the preceding section.
⇐ As g_0 is finitely $\mathcal{K}_V$-determined it is transverse to V in a punctured neighbourhood of 0. Let W be such a punctured neighbourhood with a representative of g_0 still denoted by g_0. Let $\{\eta_i\}$ be a set of vector fields in θ_V which generate $\Theta_{V,y'}$ for y' in a neighbourhood of 0 which includes W (by shrinking W if necessary). For $y \in W$, let $S = F^{-1}(g_0(y)) \cap \Sigma(F)$, which is finite. For each i let ξ_i be a lift of η_i which, by shrinking U if necessary, is defined on U. Then, $F : \mathbb{C}^{n'},S \to \mathbb{C}^{p'},g_0(y)$ is stable. Pick a subset $\{\eta_1,\ldots,\eta_r\}$ of the above set $\{\eta_i\}$ such that $\langle \eta_{1(g_0(y))},\ldots,\eta_{r(g_0(y))} \rangle$ spans a complementary subspace to

$dg_0(y)(T\mathbb{C}^p)$. Then, since $\xi_i(F) = F \circ \eta_i$, by a standard argument in e.g. Martinet [Ma1], $F: \mathbb{C}^{n'}, S \to \mathbb{C}^{p'}, g_0(y)$ is $\mathcal{A}$-equivalent as a multi-germ to $f_0 \times \mathrm{id} : \mathbb{C}^{n'}, S \to \mathbb{C}^{p'}, y$. This implies that $f_0 : \mathbb{C}^n, S \to \mathbb{C}^p, y$ is stable (f_0 is stable if and only if $f_0 \times \mathrm{id}$ is by the infinitesimal criteria of Mather [M-IV]). As y was an arbitrary point of W, f_0 is stable in a punctured neighbourhood of 0 and so is finitely $\mathcal{A}$-determined.

Conversely, if f_0 is finitely $\mathcal{A}$-determined then for y in a punctured neighbourhood W of 0, $f_0 : \mathbb{C}^n, S \to \mathbb{C}^p, y$ is stable. Hence, $F: \mathbb{C}^{n'}, S \to \mathbb{C}^{p'}, y$ is an $\mathcal{A}$-trivial unfolding of f_0. Thus, there are vector fields ξ_i, η'_i defined near S and y so that $\xi_i(F) = \eta'_i \circ F$ and $\{\eta'_{i(y)}\}$ span a subspace complementary to $\mathbb{C}^p$. Thus, $\eta'_i \in \Theta_{V,y}$. By choosing W smaller if necessary, $\{\eta_i\}$ generate $\Theta_{V,y}$ for $y \in W$. Hence, the subspace spanned by $\{\eta'_{i(y)}\}$ is contained in that spanned by $\{\eta_{i(y)}\}$. Thus, $\mathbb{C}^p$ is transverse to V at y. Thus, $\mathbb{C}^p$ is transverse to V in the punctured neighbourhood W, i.e. g_0 is transverse to V on W and hence is finitely $\mathcal{K}_V$-determined. □

Second, we relate $\mathcal{K}_V$-triviality of unfoldings of g_0 with $\mathcal{A}$-triviality of unfoldings of f_0. We let $g(x,u)$ be an unfolding of g_0 and $g_1(x,u,v) : \mathbb{C}^{p+q+r}, 0 \to \mathbb{C}^{p'+q+r}, 0$ an extension of g. We let f and f_1 denote the induced unfoldings of f_0.

Proposition 2.4: i) *If* g *is a* $\mathcal{K}_V$*-trivial unfolding (respectively* $\mathcal{K}_V$*-trivial family) then* f *is an* $\mathcal{A}$*-trivial unfolding (respectively* $\mathcal{A}$*-trivial family).*
ii) *If* g_1 *is a* $\mathcal{K}_V$*-trivial extension of* g *then* f_1 *is an* $\mathcal{A}$*-trivial extension of* f.

Proof: We give the proof of ii); that of i) is analogous (and slightly easier).

By the infinitesimal criterion, there exist germs of vector fields $\zeta_i \in \mathcal{C}_{y,u,v}\left\{\frac{\partial}{\partial y_i}\right\}$, $\chi_i \in \mathcal{C}_{y,u,v}\{\eta_i\}$, and $\gamma_i \in \mathcal{C}_{u,v}\left\{\frac{\partial}{\partial u_i}\right\}$ (with $\{\eta_i\}$ generating Θ_V) such that

$$\frac{\partial \bar{g}_1}{\partial v_i} = -\zeta_i(\bar{g}_1) - \gamma_i(\bar{g}_1) + \chi_i \circ g_1 . \tag{2.5}$$

Let $\{\xi_i\}$ denote the lifts of the $\{\eta_i\}$. If $\chi_i = \sum h_{ij}\eta_j$, let $\delta_i = \sum h'_{ij}\xi_j$ with $h'_{ij} = h''_{ij} \circ (F \times id)$. To define h''_{ij}, we note that h_{ij} is a germ defined on $\mathbb{C}^{p+q+r}$; however, $g_1 : \mathbb{C}^{p+q+r},0 \to \mathbb{C}^{p'+q+r},0$ is a germ of an immersion. Thus, $h_{ij} = g_1^* h''_{ij}$ for some h''_{ij} on $\mathbb{C}^{p'+q+r},0$. We also replace χ_i by $\chi'_i = \sum h''_{ij}\eta_j$ where η_j also denotes its trivial extension to $\mathbb{C}^{p'+q+r},0$. Then (2.5) remains valid if we replace χ_i by χ'_i since $\chi'_i \circ g_1 = \chi_i \circ g_1$. Also

(2.6) $$\delta_i\,(F \times id) = \chi'_i \circ (F \times id)\,.$$

Now, f_1 is formed from g_1 and $F \times id$ by fiber product. We make this explicit. Let

$$H_1 : \mathbb{C}^{n'+p+q+r},0 \longrightarrow \mathbb{C}^{2p'+q+r},0$$

be defined by $H_1(x',y,u,v) = (\bar{F}(x'), \bar{g}_1(y,u,v), u,v)$, and

$$H : \mathbb{C}^{n'+p+q},0 \longrightarrow \mathbb{C}^{2p'+q},0$$

by $H(x',y,u) = (F(x'), \bar{g}(y,u), u)$. Let,

$$\Delta_1 = \{(y',y',u,v) : y' \in \mathbb{C}^{p'}\},$$

$$\Delta = \{(y',y',u) : y' \in \mathbb{C}^{p'}\}\,.$$

Then, f_1 and f are the restrictions of H_1 and H

$$H_1 : H_1^{-1}(\Delta_1) \longrightarrow \Delta_1 \qquad\qquad H : H^{-1}(\Delta) \longrightarrow \Delta\,.$$

We wish to prove that $H_1 \mid H_1^{-1}(\Delta_1)$ is an $\mathcal{A}$-trivial extension of $H \mid H^{-1}(\Delta)$.

Weclaim

(2.7) $$\frac{\partial \bar{H}_1}{\partial v_i} = \left(0, \frac{\partial \bar{g}_1}{\partial v_i}\right) = -(\delta_i,\zeta_i)\,H_1 - \gamma_i(H_1) + (\chi'_i,\chi'_i) \circ H_1$$

for on the first component

$$0 = -\delta_i(F \times id) - 0 + \chi'_i \circ (F \times id),$$

and on the second component

$$\frac{\partial \bar{g}_1}{\partial v_i} = -\zeta_i(\bar{g}_1) - \gamma_i(\bar{g}_1) + \chi'_i \circ g_1$$

which follow by (2.5) and (2.6). Also,

$$\tilde{\eta}_i = \frac{\partial}{\partial v_i} + \gamma_i + (\chi'_i,\chi'_i)$$

is tangent to Δ_1; and if we let

$$\tilde{\xi}_i = \frac{\partial}{\partial v_i} + \gamma_i + (\delta_i, \zeta_i)$$

then

$$\tilde{\xi}_i(H_1) = H_1 \circ \tilde{\eta}_i .$$

Thus, $\tilde{\xi}_i$ is tangent to $H_1^{-1}(\Delta_1)$. Then, the restrictions $\tilde{\eta}_i \mid \Delta$ and $\tilde{\xi}_i \mid H_1^{-1}(\Delta_1)$ give the vector fields which provide the infinitesimal trivialization of $H_1 \mid H_1^{-1}(\Delta_1)$ as an extension of $H \mid H^{-1}(\Delta)$. □

Now we are in a position to establish the equality of codimensions before we even define the algebraic homomorphism between normal spaces. It is enough to show: 1) if g is a $\mathcal{K}_V$-versal unfolding of g_0 then the induced f is an $\mathcal{A}$-versal unfolding of g and 2) there is an $\mathcal{A}$-miniversal unfolding f of f_0 induced by an unfolding g with g $\mathcal{K}_V$-versal. For by the versality theorem, 1) implies $\mathcal{A}_e$-codim$(f_0) \leq \mathcal{K}_{V,e}$-codim$(g_0)$ while 2) implies the reverseinequality.

Then, 1) is established by

Lemma 2.8: *Let* g *be a* $\mathcal{K}_V$*-versal unfolding of* g_0*, then* f *is an* $\mathcal{A}$*-versal unfolding of* f_0.

Proof: Let f_1 be an extension of f. To prove that f is $\mathcal{A}$-versal, it is sufficient to prove that any such f_1 is an $\mathcal{A}$-trivial extension of f. If we can show that f_1 is induced by a g_1 which is an extension of g, then, by the $\mathcal{K}_V$-versality of g, g_1 is a $\mathcal{K}_V$-trivial extension of g; and by proposition 1.5, f_1 is an $\mathcal{A}$-trivial extension of f. We actually prove a variant of this where g_0, g and g_1 are replaced by related germs h_0, h and h_1, which induce f_0, f and f_1 from a larger stable unfolding so we can still apply proposition 1.5.

To define the h's, we enlarge the stable unfolding F to include explicitly all of the unfoldings under consideration. We represent F as an unfolding $F(x,w) = (\bar{F}(x,w), w)$. The unfolding $g(y,u) = (\bar{g}(y,u),u)$, $\bar{g}: \mathbb{C}^{p+q},0 \longrightarrow \mathbb{C}^{p'},0$ has the form $(y,w) = \bar{g}(y,u) = (\bar{g}'(y,u), \bar{g}''(y,u))$. Define a map $\varphi : \mathbb{C}^{p'+q},0 \longrightarrow \mathbb{C}^{p'+q},0$ by $\varphi(y,w,u) = (\bar{g}'(y,u), \bar{g}''(y,u) + w, u)$. It is easily checked that φ is a germ of a diffeomorphism, so that $F \times \mathrm{id}$ pulls back via φ to an unfolding

$$F_1(x,u,w) = (\bar{F}_1(x,u,w), u, w)$$

and that

$$\bar{F}_1(x,u,0) = \bar{f}(x,u) \qquad \text{and} \qquad \bar{F}_1(x,0,w) = \bar{F}(x,w).$$

Consider the unfolding

$$F_2(x,u,w,v) = (\bar{F}_1(x,u,w) - \bar{f}(x,u) + \bar{f}_1(x,u,v), u,w,v).$$

Then

(2.9) $\quad F_2(x,u,0,v) = (\bar{f}_1(x,u,v), u,0,v) \quad$ and $\quad F_2(x,0,w,0) = (\bar{F}(x,w), 0,w,0)$

Since F is stable, by (2.9) and the infinitesimal criterion of Mather, F_2 is stable. Then, we define $h_0: \mathbb{C}^p,0 \to \mathbb{C}^{p'+q+r},0$, $\bar{h}: \mathbb{C}^{p+q},0 \to \mathbb{C}^{p'+q+r},0$, and $\bar{h}_1: \mathbb{C}^{p+q+r},0 \to \mathbb{C}^{p'+q+r},0$ by $h_0(y) = (y,0,0,0)$, $\bar{h}(y,u) = (y,u,0,0)$, and $\bar{h}_1(y,u,v) = (y,u,0,v)$. By (2.9) we see that $\bar{h}$ pulls back F_2 to give f, $\bar{h}_1$ pulls back F_2 to give f_1 and h_1 is an extension of h. If we knew that h_1 were a $\mathcal{K}_{V''}$-trivial extension of h, where V" = D(F_2), then by proposition 1.5 we could draw the desired conclusion.

To see that it is, we define $G_0: \mathbb{C}^p,0 \to \mathbb{C}^{p'+q+r},0$ by $G_0(y) = (y, 0, 0)$ and the unfolding $G(y,u) = (\bar{G}(y,u), u)$ by $\bar{G}(y,u) = (\bar{g}(y,u),u,0)$. Then, $g_0 = \pi_* G_0$ and $\bar{g} = \pi_* \bar{G}$ for $\pi: \mathbb{C}^{p'+q+r},0 \to \mathbb{C}^{p'},0$ the projection. Thus, by proposition 1.5, G is a $\mathcal{K}_{V'}$-versal unfolding of G_0 where $V' = V \times \mathbb{C}^{q+r}$. Also, $(\varphi \times id)_* h = G$, $(\varphi \times id)_* h_0 = G_0$, and $\varphi \times id(V'') = D(F \times id) = V'$. Since $\varphi \times id$ is a diffeomorphism, h is a $\mathcal{K}_{V''}$-versal unfolding of h_0 if and only if G is a $\mathcal{K}_{V'}$-versal unfolding of G_0, which it is. Hence, h_1 is a $\mathcal{K}_{V''}$-trivial extension of h ; and thus, f_1 is an $\mathcal{A}$-trivial extension of f. □

For 2) we let $f(x,u) = (\bar{f}(x,u), u)$ denote an $\mathcal{A}$-versal unfolding of f_0 with $f: \mathbb{C}^{p+q},0 \to \mathbb{C}^{n+q},0$. We define an unfolding of $g_0(y) = y$ by $\bar{g}(y,u) = (y,u)$.

Lemma 2.10: g *is a* $\mathcal{K}_V$*-versal unfolding of* g_0*, where* $V = D(f)$.

Proof: Since

$$T\mathcal{K}_e \cdot f_0 + \langle \frac{\partial}{\partial y_1}, \dots, \frac{\partial}{\partial y_p} \rangle \supseteq T\mathcal{A}_e \cdot f_0$$

it follows that the $\mathcal{A}$-versal unfolding f is also a stable unfolding of f_0. Then, we may use f for our stable unfolding F.

Let g_1 be an extension of g, with additional parameters $v \in \mathbb{C}^r$. Define $\varphi: \mathbb{C}^{p+q+r},0 \to \mathbb{C}^{p+q+r},0$ by $\varphi(y,u,v) = (\bar{g}_1(y,u,v), v)$. As g_1 is an extension of g, $\bar{g}_1(y,u,0) = (y,u)$. Hence, φ is a germ of a diffeomorphism by the inverse function theorem. We may pull back $f \times id$ by φ to obtain an unfolding $f_1: \mathbb{C}^{n+q+r},0 \to \mathbb{C}^{p+q+r},0$. Since $\varphi(y,u,0) = (y,u,0)$, $\bar{f}_1(x,u,0) = \bar{f}(x,u)$. Note even though f_1 is a pull-back of a trivial unfolding $f \times id$, the pull-back is not in the usual sense of unfoldings; hence, the unfolding need not be an $\mathcal{A}$-trivial extension f. However, f_1 is an extension of the unfolding f which is $\mathcal{A}$-versal. Hence, f_1 is an $\mathcal{A}$-trivial extension of f by the versality theorem.

By the infinitesimal criterion, there exist vector fields of the form

$$\chi_i = \frac{\partial}{\partial v_i} + \gamma_i + \zeta_i \qquad \delta_i = \frac{\partial}{\partial v_i} + \xi_i + \zeta_i \qquad 1 \leq i \leq r$$

where $\gamma_i \in \mathcal{C}_{y,u,v}\left\{\frac{\partial}{\partial y_i}\right\}$, $\xi_i \in \mathcal{C}_{x,u,v}\left\{\frac{\partial}{\partial x_j}\right\}$, $\zeta_i \in \mathcal{C}_{u,v}\left\{\frac{\partial}{\partial u_j}\right\}$ and such that

$$\delta_i(f_1) = \chi_i \circ f_1 .$$

Thus, χ_i is f_1-liftable and

$$\frac{\partial}{\partial v_i} = \chi_i - \gamma_i - \zeta_i .$$

Consider the unfoldings h and h_1 of $h_0(y) = (y,0,0)$ with $\bar{h}(y,u) = (y,u,0)$ and $\bar{h}_1(y,u,v) = (y,u,v)$.

$$\chi_i \circ \bar{h}_1 = \chi_i, \quad \zeta_i(\bar{h}_1) = \zeta_i, \quad \gamma_i(\bar{h}_1) = \gamma_i, \quad \text{and} \quad \frac{\partial \bar{h}_1}{\partial v_i} = \frac{\partial}{\partial v_i} .$$

Hence,

$$\frac{\partial \bar{h}_1}{\partial v_i} = -\gamma_i(\bar{h}_1) - \zeta_i(\bar{h}_1) + \chi_i \circ \bar{h}_1 \qquad 1 \leq i \leq r .$$

Hence, h_1 is a $\mathcal{K}_{V'}$-trivial extension of h where $V' = D(f_1)$.

Now, $\varphi(D(f_1)) = D(f) \times \mathbb{C}^r = V \times \mathbb{C}^r$ and $\varphi(y,u,0) = (y,u,0)$. Thus, $\varphi_* h_1$ is a $\mathcal{K}_{V\times\mathbb{C}^r}$ trivial extension of $\varphi_* h$ by proposition 1.5 and hence $g_1 = \pi_*\varphi_* h_1$ is a $\mathcal{K}_V$-trivial extension of $g = \pi_*\varphi_* h$. As g_1 was an arbitrary extension of g, g is $\mathcal{K}_V$-versal. □

Now, if g is a $\mathcal{K}_V$-miniversal unfolding of g_0 on q parameters, then the induced f is an $\mathcal{A}$-versal unfolding of f_0 by lemma 2.8. Thus, by the versality theorem, $\mathcal{K}_{V,e}$-codim $(g_0) = q \geq \mathcal{A}_e$-codim (f_0). On the other hand, if f is an $\mathcal{A}_e$-miniversal unfolding of f_0, then the unfolding of g_0 defined in lemma 2.10 is $\mathcal{K}_V$-versal so the inequality is reversed. We conclude,

Theorem 1: *With the preceding notation*

$$\mathcal{A}_e\text{-codim}\,(f_0) = \mathcal{K}_{V,e}\text{-codim}\,(g_0) .$$

§3. Isomorphism of Normal Spaces

As in the preceding section, we let $f_0 : \mathbb{C}^n,0 \longrightarrow \mathbb{C}^p,0$ have a stable unfolding $F : \mathbb{C}^{n'},0 \longrightarrow \mathbb{C}^{p'},0$ with $g_0 : \mathbb{C}^p,0 \longrightarrow \mathbb{C}^{p'},0$ denoting the inclusion of $\mathbb{C}^p$. By a choice of local coordinates we may assume $F(x,u) = (\bar{F}(x,u),u) = (y,u)$ and $g_0(y) = (y, 0)$.

In this section we shall define an isomorphism between $N\mathcal{K}_{V,e} \cdot g_0$ and $N\mathcal{A}_e \cdot f_0$ when both (i.e. either) are finite dimensional. For $\zeta \in \theta(g_0)$, we may represent $\zeta = (\zeta_1, \zeta_2)$ where ζ_1 denotes the y-component and ζ_2 the u-component of ζ. We define a $\mathcal{C}_y$-linear homomorphism $\Phi : \theta(g_0) \to \theta(f_0)$ by

$$\Phi(\zeta) = -\zeta_1 \circ f_0 + d_u\bar{F}(x,0)(\zeta_2) \circ f_0$$

Theorem 2: Φ *induces an isomorphism of* $\mathcal{C}_y$*-modules*

$$\bar{\Phi} : N\mathcal{K}_{V,e} \cdot g_0 \xrightarrow{\sim} N\mathcal{A}_e \cdot f_0$$

Proof: The proof of this theorem will occupy the rest of this section.

Given a 1-parameter unfolding of g_0, which we denote by $g_t(y)$ instead of $\bar{g}(y,t)$, we can associate to it an element of $\theta(g_0)$, namely $\zeta = \frac{\partial g_t}{\partial t}\big|_{t=0}$. We shall explicitly show that $\Phi(\zeta)$ is the corresponding element of $\theta(f_0)$ obtained from the induced deformation f_t of f_0 which is defined as a fiber product

(3.1) $$f_t : X_t = \{(x,u,y) : F(x,u) = g_t(y)\} \xrightarrow{pr} \mathbb{C}^p$$

with $pr(x,u,y) = y$.

We write $g_t(y) = (g_{1t}(y), g_{2t}(y)) = (y,u)$ so that $g_{10}(y) = y$, $g_{20}(y) \equiv 0$. Then, (3.1) defines X_t by

$$\bar{F}(x,u) = g_{1t}(y) \quad \text{and} \quad u = g_{2t}(y);$$

or x and y are related by

(3.2) $$g_{1t}(y) - \bar{F}(x, g_{2t}(y)) = 0.$$

Let $H(x,y,t)$ denote the function on the left hand side of (3.2). We apply the implicit function theorem to parametrize $H^{-1}(0)$.

(3.3) $$d_yH(0,0,0) = d_yg_{10}(0) - d_u\bar{F}(0,0) \circ dg_{20}(0).$$

Since $g_{10} = \mathrm{id}$ and $g_{20} \equiv 0$, we see from (3.3) that $d_yH(0,0,0) = I$. Thus, by the implicit function theorem, we may represent $H^{-1}(0)$ as the graph of y as a function of (x,t), $y = \psi_t(x)$.

Then, $X_t = \{(x,u,y) : u = g_{2t}(y)\ ,\ y = \psi_t(x)\}$. Let $\varphi_t(x) = g_{2t} \circ \psi_t(x)$ so that $\varphi_0(x) = g_{20} \circ \psi_0(x) \equiv 0$. Also, $g_{10} = \mathrm{id}$ so for small t , g_{1t} is a germ of a diffeomorphism. Hence, by (3.2)

$$y = g_{1t}^{-1} \circ \bar{F}(x, g_{2t}(y)) .$$

Thus, by the above description of X_t and (3.1),

$$y = \psi_t(x) = g_{1t}^{-1} \circ \bar{F}(x, \varphi_t(x))$$

and so

$$f_t(x) = g_{1t}^{-1} \circ \bar{F}(x, \varphi_t(x)) . \tag{3.4}$$

Thus, by the chain rule

$$\frac{\partial f_t}{\partial t}\Big|_{t=0} = \frac{\partial g_{1t}^{-1}}{\partial t}\Big|_{t=0} \circ \bar{F}(x,\varphi_0(x)) + dg_{10}^{-1} \circ \frac{\partial \bar{F}}{\partial u}(x,\varphi_0(x)) \frac{\partial \varphi_t}{\partial t}\Big|_{t=0} . \tag{3.5}$$

From $g_{1t}^{-1} \circ g_{1t} = \mathrm{id}$ we obtain

$$\frac{\partial g_{1t}^{-1}}{\partial t}\Big|_{t=0} \circ g_{10} + dg_{10}^{-1} \frac{\partial g_{1t}}{\partial t}\Big|_{t=0} = 0 . \tag{3.6}$$

Since $g_{10} = \mathrm{id}$, (3.6) implies

$$\frac{\partial g_{1t}^{-1}}{\partial t}\Big|_{t=0} = -\frac{\partial g_{1t}}{\partial t}\Big|_{t=0}$$

Also, $\varphi_0(x) = 0$ and $\bar{F}(x,0) = f_0(x)$ so (3.5) becomes

$$\frac{\partial f_t}{\partial t}\Big|_{t=0} = -\frac{\partial g_{1t}}{\partial t}\Big|_{t=0} \circ f_0(x) + \frac{\partial \bar{F}}{\partial u}(x,0) \frac{\partial \varphi_t}{\partial t}\Big|_{t=0} . \tag{3.7}$$

Then,

$$\varphi_t = g_{2t} \circ \psi_t, \text{ or}$$

$$\varphi_t(x) = g_{2t} \circ g_{1t}^{-1} \circ \bar{F}(x, \varphi_t(x)) .$$

Hence

$$\frac{\partial \varphi_t}{\partial t}\Big|_{t=0} = \frac{\partial g_{2t}}{\partial t}\Big|_{t=0} \circ g_{10}^{-1} \circ \bar{F}(x,\varphi_0(x)) + dg_{20} \circ (-) . \tag{3.8}$$

Since g_{20}, and hence dg_{20}, equals 0, the second term vanishes. Thus, (3.8) becomes

$$\frac{\partial \varphi_t}{\partial t}\Big|_{t=0} = \frac{\partial g_{2t}}{\partial t}\Big|_{t=0} \circ f_0(x) .$$

Substituting into (3.7) yields

$$\frac{\partial f_t}{\partial t}\Big|_{t=0} = -\frac{\partial g_{1t}}{\partial t}\Big|_{t=0} \circ f_0(x) + d_u\bar{F}(x,0)\left(\frac{\partial g_{2t}}{\partial t}\Big|_{t=0}\right) \circ f_0(x) . \tag{3.9}$$

If

$$\zeta = \frac{\partial g_t}{\partial t}\Big|_{t=0} = \left(\frac{\partial g_{1t}}{\partial t}\Big|_{t=0}, \frac{\partial g_{2t}}{\partial t}\Big|_{t=0}\right) = (\zeta_1, \zeta_2)$$

then, we obtain from (3.9)

$$\frac{\partial f_t}{\partial t}\Big|_{t=0} = -\zeta_1 \circ f_0 + d_u\bar{F}(x,0)(\zeta_2) \circ f_0 . \tag{3.10}$$

We see that $\Phi(\zeta)$ is equal to the right hand side of (3.10):

$$\Phi : \theta(g_0) \longrightarrow \theta(f_0)$$

(3.11)

$$\Phi(\zeta) = -\zeta_1 \circ f_0 + d_u\bar{F}(x,0)(\zeta_2) \circ f_0$$

Next, if $\zeta \in T\mathcal{K}_{V,e} \cdot g_0$, then $\zeta = \frac{\partial g_t}{\partial t}\Big|_{t=0}$ for g_t a $\mathcal{K}_V$-trivial deformation. By proposition 2.4, f_t is an $\mathcal{A}$-trivial deformation of f_0. Thus,

$$\Phi(\zeta) = \frac{\partial f_t}{\partial t}\Big|_{t=0} \in T\mathcal{A}_e \cdot f_0 .$$

Thus,

$$\Phi(T\mathcal{K}_{V,e} \cdot g_0) \subset T\mathcal{A}_e \cdot f_0$$

and induces a $\mathcal{C}_y$-module homomorphism.

$$\bar{\Phi} : N\mathcal{K}_{V,e} \cdot g_0 \longrightarrow N\mathcal{A}_e \cdot f_0 . \tag{3.12}$$

We now show this is an isomorphism.

Given $\xi \in \theta(f_0)$, then $\xi = \frac{\partial f_t}{\partial t}\Big|_{t=0}$ with f_t induced, up to $\mathcal{A}$-equivalence, by an unfolding g_t. Thus, g_t induces f'_t with f_t $\mathcal{A}$-equivalent to f'_t, say $f_t = \psi_t \circ f'_t \circ \varphi_t$ with $\psi_0 =$ id, $\varphi_0 =$ id. We compute

$$\frac{\partial f_t}{\partial t}\Big|_{t=0} = \frac{\partial f'_t}{\partial t}\Big|_{t=0} + \frac{\partial \psi_t}{\partial t}\Big|_{t=0} \circ f'_0 - df'_0\left(\frac{\partial \varphi_t}{\partial t}\Big|_{t=0}\right)$$

$$= \frac{\partial f'_t}{\partial t}\Big|_{t=0} + \eta \circ f_0 - \xi(f_0) .$$

Thus,

$$\xi = \frac{\partial f_t}{\partial t}\Big|_{t=0} \equiv \Phi(\zeta) \bmod T\mathcal{A}_e \cdot f_0 .$$

Hence, $\bar{\Phi}$ is surjective.

By Theorem 1, the spaces in (3.12) have the same dimension as vector spaces; as $\bar{\Phi}$ is surjective it is an isomorphism. □

We can now refine our earlier results relating the versality of g_0 and f_0.

Corollary 1: *With the preceding notation, let g be an unfolding of* g_0 *and let f denote the induced unfolding of* f_0. *Then, f is* $\mathcal{A}$*-versal if and only if g is* $\mathcal{K}_V$*-versal.*

Proof: The proof of the theorem shows that for each i,

$$\bar{\Phi}\left(\frac{\partial g}{\partial u_i}\right) = \frac{\partial f}{\partial u_i}$$

Hence,the corollary follows by the versality theorem and theorem 2. □

We also obtain the analog of theorem 2 for multi-germs, which follows by the same proofs except applied to multi-germs.

Let $f_0: \mathbb{C}^n,S \to \mathbb{C}^p,0$ have a stable unfolding $F : \mathbb{C}^{n'},S \to \mathbb{C}^{p'},0$ with $g_0 : \mathbb{C}^p,0 \to \mathbb{C}^{p'},0$ denoting the inclusion of $\mathbb{C}^p$. Then, Φ defined by (3.11) also defines a homomorphism for $\theta(f_0)$ denoting the module of vector fields along the multi-germ f_0. Then, Φ also induces an isomorphism in this case.

Theorem 3: i) *The multi-germ* f_0 *has finite* $\mathcal{A}$*-codimension if and only if* g_0 *has finite* $\mathcal{K}_V$*-codimension;*

ii) *in the case of* i) Φ *induces an isomorphism*

$$\bar{\Phi}: N\mathcal{K}_{V,e} \cdot g_0 \xrightarrow{\sim} N\mathcal{A}_e \cdot f_0.$$

§4 Several Consequences

We deduce consequences of the main theorems for: a) placing Mond's formula in a more general context as an analogue of Milnor's formula but for nonlinear sections of nonisolated hypersurface singularities and b) verifying that a method for computing the versality discriminant of an unfolding of a hypersurface singularity (given in [DG] for the Pham example) is valid in general.

Nonlinear Sections of Hypersurface Singularities

Let $V,0 \subset \mathbb{C}^m,0$ be a hypersurface germ and let $g_0 : \mathbb{C}^p,0 \longrightarrow \mathbb{C}^m,0$ be a germ of an immersion. We can define two numbers associated to the nonlinear section g_0, a number defined algebraically, which measures the codimension of g_0, and a number defined geometrically, which is the analogue of the Milnor number for $g_t(\mathbb{C}^p) \cap V$ with g_t a perturbation of g_0. If we ask when these two numbers are equal, it turns out that not only can Mond's formula be interpreted as an equality of these numbers but, in this context, it is related to other formulas which involve seemingly unrelated numbers such as the multiplicity of the discriminant for a versal deformation and a special case of Greuel's and Lê's formula for the Milnor number of isolated complete intersection singularities [G], [L].

The algebraically defined number associated to g_0 is its $\mathcal{K}_V$-codimension

$$\nu_{alg}(g_0) = \mathcal{K}_{V,e}\text{-codim}(g_0)$$

For this number to be finite we must assume that g_0 is transverse to V in a punctured neighborhood of 0. For the geometrically defined number, we consider a one-parameter family of germs g_t such that g_t is transverse to V for $t \neq 0$. Here we have to use a weaker notion of transversality than that used in § 1, i.e. we choose a Whitney stratification of V with the property that all $\eta \in \theta_V$ are tangent to the strata and require transversality to all of the strata. Then the geometric number which is the analogue of the Milnor number is

$$\nu_{geom}(g_0) = |\chi(g_t(\mathbb{C}^p) \cap V \cap B_\varepsilon) - 1|.$$

Here $\chi(g_t(\mathbb{C}^p) \cap V \cap B_\varepsilon)$ is the topological Euler characteristic, B_ε is a ball about 0 of radius ε and ε and t have to chosen appropriately small. This geometric number can be shown to be well-defined. Mond's formula and other related formulas suggest the following.

BASIC QUESTION : Suppose both V and g_0 are weighted homogeneous for the same weights on $\mathbb{C}^m$. When do we have the analogue of Milnor's formula, namely, when does (4.1) hold?

(4.1) $$v_{alg}(g_0) = v_{geom}(g_0)$$

We consider some cases where it is presently known to hold.

1) Let $V = D(F) = \text{image}(F)$ where $F: \mathbb{C}^n,0 \longrightarrow \mathbb{C}^{n+1},0$ is a stable germ, and let $g_0: \mathbb{C}^3,0 \longrightarrow \mathbb{C}^{n+1},0$ denote a germ of an immersion transverse to F with f_0 the pullback. By theorem 2

$$\mathcal{A}_e\text{-codim}(f_0) = \mathcal{K}_{V,e}\text{-codim}(g_0) = v_{alg}(g_0).$$

If g_t is a family such that $g_t(\mathbb{C}^3)$ is transverse to V for $t \neq 0$, then by the proof of proposition 2.3, the pull-back family f_t is stable for $t \neq 0$. Then $f_t(\mathbb{C}^2) \cap B_\varepsilon = g_t(\mathbb{C}^3) \cap V \cap B_\varepsilon$. Thus, $v_{geom}(g_0) = |\chi(f_t(\mathbb{C}^2) \cap B_\varepsilon) - 1|$. Thus, by the result of de Jong and van Straten [JS], (4.1) holds when g_0 and F are weighted homogeneous for the same weights on $\mathbb{C}^{n+1}$.

2) Let $V = D(F)$ where $F: \mathbb{C}^{n+q},0 \longrightarrow \mathbb{C}^{1+q},0$ is a versal unfolding of a weighted homogeneous hypersurface singularity defined by f_0 (here $q = \tau - 1$). Also, let $g_0: \mathbb{C},0 \longrightarrow \mathbb{C}^{1+q},0$ denote the germ $g_0(y) = (y,0)$. Then, g_0 is transverse to F with f_0 the pullback. We saw in example 1.3 that

$$\mathcal{A}_e\text{-codim}(f_0) = \mathcal{K}_{V,e}\text{-codim}(g_0) \ (= v_{alg}(g_0)) = \tau - 1$$

If g_t is transverse to V for t small $\neq 0$ then $g_t(\mathbb{C}) \cap V \cap B_\varepsilon$ consists of m(V) points, where m(V) denotes the multiplicity of V. However, $m(V) = \mu$; see e.g. [T]. Hence, $v_{geom}(g_0) = \mu - 1$. Since f_0 is weighted homogeneous, $\mu = \tau$ so again we have equality in (1.4).

3) Let $V = D(F) = \text{image}(F)$ where $F: \mathbb{C}^n,0 \longrightarrow \mathbb{C}^{n+1},0$ is the stable unfolding of the germ $f_0(x) = (x^n,x^m)$ with $(n,m) = 1$. Likewise, let $g_0: \mathbb{C}^2,0 \longrightarrow \mathbb{C}^{n+1},0$ denote the germ of an immersion $g_0(y_1,y_2) = (y_1,y_2,0)$ transverse to F with f_0 the pullback. Then, a simple calculation shows that $\mathcal{A}_e\text{-codim}(f_0) = \delta(C)$ where C is the image curve $f_0(\mathbb{C})$ defined by $y_1{}^m - y_2{}^n = 0$. Also, f_0 can be deformed to a stable germ f_t so that the image curve $f_t(\mathbb{C})$ has $\delta(C)$ double points and $f_t(\mathbb{C}) \cap B_\varepsilon = g_t(\mathbb{C}^2) \cap V \cap B_\varepsilon$ for g_t the deformation of g_0 inducing f_t. Hence,

$$v_{geom}(g_0) = |\chi(f_t(\mathbb{C}) \cap B_\varepsilon) - 1| = |(1 - \delta) - 1| = \delta(C).$$

Lastly we consider a hypersurface which is not the discriminant of a stable germ.

4) Let V,0 be an isolated hypersurface singularity defined by a weighted homogeneous germ $f_0 : \mathbb{C}^n,0 \longrightarrow \mathbb{C},0$. Let $g_0: \mathbb{C}^{n-1},0 \longrightarrow \mathbb{C}^n,0$ be a germ of an immersion which is weighted homogeneous for the same weights and for which $g_0(\mathbb{C}^{n-1})$ is transverse to V in a punctured neighbourhood of 0. By a weighted homogeneous change of coordinates we may assume that g_0 is a linear embedding $g_0(x_1, \ldots ,x_{n-1}) = (x_1, \ldots ,x_{n-1},0)$. Now, θ_V is generated by $\{\zeta_{ij} = \frac{\partial f_0}{\partial x_j}\frac{\partial}{\partial x_i} - \frac{\partial f_0}{\partial x_i}\frac{\partial}{\partial x_j}$, e} where e is the Euler vector field. Thus,

$$N\mathcal{K}_{V,e}\cdot g_0 = \mathcal{C}_{X'}\left\{\frac{\partial}{\partial x_i}\right\} / (\mathcal{C}_{X'}\left\{\frac{\partial}{\partial x_j}, j= 1,\ldots , n-1\right\} + \mathcal{C}_{X'}\left\{\zeta_{ij}, e\right\}$$

$$= \mathcal{C}_{X'}\left\{\frac{\partial}{\partial x_n}\right\} / (\mathcal{C}_{X'}\left\{\frac{\partial f_0}{\partial x_i}\cdot\frac{\partial}{\partial x_n}, j= 1,\ldots , n-1\right\}$$

$$(\text{since } e \circ g_0 = \sum_{j=1}^{n-1} x_j\cdot\frac{\partial}{\partial x_j})$$

$$= \mathcal{C}_{X'} / (\frac{\partial f_0}{\partial x_i}|_{x_n=0}, j= 1,\ldots , n-1).$$

Therefore,

$$\mathcal{K}_{V,e}\text{-codim}(g_0) = \mu(f_0|\mathbb{C}^{n-1}).$$

On the other hand, the assumption of transversality of $\mathbb{C}^{n-1}$ to V off of 0 implies that h = $(x_n,f_0) : \mathbb{C}^n,0 \longrightarrow \mathbb{C}^2,0$ defines an isolated weighted homogeneous complete intersection singularity. For $g_t(x') = (x', t)$, $g_t(\mathbb{C}^{n-1})$ is transverse to V for small $t \neq 0$. Thus, $h^{-1}(0,t) = g_t(\mathbb{C}^{n-1}) \cap V$ is a Milnor fiber of h so that $\nu_{geom}(g_0) = \mu(h) = \mu(f_0|\mathbb{C}^{n-1})$ (e.g. by a result of Greuel and Lê, [G] [L], $= \dim_{\mathbb{C}}\mathcal{J}(h)$, where $\mathcal{J}(h)$ is the Jacobian algebra of h, and by direct computation,

$$\mathcal{J}(h) = \mathcal{C}_X/(\frac{\partial f_0}{\partial x_i}|_{x_n=0}, j= 1,\ldots , n-1)).$$

Again (1.4) holds.

Versality Discriminant

Let $f_0 : \mathbb{C}^n,0 \longrightarrow \mathbb{C},0$ be a weighted homogeneous isolated hypersurface singularity. We can assign weights to $\theta(f_0)$ via $wt(\frac{\partial}{\partial x_i}) = -wt(x_i)$ and this induces weights on $N\mathcal{A}_e \cdot f_0 = N\mathcal{K}^+_e \cdot f_0$ ($\mathcal{K}^+$ is the usual action of $\mathcal{K}$ together with $\mathbb{C}$ acting by translation on $\mathbb{C}$). We let $N\mathcal{A}_e \cdot f_{0(<m)}$ denote the terms of weight < m. For a given m, let $\{\varphi_i\}_{i=1}^q$ be a basis for $N\mathcal{A}_e \cdot f_{0(<m)}$ and consider the unfolding

$$F(x,u) = (\bar{F}(x,u), u) = (f_0(x) + \sum_{i=1}^q u_i\varphi_i, u)$$

The versality discriminant for F consists of $z = (y,u) \in \mathbb{C}^{1+q}$ such that if $S = F^{-1}(y,u) \cap \Sigma(F)$ (recall $\Sigma(F)$ is the critical set of F) then $F : \mathbb{C}^{n+q},S \rightarrow \mathbb{C}^{1+q},z$ is not infinitesimally stable. Understanding the versality discriminant of F is a basic step in understanding the structure of F and determining whether, e.g., it is topologically versal.

In [DG] a procedure was given for computing the versality discriminant for the Pham example. We show here that this procedure works for all such unfoldings described above. Let $\{\bar{\varphi}_i\}_{i=1}^r$ be a basis for $N\mathcal{A}_e \cdot f_{0(\geq m)}$. Also, let

$$e = d.y.\frac{\partial}{\partial y} - \sum_{j=1}^n a_j x_j \frac{\partial}{\partial x_j} \qquad \text{with } a_j = wt(x_j) \text{ and } d = wt(y).$$

Then,

$$e(\bar{F}) \stackrel{def}{=} d.\bar{F}.\frac{\partial}{\partial y} - \sum_{j=1}^n a_j x_j \cdot \frac{\partial \bar{F}}{\partial x_j} = \sum_{j=1}^q wt(u_j) \cdot u_j \cdot \varphi_j$$

Since $\{\varphi_i , \bar{\varphi}_j\}$ is a basis for $N\mathcal{A}_e \cdot f_0$, by the preparation theorem we may write

(4.2)

$$\varphi_i \cdot e(\bar{F}) = \sum_{j=1}^q h_{ij}(u) \cdot \varphi_j + \sum_{j=1}^r g_{ij}(u) \cdot \bar{\varphi}_j \mod \left(\frac{\partial \bar{F}}{\partial x_1}, \dots, \frac{\partial \bar{F}}{\partial x_n}\right)$$

$$\bar{\varphi}_i \cdot e(\bar{F}) = \sum_{j=1}^q h'_{ij}(u) \cdot \varphi_j + \sum_{j=1}^r g'_{ij}(u) \cdot \bar{\varphi}_j \mod \left(\frac{\partial \bar{F}}{\partial x_1}, \dots, \frac{\partial \bar{F}}{\partial x_n}\right)$$

Let H be the $(\tau-1)\times r$ matrix with entries

$$H = \left(\underset{q\times r}{g_{ij}(u)} \mid \underset{r\times r}{y\cdot\delta_{ij} - g'_{ij}(u)}\right)$$

Let W be the variety defined by the vanishing of the $r\times r$ minors of H. For the Pham example [DG], it is shown that this yields the versality discriminant. This is in fact true in general:

Proposition 4.3: W *is the versality discriminant of* F.

Proof: The proof is a consequence of the proof of proposition 2.3 together with the construction due to Saito [S] of the generators of θ_V for $V = D(F_1)$ with F_1 the versal unfolding of f_0 (see example 1.3)

$$F_1(x,u,v) = \left(f_0(x) + \sum_{i=1}^{q} u_i\cdot\varphi_i + \sum_{j=1}^{r} v_j\cdot\bar{\varphi}_j,\ u,\ v\right)$$

By the proof of of proposition 2.3, $(y,u) \in \mathbb{C}^{1+q} \subset \mathbb{C}^{1+q+r}$ belongs to W exactly if $\mathbb{C}^{1+q}$ fails to be transverse to $V = D(F_1)$ at $(y,u,0)$. Recall that transversality holds at (y,u) if

(4.4) $$\mathbb{C}^{1+q} + \langle \eta_{0(y,u)}, \ldots, \eta_{\tau-1(y,u)} \rangle = \mathbb{C}^{1+q+r}$$

where $\{\eta_i\}$ denote the set of generators for θ_V constructed by Saito. However, note that (4.2) implies

(4.5)
$$d\bar{F}\cdot\varphi_i\cdot\frac{\partial}{\partial y} = \sum_{j=1}^{q} h_{ij}(u)\cdot\varphi_j + \sum_{j=1}^{r} g_{ij}(u)\cdot\bar{\varphi}_j \mod\left(\frac{\partial\bar{F}}{\partial x_1},\ldots,\frac{\partial\bar{F}}{\partial x_n}\right)$$

$$d\bar{F}\cdot\bar{\varphi}_i\cdot\frac{\partial}{\partial y} = \sum_{j=1}^{q} h'_{ij}(u)\cdot\varphi_j + \sum_{j=1}^{r} g'_{ij}(u)\cdot\bar{\varphi}_j \mod\left(\frac{\partial\bar{F}}{\partial x_1},\ldots,\frac{\partial\bar{F}}{\partial x_n}\right)$$

It follows from Saito's construction (1.3) and (4.5) that

$$\eta_k = (y-h_{kk})\cdot\frac{\partial}{\partial u_k} + \sum_{j\neq k} h_{kj}(u)\cdot\frac{\partial}{\partial u_j} + \sum_{j=1}^{r} g_{kj}(u)\cdot\frac{\partial}{\partial v_j} \qquad 1 \leq k \leq q$$

$$\eta_{q+k} = (y-g'_{kk})\cdot\frac{\partial}{\partial v_k} + \sum_{j=1}^{q} h_{kj}(u)\cdot\frac{\partial}{\partial u_j} + \sum_{j\neq k} g_{kj}(u)\cdot\frac{\partial}{\partial v_j} \qquad 1 \leq k \leq r$$

together with

$$\eta_0 = d.y.\frac{\partial}{\partial y} - \sum_{j=1}^{q} b_j\cdot u_j\cdot\frac{\partial}{\partial u_i} + \sum_{j=1}^{r} c_j\cdot v_j\frac{\partial}{\partial v_j}$$

are generators for V. Hence, (4.4) holds exactly when the v-components of the vectors $\eta_i | \mathbb{C}^{1+q}$ span $\mathbb{C}^r$. These components give (up to signs in columns) exactly the matrix H (since $\eta_0 | \mathbb{C}^{1+q}$ has v-component = 0). Thus, (4.4) fails exactly when the rank of H < r. □

Bibliography

A Arnold, V. I. *Wave front evolution and equivariant Morse lemma*, Comm. Pure App. Math. **29** (1976), 557-582.

B Bruce, J.W. *Functions on Discriminants*, Jour. London Math. Soc. **30** (1984), 551- 567

D1 Damon, J. a) *The unfolding and determinacy theorems for subgroups of $\mathcal{A}$ and $\mathcal{K}$*, Proc. Sym. Pure Math. **40** (1983) 233-254.
b) *The unfolding and determinacy theorems for subgroups of $\mathcal{A}$ and $\mathcal{K}$*, Memoirs A.M.S. 50, no. 306 (1984).

D2 ------ *Deformations of sections of singularities and Gorenstein surface singularities*, Amer. J. Math. **109** (1987) 695-722.

DG Damon, J. and Galligo, A. *Universal Topological Stratification for the Pham Example*, preprint.

DP du Plessis, A. *On the genericity of topologically finitely determined map germs*, Topology **21** (1982), 131- 156

G Greuel, G-M. *Der Gauss-Manin Zushammenhang isolierter Singularitaten von vollstandigen Durchschnitten*, Math. Ann. **214** (1975), 235-266.

JS de Jong, T. and van Straten , D. *Disentanglement s,* These Proceedings.

L Lê D.T. *Calculation of the Milnor number of Isolated Singularities of Complete Intersection.* Func. Analysis and Appl. **8** (1974), 127-131.

M Mather, J. *Stability of C^∞-mappings:*
II. Infinitesimal stability implies stability, Ann. of Math. (2) **89** (1969), 254-291;
III. Finitely determined map germs, Inst. Hautes Etudes Sci. Publ. Math. **36** (1968), 127-156;
IV. Classification of stable germs by **R**-*algebras*, Inst. Hautes Etudes Sci. Publ. Math. **37** (1969), 223-248.
V. *Transversality*, Advances in Math. **4** (1970), 301-336.

M2 Mather, J. *How to stratify mappings and jet spaces*, in Singularités d'Applications Differentiables, Plans-sur-Bex, Lectures Notes in Math., vol. 535, Springer-Verlag, Berlin and New York, (1975), 128-176

Ma1 Martinet, J. *Deploiements versels des applications differentiables et classification des applications stables*, Singularités d'Applications Differentiables, Plans-Sur-Bex, Springer Lecture Notes 535 (1975) 1-44.

Ma2 ------ *Deploiements stables des germs de type fini et determination fini des applicationsdifferentiables* (1976), preprint.

MM Marar, W. and Mond, D. *Multiple point schemes for corank 1 maps*, Jour. London Math. Soc. **39** (1989), 553-567

Mo1 Mond, D. *On the classification of of germs of maps from* $\mathbb{R}^2$ *to* $\mathbb{R}^3$, Proc. London Math. Soc. (3) **50** (1985), 333 - 369.

Mo2 Mond, D. *Some remarks on the geometry and classification of germs of maps from surfaces to 3-space*, Topology **26** (1987), 361- 383.

Sa Saito, K. *Theory of logarithmic differential forms and logarithmic vector fields*, J. Fac, Sci. Univ. Tokyo Sect. Math. **27** (1980), 265-291.

T Terao, H. *Discriminant of a holomorphic map and logarithmic vector fields*, J. Fac. Sci. Univ. Tokyo Sect. Math. **30** (1983), 379- 391.

Te Teissier, B. *Cycles évanescents, sections planes, et conditions de Whitney*, Singularités à Cargèse, Asterisque 7,8 (1973), 285–362.

Department of Mathematics
Universty of North Carolina
Chapel Hill
North Carolina 27599
U S A

DIFFERENTIAL FORMS AND HYPERSURFACE SINGULARITIES

Alexandru Dimca

One of the main tools in studying an isolated hypersurface singularity $(X,0) \subset (\mathbb{C}^n,0)$ is the use of the (holomorphic) differential forms in the language of the Gauss–Manin connection [B], [Ma], [G]. This language (in the more refined version coming from the theory of $\mathcal{D}$–modules) has also been used to describe the (mixed) Hodge filtration on the cohomology $H^{n-1}(F)$ of the Milnor fiber of $(X,0)$, see [SS].

In this approach the differential forms are gradually replaced by some more abstract objects and one looses much of the possibility of explicit computations which is usually associated with the differential forms. For instance, one is able in this way to compute the Jordan normal form of the monodromy operator T acting on $H^{n-1}(F)$ see [Sk1] , but one is unable to describe explicit bases for $H^{n-1}(F)$ in terms of differential forms, with the exception of the weighted homogeneous singularities [OS], [D2].

In this paper we try to understand explicitly the cohomology of the complement $B_\varepsilon \setminus X$ of a good representative X for $(X,0)$ in a small open ball B_ε , in terms of differential forms on $B_\varepsilon \setminus X$. This cohomology can be identified essentially to the eigenspace in $H^{n-1}(F)$ corresponding to the eigenvalue 1 of the monodromy operator T and hence our problem is part of the unsolved problem mentionned above.

Due to a theorem of Grothendieck, we can work only with meromorphic forms on B_ε having poles along X . The complex of these meromorphic forms has a natural polar filtration given by the order of poles along X .

This filtration gives rise to a spectral sequence which is the main technical object of interest for us. We discuss various properties of the E_2 and E_3 terms of this spectral sequence and give conditions for degeneracy at these stages.

In the final sections we treat in detail the curve singularities and the $T_{p,q,r}$ surface singularities as well as their double suspensions. This leads to the next remarkable fact. The polar filtration induced on $H^n(B_\varepsilon \setminus X)$ is related to some (naturally associated) Hodge filtration, but in general these two filtrations are different, see (2.5) and (5.4, ii).

As main applications of our technique (the study of the spectral sequence and the explicit description of $H^n(B_\varepsilon \setminus X)$ in terms of differential forms) we mention:

(i) new formulas for the Euler characteristic of the Milnor fiber (and of the associated weighted projective hypersurface) of a weighted homogeneous polynomial with a 1–dimensional singular locus [D2], Prop. (3.19).

(ii) a better understanding of the dependence of the Betti numbers for hypersurfaces in $\mathbb{P}^n$ with isolated singularities on the position of these singularities with respect to some linear systems [D3].

In the present paper we use some of our results in [D2], [D3] and, conversely, we complete and improve some of our results there.

For instance, (3.4) and (3.5) below give larger classes of transversal singularity types for which the Euler characteristic formula in Prop. 3.19 [D2] holds. In the same time, (3.4, ii) shows that it is enough to take in this formula $m = n + 2$ for all these classes of transversal singularities, a fact which is quite important for numerical computations.

However, there are still a lot of provoking open questions, see (2.11), (3.3), (3.6), (4.5) and an obscure relation with some results by Arnold and Varchenko to clarify, see (4.7), (4.9).

ACKNOWLEDGEMENT

I would like to express my gratitude to Professor F. Hirzebruch, and to the Max–Planck–Institut für Mathematik in Bonn for material support and for a really stimulating and pleasant mathematical (and not only!) atmosphere.

§ 1. Topological and MHS preliminaries

Let $X : f = 0$ be an isolated hypersurface singularity at the origin of $\mathbb{C}^n$, with $n \geq 2$. Let $K = X \cap S_\varepsilon$ be the associated link, where $S_\varepsilon = \partial \bar{B}_\varepsilon$ and $B_\varepsilon = \{x \in \mathbb{C}^n; |x| < \varepsilon\}$ for $\varepsilon > 0$ small enough. Recall the well–known result of Milnor [M].

(1.1) PROPOSITION

(i) The pair $(\mathbb{C}^n, X)$ has a conic structure at the origin, i.e. there exists a homeomorphism $(B_\varepsilon, B_\varepsilon \cap X) \simeq C(S_\varepsilon, K)$.

(ii) For $n = 2$, K is a disjoint union of circles S^1, one for each irreducible component of X.

(iii) For $n > 2$, K is a $(n-3)$–connected manifold of dimension $2n-3$.

In this paper we are interested in the next (local) cohomology groups, always with $\mathbb{C}$–coefficients:

$$(1.2) \qquad H_0^k(X) = H^k(X, X\setminus\{0\}) \xleftarrow[\sim]{\partial} \tilde{H}^{k-1}(X\setminus\{0\}) \simeq \tilde{H}^{k-1}(K)$$

$$H^k(B_\varepsilon\setminus X) \simeq H^k(S_\varepsilon\setminus K) \simeq H_{2n-1-k}(S_\varepsilon, K) \xrightarrow[\sim]{\partial} \tilde{H}_{2n-2-k}(K)$$

(all the indicated isomorphisms being straightforward).

There is a Gysin sequence relating these groups

$$(1.3) \quad \dots \longrightarrow H^k(B_\varepsilon\setminus\{0\}) \xrightarrow{j^*} H^k(B_\varepsilon\setminus X) \xrightarrow{R} H^{k-1}(X\setminus\{0\}) \xrightarrow{\delta} H^{k+1}(B_\varepsilon\setminus\{0\}) \longrightarrow$$

where $j : B_\varepsilon\setminus X \longrightarrow B_\varepsilon\setminus\{0\}$ is the inclusion and R is the Poincaré (or Leray) residue map.

In particular, for $n > 2$ we get an isomorphism

$$(1.4) \qquad H^n(B_\varepsilon\setminus X) \xrightarrow[\sim]{R} H^{n-1}(X\setminus\{0\}) = H^{n-1}(K)$$

while for $n = 2$ we get an exact sequence

$$(1.4') \qquad 0 \longrightarrow H^2(B_\varepsilon\setminus X) \xrightarrow{R} H^1(X\setminus\{0\}) \longrightarrow \mathbb{C} \longrightarrow 0 .$$

By the work of Deligne [De], Durfee [Df] and Steenbrink [S3] the cohomology group $H^{n-1}(K)$ has a MHS (mixed Hodge structure) of weight $\geq n$ (i.e. $W_{n-1}H^{n-1}(K) = 0$).

Using (1.2), (1.4) and (1.4′) we may transport this MHS on $H_0^n(X)$ and $H^n(B_\varepsilon\setminus X)$ respectively, such that $J^{-1} : H_0^n(X) \longrightarrow H^{n-1}(K)$ becomes a morphism of type (0,0) while R becomes a morphism of type $(-1,-1)$ as usual [S4].

(1.5) EXAMPLES

(i) Curve singularities $(n = 2)$. Using essentially [Df], Example (3.12) it follows that $H^1(K)$ is in this case pure of type (1,1).

(ii) Surface singularities $(n = 3)$. Let $(\tilde{X},D) \longrightarrow (X,0)$ be the resolution of the singularity $(X,0)$ with exceptional divisor $D = \cup D_i$, D_i smooth and intersecting each other transversally. Then Example (3.13) in [Df] tells that the only (possibly) nonzero Hodge numbers of $H^2(K)$ are the next: $h^{2,2}$ = number of cycles in D and $h^{2,1} = h^{1,2} = \sum_i g(D_i)$, where $g(D_i)$ denotes the genus of the irreducible component D_i of D. In particular, if $\dim H^2(K) = 1$ it follows that the only nonzero Hodge number is $h^{2,2} = 1$. This holds for instance for the $T_{p,q,r}$ surface singularities, defined by the equation

$$f = xyz + x^p + y^q + z^r = 0 \quad \text{for} \quad \frac{1}{p} + \frac{1}{q} + \frac{1}{r} < 1 .$$

Note that by duality [Df], one has for such singularities

$$h^{0,0}(H^1(K)) = h^{2,2}(H^2(K)) = 1 .$$

(iii) $(X,0)$ is weighted homogeneous. In this case $H^{n-1}(K)$ is pure of weight n and the computation of the corresponding Hodge numbers follows from [S1].

Consider next the Milnor fibration associated to f

$$F \longrightarrow S_\varepsilon \setminus K \longrightarrow S^1$$

and the corresponding Wang sequence [M]:

$$(1.6) \qquad 0 \longrightarrow H^{n-1}(S_\varepsilon \setminus K) \longrightarrow H^{n-1}(F) \xrightarrow{T-I} H^{n-1}(F) \longrightarrow H^n(S_\varepsilon \setminus K) \longrightarrow 0$$

where T denotes the monodromy operator.

Now $H^n(S_\varepsilon \setminus K) = H^n(B_\varepsilon \setminus X)$ has a MHS by the above discussion, Steenbrink [S2] and Varchenko [V1] have constructed MHS on $H^{n-1}(F)$ but since T is not a MHS morphism, we cannot use the sequence (1.6) to compute the MHS on $H^n(B_\varepsilon\setminus X)$. However T_s, the semisimple part of T, is a MHS morphism and let $h_\lambda^{p,q}(F)$ denote the (p,q) Hodge number of the sub MHS structure $\ker(T_s-\lambda I)=H^{n-1}(F)_\lambda\subset H^{n-1}(F)$. A slight variation of the sequence (1.6), namely

$$(1.6') \qquad H_0^{n-1}(X)\longrightarrow H_c^{n-1}(F)\xrightarrow{\ j\ } H^{n-1}(F)\longrightarrow H_0^n(X)\longrightarrow 0$$

it is known to be a MHS sequence, see [S3], p. 521. Since $T-I=j\,\mathrm{Var}$, where $\mathrm{Var}: H^{n-1}(F)\longrightarrow H_c^{n-1}(F)$ is the variation map, it follows that any element in $\operatorname{coker} j=H_0^n(X)$ can be represented by some element in $H^{n-1}(F)_1$.

Hence we have the next result

$$(1.7) \qquad h^{p,q}(H_0^n(X))\le h_1^{p,q} \quad \text{for all } p \text{ and } q.$$

(1.8) EXAMPLE

For the $T_{p,q,r}$ surface singularities one has $h_1^{1,1}=h_1^{2,2}=1$ according to [S2], p. 554. Hence it is not true that the inequalities in (1.7) are equalities.

Finally we recall some facts about the double suspension. This is the process of passing from the singularity $X: f=0$ in $\mathbb{C}^n$ to the singularity $\bar{X}:\bar{f}=0$ in $\mathbb{C}^{n+2}$, with

$$\bar{f}=f(x)+t_1^2+t_2^2.$$

Using the Thom–Sebastiani formula for Hodge numbers [SS], it follows that

$$(1.9) \qquad h_\lambda^{p,q}(F)=h_\lambda^{p+1,q+1}(\bar{F})$$

for any p,q and eigenvalue λ of $T=\bar{T}$. Here $\bar{F}$ (resp. $\bar{T}$) denotes the Milnor fiber (resp. monodromy operator) of the singularity $(\bar{X},0)$.

Note that under the identification $H^{n-1}(F)\simeq H^{n+1}(\bar{F})$ one has $T-I=\bar{T}-I$ and hence $\operatorname{coker} j\simeq\operatorname{coker}\bar{j}$, where $\bar{j}$ is the morphism in the sequence (1.6′) corresponding to

$(\overline{X},0)$. In this way we get the next equality

$$(1.10) \qquad h^{p,q}(H_0^n(X)) = h^{p+1,q+1}(H_0^{n+2}(\overline{X})) .$$

In conclusion, all these invariants behave nicely with respect to the double suspension.

§ 2. Definition and first properties of the spectral sequence

Let $\Omega^{\cdot}$ denote the stalk at the origin of the (holomorphic) de Rham complex on $\mathbb{C}^n$. Let $\Omega^{\cdot}_f$ be the localization of the complex $\Omega^{\cdot}$ with respect to the multiplicative system $\{f^s ; s \geq 0\}$.

Since $B_\varepsilon \setminus X$ is a Stein manifold, Grothendieck Theorem (Thm. 2 in [Gk]) and an obvious direct limit argument give the next result.

(2.1) PROPOSITION

$$H^{\cdot}(B_\varepsilon \setminus X) = H^{\cdot}(\Omega^{\cdot}_f).$$

Consider the polar filtration F on $\Omega^{\cdot}_f$ defined as follows:

$$F^s\Omega^j_f = \left\{ \frac{\omega}{f^{j-s}} ; \omega \in \Omega^j \right\} \text{ for } j-s \geq 0 \text{ and}$$

$$F^s\Omega^j_f = 0 \text{ for } j-s < 0 \text{, where } s \in \mathbb{Z} .$$

By the general theory of spectral sequences we get an E_1–spectral sequence $(E_r(X,0),d_r)$ converging to $H^{\cdot}(B_\varepsilon \setminus X)$ and such that

$$E_1^{s,t}(X,0) = H^{s+t}(F^s\Omega^{\cdot}_f/F^{s+1}\Omega^{\cdot}_f) .$$

This E_1–term can be described more explicitly as follows ([D2], Lemma (3.3)).

(2.2) LEMMA

The nonzero terms in $E_1(X,0)$ are the following:

(i) $E_1^{s,0} = \Omega^s$ for $s = 0, \dots, n$;

(ii) $E_1^{s,1} = \Omega^s_X$ for $s = 0, \dots, n-3$, there is an exact sequence $0 \longrightarrow \Omega^{n-2}_X \xrightarrow{u} E_1^{n-2,1} \xrightarrow{v} K_f \longrightarrow 0$ and $E_1^{n-1,1} = \Omega^n / f\,\Omega^n$;

(iii) $E_1^{n-t-1,t} = K_f$, $E_1^{n-t,t} = \Omega^n_X = T_f$ for $t \geq 2$.

Here $\Omega^k_X = \Omega^k / (f\,\Omega^k + df \wedge \Omega^{k-1})$ is the stalk at the origin of the sheaf of k–differential forms on (X,0) [L] and T_f is just a simpler notation for Ω^n_X , recalling the relation with the Tjurina algebra of the singularity f . And K_f is defined by

$$K_f = \{ [\omega] \in \Omega^{n-1}_X \; ; \; df \wedge \omega = f \cdot h \cdot \omega_n \}$$

for some analytic germ $h \in \mathcal{O}_n$ and with $\omega_n = dx_1 \wedge \dots \wedge dx_n$, the standard "volume form". If $M_f = \mathcal{O}_n / J_f$ is the Milnor algebra of the singularity f , $J_f = \left[\frac{\partial f}{\partial x_1}, \dots, \frac{\partial f}{\partial x_n} \right]$ being the Jacobian ideal of f , it is easy to see that one can identify K_f with the ideal $(f)^\perp$ in M_f consisting of the elements annihilated by f in M_f . This identification is given explicitly by

$$K_f \ni [\omega] \longmapsto [h] \in (f)^\perp$$

with ω and h as in the definition of K_f . The morphisms u and v above are given by the next formulas.

$$u([\alpha]) = \left[\frac{df}{f} \wedge \alpha \right] \text{ and } v\left(\left[\frac{\omega}{f} \right] \right) = [\omega] \; .$$

The differentials $d_1^t : E_1^{n-1-t,t} \longrightarrow E^{n-t,t}$ (for $t \geq 2$) can be described easily using this notations, namely

$$d_1^t[\omega] = [d\omega - th\,\omega_n] \tag{2.3}$$

for ω and h as above.

Now we describe ker d_1^1 and coker d_1^1 in more familiar terms.
Let $A = \mathrm{im}(u)$ and note that

$$\tilde{T}_f = E_1^{n-1,1}/d_1^1(A) = \Omega^n/(f\,\Omega^n + df \wedge d\Omega^{n-2})$$

is a μ-dimensional vector space over $\mathbb{C}$, where $\mu = \dim M_f =$ the Milnor number of f, see [Ma], p. 416. Consider now the induced map by d_1^1, namely

$$\tilde{d}_1^1 : K_f = E_1^{n-2,1}/A \longrightarrow E_1^{n-1,1}/d_1^1(A) = \tilde{T}_f .$$

Note that $\tilde{d}_1^1$ is given again by the formula (2.3) with $t = 1$, but the right hand side class is in $\tilde{T}_f$ this time and not in T_f.

(2.4) <u>PROPOSITION</u>

(i) $$E_2^{n-2,1} = \begin{cases} \ker \tilde{d}_1^1 & \text{for} \quad n > 2 \\ \ker \tilde{d}_1^1 \oplus \mathbb{C} < \frac{df}{f} > & \text{for} \quad n = 2 \end{cases}$$

(ii) $$E_2^{n-1,1} = \operatorname{coker} \tilde{d}_1^1 .$$

<u>PROOF</u>

One clearly has $E_2^{n-1,1} = \operatorname{coker} d_1^1 = \operatorname{coker} \tilde{d}_1^1$ and hence we have to prove only the first claim. We treat only the case $n > 2$, the other one being similar.

Note that $E_2^{n-1,1} = \ker d_1^1/B$, with

$$B = \left\{ \left[\frac{df \wedge d\beta}{f} \right] ; \beta \in \Omega^{n-3} \right\} .$$

On the other hand

$$\ker \tilde{d}_1^1 = \frac{\ker d_1^1 + A}{A} = \frac{\ker d_1^1}{A \cap \ker d_1^1} .$$

So it is enough to show that $B = A \cap \ker d_1^1$. Let $\omega = \frac{df}{f} \wedge \alpha$ be in $\ker d_1^1$. Then it follows that $df \wedge d\alpha = f \cdot \gamma$ for some $\gamma \in \Omega^n$. Consider now $d\alpha$ as an element in $H^0(X, d\Omega_X^{n-2})$. The above relation shows that $(d\alpha)_x = 0$ for any $x \in X\setminus\{0\}$ and hence $d\alpha$ has the support contained in $\{0\}$. But the cohomology group $H^0_{\{0\}}(X, d\Omega_X^{n-2})$ is trivial by [L], p. 159 and hence $d\alpha = 0$. Using the exactness of the de Rham complex $(\Omega_X^{\cdot}, d)$ at position $(n-2)$ [L] loc. cit. it follows that $\alpha = d\beta$ and hence $\ker d_1^1 \cap A \subset B$. Since the converse inclusion is trivial, we have got the result.

Again by exactness of de Rham complexes we have that the only possibly nonzero E_2-terms are $E_2^{0,0} = E_2^{0,1} = \mathbb{C}$ and $E_2^{n-1-t,t}$, $E_2^{n-t,t}$ for $t \geq 1$, i.e. our spectral sequence is essentially situated on two semilines: $s+t = n-1$, $t \geq 1$ and $s+t=n$, $t \geq 1$.

Note that on $H^k(B_\varepsilon\setminus X)$ we have now <u>two decreasing filtrations</u>:

(i) the filtration F coming from the polar filtration on $\Omega_f^{\cdot}$, namely

$$F^sH^k(B_\varepsilon\setminus X) = \operatorname{im}\{H^k(F^s\Omega_f^{\cdot}) \longrightarrow H^k(\Omega_f^{\cdot}) = H^k(B_\varepsilon\setminus X)\}$$

(ii) the Hodge filtration F_H which is part of the MHS on $H^k(B_\varepsilon\setminus X)$ coming from the MHS on $H^{k-1}(X\setminus\{0\}) = H^{k-1}(K)$ as explained in the first section (for $k = n$).

(2.5) PROPOSITION

$$F^sH^n(B_\varepsilon\setminus X) \supset F_H^{s+1}H^n(B_\varepsilon\setminus X) \text{ for any } s \text{ and } F^0 = F_H^1 = H^n(B_\varepsilon\setminus X).$$

PROOF

Any isolated hypersurface singularity $(X,0)$ can be put on a projective hypersurface $V \subset \mathbb{P}^n$ of degree N arbitrarily large [B]. Let a be the only singular point of V and such that $(V,a) \simeq (X,0)$. Consider the diagram (we assume $n > 2$ but the case $n = 2$ is similar!)

$$\begin{array}{ccccccc}
H_0^{n-1}(V^*) & \xrightarrow{\delta} & H_a^n(V) & \longrightarrow & H_0^n(V) & \longrightarrow & 0 \\
\uparrow & & \wr\uparrow \delta & & & & \\
R \;\Big|\; \wr & & H^{n-1}(X\setminus\{0\}) & & & & \\
\Big| & & \wr\uparrow R & & & & \\
H^n(U) & \xrightarrow{\rho} & H^n(B_\varepsilon\setminus X) & & & &
\end{array}$$

where $U = \mathbb{P}^n\setminus V$, $V^* = V\setminus\{a\}$, $H_0^n(V)$ denote the primitive cohomology of V and we identify B_ε with a small neighbourhood W of a in $\mathbb{P}^n$ and X with $W \cap V$. For more details see [D3].

For $N = \deg V$ large enough, it is known that $H_0^n(V) = 0$ [Sk2], [D2]. Since the Poincaré residue maps R are both isomorphisms of MHS of type $(-1,-1)$, while the morphisms δ are of type $(0,0)$, it follows that ρ is also a morphism of type $(0,0)$ (in fact ρ is induced by the inclusion $B_\varepsilon\setminus X = W\setminus V \hookrightarrow U$ and hence it is natural to expect type $(0,0)$!). It follows that

$$F_H^{s+1}H^n(B_\varepsilon\setminus X) = \rho\,(F_H^{s+1}H^n(U)) .$$

The cohomology group $H^n(U)$ has also a polar filtration F in addition to its Hodge filtration F_H, see [D2].

Moreover, it is clear that

$$\rho\,(F^sH^n(U)) \subset F^sH^n(B_\varepsilon\setminus X) .$$

The result now follows from the corresponding result for the filtrations F and F_H on $H^\cdot(U)$ proved in [DD].

(2.6) COROLLARY

Any cohomology class in $H^n(B_\varepsilon\setminus X)$ can be represented by a meromorphic n–form having a pole along X of order at most n.

(2.7) REMARK

Y. Karpishpan has given another proof for the inclusion in (2.5) and has also shown that on the other cohomology group $H^{n-1}(B_\varepsilon\setminus X)$ there is a reverse inclusion, see [Ka].

Moreover, on this lower cohomology group

$$H^{n-1}(B_\varepsilon \setminus X) \simeq H^{n-1}_{\{0\}}(\mathbb{C}_X)$$

there is another "Hodge type" filtration

$$H^1_{\{0\}}(d\Omega_X^{n-3}) \subset H^2_{\{0\}}(d\Omega_X^{n-4}) \subset \dots \subset H^{n-1}_{\{0\}}(\mathbb{C}_X)$$

see [L], p. 159.

It seems to be an interesting question to compare this filtration to the polar filtration.

Next we investigate the behaviour of the spectral sequence $(E_r(X),d_r)$ with respect to the double suspension.

First we look at the E_1-term. It is convenient to work with an "approximation" of this term, which forgets the difference between the case $t = 1$ and $t \geq 2$. Namely we define for all $t \in \mathbb{Z}$

$$\hat{E}_1^{n-1-t,t} = K_f \,, \quad \hat{E}_1^{n-t,t} = T_f$$

and let the differential $\hat{d}_1^t : \hat{E}_1^{n-1-t,t} \longrightarrow \hat{E}_1^{n-t,t}$ be given by the formula (2.3).

Let $\hat{\bar{E}}_1^{s,t}$ denote the corresponding spaces for the singularity $\bar{f}$.

Consider also the differential forms

$$\bar{\omega}_2 = dt_1 \wedge dt_2 \text{ and } \gamma = \tfrac{1}{2}(t_1dt_2 - t_2dt_1)\,.$$

Note that one has

$$d\gamma = \bar{\omega}_2 \text{ and } d(t_1^2 + t_2^2) \wedge \gamma = (t_1^2 + t_2^2) \cdot \bar{\omega}_2\,.$$

(2.8) PROPOSITION

The diagram

$$\begin{array}{ccc}
\hat{E}_1^{\,n-1-t\,,\,t} & \xrightarrow{\ \hat{d}_1^{\,t}\ } & \hat{E}_1^{\,n-t,t} \\
\varphi \big\downarrow \wr & & \wr \big\downarrow \psi \\
\hat{\bar{E}}^{\,n-t,t+1} & \xrightarrow{\ \hat{d}_1^{\,t+1}\ } & \hat{\bar{E}}^{\,n+1-t,t+1}
\end{array}$$

with $\varphi(\alpha) = \alpha \wedge \bar{\omega}_2 + (-1)^n \beta \wedge \gamma$ (where β is determined by $df \wedge \alpha = f \cdot \beta$) and $\psi(\varepsilon) = \varepsilon \wedge \bar{\omega}_2$ is commutative for all $t \in \mathbb{Z}$. Moreover φ and ψ are linear isomorphisms.

PROOF

First note that $\varphi(\alpha) \in K_{\bar{f}}$ since $df \wedge \varphi(\alpha) = \bar{f}\beta \wedge \bar{\omega}_2$. The commutativity follows by a direct computation. And φ and ψ are isomorphisms since the Milnor and the Tjurina algebras of f and $\bar{f}$ are isomorphic.

We can next define (for any $t \in \mathbb{N}$) $\hat{E}_2^{\,n-1-t,t} = \ker \hat{d}_1^{\,t}$, $\hat{E}_2^{\,n-t,t} = \operatorname{coker} \hat{d}_1^{\,t}$ and similarly for the singularity $\bar{f}$ the spaces $\hat{\bar{E}}_2^{\,s,t}$. We get from (2.8) a diagram

(2.9)

$$\begin{array}{ccc}
\hat{E}_2^{\,n-1-t\,,\,t} & \xrightarrow{\ \hat{d}_2^{\,t}\ } & \hat{E}_2^{\,n-t+1,t-1} \\
\bar{\varphi} \big\downarrow \wr & & \big\downarrow \wr \bar{\psi} \\
\hat{\bar{E}}_2^{\,n-t,t+1} & \xrightarrow{\ \hat{d}_2^{\,t+1}\ } & \hat{\bar{E}}_2^{\,n-t+2,t}
\end{array}$$

where the isomorphisms $\bar{\varphi}$, $\bar{\psi}$ are induced by φ, ψ and the differentials $\hat{d}_2$ are induced by the differentials d_2 in the spectral sequences $E_r(X)$ and $E_r(\bar{X})$.

(2.10) PROPOSITION

The diagram (2.9) is commutative up–to the factor $(t-1)t^{-1}$ for all $t \geq 2$.

PROOF

For $t \geq 2$, to say that $[\alpha] \in K_f$ is in $\ker d_1^t$ means that (possibly after choosing another representant α of the class $[\alpha]$!)

$$d\left[\frac{\alpha}{f^t}\right] = \frac{\beta}{f^{t-1}} \text{ for some } \beta \in \Omega^n .$$

A direct computation shows that

$$d\left[\frac{\varphi(\alpha)}{\overline{f}^{t+1}}\right] = (1-\tfrac{1}{t})\frac{\beta \wedge \overline{\omega}_2}{\overline{f}^t} + (-1)^n d\left[\frac{\beta \wedge \overline{\gamma}}{t\overline{f}^t}\right].$$

But this clearly implies that

$$\hat{d}_2^{t+1}(\overline{\varphi}(\alpha)) = (1-\tfrac{1}{t})\,\overline{\psi}(\hat{d}_2^{t}(\alpha)) .$$

(2.11) REMARK

Let $\left\{\alpha_i f^{-t_i}\right\}_{i\in I}$ be a basis for $H^{n-1}(B_\varepsilon \backslash X)$. Then it is obvious that the classes $\left\{\varphi(\alpha_i)\overline{f}^{-t_i-1}\right\}_{i\in I}$ form a basis for $H^{n+1}(\overline{B}_\varepsilon \backslash \overline{X})$, where $\overline{B}_\varepsilon = \{\overline{x} \in \mathbb{C}^{n+2}; |\overline{x}| < \varepsilon\}$ is a small ball in $\mathbb{C}^{n+2}$.

The similar statement for the top groups $H^n(B_\varepsilon \backslash X)$ and $H^{n+2}(\overline{B}_\varepsilon \backslash \overline{X})$ is still open, see (5.4, i) below.

(2.12) REMARK

As the referee has pointed out, an explicit connection between our complex $\Omega_f^{\cdot}$ and the complex of $\mathcal{D}$–modules $\Omega^{\cdot}[D]$ considered usually in the study of the Gauss–Manin system (see for instance [SS]) can be established as follows.

The projection $\Omega_f^{\cdot} \longrightarrow \Omega_f^{\cdot}/\Omega^{\cdot}$ is a quasi–isomorphism and the quotient $\Omega_f^{\cdot}/\Omega^{\cdot}$ in turn can be identified (up to a shift) to the quotient $\Omega^{\cdot}[D]/t\Omega^{\cdot}[D]$.

§ 3. Some results on the E_2 and E_3 terms

It was shown in [D2] that the spectral sequence $(E_n(X),d_r)$ degenerates at E_2 if and only if $(X,0)$ is a weighted homogeneous singularity and that in this case everything can be computed quite explicitly.

We assume from now on that this is not the case and hence, according to Saito's Theorem [St] we have $f \notin J_f$.

(3.1) LEMMA

Let $m \subset T_f$ (resp. $m \subset \tilde{T}_f$) denote the subspace corresponding to the classes of differential forms $h\,\omega_n$ with $h \in \mathcal{O}_n$ such that $h(0) = 0$. Then $\operatorname{im}(d_1^t) \subset m$ for any $t \geq 1$. (For $t = 1$ the statement refers of course to $\tilde{d}_1^1$).

PROOF

Let $\alpha \in K_f$. Then the relation $df \wedge \alpha = f \cdot h \cdot \omega_n$ can be written as $D(f) = h \cdot f$ where D is the derivation of $\mathcal{O}_n$ given by

$$D = \sum a_i \frac{\partial}{\partial x_i}$$

where a_i are the coefficients of the monomials $dx_1 \wedge \ldots \wedge \widehat{dx_i} \wedge \ldots \wedge dx_n$ in α (with suitable signs). To prove that $d_1^t[\alpha] \in m$ it is enough to show that

$$\operatorname{Trace}(D) = \sum \frac{\partial a_i}{\partial x_i}(0) = 0$$

since $h(0) = 0$ by Saito's Theorem. When $\operatorname{ord}(f) \geq 3$ this follows directly from [SW]. When $\operatorname{ord}(f) = 2$ we can write by the Splitting Lemma

$f = g(u_1, \ldots, u_k) + u_{k+1}^2 + \ldots + u_n^2$ with $\operatorname{ord}(g) \geq 3$. Then any element from K_f (thought as a derivation) may be obtained as follows. Let $\tilde{D}$ be a derivation of $\mathbb{C}\{u_1, \ldots, u_k\}$ such that $\tilde{D}(g) = h \cdot g$. Then the derivation

$$D = \tilde{D} + \frac{h}{2} \sum_{j=k+1,n} u_j \frac{\partial}{\partial u_j}$$

satisfies $D(f) = h \cdot f$ and $\mathrm{Trace}(D) = \mathrm{Trace}(\tilde{D}) = 0$. To see that the correspondence $\tilde{D} \longrightarrow D$ sets up an isomorphism $K_g \longrightarrow K_f$, recall the identification $K_f \simeq (f)^{\perp}$.

(3.2) <u>LEMMA</u>

For $t >> 0$ one has

$$\dim E_2^{n-1-t,t} = \dim E_2^{n-t,t} \leq \mathrm{codim}((f) + (f)^{\perp})$$

where the codimension is taken with respect to the Milnor algebra M_f.

<u>PROOF</u>

Using the identification $M_f \simeq \Omega^n/df \wedge \Omega^{n-1}$ we have a canonical projection $\rho : M_f \longrightarrow T_f$ with $\ker \rho = (f)$. Recall the identification $K_f \simeq (f)^{\perp}$ and let $\bar{K} = \rho((f)^{\perp})$.

Let $\bar{S}$ be a complement of the vector subspace $\bar{K}$ in T_f. And let $(f)^{\perp} = ((f)^{\perp} \cap (f)) + L$ be a direct sum decomposition of $(f)^{\perp}$.

Then $\dim L = \dim \bar{K} = \ell$.

For $t \geq 2$, the differential $d_1^t : (f)^{\perp} \longrightarrow T_f$ has a block decomposition (corresponding to the above decompositions) of the form

$$d_1^t \sim \left[\begin{array}{c|c} A - tI & B \\ \hline C & D \end{array}\right]$$

where A is an $\ell \times \ell$ – matrix . For $e_1, \dots, e_\ell$ a basis for L we let $\rho(e_1), \dots, \rho(e_\ell)$ be a basis for $\bar{K}$ and that is why the identity matrix I occurs above.

It is clear that for $t >> 0$, the matrix $A_t = A - tI$ is invertible and hence $\mathrm{rank}\, d_1^t \geq \ell$, which is equivalent to our claim.

(3.3) QUESTION

With the above notations it is easy to see that $\operatorname{rank} d_1^t = \ell$ for all $t >> 0$ if and only if $D = 0$ and $CA^kB = 0$ for all $k \geq 0$. Are these conditions satisfied for any singularity f? (These conditions say essentially that the behaviour at infinity of the E_1–term is as simple as possible).

(3.4) PROPOSITION

The next statements are equivalent.

(i) The E_3–term of the spectral sequence $E_r(X)$ is finite (i.e. has finitely many non zero entries).

(ii) $E_3^{n-t,t} = 0$ for $t > n$ and $E_3^{n-1-t,t} = 0$ for $t > n+1$

(iii) $f^2 \in J_f$ and $\operatorname{rank} d_1^t = 2\tau - \mu$ for all $t >> 0$, where $\tau = \tau(f) =$ the Tjurina number of f and $\mu = \mu(f) =$ the Milnor number of f.

PROOF

(i) ⇒ (iii). If $f^2 \notin J_f$, then one has

$$\operatorname{codim}((f) + (f)^\perp) < \operatorname{codim}(f)^\perp = \mu - \tau .$$

Hence for $t >> 0$ one has $\dim E_2^{n-1-t,t} < \mu - \tau$. Let $V \subset \mathbb{P}^n$ be a projective hypersurface having just one singular point a and such that $(V,a) \simeq (X,0)$. Then the spectral sequence associated to V has a finite E_3 – term by (i) and Theorem 3.9 in [D2]. Using the computation of the Euler characteristic of V as in the proof of (3.19) [D2], one gets

$$\dim E_2^{n-1-t,t} + \dim E_1^{n-t,t-1} = \mu$$

for all $t >> 0$. This is a contradiction since $\dim E_1^{n-t,t-1} = \tau$.

In the same way one gets a contradiction if $\operatorname{rank} d_1^t > 2\tau - \mu$ for $t >> 0$. Note that $\operatorname{rank} d_1^t$ becomes constant for $t >> 0$ and the case $\operatorname{rank} d_1^t < 2\tau - \mu$ is excluded by (3.2).

(iii) ⇒ (i) Recall the notations from the proof of (3.2). Let $S \subset M_f$ be a vector subspace such that $\rho(S) = \overline{S}$ and $S + (f)^{\perp} = M_f$ is a direct sum. We may think of B as a linear map $(f) \longrightarrow \overline{K}$ and of A_t as a linear map $L \longrightarrow \overline{K}$. Then $\ker d_1^t = < u - A_t^{-1}Bu \,;\ u \in (f) >$. It is clear that $\lim A_t^{-1}Bu = 0$ for $t \longrightarrow \infty$ and hence $\ker d_1^t$ converges to (f) in the corresponding grassmannian.

We can identify S with $\ker d_1^t$ via the obvious maps

$$S \ni a \longmapsto a \cdot f \in (f) \longmapsto af - A_t^{-1}B(af) \in \ker d_1^t .$$

And the composition

$$S \hookrightarrow M_f \xrightarrow{\ \ell\ } T_f \longrightarrow \operatorname{coker} d_1^t$$

gives again an isomorphism.

Via these two isomorphisms we regard d_2^t as an endomorphism of S. This endomorphism can be described explicitly as follows: $d_1^t(af - A_t^{-1}B(af)) = 0$ means that $(af - A_t^{-1}B(af)) \cdot \omega_n = df \wedge \alpha$ and $d\alpha - tA_t^{-1}B(af) \cdot \omega_n = df \wedge \eta + \lambda f \omega_n$ for some $\alpha, \eta \in \Omega^{n-1}$ and $\lambda \in \mathcal{O}_n$. But then one has

$$d\left[\frac{\alpha}{f^t}\right] = \frac{\lambda - ta}{f^{t-1}}\,\omega_n - \frac{d\beta}{(t-1)f^{t-1}} + d\left[\frac{\beta}{(t-1)f^{t-1}}\right].$$

It follows that $d_2^t : S \longrightarrow S$ has a matrix of the next form

$$-tI + P + (t-1)^{-1}Q$$

for some fixed matrices P and Q.

From this formula it is clear that d_2^t is an isomorphism for $t >> 0$ and hence the E_3–term is finite.

(i) & (iii) ⇒ (ii) Let $s = \max\{t, d_2^t$ is not an isomorphism$\}$. Using the projectivization V as above we get $\dim E_2^{n-1-s,s} = \mu - \tau$. Note that $\operatorname{rank} d_1^t \le 2\tau - \mu$ for all t. It follows that $\dim E_2^{n+1-s,s-1} \ge \mu - \tau$. Since d_2^s is not an isomorphism, it follows that $E_3^{n+1-s,s-1} \neq 0$.

But one clearly has $E_3^{n+1-s,s-1} = E_\infty^{n+1-s,s-1}$ by the definition of s.

Hence $E_\infty^{n+1-s,s-1} \neq 0$ which is possible according to Proposition (2.5) only for $s-1 \leq n$. Finally (ii) $\Rightarrow$ (i) is obvious and this ends the proof.

(3.5) EXAMPLES

(i) Singularities f with $\mu - \tau = 1$.

The ideal (f) in M_f is 1–dimensional and $f^2 \in J_f$. Moreover $\operatorname{rank} d_1^t = \tau - 1$ by (3.1) and (3.2) for $t >> 0$ and hence all these singularities fulfill the condition (iii) in (3.4).

(ii) Semiweighted homogeneous singularities of the form $f = f_0 + f'$ with f_0 weighted homogeneous of type $(w_1, \dots, w_n; N)$ (and defining an isolated singularity at the origin) and f' containing only monomials of degree $> \max(N, (n-1)N - 2\Sigma w_i)$ with respect to the given weights $\underline{w} = (w_1, \dots, w_n)$.

Consider the usual filtration G on $\Omega^{\cdot}$ given by $\deg(x_i) = \deg(d\,x_i) = w_i$ and note that there are induced filtrations G on K_f and T_f. The differentials d_1^t are all compatible with these filtrations G.

A more subtle point is that the identification $K_f \simeq (f)^{\perp}$ is compatible with the filtrations, if we consider $(f)^{\perp} \subset M_f = \Omega^n / df \wedge \Omega^{n-1}$ with the filtration induced by that on Ω^n. This follows from the fact that the morphism

$$\theta = df \wedge : \Omega^{n-1} \longrightarrow \Omega^n$$

is strictly compatible with the filtration G, i.e. $\theta(G^s\Omega^{n-1}) = G^{s+N}\Omega^n \cap \operatorname{im}\theta$. This result is mentioned in [AGV], p. 211–212 and can be easily proved.

Recall now that the hessian of f, namely

$$\operatorname{hess}(f) = \det \left[\frac{\partial^2 f}{\partial x_i \partial x_j}\right]_{i,j=1,n}$$

generates the minimal ideal in M_f [AGV], p. 102. Clearly $\operatorname{hess}(f)$ has filtration order $\operatorname{ord}(\operatorname{hess}(f))$ exactly $nN - \Sigma w_i$. Recalling the notations from the proof of (3.4), it follows that S can be generated by elements with order $\leq \operatorname{ord}(\operatorname{hess}(f)) - N$.

Note that $\rho : M_f \longrightarrow T_f$ induces an isomorphism at the graded pieces

$G^sM_f/G^{s+1}M_f \longrightarrow G^sT_f/G^{s+1}T_f$ for $s \leq \mathrm{ord}(\mathrm{hess}(f)) - N$ (use the restriction on f'!).
It follows that

$$\dim \operatorname{coker} d_1^t \geq \dim S = \mu - \tau \text{ for all } t \geq 2 .$$

Since for $t >> 0$, one has also the converse inequality by (3.2), it follows that these singularities f satisfy the second condition in (iii) in (3.4). The first condition i.e. $f^2 \in J_f$ follows again from the assumption on f'.

(iii) Curve singularities with Newton nondegenerate equations

The condition $f^2 \in J_f$ follows now from the Briançon–Skoda Theorem [BS]. And the argument in (ii) above based on filtrations can be repeated since in this case the morphism θ is strictly compatible with the Newton filtrations on $\Omega^\cdot$ by Kouchnirenko results [K], Thm. 4.1. ii.

(iv) Singularities with $\mu - \tau = 2$ and $d_1^t(m^2 \cap (f)^\perp) \subset m^2T_f$, where m denotes the maximal ideal in M_f.

These singularities satisfy $(f)^\perp \supset m^2$ (in particular $f^2 \in J_f$) and an argument similar (and simpler) to that in (ii) shows that they fulfill the condition (iii) in (3.4).

However, note that the apparently natural condition on d_1^t above is not satisfied by all the singularities. It fails for instance for the bimodal singularities

$$Q_{k,i} : f = x^3 + yz^2 + x^2y^k + by^{3k+i}$$

with $k > 1$, $i > 0$ and $b = b_0 + b_1y + \dots + b_{k-1}y^{k-1}$ where $b_0 \neq 0$. To see this, one can use the relations among f, $\frac{\partial f}{\partial x}$, $\frac{\partial f}{\partial y}$ and $\frac{\partial f}{\partial z}$ listed by Scherk in [Sk1], p. 75.

(3.6) Remark

It was recently shown by the author [D4] that the spectral sequence $(E_r(X),d_r)$ degenerates at a finite step $s(X)$ for any hypersurface singularity (even nonisolated!). It is however an interesting open problem to give nice estimates for the number $s(X)$. For instance we conjecture that $s(X) \leq n + 1$.

§ 4 Plane curve singularities and their double suspensions

We consider in this section isolated curve singularities $X : f = f_1 \dots f_p = 0$ in $\mathbb{C}^2$ having p branches.

(4.1) PROPOSITION

(i) $$H^1(B_\varepsilon \setminus X) = \mathbb{C}\left\langle \frac{df_1}{f_1}, \dots, \frac{df_p}{f_p} \right\rangle$$

(ii) $H^2(B_\varepsilon \setminus X) = \mathbb{C} < \omega_1, \dots, \omega_{p-1} >$ where $\omega_i = df_i \wedge df_{i+1} / f_i f_{i+1}$ for $i = 1, \dots, p-1$.

PROOF

(i) Let $H : y_1 \dots y_p = 0$ be the union of the coordinate hyperplanes in $\mathbb{C}^p$ and let $\tilde{f} = (f_1, \dots, f_p) : B_\varepsilon \setminus X \longrightarrow \mathbb{C}^p \setminus H$ be the obvious map. It is known that

$$H^1(\mathbb{C}^p \setminus H) = \mathbb{C}\left\langle \frac{dy_1}{y_1}, \dots, \frac{dy_p}{y_p} \right\rangle$$

and that the induced map

$$H_1(\tilde{f}) : H_1(B_\varepsilon \setminus X) \longrightarrow H_1(\mathbb{C}^p \setminus H)$$

is an epimorphism (for the corresponding statement at π_1-level see if necessary [D1], Lemma (2.2)).

Since these two homology groups have the same rank p (use (1.2!) it follows that $H_1(\tilde{f})$ and $H^1(\tilde{f})$ are isomorphisms.

(ii) By (1.2) we know that $b_2(B_\varepsilon \setminus X) = p - 1$ (b_2 being the second Betti number) and hence it is enough to show that $\omega_1, \dots, \omega_{p-1}$ are linearly independent.

By (1.4′) it is enough to show that $R\omega_1, \dots, R\omega_{p-1}$ are linearly independent. For each branch $X_i : f_i = 0$ choose a normalization $\varphi_i : (\tilde{X}_i, 0) = (\mathbb{C}, 0) \longrightarrow (X_i, 0)$ and note that

$$\varphi = \coprod_i \varphi_i : \coprod_i (\tilde{X}_i \setminus \{0\}) \longrightarrow \coprod_i X_i \setminus \{0\} = X \setminus \{0\}$$

is a homeomorphism.

Hence we get an identification

$$H^1(X\setminus\{0\}) \xrightarrow{\varphi^*} \oplus H^1(\tilde{X}_i\setminus\{0\}) = \mathbb{C}^p .$$

Let us compute $\varphi^* R(\omega_i) = (a_1, \dots, a_p) \in \mathbb{C}^p$. When computing the component a_j one can replace the Poincaré residue map R (along $X\setminus\{0\}$) with the Poincaré residue map R_j (along $X_j\setminus\{0\}$) and this gives $a_j = \varphi_j^* R_j(\omega_i)$.

It follows that $a_j = 0$ for $j \neq i, i+1$ and $a_i = -a_{i+1} = (X_i,X_{i+1})_0$ = the intersection multiplicity of the branches X_i and X_{i+1}. Indeed

$$a_i = \varphi_i^* \left[\frac{df_{i+1}}{f_{i+1}}\right] = m \left[\frac{dt}{t}\right]$$

if $f_{i+1}(\varphi_i(t))$ has order m in t. But this order m is precisely $(X_i,X_{i+1})_0$, see for instance [BK], p. 411.

From this computation it follows that $\varphi^*(R(\omega_i))$ for $i = 1, \dots, p-1$ are linearly independent and this ends the proof.

(4.2) COROLLARY

The nonzero terms of the limit E_∞ of the spectral sequence $E_r(X)$ associated to the plane curve singularity $(X,0)$ are the following: $E_\infty^{0,0} = \mathbb{C}\langle 1 \rangle$, $E_\infty^{0,1} = \mathbb{C}\left\langle \frac{df_1}{f_1}, \dots, \frac{df_p}{f_p} \right\rangle$ and $E_\infty^{1,1} = \mathbb{C}\langle \omega_1, \dots, \omega_{p-1} \rangle$.

(4.3) COROLLARY (compare to (3.4)).

For plane curve singularities $(X,0)$ the next two statements are equivalent:

(i) The spectral sequence $E_r(X)$ degenerates at E_3;

(ii) The E_3–term of the spectral sequence $E_r(X)$ is finite.

PROOF

Clearly we have to show only (ii) $\Rightarrow$ (i). By the proof of (3.4), the condition (ii) implies that $\operatorname{rank} d_1^t \leq 2\tau - \mu$ for any $t \geq 2$.

Let $a_t = \dim E_2^{1-t,t}$, $b_t = \dim E_2^{2-t,t}$ and note that

α. $a_t = b_t \geq \mu - \tau$ for any $t \geq 2$.

β. $a_1 = p$ since $\ker d_1^1 = E_\infty^{0,1}$

γ. $b_1 = \mu - \tau + p - 1$ by (2.4).

Consider the number

$$s = \min\{t \geq 2,\ d_2^t \text{ is not injective}\} \in \mathbb{N} \cup \{\infty\}.$$

If $s = \infty$, i.e. all the differentials d_2^t are injective it is clear that the spectral sequence $E_r(X)$ degenerates at E_3.

If $2 \leq s < \infty$, then it follows using $\alpha.$, $\beta.$ and $\gamma.$ and (4.2) that

$$0 \neq \ker d_2^s = E_3^{1-s,s} = E_\infty^{1-s,s}$$

in contradiction with (4.2).

To investigate the spectral sequence $E_r(X,0)$ for the double suspension of our curve singularity we need the next result.

(4.4) LEMMA

Assume that $(X,0)$ satisfies one of the following conditions:

(a) $\mu - \tau = 1$ or $\mu - \tau = 2$ and $d_1^t(m^2 \cap (f)^\perp) \subset m^2 T_f$;

(b) $(X,0)$ is semi weighted homogeneous;

(c) $(X,0)$ has a Newton nondegenerate equation $f = 0$.

Consider the diagram

$$\begin{array}{cccc}
\tilde{d}_1^1 & \tilde{T}_f = E_1^{1,1} & \longrightarrow\!\!\!\!\rightarrow & E_2^{1,1} \\
K_f & \downarrow & & \downarrow \\
\hat{d}_1^1 & T_f = \hat{E}_1^{1,1} & \longrightarrow\!\!\!\!\rightarrow & \hat{E}_2^{1,1}
\end{array}$$

Then:

(i) The elements $[\omega_1], \dots ,[\omega_{p-1}]$ are linearly independent in $\hat{E}_2^{1,1}$.

(ii) There is a direct sum decomposition

$$\hat{E}_1^{1,1} = \bar{S} + \operatorname{im} \hat{d}_1^1 + \mathbb{C} < \omega_1, \dots ,\omega_{p-1} > .$$

In particular $\dim (\ker \hat{d}_1^1) = \mu - \tau + p - 1$. (The definition of $\bar{S}$ will be given in the proof).

<u>PROOF</u>

(i) We have to show that a relation

$$\Sigma\, c_i \omega_i = \alpha + \frac{df}{f} \wedge \beta + d \left[\frac{\gamma}{f} \right]$$

implies $c_1 = \dots = c_p = 0$.

Taking residue R_j along $X_j \setminus \{0\}$ we get

$$-m_{j-1} c_{j-1} + m_j c_j = 0 \text{ with } m_k = (X_k, X_{k+1})_0 .$$

These relations for $j = 1, \dots ,p-1$ (with $c_0 = m_0 = 0$) clearly give $c_1 = \dots = c_{p-1} = 0$.

(ii) In the case (a) we take $\bar{S} = < 1 >$, resp. $\bar{S} = < 1,\ell >$, with ℓ a generic linear form. In the cases (b) and (c) we take S and $\bar{S}$ as in the proof of (3.4) and in

Examples (3.5. ii, iii).

Note that all the elements in $\bar{S}$ have orders < order(f) , while all the elements ω_i have orders equal to order (f) , since we can write

$$\omega_i = (f_1 \cdots \hat{f}_i \hat{f}_{i+1} \cdots f_p) df_i \wedge df_{i+1}/f .$$

This remark combined with (i) shows that the sum in (ii) is indeed direct.

(4.5) QUESTION

Is it true that $\dim(\ker \hat{d}_1^1) = \mu - \tau + p - 1$ for any plane curve singularity?

Let now $\bar{X} : \bar{f} = 0$ be the double suspension in $\mathbb{C}^4$ of the curve singularity $X : f = 0$ in $\mathbb{C}^2$.

(4.6) PROPOSITION

Assume that (X,0) satisfies one of the conditions in (4.4). Then the spectral sequence $(E_r(\bar{X}), d_r)$ degenerates at E_3 and the limit term E_∞ is described explicitly as follows

$$E_\infty^{0,0} = \mathbb{C} < 1 > , \quad E_\infty^{0,1} = \mathbb{C} \left\langle \frac{d\bar{f}}{\bar{f}} \right\rangle$$

$$E_\infty^{1,2} = \mathbb{C} \left\langle \frac{\varphi(\alpha_1)}{\bar{f}^2}, \dots, \frac{\varphi(\alpha_{p-1})}{\bar{f}^2} \right\rangle \text{ where } \alpha_i = f \cdot \left[\frac{df_i}{f_i}\right] \text{ and}$$

$$E_\infty^{2,2} = \mathbb{C} \left\langle \frac{\psi(\beta_1)}{\bar{f}^2}, \dots, \frac{\psi(\beta_{p-1})}{\bar{f}^2} \right\rangle \text{ where } \beta_i = f \cdot \omega_i .$$

PROOF Use (4.4) and (2.10).

(4.7) REMARK

For $\beta \in \mathbb{C}$ consider the vector space $D(f,\beta) = \Omega^n/(df \wedge d\Omega^{n-2} + K(f,\beta))$ with $K(f,\beta) = \mathbb{C} < d\alpha + \beta(df \wedge \alpha)f^{-1}$; for $\alpha \in K_f >$. These vector spaces were investigated by Arnold [A] and Varchenko [V2], who have evaluated $\dim D(f,\beta)$ in terms of other numerical invariants of the singularity f.

One has clearly an epimorhism $D(f,-t) \longrightarrow E_2^{n-t,t}$ for any positive integer $t \geq 1$. In the curve case one has even an isomorphism

$$D(f,-1) \xrightarrow{\sim} E_2^{1,1}$$

since both vector spaces have dimension $\mu - \tau + p - 1$ by Arnold [A], Varchenko [V2] and our results above (we need only $\dim \ker \tilde{d}_1^1 = p - 1!$).

It follows that for any plane curve singularity f one has

$$(4.8) \qquad (f)\omega_2 \subset df \wedge d\Omega^0 + K(f,-1).$$

The vector spaces $D(f,\beta)$ for $\beta = -p/q$ a (negative) rational number can be related to similar spectral sequences converging to

$$H^{n-1}(F)_\lambda = \ker(T - \lambda I)$$

for $\lambda = \exp(2\pi i p/q)$.

However the deeper relations between these two points of view are not at all clear to the author. In particular, one may ask

(4.9) QUESTION

What is the higher dimensional analogue of (4.8)?

§ 5. $T_{p,q,r}$ – singularities and their double suspensions

Let $X : f = xyz + x^p + y^q + z^r = 0$ $\left[\frac{1}{p} + \frac{1}{q} + \frac{1}{r} < 1\right]$ be a $T_{p,q,r}$ surface singularity. These singularities play an important role in the classification of singularities. They are unimodal in Arnold sense, see [AGV], p. 246 and, on the other hand, they are the surface cusp singularities which embed in codimension 1 [L], p. 17.

They are interesting for us since they (or rather their double suspension) give counterexamples to some "natural" conjectures. All the explicit computations in this section are based on the computations done by Scherk in his thesis [Sk1], p. 53 (when computing the Gauss–Manin connection of a $T_{p,q,r}$ – singularity). It is well–known that

$$\mu = \tau + 1 = p + q + r - 1 \text{ and}$$

$M_f = \mathbb{C} < 1, x, \dots, x^{p-1}, y, \dots, y^{q-1}, z, \dots, z^{r-1}, f >$, $(f)^{\perp} = m =$ the maximal ideal in M_f

$T_f = \mathbb{C} < 1, x, \dots, x^{p-1}, y, \dots, y^{q-1}, z, \dots, z^{r-1} > \omega_3$ with $\omega_3 = dx \wedge dy \wedge dz$.

Let $s = 1 - \frac{1}{p} - \frac{1}{q} - \frac{1}{r}$ and $\lambda = 1 + pqrx^{p-3}y^{q-3}z^{r-3}$. To avoid discussion of some special cases, we assume that $\min(p,q,r) \geq 3$ and then λ is an invertible element in $\mathcal{O}_3$.

In particular, the elements

$$x\lambda, \dots, x^{p-1}\lambda, y\lambda, \dots, y^{q-1}\lambda, z\lambda, \dots, z^{r-1}\lambda, f\lambda$$

give a basis for $(f)^{\perp}$.

Using [Sk1] one may derive the next relations among f , $f_x = \frac{\partial f}{\partial x}$, $f_y = \frac{\partial f}{\partial y}$ and $f_z = \frac{\partial f}{\partial z}$.

$$\begin{aligned}(A_x) : x\lambda f = & \left[\tfrac{1}{p}x^2 + qrsy^{q-2}z^{r-2} + qrx^{p-1}y^{q-3}z^{r-3}\right]f_x + \\ & + \left[\tfrac{1}{q}xy + rsz^{r-1} + prx^{p-2}y^{q-2}z^{r-3}\right]f_y + \\ & + \left[\tfrac{1}{r}xz + sxz + pqx^{p-2}y^{q-3}z^{r-3}\right]f_z\end{aligned}$$

and two similar equations (A_y) and (A_z) obtained from (A_x) by permuting cyclically the letters x,y,z and p,q,r .

And another (even more tedious!) relation

$$\begin{aligned}(B) : \lambda f^2 = & \left[\tfrac{1}{p}xf + \tfrac{s}{p}x^2yz + qrs^2y^{q-1}z^{r-1} + qrsx^{p-1}y^{q-2}z^{r-2} + \right. \\ & \left. + qrx^{p-2}y^{q-3}z^{r-3}f\right]f_x + \left[\tfrac{1}{q}yf + \tfrac{s}{q}xy^2z - rs^2yz^r + \right. \\ & \left. + prsx^{p-2}y^{q-1}z^{r-2} + prx^{p-3}y^{q-2}z^{r-3}f\right]f_y + \\ & + \left[\tfrac{1}{r}zf + \tfrac{s}{r}xyz^2 + s^2xyz^2 + pqsx^{p-2}y^{q-2}z^{r-1} + pqx^{p-3}y^{q-3}z^{r-2}f\right]f_z\end{aligned}$$

It follows from (B) that $\hat{d}_1^t(\lambda f) = 0$ for all $t \geq 1$. Moreover, $\operatorname{im} \hat{d}_1^t \subset m$ by Lemma (3.1) and using (A_x) , (A_y) and (A_z) it follows that

$$\hat{d}_1^t(m^k \cap (f)^{\perp}) \subset m^k \text{ for all } k \geq 1 .$$

We want to show that the map $(f)^{\perp}/(f) \longrightarrow m$ ($m \subset T_f$ the "maximal ideal" as in (3.1)) induced by $\hat{d}_1^t$ is an isomorphism. By the above remark, it is enough to show that the graded pieces

$$d(k) : \frac{m^k \cap (f)^{\perp} + (f)}{m^{k+1} \cap (f)^{\perp} + (f)} \longrightarrow \frac{m^k}{m^{k+1}}$$

are isomorphisms for all $k \geq 1$.

Using (A_x) we get

$$d(k)[\lambda x^k] = (1 + \frac{k}{p} - t)[x^k]$$

for any $k = 1, \ldots, p-1$. Using (A_y) and (A_z) we get similar formulas for $d(k)[\lambda y^k]$ and $d(k)[\lambda z^k]$.

These formulas clearly prove our claim.

Hence $\hat{E}_2^{2-t,t} = \mathbb{C} < \lambda f >$, $\hat{E}_2^{3-t,t} = \mathbb{C} < 1 >$ for all $t \geq 1$.

Using the proof of (3.4) to identify $\tilde{d}_2^t : \mathbb{C} \longrightarrow \mathbb{C}$ to the multiplication with a constant $c(t)$, one can compute

$$c(t) = 3 - 2s - t + \frac{2s^2 - 2s + 1}{t - 1} .$$

In particular $c(t) \neq 0$ for any $t \geq 2$. These computations imply the next result.

(5.1) PROPOSITION

(i) There exists a differential form $\alpha \in K_f$ such that $\tilde{d}_1^1(\alpha) = 0$.

(ii) The spectral sequence $(E_r(X,0), d_r)$ associated to the T_{pqr} surface singularity degenerates at E_3 and the nonzero terms of the limit are $E_\infty^{0,0} = \mathbb{C} < 1 >$, $E_\infty^{0,1} = \mathbb{C}\langle \frac{df}{f} \rangle$, $E_\infty^{1,1} = \mathbb{C}\langle \frac{\alpha}{f} \rangle$ and $E_\infty^{2,1} = \mathbb{C}\langle \frac{xyz\omega_3}{f} \rangle$ with $\omega_3 = dx \wedge dy \wedge dz$.

PROOF

The above computations show that $E_3^{2-t,t} = 0$ for all $t \geq 2$. Since we know that $b_2(B_\varepsilon \setminus X) = 1$ in this case (recall 1.5. ii), it follows that $E_\infty^{1,1} = \mathbb{C} < \frac{\alpha}{f} > = \ker d_1^1$.

Consider now the projection $\sigma : \tilde{T}_f \longrightarrow T_f$ and note that $xyz\omega_3$ generates $\ker \sigma$.

Since $\dim \ker \tilde{d}_1^1 = \dim \ker \hat{d}_1^1 = 1$, it follows that $xyz\omega_3$ is not in $\mathrm{im}(\tilde{d}_1^1)$.

Hence $E_2^{2,1}$ is spanned by the classes ω_3 and $xyz\omega_3$. Since d_2^2 kills ω_3 by the above computation of $c(t)$, it follows that

$$E_3^{2,1} = E_\infty^{2,1} = \mathbb{C} < xyz\omega_3/f > .$$

(5.2) REMARKS

(i) Since $\ker \tilde{d}_1^1 = \ker \hat{d}_1^1$, it follows that the form α which occurs in (5.1) is precisely the 2–form associated in an obvious way to the relation (B) above.

(ii) We would like to stress the fact that the computation of the Gauss–Manin connection for the $T_{p,q,r}$ surface singularities in [Sk1] or [SS] gives no indication on the explicit 3–form generating $H^3(B_\varepsilon \setminus X)$.

Let now $(\bar{X},0) \subset (\mathbb{C}^5,0)$ be the double suspension of the T_{pqr} surface singularity $(X,0)$.

(5.3) PROPOSITION

The spectral sequence $(E_r(\bar{X}),d_r)$ degenerates at E_3 and the nonzero terms of the limit are the following

$$E_\infty^{0,0} = \mathbb{C} < 1 >,\ E_\infty^{0,1} = \mathbb{C} \langle \frac{d\bar{f}}{\bar{f}} \rangle,\ E_\infty^{2,2} = \mathbb{C} \langle \frac{\varphi(\alpha)}{\bar{f}^2} \rangle$$

and $E_\infty^{4,1} = \mathbb{C} \langle \frac{\omega_5}{\bar{f}} \rangle$ with $\omega_5 = dx \wedge dy \wedge dz \wedge dt_1 \wedge dt_2$.

PROOF

Since we know that $\dim E_\infty^{2,2} = 1 = b_4(\overline{B}_\varepsilon\backslash\overline{X})$ by (2.11), it follows that $\tilde{d}_1^1$ is injective and hence $E_2^{4,1} = \mathbb{C}\langle\frac{\omega_5}{f}\rangle$, as coker $\tilde{d}_1^1$ should be 1–dimensional and we use also (3.1).

Next all $E_2^{4-t,t}$ and $E_2^{5-t,t}$ for $t \geq 2$ are 1–dimensional by the above properties of the T_{pqr} surface singularity $(X,0)$ and (2.8).

Using (2.10) it follows that d_2^t are isomorphisms for all $t \geq 3$ and this clearly ends the proof.

(5.4) REMARKS

(i) It is easy to see that one has the next equality of classes in $H^5(\overline{B}_\varepsilon\backslash\overline{X})$:

$$\left[\frac{\omega_5}{f}\right] = -2\left[\frac{1}{p}+\frac{1}{q}+\frac{1}{r}\right]\left[\frac{xyz\omega_5}{f^2}\right].$$

Hence in this case again ψ induces an explicit basis, compare with (2.11).

(ii) Using (5.3) one gets

$$H^5(\overline{B}_\varepsilon\backslash\overline{X}) = F^4H^5(\overline{B}_\varepsilon\backslash\overline{X}) \supsetneq F_H^5H^5(\overline{B}_\varepsilon\backslash\overline{X}) = 0\,.$$

The last equality comes from the fact that $H^5(\overline{B}_\varepsilon\backslash\overline{X})$ has a Hodge structure of type (4,4) by (1.5. ii) and (1.10).

This shows that the inclusions in Prop. 2.5 may be strict and hence the filtration F is a (subtler and more difficult to compute) filtration different from the Hodge filtration $F_H^{\bullet}$ on $H^n(B_\varepsilon\backslash X)$.

See also the paper by Karpishpan [Ka] for a different example.

References

[A] Arnold, V.I.: Normal forms of Poisson structures and of other powers of volume forms, Tr. Sem. I.G. Petrovskogo, 12, 1–15 (1985).

[AGV] Arnold, V.I., Gusein–Zade, S.M., Varchenko, A.N.: Singularities of Differentiable Maps, vol. I, Monographs in Math. 82, Birkhäuser 1985.

[BS] Briançon, J., Skoda, H.: Sur la cloture intégrale d'un idéale de germes de fonctions holomorphes en un point de $\mathbb{C}^n$, C.R. Acad. Sci. Paris 278, 949–951 (1974).

[B] Brieskorn, E.: Die Monodromie der isolierten Singularitäten von Hyperflächen, Manuscripta math. 2, 103–161 (1970).

[BK] Brieskorn, E., Knörrer, H.: Plane Algebraic Curves, Birkhäuser 1986.

[De] Deligne, P.: Theorie de Hodge II, III, Publ. Math. IHES 40, 5–58 (1971) and 44, 5–77 (1974).

[DD] Deligne, P., Dimca, A.: Filtrations de Hodge et par l'ordre du pôle pour les hypersurfaces singulières Ann. Sci. Ec. Norm. Sup. (to appear).

[D1] Dimca, A.: On analytic abelian coverings, Math. Ann. 279, 501–515 (1988).

[D2] Dimca, A.: On the Milnor fibrations of weighted homogeneous polynomials, Compositio Math. (to appear).

[D3] Dimca, A.: Betti numbers of hypersurfaces and defects of linear systems, Duke Math. J. 60, 285–298 (1990).

[D4] Dimca, A.: On the de Rham cohomology of a hypersurface complement, Amer. J. Math. (to appear).

[Df] Durfee, A.H.: Mixed Hodge structures on punctured neighborhoods, Duke Math. J. 50, 1017–1040 (1983).

[G] Greuel, G.–M.: Der Gauss–Manin–Zusammenhang isolierter Singularitäten von vollständigen Durchschnitten, Math. Ann. 214, 235–266 (1975).

[Gk] Grothendieck, A.: On the de Rham cohomology of algebraic varieties, Publ. Math. IHES 29, 351–358 (1966).

[Ka] Karpishpan, Y.: Pole order filtration on the cohomology of algebraic links (preprint MIT).

[K] Kouchnirenko, A.G.: Polyèdres de Newton et nombres de Milnor, Invent. Math. 32, 1–31 (1976).

[L] Looijenga, E.J.N.: Isolated Singular Points on Complete Intersections, London Math. Soc. Lecture Note Series 77, Cambridge Univ. Press, 1984.

[Ma] Malgrange, B.: Intégrales asymptotiques et monodromie, Ann. Sci. Ecole Norm. Sup. 7, 405–430 (1974).

[M] Milnor, J.: Singular Points of Complex Hypersurfaces. Ann. of Math. Studies 61, Princeton Univ. Press, 1968.

[OS] Orlik. P., Solomon, L.: Singularities I, Hypersurfaces with an isolated singularity, Adv. in Math. 27, 256–272 (1978).

[St] Saito, K.: Quasihomogene isolierte Singularitäten von Hyperflächen, Invent. Math. 14, 123–142 (1971).

[SW] Scheja, G., Wiebe, H.: Über Derivationen in isolierten Singularitäten auf vollständigen Durchschnitten, Math. Ann. 225, 161–171 (1977).

[Sk1] Scherk, J.: Isolated Singular Points and the Gauss–Manin Connection, D. Ph. Thesis, Oxford, 1977.

[Sk2] Scherk, J.: On the monodromy theorem for isolated hypersurface singularities, Invent. Math. 58, 289–301 (1980).

[SS] Scherk, J., Steenbrink, J.H.M.: On the mixed Hodge structure on the cohomology of the Milnor fiber, Math. Ann. 271, 641–665 (1985).

[S1] Steenbrink, J.H.M.: Intersection form for quasihomogeneous singularities, Compositio Math. 34, 211–223 (1977).

[S2] Steenbrink, J.H.M.: Mixed Hodge structure on the vanishing cohomology. In:

Holm, P. ed.: Real and Complex Singularities, pp. 525–563, Oslo 1976, Alphen aan de Rijn Sijthoff, Noordhoff 1977.

[S3] Steenbrink, J.H.M.: Mixed Hodge structures associated with isolated singularities, Proc. Symp. Pure Math. 40, Part 2, pp. 513–536 (1983).

[S4] Steenbrink, J.H.M.: Mixed Hodge Structures and Singularities (book to appear).

[V1] Varchenko, A.N.: Asymptotic Hodge structure in the vanishing cohomology, Math. USSR Izv. 18, 469–512 (1982).

[V2] Varchenko, A.N.: Local classification of volume forms in the presence of a hypersurface, Funct. Anal. Appl. 19 (4), 269–276 (1985).

Max–Planck–Institut
für Mathematik
Gottfried–Claren–Str. 26
5300 Bonn 3
FRG

Local Reflexional and Rotational Symmetry in the Plane

P.J.Giblin and F.Tari

§1. Introduction

The local reflectional symmetry of plane curves has been studied in a number of articles [B], [B-G1], [B-G-G], [G], [G-B]. The basic idea is to consider a smooth parametrised curve $\gamma : \mathbf{I} \longrightarrow \mathbf{R}^2$ where $\mathbf{I}$ is an open interval of $\mathbf{R}$ or $\mathbf{I}=S^1$. We normally assume γ is an embedding. We look for pairs of parameter values (t_1, t_2) for which there is some line ℓ with reflexion in ℓ taking $\gamma(t_1)$ and its tangent line to $\gamma(t_2)$ and its tangent line (Fig.1). Thus ℓ is an "infinitesimal axis of symmetry" for γ.

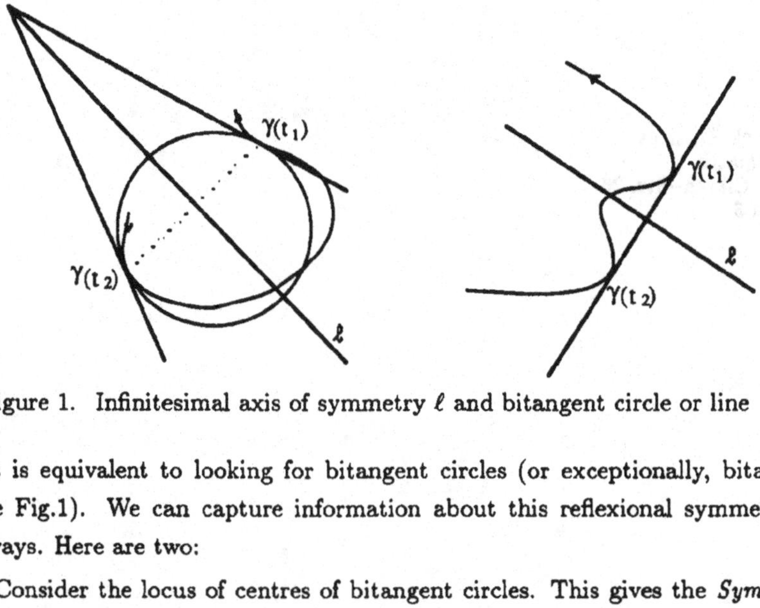

Figure 1. Infinitesimal axis of symmetry ℓ and bitangent circle or line

This is equivalent to looking for bitangent circles (or exceptionally, bitangent lines -see Fig.1). We can capture information about this reflexional symmetry in several ways. Here are two:

(1) Consider the locus of centres of bitangent circles. This gives the *Symmetry Set* (SS) of the curve γ.

(2) Consider the locus of lines ℓ as a set in the dual plane. This gives the dual of the symmetry set (for ℓ in fact is always tangent to the symmetry set at the centre of the circle).

Some information on the symmetry set can be obtained by direct arguments [G-B] but by far the most fruitful approach is to regard it as part of the *full bifurcation set* of the family of the distance-squared functions on γ [B-G1]. That is, we consider the family $F : \mathbf{I} \times \mathbf{R}^2 \longrightarrow \mathbf{R}$ given by $F(t, x) = \|\gamma(t) - x\|^2$, and define the full bifurcation set

$$B(F) = \{x \in \mathbf{R}^2 : F(-,x) \text{ has a degenerate singularity at some } t, \\ \text{or two singularities at } t_1, t_2 \text{ with } F(t_1,x) = F(t_2,x)\}$$

This is precisely the union of the evolute E and the symmetry set SS of γ : $B(F) = E \cup SS$. This situation is ideal in the sense it enables us to apply the theory of versal unfoldings, normal forms etc to the structure of the symmetry set [B-G1].

In this paper we explore the dual of the symmetry set of a plane curve and of 1-parameter families of such by two methods. The first, suggested by Bill Bruce, identifies it as a bifurcation set and the second identifies it as a discriminant: the set of critical values of a map. In both cases we can again apply standard techniques of singularity theory. The second method brings out the connexion with maps $\mathbf{R}^2 \longrightarrow \mathbf{R}^2$ symmetric under reflexion in a line, studied in [B-G2]. We shall mostly state results and give methods and examples. Full details of proofs are in [T].

The first suggestion that the dual of the symmetry set might be interesting came from computer pictures of the symmetry set drawn by Richard Morris. The four-cusped moth, which figures prominently in the list of symmetry sets of 1-parameter families of curves, appeared to have two inflexions. This would make its dual a closed curve with two cusps and four inflexions, that is a "lips" curve but definitely not the lips we get as the apparent coutour of a surface [A2], [B2], [Mc], which has two inflexions. So the hunt was on.

It is natural to look for a corresponding theory to investigate local *rotational* symmetry of plane curves, and we present one possible approach in §3. The basic idea here is to look for centres of local rotational symmetry in the sense of centres C for which there is a rotation about C taking a point $\gamma(t_1)$, together with its tangent line and its *centre of curvature*, to $\gamma(t_2)$ together with its tangent line and its centre of curvature. It turns out that, at least in some cases, the locus of such centres C (which we call the *rotational symmetry set* (RSS)) can be described locally as the set of critical values (discriminant) of a map. It is a very striking fact that the local structure of the rotational symmetry sets, including that in 1-parameter families, closely resembles the local structure of the duals of the symmetry set.

Again we are much indebted here to the inspirational computer pictures of Richard Morris , which constantly suggested new things to prove as well as illustrating those already proved. We include a brief word here about the technique for drawing the rotational symmetry set of a given curve. This technique led to a fast algorithm for drawing the symmetry set.

As detailed in §3 , we are essentially looking for pairs of points (t_1, t_2) on the curve γ at which the curvatures are equal: $\kappa(t_1) = \kappa(t_2)$. A simple construction give the centre C for every such pair. Drawing the graph of curvature (Fig.2) we are looking for all pairs of points at the same level on the graph. This is done by starting at a maximum or minimum (corresponding to a vertex of γ) and working from there, keeping track of when one value, t_1 or t_2, has to turn round because the other has reached another maximum or minimum (Fig.2(*ii*)). Eventually these pairs (t_1, t_2) converge on another maximum or minimum and end with a pair of the form (t_1, t_1). There are also some closed paths in the set $S = \{(t_1, t_2) : \kappa(t_1) = \kappa(t_2)\}$: these have no natural starting point. Such a path is suggested in (Fig.2 (*iii*)).

A somewhat similar idea works for the symmetry set: starting at a vertex of γ we calculate pairs of points of contact of bitangent circles which diverge from this vertex and keep track of them until they converge at another vertex. Again there are exceptional, closed pieces which do not arise this way. Both algorithms have been fully implemented in Fortran by Richard Morris [M].

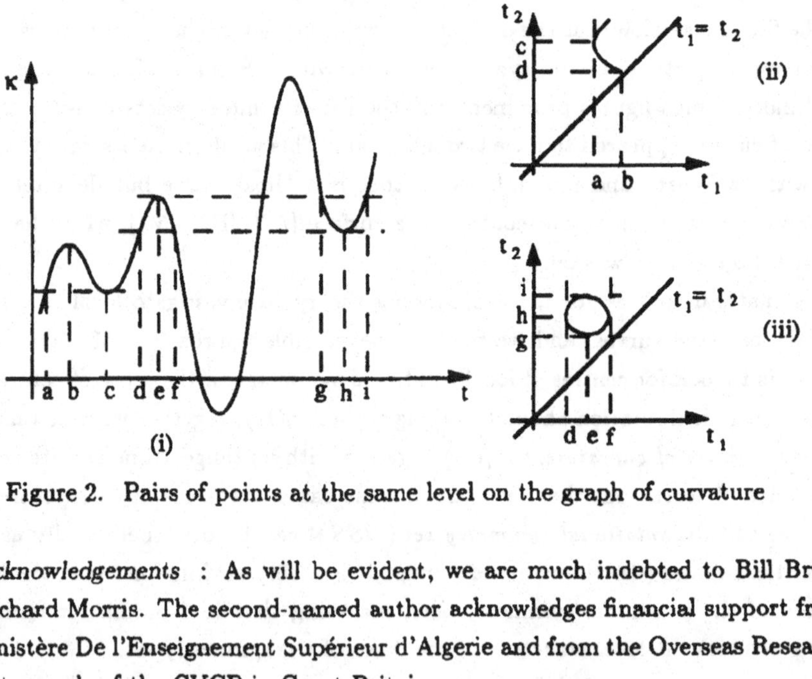

Figure 2. Pairs of points at the same level on the graph of curvature

Acknowledgements : As will be evident, we are much indebted to Bill Bruce and Richard Morris. The second-named author acknowledges financial support from the Ministère De l'Enseignement Supérieur d'Algerie and from the Overseas Research Student awards of the CVCP in Great Britain.

§2. Duals of symmetry sets of plane curves

Given a curve in the plane, at each point there is a tangent line (or several tangent lines). Each tangent line is represented by a point in the dual plane and the set of all these points is called the dual curve. An inflexion on the curve corresponds to a cusp on the dual curve and vice-versa. When the original curve is smooth and generic the dual has the same singularities as generic wave fronts. It also has the same transitions as the propagations of wave fronts when considering generic deformation of the curve. (See [A2] ch.8 for an illustration of these.)

However, the duals of symmetry sets of generic plane curves and families of such curves cannot be deduced from these results; for example the symmetry set in the case of a "biosculating circle" (a circle osculating γ at two points) is either an isolated point or two cusps with the same origin [B-G1, ex.4.4].

In this section two methods of studing the duals of symmetry sets of plane curves and their changes in generic 1-parameter families are described . The first method was suggested by J.W.Bruce. The idea is to define the dual of the symmetry set as the bifurcation set of a germ or a bi-germ of a map $\mathbf{R} \longrightarrow \mathbf{R}^2$ and use stratified Morse functions to describe the changes which occur in generic 1-parameter families of bifurcation sets in the plane [B-G1,§4]. The case of duals of symmetry sets of surfaces is studied in [B-W].

In the second method we define the dual of the symmetry set locally as the discriminant of a map from the plane to the plane or a discriminant of a symmetric map from the plane to the plane. This highlights the connection with the rotational symmety set RSS^- defined in §3.

The dual of the symmetry set as a bifurcation set

For a smooth curve γ, the circles whose centres give the symmetry set are tangent to the curve at two different places or have higher contact with the curve at a single point (Fig.3).

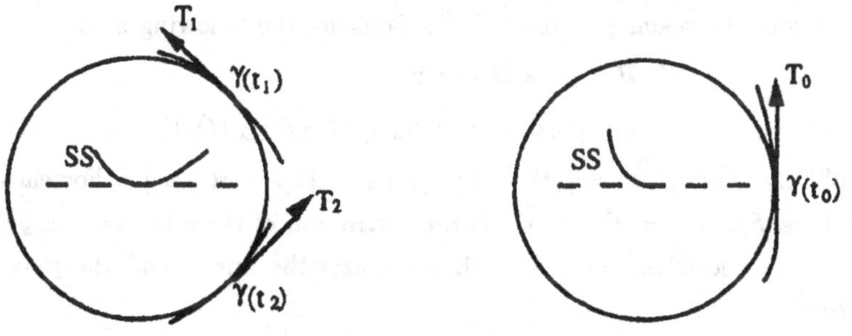

Figure 3.

The tangent to the symmetry set in the first case has the direction of $T_2 - T_1$ (Fig.3,right). When the circle has exactly 4-point contact with the curve the symmetry set has an end point and the (one-sided) tangent line has the direction of the normal to the curve (Fig.3,left). In both cases the tangent line to the symmetry set is an infinitesimal axis of symmetry to the curve (see §1). If the curve is locally folded up, *i.e.*, taken by the map (x, y^2) in the coordinate system with the x-axis the tangent line to the symmetry set and the y-axis the normal to it, the result is two tangential pieces of curve or a single singular curve (Fig.4). These are unstable when considered as bigerms or germs $\mathbf{R} \longrightarrow \mathbf{R}^2$. Thus each line ℓ in the plane can be chosen as an x-axis, and the map $\mathbf{R}^2 \longrightarrow \mathbf{R}^2$ corresponding to $(x,y) \mapsto (x,y^2)$ when ℓ=x-axis can be applied to γ. The dual of the symmetry set consists of those ℓ giving unstable germs or bigerms.

It is easy to visualise the unstable lines in the following example. Let $\gamma(y) = (y^2+y^3, y)$, the reflexion in the x-axis takes γ to (y^2+y^3, y^2) which is a cusp. When we move the axis of reflexion parallel to the x-axis and reflect γ with respect to the line $y = u$, the resulting curve is $(y^2+y^3, (y-u)^2)$. The germ $(y^2+y^3, (y-u)^2)$ is a versal unfolding of the cusp.

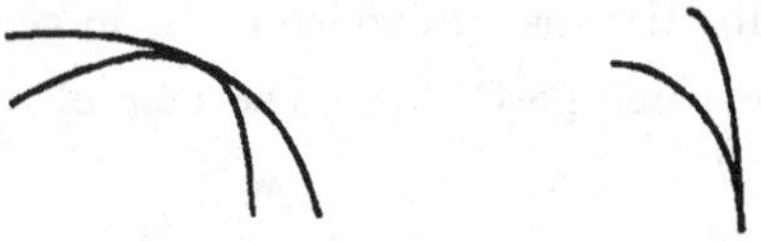

Figure 4.

Let L be the set of all oriented lines in the plane. There is a 1-1 correspondence between L and $S^1 \times \mathbf{R}$ since each line is the set of points x satisfying $x.u = \lambda$ for some $(u,\lambda) \in S^1 \times \mathbf{R}$. Suppose we are given a smooth curve γ and a line $\ell = (u,\lambda)$ (Fig.5). Let $p_{(u,\lambda)}(t)$ and $d_{(u,\lambda)}(t)$ be the orthogonal projection of $\gamma(t)$ on ℓ and the distance of $\gamma(t)$ to ℓ respectively. It is easy to check that :

$d_{(u,\lambda)}(t) = \lambda - \gamma(t).u$

$p_{(u,\lambda)}(t) = \gamma(t) + (\lambda - \gamma(t).u)u,$

where . denotes the scalar product in $\mathbf{R}^2$. Consider the following map

$$F : \mathbf{R} \times S^1 \times \mathbf{R} \longrightarrow \mathbf{R}^2$$

$$(t,u,\lambda) \longmapsto p_{(u,\lambda)}(t) + d^2_{(u,\lambda)}(t).u$$

More explicitly : $F(t,u,\lambda) = \gamma(t) + (\lambda - \gamma(t).u)[1 + (\lambda - \gamma(t).u)]u$. For each (u,λ), $F_{(u,\lambda)}$ (where $F_{(u,\lambda)}(t) = F(t,u,\lambda)$) is the restriction of the fold map (x,y^2) to the curve γ in the coordinate system with the x-axis the line ℓ and the y-axis a line parallel to u.

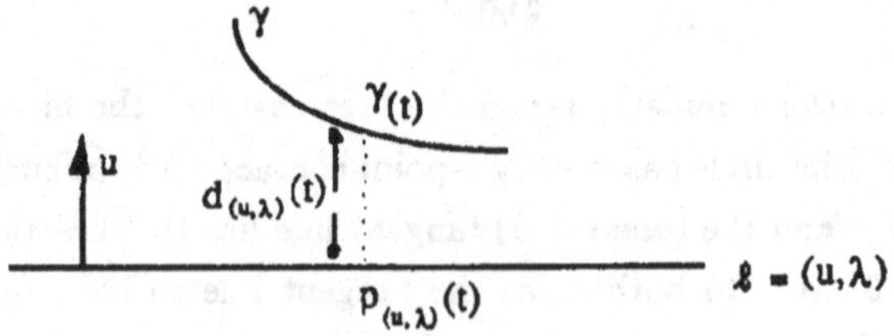

Figure 5.

2.1. Definition : *The bifurcation set of the map F, denoted by $B(F)$, is the set of points (u, λ) where $F_{(u,\lambda)}$ is locally unstable as a map $\mathbf{R} \longrightarrow \mathbf{R}^2$, with respect to smooth changes of coordinates in source and target.*

It is a staightforward exercise to verify the following.

2.2. Proposition : *The bifurcation set of the map F is the union of the dual of the symmetry set and the dual of the evolute of the curve γ.*

In order to study the singularities occurring in the duals of symmetry sets one has to consider the bi-germ (F_1, F_2) associated to the two pieces of curves γ_1 and γ_2 which locally give the symmetry set (Fig.3,right) or the uni-germ F associated to a single piece of curve γ with a vertex (Fig.3,left). For generic 1-parameter families of curves (γ_s) we consider the 'big' family of germs:

$$\tilde{F} : \mathbf{R} \times S^1 \times \mathbf{R} \times \mathbf{R} \longrightarrow \mathbf{R}^2$$
$$(t, u, \lambda, s) \longmapsto \mathbf{F}_s(t, u, \lambda)$$

and the 'big' bifurcation set $B(\tilde{F})$. The individual bifurcation sets are recovered by taking the intersection of $B(\tilde{F})$ with the fibres of generic functions on $B(\tilde{F})$ [A1],[B1],[B-G1].

Following the notation in [B-G1] the cases of interest for the study of the duals of symmetry sets and their 1-parameter families are : inflexion and higher inflexion on the symmetry set, A_2^2, A_3 and the A_4. (Here A_k is Arnold's notation for the singularity of the distance-squared function, and A_2^2 refers to two A_2 singularities at the same level (biosculating circle). Thus A_3 stands for a vertex, and A_4 for a higher vertex, on γ.)

2.3. Proposition : *The dual of the symmetry set undergoes the following transitions in generic 1-parameter families of curves:*

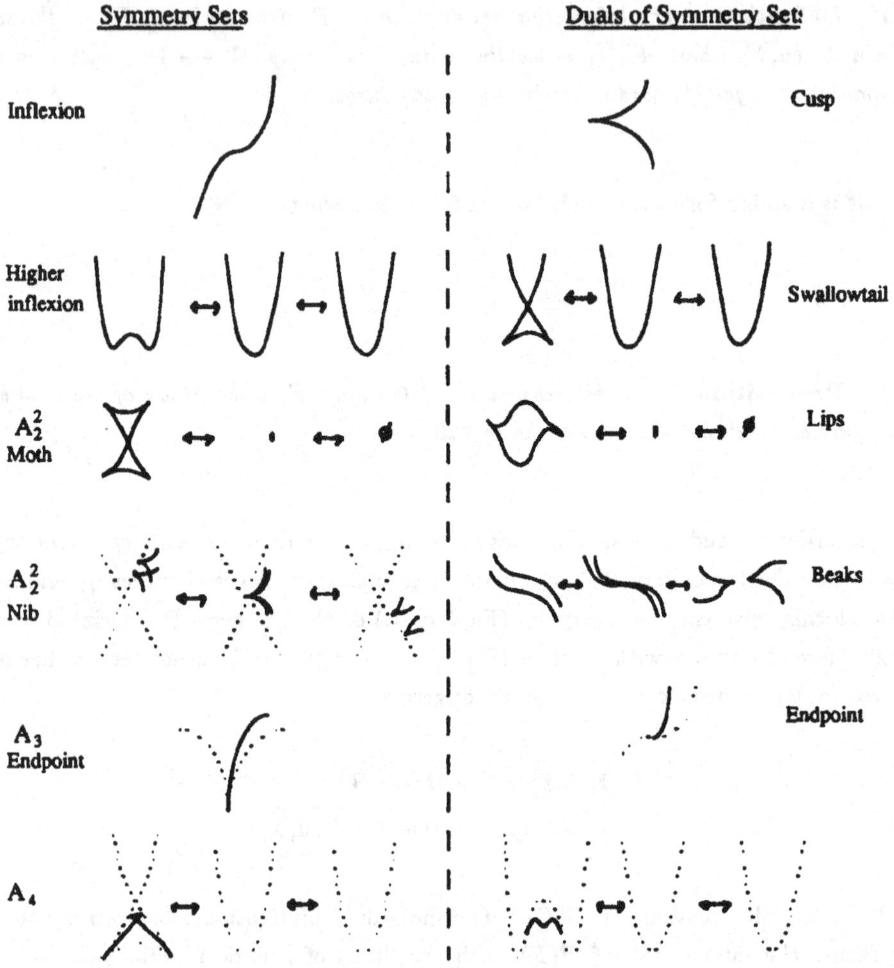

Figure 6. Transitions on 1-parameter families of duals of symmetry sets

2.4. **Remarks :**

1. The information needed to draw these pictures comes by putting together the previously found transitions on symmetry sets [B-G1] and the new information on duals. Thus cusps correspond to inflexions and vice-versa.

2. The lips in the lips-transition on the dual of the symmetry set have four inflexions and a common tangent line at two different points. This contrasts with the generic lips in projections of surfaces [B2] which has two inflexions only.

The dual of the symmetry set as a discriminant

Let γ be a smooth embedded curve and G the following map

$$G : \mathbf{R} \times \mathbf{R} \longrightarrow L$$
$$(t_1, t_2) \longmapsto \ell(t_1, t_2),$$

where $\ell(t_1, t_2)$ is the perpendicular bisector of the segment $[\gamma(t_1), \gamma(t_2)]$ (Fig.7(*i*)). Of course, if $t_1 = t_2$ then ℓ is the normal at $\gamma(t_1)$..For calculations we write a line ℓ in a chosen coordinate system as the set of points (x, y) with $y = ax + b$ and identify ℓ with (a, b) so that $L = \mathbf{R}^2$. (We shall avoid lines parallel to the y-axis in our calculations.)

As in the first method we distinguish two cases :

Case 1.

The symmetry set is locally obtained from two pieces of curve γ_1 and γ_2. We can write $\gamma_1(x) = (f_1(x), x)$ and $\gamma_2(x) = (f_2(x), x)$ (Fig.7(*ii*)).

The line $\ell(x_1, x_2) = (-a(x_1, x_2), b(x_1, x_2))$ can be expressed in terms of f_1, f_2, x_1 and x_2 :

$$a(x_1, x_2) = \frac{f_1(x_1) - f_2(x_2)}{x_1 - x_2},$$
$$b(x_1, x_2) = \tfrac{1}{2}(x_1 + x_2) + \tfrac{1}{2}a(x_1, x_2)(f_1(x_1) + f_2(x_2)).$$

The map germ

$$G : \mathbf{R}^2 \longrightarrow \mathbf{R}^2$$
$$(x_1, x_2) \longmapsto (-a(x_1, x_2), b(x_1, x_2))$$

is smooth.

2.5. Proposition : *The discriminant of the map G is locally the dual of the symmetry set of the curve γ.*

Thus we expect in the codimension ≤ 1 cases the occurrence of stable cusps, swallowtails, lips and beaks transitions [R]. This is indeed the case, as can be seen from Proposition (2.3).

Case 2

The symmetry set here is locally obtained from a neighbourhood of a point on a single curve γ. The point is generically a vertex but higher vertices can occur in 1-parameter families of curves. If we write γ locally in the form $\gamma(x) = (f(x), x)$ where f is a smooth function (Fig.7(*iii*)) then:

$$a(x_1, x_2) = \frac{f(x_1) - f(x_2)}{x_1 - x_2}$$
$$b(x_1, x_2) = \tfrac{1}{2}(x_1 + x_2) + \tfrac{1}{2}a(x_1, x_2)(f(x_1) + f(x_2))$$

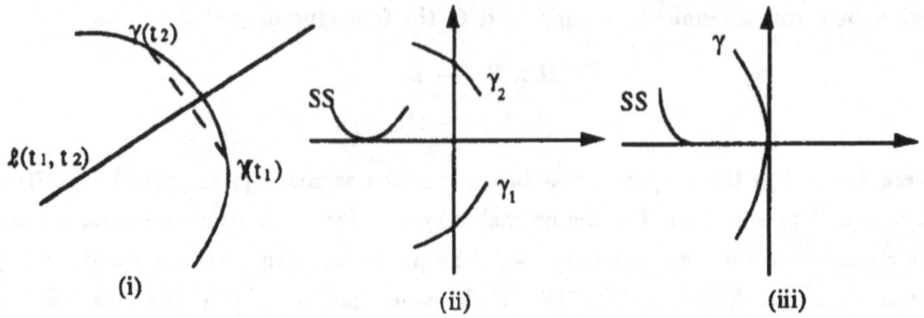

Figure 7.

The map germ

$$G : \mathbf{R}^2 \longrightarrow \mathbf{R}^2$$
$$(x_1, x_2) \longmapsto (-a(x_1, x_2), b(x_1, x_2))$$

is a smooth symmetric map with respect to reflexion in the diagonal $\Delta = \{(x, x), x \in \mathbf{R}\}$.

The symmetric germs $\mathbf{R}^2 \longrightarrow \mathbf{R}^2$ are the invariant germs of the action of $\mathbf{Z}_2$ on the source, where the group $\mathbf{Z}_2$ is generated by reflexion in the diagonal Δ. The equivariant change of coordinates in the source $(x, y) \mapsto (x + y, x - y)$ transforms symmetry with respect to Δ to symmetry with respect to the x-axis.

A classification of invariant germs up to equivariant change of coordinates in the source ($\mathbf{Z}_2$ acting by reflexion in the x-axis), and any change of coordinates in the target, can be deduced from the classification of germs of projections of surfaces with boundary ([B-G2], remarks following Theorem 1.2). All that is needed is to replace in the list of normal forms obtained in ([B-G2], Table 5.1) y by y^2. This yields the following table of normal forms of symmetric map germs of codimension ≤ 1, where we also give the codimension and the name of the corresponding boundary singularity.

2.6. Table of normal forms ($\epsilon = \pm 1$)

Normal form	Name of the boundary singularity	Codimension
I. (x, y^2)	Submersion	0
II. $(x, xy^2 + y^4)$	Semi-fold	0
III. $(x, xy^2 + y^6)$	Semi-cusp	1
IV. $(x, x^2y^2 + \epsilon y^4)$	Semi-lips ($\epsilon = +1$) Semi-beaks ($\epsilon = -1$)	1
V. $(y^2 + x^3, x^2)$	Boundary cusp	1

2.7. Proposition : *(i) The discriminant of the germ G is locally the union of the dual of the symmetry set and the dual of the evolute the curve γ.*

(ii) The germ G is equivalent in the equivariant sense to the germ $(x, xy^2 + y^4)$ at an A_3 on the curve and to $(x, x^2y^2 + y^4)$ at an A_4.

The proof of this proposition is a computational exercise. In (*ii*) it is possible to write the changes of coordinates explicitly.

The germ $(x, xy^2 + y^4)$ is stable and its discriminant is drawn in Fig 8(*i*) .The germ $(x, x^2y^2 + y^4)$ is of codimension 1. Generic sections of the discriminant of its versal unfolding are as in Fig.8(*ii*).

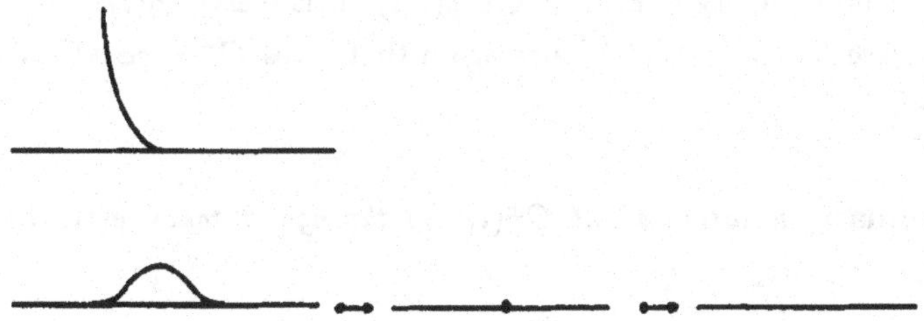

Figure 8.

Taking into account the presence of inflexions on the duals, confirms the pictures of the transitions on the dual of symmetry set and evolute at an A_3 and A_4 shown in Fig.6.

§3. Rotational symmetry sets

The centre maps

Let $\gamma : \mathbf{I} \longrightarrow \mathbf{R}^2$ be, as usual, an embedded smooth curve, where $\mathbf{I}$ is either an open interval of $\mathbf{R}$ or else the unit circle. Consider two points $\gamma(t_1), \gamma(t_2)$ at which the (unit) tangent vectors are $T(t_1)$, $T(t_2)$ respectively. We seek two points $C^{\pm} = C^{\pm}(t_1, t_2)$ which are the centres of rotation taking $\gamma(t_1)$ to $\gamma(t_2)$ and $T(t_1)$ to $\pm T(t_2)$. Hence in each case the tangent *line* at $\gamma(t_1)$ is taken to that at $\gamma(t_2)$. Using the complex numbers $\mathbf{C}$ to parametrize $\mathbf{R}^2$ and writing θ for the angle of rotation, we have

$$\gamma(t_2) - C^{\pm} = e^{i\theta}(\gamma(t_1) - C^{\pm}) \quad \text{and} \quad T_2 = \pm T_1 e^{i\theta}.$$

It follows that the *centre maps* are given by

$$C^{\pm}(t_1, t_2) = \frac{\gamma(t_2)T(t_1) \mp \gamma(t_1)T(t_2)}{T(t_1) \mp T(t_2)} \in \mathbf{C}, \tag{1}$$

provided $T(t_1) \neq \pm T(t_2)$.

An interesting limiting case occurs for C^+ when t_1 and t_2 both tend to the same value t. Then it is not hard to show that $C^+(t_1,t_2) \longrightarrow e(t)$, where $e(t)$ is the centre of curvature of γ at $\gamma(t)$. (Thus $e(t) = \gamma(t) + \frac{1}{\kappa(t)}N(t)$, N being the unit normal and κ the curvature of γ). Thus we define

$$C^+(t,t) = e(t); \tag{2}$$

the resulting C^+ is still smooth. On the other hand there is no point in extending C^+ to parallel tangents, *i.e.*, $T(t_1) = T(t_2)$ and $\gamma(t_1) \neq \gamma(t_2)$, for that merely gives $C^+ = \infty$.

There is no difficulty interpreting $C^-(t,t)$: it is merely $\gamma(t)$. We do not extend C^- to the case $T(t_1) = -T(t_2)$ since again that gives $C^- = \infty$. We can now check the following.

3.1. Lemma : *Rotation about $C^\pm(t_1,t_2)$ through θ takes $e(t_1)$ to $e(t_2)$ if and only if $\kappa(t_1) = \pm\kappa(t_2)$.*

3.2. Definition : *The Rotational Symmetry Set (RSS) consists of two parts, RSS^+ and RSS^-:*

$$RSS^\pm = \{C^\pm(t_1,t_2) : \kappa(t_1) = \pm\kappa(t_2)\},$$

where we use C^+ in the form extended by (2). (See Fig.9 for an example of RSS^+).

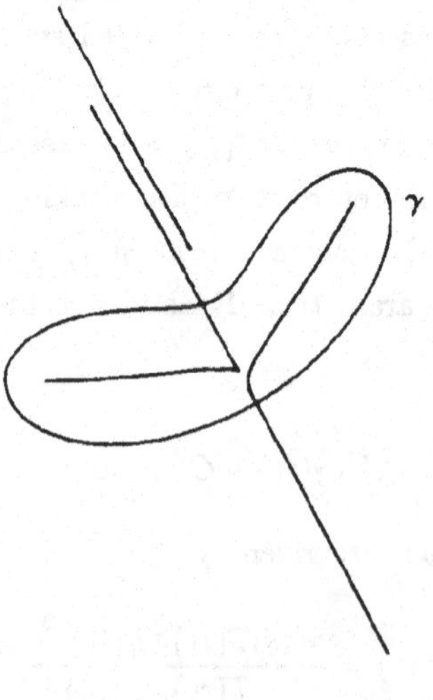

Fig. 9. A rotational symmetry set (RSS^+)

3.3. Remarks :

(1) If $C^+(t_1,t_2) \in RSS^+$ for sequences of points $t_1 \to t$, $t_2 \to t$ then $\gamma(t)$ is a *vertex* of γ : $\kappa'(t) = 0$. Thus RSS^+ contains the centre of curvature at each vertex of γ. Note that the symmetry set (§2 Fig.3) also contains these points.

(2) $C^-(t,t) \in RSS^-$ requires $\kappa(t) = 0$, *i.e.*, γ has an inflexion at $\gamma(t)$, and then $C^-(t,t) = \gamma(t)$. Thus RSS^- contains all the inflexion points of γ.

(3) The angle of rotation θ has been suppressed above. We have not so far attempted to include it in a coherent theory.

For the time being, let us consider RSS^+. Not only is RSS^+ the image by C^+ of $\{(t_1,t_2) : \kappa(t_1) = \kappa(t_2)\}$, but we have the following result. The proof is straightforward.

3.4. Proposition : *The set of critical values of C^+ is precisely $RSS^+ \cup SS \cup E$, the union of RSS^+, the symmetry set and the evolute of γ. Note that the last arises as the image of the diagonal $\{(t,t)\}$ under C^+.*

The simple case where RSS^+ does not cross the symmetry set or the evolute

Suppose that $p_0 = C^+(t_1^0,t_2^0) \in RSS^+$. The situation is considerably simplified when $p_0 \notin SS$ and $p_0 \notin E$ (so $t_1^0 \neq t_2^0$), for then RSS^+ is locally the critical locus of C^+ : a map from the plane to the plane. In fact it is not hard to find the conditions for the map C^+ to have a fold, cusp, swallowtail, lips and beaks singularity at p_0. (For terminology, see [R].) These are expressed in terms of the successive derivatives of κ (with respect to arclength) at the two points t_1^0 and t_2^0 as follows.

Fold conditions : $\kappa(t_1^0) = \kappa(t_2^0)$, $\kappa'(t_1^0) \neq \kappa'(t_2^0)$.

Cusp conditions : $\kappa(t_1^0) = \kappa(t_2^0)$, $\kappa'(t_1^0) = \kappa'(t_2^0)$, $\kappa''(t_1^0) \neq \kappa''(t_2^0)$.

Swallowtail conditions : $\kappa(t_1^0) = \kappa(t_2^0)$, $\kappa'(t_1^0) = \kappa'(t_2^0)$, $\kappa''(t_1^0) = \kappa''(t_2^0)$, $\kappa'''(t_1^0) \neq \kappa'''(t_2^0)$.

Lips conditions : $\kappa(t_1^0) = \kappa(t_2^0)$, $\kappa'(t_1^0) = \kappa'(t_2^0) = 0$, $\kappa''(t_1^0)\kappa''(t_2^0) < 0$.

Beaks conditions : $\kappa(t_1^0) = \kappa(t_2^0)$, $\kappa'(t_1^0) = \kappa'(t_2^0) = 0$, $\kappa''(t_1^0)\kappa''(t_2^0) > 0$.

For the last two cases there is an additional condition needed to ensure that the map C^+ is equivalent to the normal form of the swallowtail.

Consider the third case. A family of smooth curves $\{\gamma_u\}$ with $\gamma_0 = \gamma$ will give rise to a family of maps $\{C_u^+\}$ and C_0^+ has swallowtail singularity. By writing down the condition for the family $\{C_u^+\}$ to versally unfold this singularity we find that, for a generic family $\{\gamma_u\}$, the RSS^+ does indeed undergo a swallowtail transition (Fig.10).

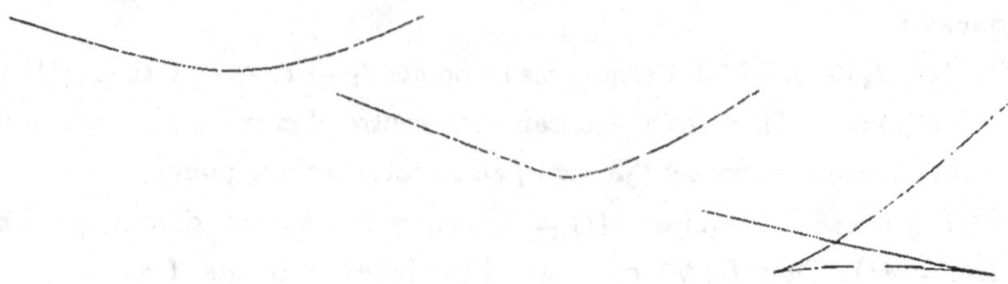

Fig. 10. Swallowtail transition on rotational symmetry set.

Computer pictures.

We continue to suppose $p_0 \notin SS$, $p_0 \notin E$ as above. It is easier to see what is going on geometrically in the case of a lips or beaks transition on RSS^+ for a family of plane curves. Here, the set $S = \{(t_1, t_2) : \kappa(t_1) = \kappa(t_2)\}$ is, at the moment of transition, itself singular, and it undergoes a Morse transition.

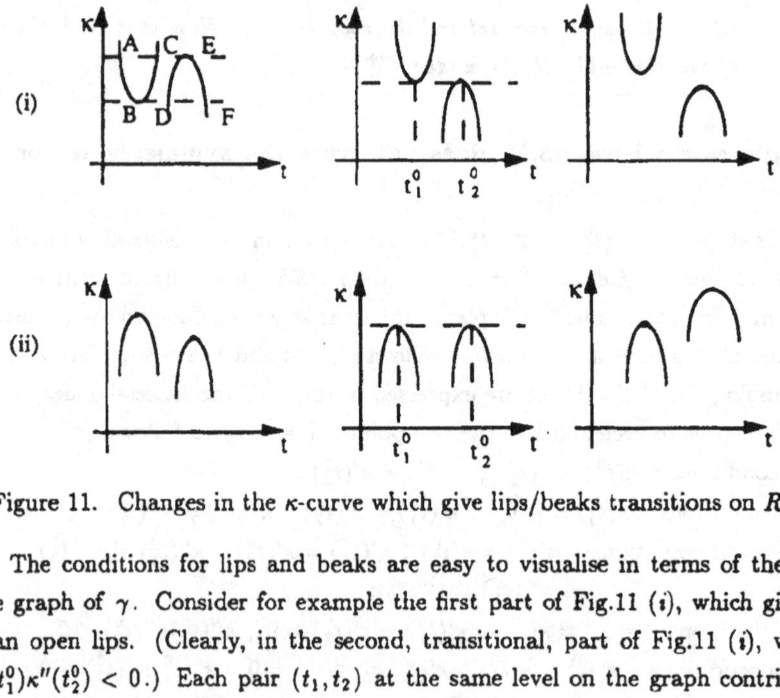

Figure 11. Changes in the κ-curve which give lips/beaks transitions on RSS^+

The conditions for lips and beaks are easy to visualise in terms of the curvature graph of γ. Consider for example the first part of Fig.11 (i), which gives rise to an open lips. (Clearly, in the second, transitional, part of Fig.11 (i), we have $\kappa''(t_1^0)\kappa''(t_2^0) < 0$.) Each pair (t_1, t_2) at the same level on the graph contributes a point $C^+(t_1, t_2)$ to RSS^+. For instance the points on the arc AB of the graph are paired with those on the arc EF. Looking at $\kappa'(t)$ along AB (it goes from < 0 to 0) and along EF (it goes from 0 to < 0), it is clear that there will be some pair of points (t_1, t_2) where $\kappa'(t_1) = \kappa'(t_2)$, and this gives a cusp on RSS^+ (see the conditions above). Similarly BC and DE give a cusp somewhere; for sufficiently small perturbations from the transition state there will be just two cusps altogether. On the other hand pairs B,D; B,F; A,E; C,E all give *inflexions* on RSS^+; the inflexion condition is $\kappa'(t_1)$ or $\kappa'(t_2) = 0$. Thus RSS^+ has four inflexions and two cusps. Again, checking versality conditions, we find that a generic family of curves passing through a situation satisfying the lips/beaks conditions above does give a lips/beaks transition on RSS^+, but with the unusual feature of four inflexions on the lips. Notice that this is similar to the situation for the duals of symmety sets §2. (See Fig.12 and compare with Fig.6.)

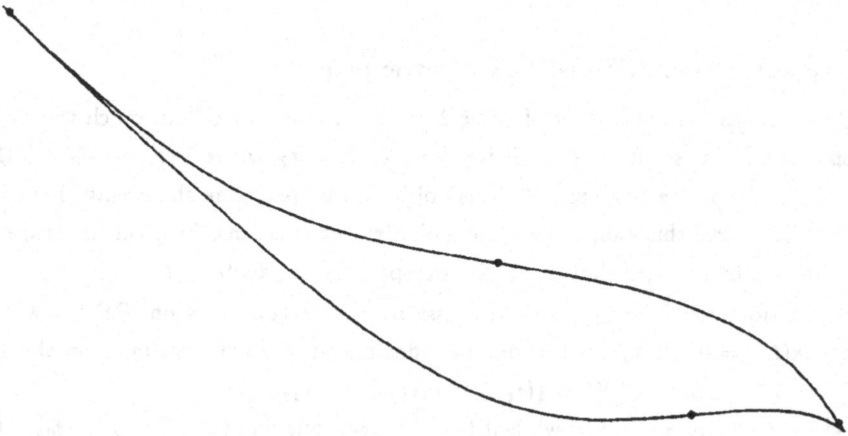

Fig. 12. Computer picture of a lips on the rotational symmetry set.

When $C^+(t_1^0, t_2^0)$ belongs to RSS^+ and also to SS or E (or both), the map C^+ tends to be much more degenerate than is the case for swallowtail, lips or beaks as above (in some cases C^+ is of corank 2), and this makes it harder to identify rigorously how RSS^+ behaves in a 1-parameter family . However, in [T], all the cases are explored and treated by informal genericity arguments and by putting together information found previously by studying the symmetry set and the rotational symmetry set. For example, as the symmetry set passes the moth transition, the combined $RSS^+ \cup SS$ appears to undergo the transition of Fig.13.

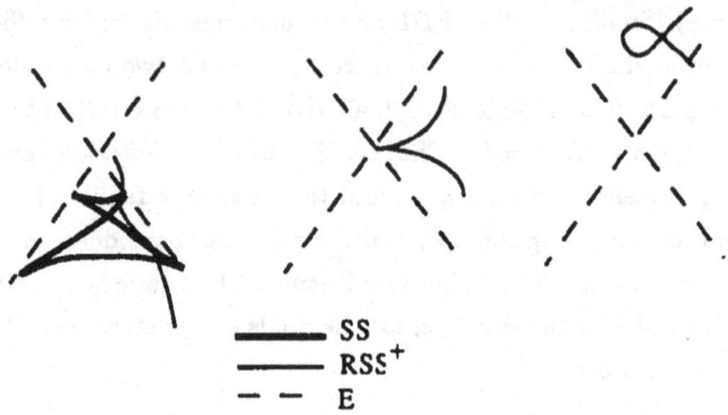

Figure 13. Transition on RSS^+, SS, E at an A_2^2

RSS^- and the connexion with symmetric maps

RSS^- arises as part of the set of critical values of the map C^- in much the same way that RSS^+ arises from C^+. If we consider $t_1 \neq t_2$ where $\kappa(t_1) = -\kappa(t_2)$ the point $C^-(t_1, t_2)$ is the same as $C^+(t_1, t_2)$ obtained by reversing the orientation of γ near t_2. Of course this cannot be done globally, but it means the *local* structure of the RSS^- is the same as that of RSS^+ except close to points (t, t).

So consider say $t = t_0$ on γ and assume that $C(t_0, t_0)$ is on RSS^-, which requires $\kappa(t_0) = 0$. If t_0 is an ordinary inflexion on γ ($\kappa'(t_0) \neq 0$), then the set

$$S^- = \{(t_1, t_2) : \kappa(t_1) = -\kappa(t_2)\}$$

is smooth close to $p_0 = (t_0, t_0)$, while if t_0 is a higher inflexion ($\kappa'(t_0) = 0, \kappa''(t_0) \neq 0$) then S^- has an isolated point at p_0. The image $RSS^- = C^-(S^-)$ is a smooth curve with an endpoint at $\gamma(t_0)$ in the first case and is merely $\{\gamma(t_0)\}$ in the second (Fig.14).

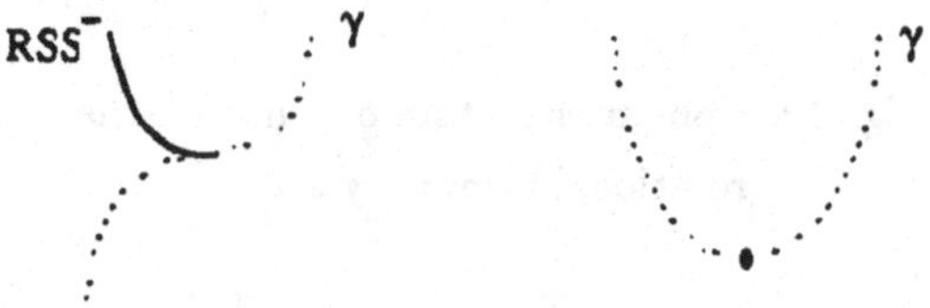

Figure 14. RSS^- near an inflexion and higher inflexion on γ

Note that the map $C^-(: \mathbf{R} \times \mathbf{R} \longrightarrow \mathbf{C} = \mathbf{R}^2)$ is *symmetric* with respect to interchange of variables in the source : $C^-(t_1, t_2) = C^-(t_2, t_1)$. Of course, the same goes for C^+, but when we examine C^+ close to a point (t_0, t_0) where $C^+(t_0, t_0) \in RSS^+$ we find C^+ is very degenerate (see above). On the other hand with C^-, the critical set near a point $p_0 = (t_0, t_0)$ with $\kappa(t_0) = 0$ consists precisely of S^- and the diagonal $\Delta = \{(t, t)\}$. The images $C^-(S^-)$ is RSS^- and $C^-(\Delta)$ is the curve $\gamma(I)$ itself. This means that we can in principle study the pair $(RSS^-, \gamma(I))$ using the classification of symmetric maps found in [B-G2] (see (2.6) above). We find the following:

3.5. Proposition : *(i) The pair $(RSS^-, \gamma(I))$ has locally the stable structure of Fig.15 (i) when $\kappa'(t_0) \neq 0$, $\kappa(t_0) = 0$, $\kappa''(t_0) \neq 0$ (ordinary inflexion on the curve). (Case II in Table(2.6).)*

(ii) In a generic family of curves (γ_u) with $\gamma = \gamma_0$ having a point t_0 where, $\kappa'(t_0) \neq 0$, $\kappa(t_0) = \kappa''(t_0) = 0$, $\kappa'''(t_0) \neq 0$, the pair $(RSS^-, \gamma(I))$ undergoes a transition as in Fig.15 (ii). (Case III in Table(2.6).)

(iii) In a generic family of curves (γ_u) with $\gamma = \gamma_0$ having a higher inflexion at t_0, $\kappa(t_0) = \kappa'(t_0) = 0, \kappa''(t_0) \neq 0$, the pair $(RSS^-, \gamma(I))$ undergoes a transition as in Fig.15 (iii). (Case IV in Table(2.6). In fact, the corresponding semi-beaks transition fails to occur in the geometrical setting.)

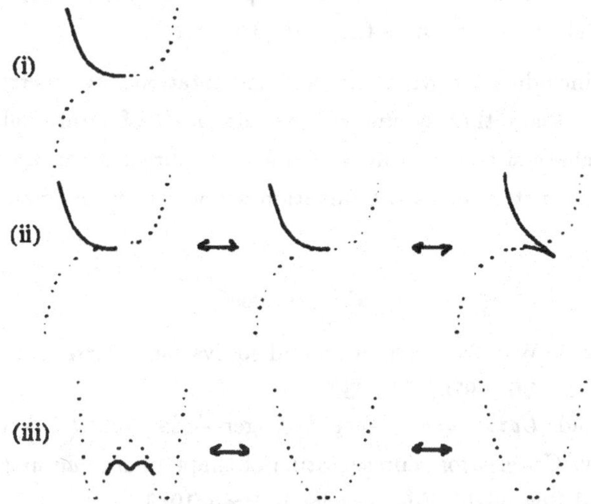

Figure 15. Transitions on 1-parameter families of RSS^-

3.6. **Remark** : The last transition of proposition (3.5 *ii*) does not occur as a transition on a generic family of duals of symmetry sets. This is an exception to the general rule that duals of symmetry sets behave similarly to rotational symmetry sets.

Dual of the rotational symmetry set

In order to study the duals of RSS^+ and RSS^- we need to identify the tangent line to either of these sets. A short calculation produces the following striking result.

3.7. Proposition : *The tangent line to RSS^+ at $C^+(t_1, t_2)$ or RSS^- at $C^-(t_1, t_2)$ is the perpendicular bisector of the line joining the centres of curvature $e(t_1)$, $e(t_2)$ at $\gamma(t_1)$, $\gamma(t_2)$ (Fig.16).*

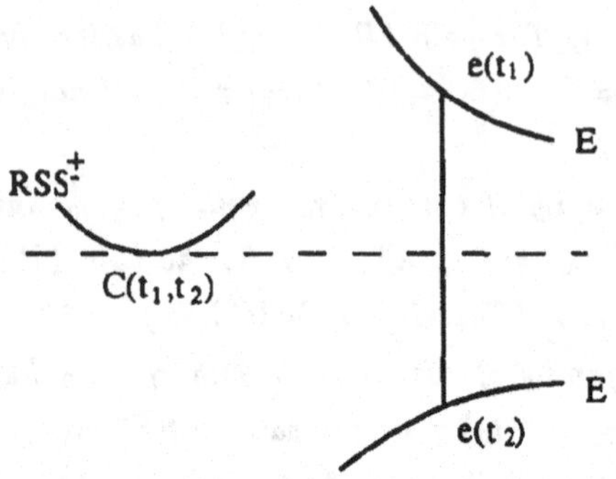

Figure 16. Two pieces of the evolute and the tangent to $RSS^{\pm}$

Note that this perpendicular bisector does passes through $C^+(t_1, t_2)$ or $C^-(t_1, t_2)$ since a rotation about one of these takes $e(t_1)$ to $e(t_2)$.

Of course since duals of symmetry sets and rotational symmetry sets behave in a similar way we know that symmetry sets and duals of rotational symmetry sets must likewise behave similarly. But we know of no direct description of the dual of the rotational symmetry set as a bifurcation set which would make the connexion explicit.

References

[A1] V.I.Arnold. Wavefront evolution and equivariant Morse lemma. *Comm. Pure. App. Math.* **29** (1976), 557-582.

[A2] V.I.Arnold. *Catastrophe theory*, Springer-Verlag 1984 (2nd edition 1986).

[B] M.Brady. Criteria for representation of shape in *Human and Machine Vision.* Beck and Rosenfield, Eds. Academic Press 1983.

[B1] J.W.Bruce. Generic functions on semi-algebraic sets, *Quart. J. Math. Oxford* (2) **37** (1986), 137-165.

[B2] J.W.Bruce. Geometry Of Singular Sets, *Math.Proc.Camb.Phil.Soc.*, **106** (1989), 495-509.

[B-G1] J.W.Bruce and P.J.Giblin. Growth, Motion and 1-parameter families of symmetry sets, *Proc. Roy .Soc. Edinburgh* **104A** (1986), 179-204.

[B-G2] J.W.Bruce and P.J.Giblin. Projections of surfaces with boundary, *Proc.London Math.Soc.*, **60** (1990), 392-416.

[B-G-G] J.W.Bruce, P.J.Giblin and C.G.Gibson. Symmetry Sets, *Proc. Roy. Soc. Edinburgh* **101A** (1985), 163-186.

[B-W] J.W.Bruce and T.Wilkinson. Folding maps and focal sets. *These proceedings.*

[G] P.J.Giblin. Local symmetry in the plane: Experiment and Theory in *Computers in Geometry and Topology*, Ed. M.C.Tangora. *Lecture Notes in Pure and Applied Mathematics* **114**(1989), 131-149. Pub. Marcel Dekker.

[G-B] P.J.Giblin and S.A.Brassett. Local symmetry of plane curves, *Amer. Math. Monthly.* 92 (1985),689-707.

[Mc] C.McCrory. Profiles of surfaces. Preprint, University of Warwick, 1981.

[M] R.J.Morris. Ph.D thesis, University of Liverpool, 1990.

[R] J.H Rieger. Families of maps from the plane to the plane. *J.London Math.Soc* (2), **36** (1987), 351-369.

[T] F.Tari, Some applications of singularity theory to the geometry of curves and surfaces, Ph.D thesis, University of Liverpool, 1990.

Department of Pure Mathematics

The University of Liverpool

P.O. Box 147

Liverpool L69 3BX

The Intersection Form of a Plane Isolated Line Singularity

V.V. Goryunov*

Abstract. Consider an isolated singularity of a real analytic function in two variables. There exists a method to calculate its intersection matrix based on consideration of a real level of some special perturbation of the function [3]. In this paper we extend this method to the case of functions with a smooth curve as a singular set.

Let $f:(\mathbb{R}^2,0)\to(\mathbb{R},0)$ be a germ of a real analytic function with a smooth critical one-dimensional set. We suppose f to be an isolated line singularity [6], i.e. its restriction to a germ of any transversal to the critical curve not passing through 0 has a Morse singularity. We complexify f and consider its stabilization $f(x,y)+z_1^2+\cdots+z_n^2$ on $\mathbb{C}^{n+2}$. A non-singular level of the stabilization in a sufficiently small ball centered at 0 is homotopy equivalent to a wedge of $(n+1)$-dimensional spheres [6]. So the only non-trivial homology group of this level is H_{n+1}. Our aim is to build up a Dynkin diagram of an intersection form on H_{n+1}. We also introduce a way to choose a distinguished basis in homology of a line singularity.

1. The main result. Suppose that coordinates x,y on the plane $\mathbb{R}^2$ are such that $0x$ is the critical set of the germ f. As $f(0)=0$ we can write $f=y^2h$ for some analytic function h. We perturb $f:\tilde{f}=y^2\tilde{h}$. The perturbed function, defined on a small ball $B\subset\mathbb{C}^2$, is again critical on $0x$. For general $\tilde{h}$ it has at general points of $0x$ an A_∞-singularity (equivalent to y^2 by a change of coordinates). At other points of $0x$, determined by $\tilde{h}=0$, there are D_∞-singularities ($\sim xy^2$). And at points outside $0x$ $\tilde{f}$ has Morse singularities.

We suppose all the point singularities of $\tilde{f}$ to be in $\mathbb{R}^2$. Moreover let all the saddle Morse points of $\tilde{f}$ lie on the real zero level X_0 of $\tilde{f}$, i.e. on the level of non-isolated singularities. Such a perturbation of f will be called *special.*

Let us consider a division of $B\cap\mathbb{R}^2$ by X_0. For every natural n we shall associate an integer bilinear form q_n to this division. The form will be symmetric for n odd and skew symmetric for n even. There also will be a periodicity: $q_{n+2k}=(-1)^kq_n$. The goal of this paper is to show that the form we are going to define is exactly an intersection form on H_{n+1} - the $(n+1)$-homology of the nonsingular complex level of the stabilized function $f(x,y)+z_1^2+\cdots+z_n^2$.

* The author is very thankful to the Mathematics Institute of the University of Warwick for its kind hospitality and support during the Symposium on Singularity Theory and its Applications when this work was completed.

First of all we indicate a formal basis of a lattice on which the form will be defined (in section 4 we shall see that there is a natural correspondence between this basis and some distinguished basis of vanishing cycles in H_{n+1}). We associate:

a formal generator m_j^0 to every saddle point s_j of $\tilde{f}$;

a generator m_j^+ to every component U_j^+ of the complement of X_0 on which $\tilde{f} > 0$ and which is relatively compact in $B \cap \mathbb{R}^2$;

m_j^- to the similar component U_j^- on which $\tilde{f} < 0$;

w_j - to every D_∞-point p_j of $\tilde{f}$;

c_j^+ - to every relatively compact segment ℓ_j^+ of the real non-isolated singular set of $\tilde{f}$ on which $\tilde{h} > 0$;

c_j^- - to the similar segment ℓ_j^- on which $\tilde{h} < 0$.

We recall several numbers from [3]:

$\nu_{0-}(j,i)$ - the number of the vertices of the curvilinear polygon U_i^- coinciding with the point s_j.

$\nu_{0+}(k,j)$ - the same number for U_k^+.

$\nu_{+-}(k,i)$ - the number of common sides of polygons U_k^+ and U_i^-.

Now we define the bilinear form q_n on the integer lattice generated by the formal generators $m_j^+ + c_j^-$: q_{odd} - symmetric, q_{even} - skew symmetric.

We begin with the squares. We need define them only in the symmetric case. We set:

$$q_n(w_j,w_j) = 4\cdot(-1)^{(n+1)/2},$$

$$q_n(a,a) = 2\cdot(-1)^{(n+1)/2} \text{ for any formal generator } a \neq w_j.$$

Let $\alpha = (-1)^{n(n-1)/2}$ and $\beta = (-1)^{n(n+1)/2}$. For example, $\alpha = \beta = 1$ if $n = 0$; $\alpha = 1, \beta = -1$ if $n = 1$. We set:

$$q_n(\overset{0}{m}_j, \overset{-}{m}_i) = \alpha\nu_{0-}(j,i),$$

$$q_n(\overset{+}{m}_k, \overset{0}{m}_j) = \alpha\nu_{0+}(k,j),$$

$$q_n(\overset{-}{m}_i, \overset{+}{m}_k) = \beta\nu_{+-}(k,i).$$

To define the other values of q_n on the basic elements we consider a model example. It will present all the possibilities of the mutual disposition of the sets in the division of $B \cap \mathbb{R}^2$ by X_0 in a neighbourhood of the critical axis $0x$. We shall code the q_n-values in a graph: its vertices represent the formal generators (the correspondence will be evident from the position) and each weighted oriented edge $a \overset{k}{\to} b$ means that $q_n(a,b) = k$.

The model example is given in Fig. 1 where each two-dimensional component of the division is labelled by the sign of $\tilde{f}$ on it. The graph is built up with an assumption that all these labelled components are relatively compact. If any of them is not, the corresponding vertex of the graph must be omitted with all its edges. The orientations of the critical axis and of the plane are essential. The weight number δ is 0 for n even and $2\cdot(-1)^{(n+1)/2}$ for n odd (so we leave an edge with weight δ non-oriented). The values $q_n(\overset{-}{m}_i, \overset{+}{m}_k)$ are not indicated.

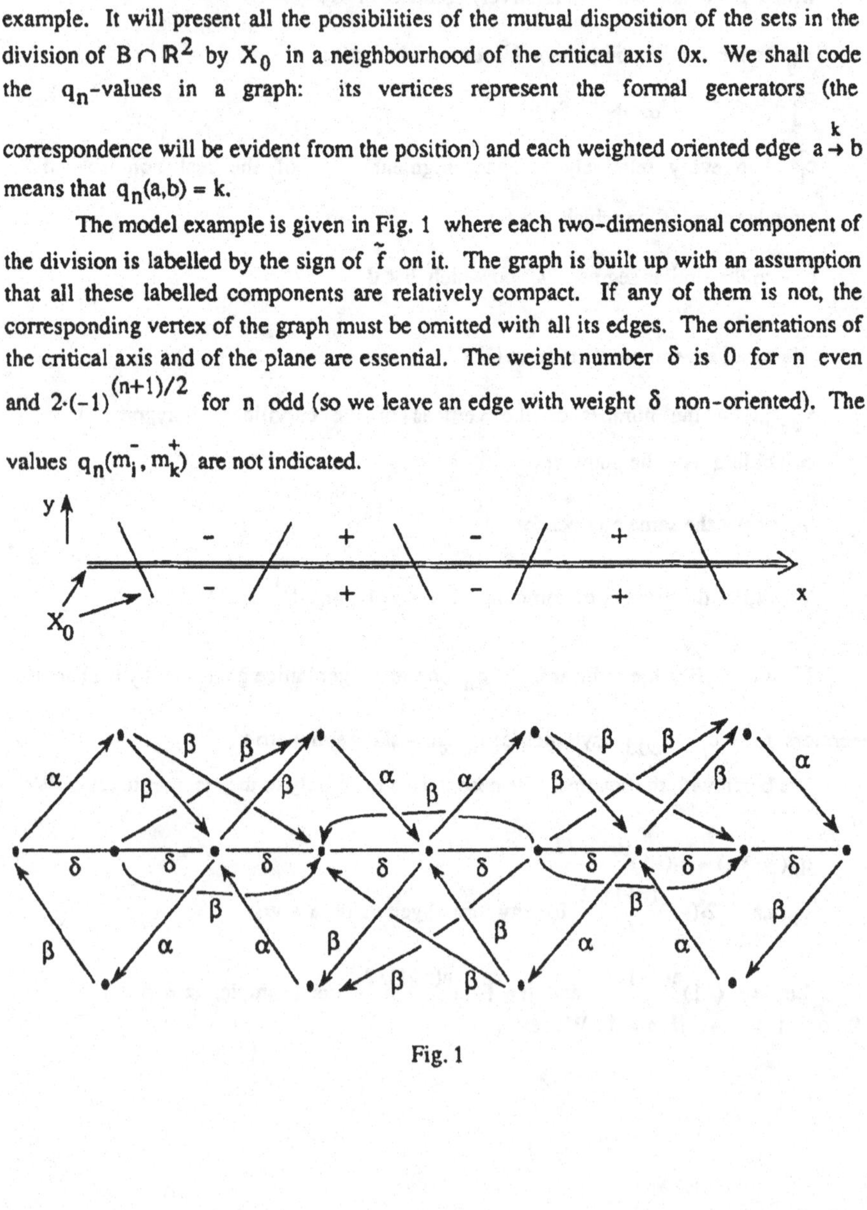

Fig. 1

The main result of this paper is

Theorem. Let q_n be a formal bilinear form defined by a special perturbation $\tilde{f}$ of a real plane isolated line singularity f. Then there is a basis in integer homology H_{n+1} of a non-singular complex level of a stabilization $f(x,y)+z_1^2+\cdots+z_n^2$ in which the intersection form coincides with q_n.

The basis just mentioned is a distinguished one (in the sense extended for line singularities - see section 3).

2. *Examples.* The beginning of the classification of the line singularities was obtained by Siersma in [6]. His list of simple functions f on $\mathbb{C}^2$ is in one-to-one correspondence with Arnol'd's list of simple functions h on $\mathbb{C}^2$ with a boundary $y = 0$ in [1] : $f = y^2h$. We give special $\mathbb{R}^2$-divisions for all these singularities and the corresponding Dynkin diagrams for $n = 0$ and $n = 1$. In the skew symmetric case $(n=0)$ all the edges have weight 1 and we omit it. In the symmetric case $(n = 1)$ the fact $q_1(a,b) = k$ is illustrated by the non-oriented edge of multiplicity $|k|$: continuous if $k > 0$ and dotted if $k < 0$. (-4)-vertices are black, (-2)-vertices are white.

$J_{k+1,\infty} : y^2(y+x^{k+1}) \sim A_k$

X_0

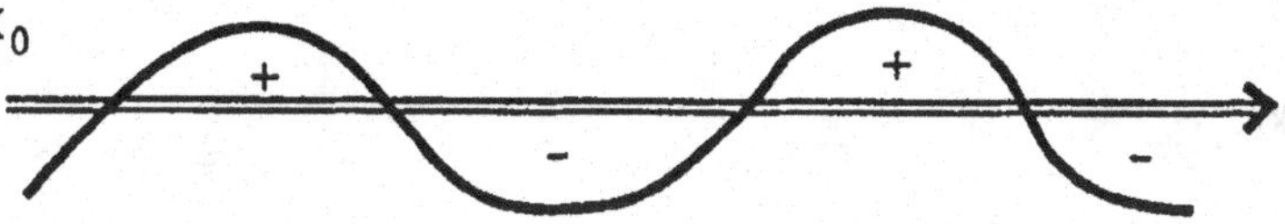

$n = 0$:

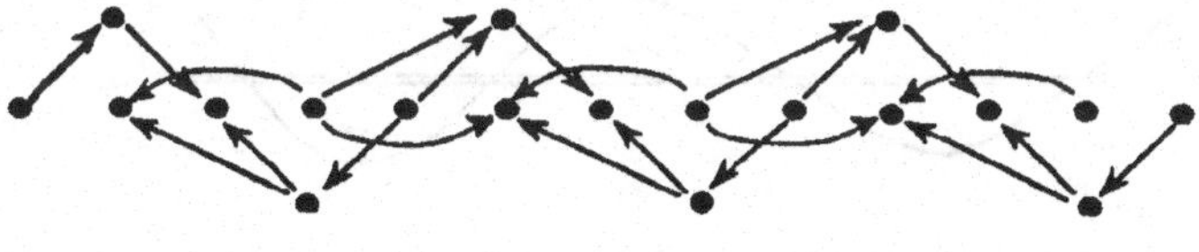

or

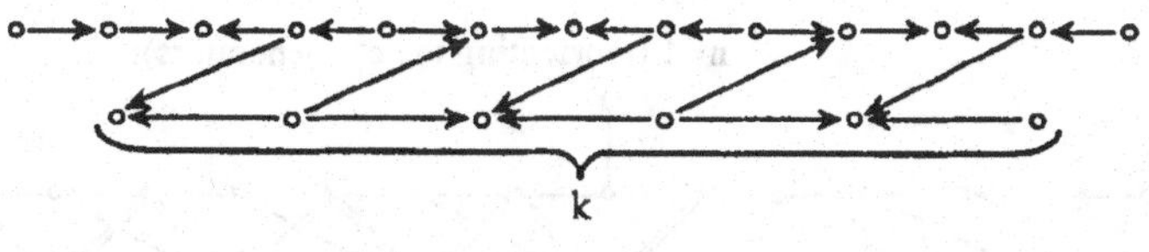

n = 1:

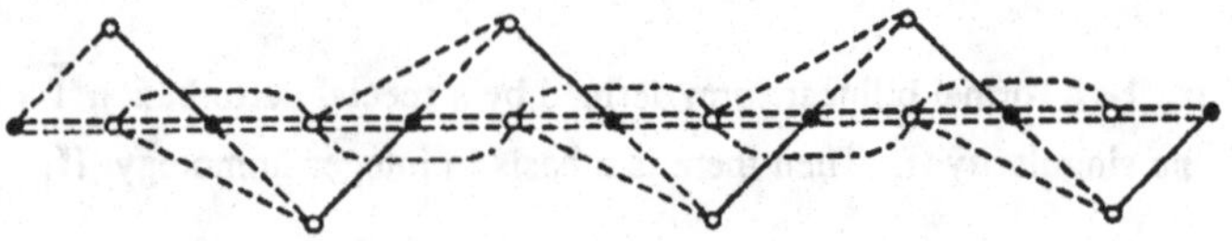

or changing the orientations of the $c_j^{\pm}$-generators

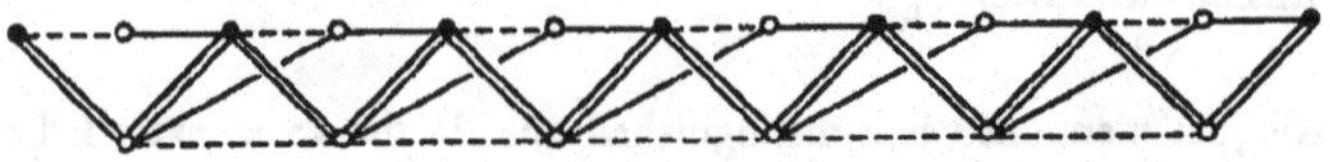

$T_{2,k+2,\infty} : y^2(x^2+y^k) \sim B_k$

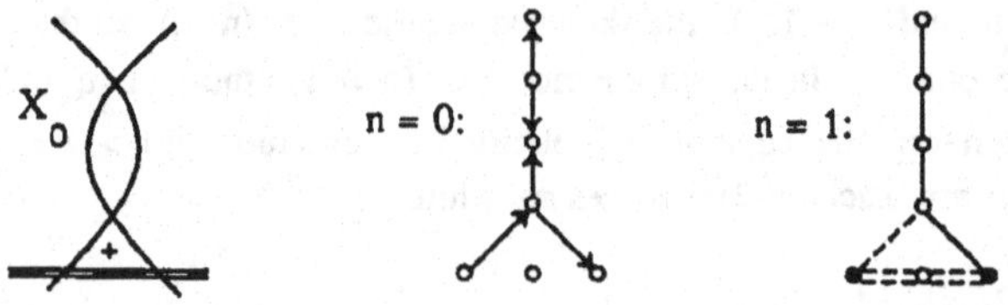

$Z_{k-2,\infty} : y^2(xy+x^k) \sim C_k$

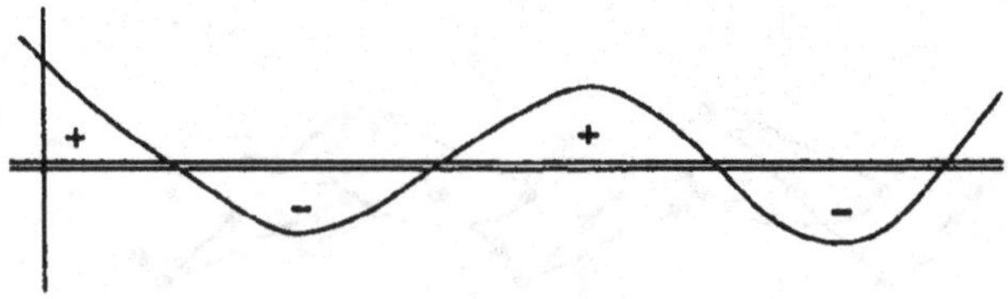

We must add one vertex to the A_{k-1}-diagram

n=0:

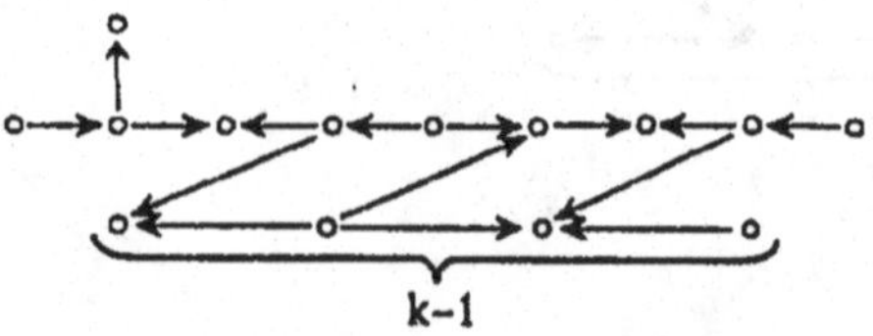

n= 1 (reorienting the $c_j^{\pm}$-generators):

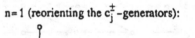

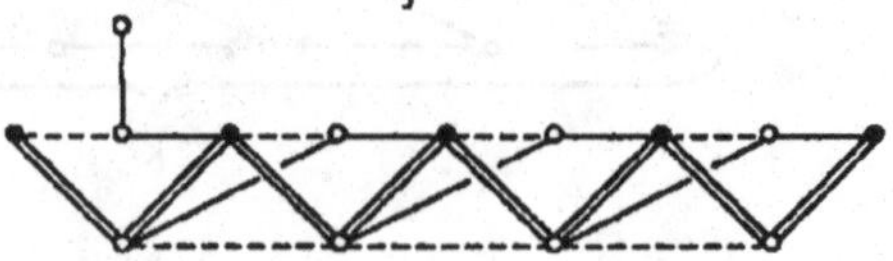

$W_{1,\infty} : y^2(y^2+x^3) \sim F_4$

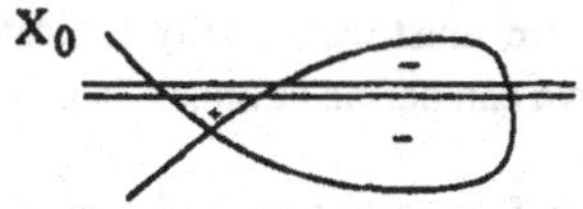

n=0:

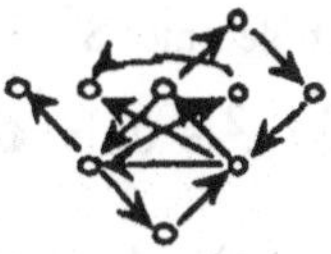

n=1 (reoreinting the $\overset{\pm}{c}$):

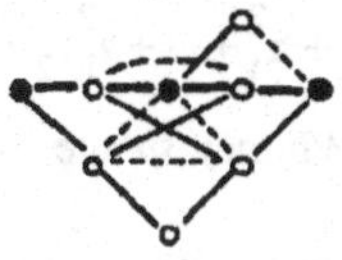

3. *A distinguished basis for a non-isolated singularity.* Here we consider a more general case than before. During this section f will be a germ of a holomorphic function on $(\mathbb{C}^k,0)$ with a one-dimensional complete intersection γ as a critical set. We again suppose that a restriction of f to a germ of any transversal to γ not passing through zero has a Morse singularity.

If $g_1 = \cdots = g_{k-1} = 0$ are equations of γ and $f(0) = 0$ then we can write

$$f = \sum_{i,j=1}^{k-1} h_{ij} g_i g_j$$

for some holomorphic functions $h_{ij} = h_{ji}$ [5]. Let us perturb f by changing g and h: $\tilde{f} = \sum \tilde{h}_{ij}\tilde{g}_i\tilde{g}_j$. For a generic perturbation the situation is as in the case of line singularities. The critical curve $\tilde{\gamma} : \tilde{g} = 0$ is smooth. $\tilde{f}$ is A_∞-singular almost everywhere on $\tilde{\gamma}$. At the points of $\tilde{\gamma}$ determined by $\det(\tilde{h}_{ij}) = 0$ $\tilde{f}$ has D_∞-singularities and several points out of j are Morse points for $\tilde{f}$ (the values of $\tilde{f}$ at these points are distinct and non-zero).

Non-critical levels of f and $\tilde{f}$ (intersected with a small ball B in $\mathbb{C}^k$ centered at 0) are diffeomorphic. If $\tilde{f}$ has no D_∞-points and $k > 2$ these levels are homotopy equivalent to a wedge of one (k-2)-dimensional sphere and $N(A_1) + \mu(\gamma)$ copies of S^{k-1} ($N(A_1)$ is a number of Morse points of $\tilde{f}$, $\mu(\gamma)$ is a Milnor number of γ) [7]. If $\tilde{f}$ has a D_∞-singularity its non-critical level is homotopy equivalent to a wedge of $N(A_1) + \mu(\gamma) + 2N(D_\infty) - 1$ spheres S^{k-1} [7].

We consider the case when $N(D_\infty) > 0$. For this case we shall define an analogue of a distinguished basis for the middle-dimensional homology of a Milnor fibre of a function with an isolated singularity. We denote $Y_t = \tilde{f}^{-1}(t) \cap B$.

Let us fix a non-critical value t_* of $\tilde{f}$. Let $\omega_0,\ldots,\omega_{N(A_1)}$ be a system of paths on $\mathbb{C}^1$ from t_* to all the critical values of $\tilde{f}$. We suppose these paths have no mutual or self-intersections. They are numerated clockwise according to the order of their going out from t_*.

The motion from t_* along the path leading to a non-zero critical value determines a usual vanishing sphere S^{k-1} on Y_{t_*}.

Approaching the value zero we must canonically indicate $\mu(\gamma) + 2N(D_\infty) - 1$ vanishing cycles. $N(D_\infty)$ of them we can define at once. In a neighbourhood of a D_∞-point we choose coordinates in which $\tilde{f}$ is written as $xy^2 + z_1^2 + \cdots + z_{k-2}^2$. Without any loss of generality suppose that we are approaching zero by the path $\varepsilon \in \mathbb{R}$, $\varepsilon > 0$, $\varepsilon \to 0$. Then a local level $xy^2 + z_1^2 + \cdots + z_{k-2}^2 = \varepsilon$ is contractible onto $S^{k-1} : x = \bar{y}^2$, $z_i \in \mathbb{R}$. We call this sphere a *Whitney sphere* as the set $xy^2 + z^2 = 0$ in $\mathbb{C}^3$ is a Whitney umbrella.

In order to determine the remaining $\mu(j) + N(D_\infty) - 1$ vanishing spheres we notice that this number is exactly the number of Morse critical points of a function $\Delta = \det(\tilde{h}_{ij})$ on $\tilde{\gamma}$ [4].

Let us consider another copy of $\mathbb{C}$ - the set of values of Δ. We join the noncritical value 0 with all critical values of $\Delta \mid_{\tilde{\gamma}}$ by a system of paths on $\mathbb{C}$ analogous to the system ω_i. The Δ-preimage on $\tilde{\gamma}$ of each path is a segment joining a pair of D_∞-points. Let ℓ be one of these segments. The way to determine a vanishing cycle by ℓ is as follows.

Consider two small disks D_1 and D_2 on γ centered at the ends of ℓ. Take a transversal to $\tilde{\gamma}$ at some point of $\ell \setminus (D_1 \cup D_2)$. For small ε its intersection with a level $\tilde{f} = \varepsilon$ is homotopy equivalent to S^{k-2}. The set Π of all these spheres is a part of our future cycle. Other parts we construct locally in neighbourhoods of the ends of ℓ.

Without any loss of generality we may consider a non-singular level of $\tilde{f}$ in a neighbourhood of each end as the set $\tilde{f} = xy^2 + z_1^2 + \cdots + z_{k-2}^2 = \varepsilon$ with real $\varepsilon > 0$. In this neighbourhood we may take a real positive semi-axis of $0x$ as ℓ. Continue Π by a set $x, y, z_i \in \mathbb{R}$ on the local level $\tilde{f} = \varepsilon$. Then add the half of the Whitney sphere $x = \bar{y}^2$, $z_i \in \mathbb{R}$ with $\operatorname{Im} y \geq 0$ or $\operatorname{Im} y \leq 0$.

The same procedure is done near another end of ℓ. If we agree orientations of Π and the hemispheres we get a vanishing cycle which is S^{k-1}. We call this cycle *composed.*

Slightly modifying the arguments of [7] it is easy to show that all the cycles we constructed here organize a basis in $H_{k-1}(Y_{t_*}, \mathbb{Z})$. This basis we call *distinguished.*

Remark. There is an unpleasant moment in constructing a composed cycle when we have no canonical choice of Whitney hemispheres. The correction may be done, for example, thus. Choose orientations of Whitney spheres, ℓ and the "transversal" sphere S^{k-2}. Orient Π as $\ell \times S^{k-2}$ and take the hemispheres with their own orientations. Then there is only one way to get a cycle.

Another method is to take a cycle consisting of 2Π and all four hemispheres with suitable orientations. This cycle has a nice property: its self-intersection is zero (see section 5). But in this way we get a basis only in the rational, not integer, homology.

4. *A distinguished basis of a real function.*

We return to a real line singularity $f(x,y) = y^2 h(x,y)$ and a complexified special perturbation $\tilde{f} = y^2\tilde{h} + z_1^2 + \dots + z_n^2$ of its stabilization. The constructions of the previous section are interpreted for it in the following way.

As a fixed non-critical level we take $Y_\varepsilon = \{\tilde{f} = \varepsilon\} \subset B \cap \mathbb{C}_{n+2}$ with real $\varepsilon > 0$ smaller than the modulus of any non-zero critical value of $\tilde{f}$. Join ε with all critical values of $\tilde{f}$ by a system of paths on $\mathbb{C}$. Suppose the paths to be without mutual and self-intersections and to lie (except for their ends) in a half-plane $\operatorname{Im} u > 0$. Note that the number of paths is less than in section 3 as there are Morse saddle points on $y^2\tilde{h} = 0$. But these points are far from the critical axis. So the vanishing cycles are defined as before. The only thing we need is to define canonical orientations.

Let us fix the orientation of the real (x,y)-plane. All the local diffeomorphisms of $\mathbb{C}_{n+2}$ which we shall use are direct products of real orientation-preserving diffeomorphisms of this plane and an identical mapping of z-space. We write z^2 instead of $z_1^2 + \dots + z_n^2$ and ∂_z instead of an n-tuple of vectors $(\partial_{z_1}, \dots, \partial_{z_n})$.

We begin with Morse cycles. Here we follow [3].

In a small neighbourhood of a saddle point s_j of $y^2\tilde{h}$ we can choose coordinates x', y', z in which $\tilde{f} = x'y' + z^2$. The Morse saddle cycle m_0^j on $\tilde{f} = \varepsilon$ is defined by $x' = \bar{y}'$, $z \in \mathbb{R}^n$. At its points with $z = 0$ we take an orientation $(ix'\partial_{x'} - iy'\partial_{y'}, \partial_z)$.

The first or the third quarter of (x',y')-plane may be a part of one of the sets U_k^+ in the division of $\mathbb{R}^2$ by the curve $y^2\tilde{h} = 0$. On local Y_ε there lies a part of a Morse cycle m_k^+ vanishing in a "maximal" Morse critical point of $\tilde{f}$. This part is given by $x', y' \in \mathbb{R}$, $z \in i\mathbb{R}^n$. At its points with $z = 0$ we take the orientation $(x'\partial_{x'} - y'\partial_{y'}, i\partial_z)$.

In a similar way a part of a "minimal" Morse cycle m_k^- is represented on $Y_{-\varepsilon}$ by the real points. We orient it on $z = 0$ by $(-x'\partial_{x'} + y'\partial_{y'}, \partial_z)$. Note that the cycle m_j^0 taken from Y_ε to $Y_{-\varepsilon}$ by multiplication by i (corresponding to its transfering by the family of levels $Y_{\varepsilon e^{i\pi t}}$, $t \in [0,1]$) has on $Y_{-\varepsilon}$ at z=0-points the orientation $(ix'\partial_{x'} - iy'\partial_{y'}, i\partial_z)$.

Now we orient Whitney and composed cycles. For this purpose we additionally fix a direction of the real x-axis. In neighbourhoods of non-isolated critical points of $\tilde{f}$ we shall use only diffeomorphisms preserving the x-axis with its orientation.

Near a D_∞-point p_j of $\tilde{f}$ choose coordinates x'', y'', z in which $\tilde{f} = \pm x''y''^2 + z^2$. The cycle w_j is given on Y_ε by $x'' = \pm \bar{y}''^2$, $z \in \mathbb{R}^n$. We orient it on $z = 0$ by $(-2ix''\partial_{x''} + iy''\partial_{y''}, \partial_z)$.

D_∞-points of the x-axis are zeros of the function $\tilde{h}$ (an equivalent of $\det(\tilde{h}_{ij})$ from section 3). On any segment $\ell_j^\pm$ of the real x-axis between two neighbouring D_∞-points $\tilde{h}$ has an extremum. So any $\ell_j^\pm$ determines a vanishing composed cycle $c_j^\pm$ in the distinguished basis of $H_{n+1}(Y_\varepsilon, \mathbb{Z})$. We shall define its orientation in the middle part Π. The choice of two Whitney hemispheres for the composed cycle is done in a way to get a cycle with the orientations coming from the corresponding Whitney spheres.

For a neighbourhood of an interior point of ℓ_j^+ take coordinates x''', y''', z in which $\tilde{f} = y'''^2 + z^2$. Then locally Π is represented by a real part of $\tilde{f} = \varepsilon$. At a point with $y''' > 0$, $z''' = 0$ we orient Π by $(\partial_{x'''}, \partial_z)$. In a neighbourhood of the "left" end of ℓ_j^+, where locally $\tilde{f} = x''y''^2 + z^2$, Π is represented by the real points of Y_ε for which $x'' \geq y''^2$. The Whitney hemisphere here is: $x'' = \bar{y}''^2$, $\operatorname{Im} y'' \leq 0$, $z \in \mathbb{R}^n$. For the "right" end of ℓ_j^+ we have $\tilde{f} = -x''y''^2 + z^2$, $\Pi = \operatorname{Re} Y_\varepsilon \cap \{x'' \leq -y''^2\}$ and the Whitney hemisphere is: $x'' = -\bar{y}''^2$, $\operatorname{Im} y'' \geq 0$, $z \in \mathbb{R}^n$.

In the same way for an interior point of ℓ_j^- we make $\tilde{f} = -y'''^2 + z^2$ with Π on $\tilde{f} = \varepsilon$ represented by real x''', z and pure imaginary y'''. For a point with $\mathrm{Im}\, y''' < 0$, $z = 0$ we define an orientation of Π by $(\partial_{x'''}, \partial_z)$. Near the "left" end of $\ell_j^- : \tilde{f} = -x''y''^2 + z^2$, $\Pi = \{x'' \in \mathbb{R}, y'' \in i\mathbb{R}, z \in \mathbb{R}^n, x'' \geq -y''^2\}$ and the Whitney hemisphere $\{x'' = -\bar{y}''^2, \mathrm{Re}\, y'' \geq 0, z \in \mathbb{R}^n\}$ on Y_ε. For the "right" end of $\ell_j^- : \tilde{f} = x''y''^2$, $\Pi = \{x'' \in \mathbb{R}, y'' \in i\mathbb{R}, z \in \mathbb{R}^n, x'' \leq y''^2\}$ and the Whitney hemisphere $\{x'' = \bar{y}''^2, \mathrm{Re}\, y'' \leq 0, z \in \mathbb{R}^n\}$ on $\tilde{f} = \varepsilon$.

Thus, as was promised in section 1, we have associated the elements of the distinguished basis of vanishing cycles to the sets of the division of $\mathbb{R}^2$ by the curve $X_0 : y^2\tilde{h} = 0$.

5. *Calculations of intersections.*

1°. Morse cycles.

The fact that the restriction of the intersection form (,) to the sublattice of H_{n+1} spanned by the Morse cycles is given exactly by the formulas of section 1 is a consequence of [3] and [2]. The only slightly non-trivial moment is to show that nothing changes for the cycles $m_k^\pm$ associated to the set $U_k^\pm$ which are partially bounded by the critical axis. But this is easily checked by a method from [3].

So we need only consider what happens near the critical axis.

2°. Selfintersection of a Whitney cycle.

Let w' be a deformation of a Whitney cycle $w : x = \bar{y}^2, z \in \mathbb{R}^n$ on complex $Y_\varepsilon : xy^2 + z^2 = \varepsilon$. Assume that w and w' intersect transversally and $w \cap w'$ has no points differing only by a sign of y and no points with $y = 0$. Consider a covering $\pi : Y_\varepsilon \to V = \{xv + z^2 = \varepsilon\}$, $v = y^2$. $\pi(w)$ is an ordinary Morse sphere S on V. $\pi(w')$ is homologous to $2S$. So an intersection number $(\pi(w), \pi(w'))$ is zero for n even and $4(-1)^{(n+1)/2}$ for n odd. Within the assumptions on w' there is exactly one point of $w \cap w'$ over any point of $\pi(w) \cap \pi(w')$. The intersection numbers of $\pi(w)$ with $\pi(w')$ and w with w' in these corresponding points coincide. So $w^2 = (w, w') = = (\pi(w), \pi(w')) = 4(-1)^{(n+1)/2}$ for n odd.

3°. Whitney-Morse intersections.

Let a Whitney cycle w be associated to a D_∞-point p and a Morse cycle $m^\pm$ to a set $U^\pm$ of our division of $\mathbb{R}^2$. The intersection of w and $m^\pm$ is non-empty iff p is one of the corners of $U^\pm$. For example, consider a local model for w as in 2° and a part of m^+ defined on $xy^2+z^2=\varepsilon$ by $x,y \in \mathbb{R}$, $y>0$, $z \in i\mathbb{R}^n$. $w \cap m^+$ is one point: $x=\sqrt{\varepsilon}$, $y=\varepsilon\sqrt{\varepsilon}$, $z=0$. w is oriented by $(-2ix\partial_x + iy\partial_y, \partial_z)$, m^+ by $(2x\partial_x - y\partial_y, i\partial_z)$. Put these systems together in the same order. The orientation we get differs from the complex orientation of Y_ε by $(-1)^{n(n+1)/2}$.

In the same way one can show that the intersections of w with the other variants of $m^\pm$ are given by the form q_n defined in section 1. The calculation of intersection with m^- is better carried out on $Y_{-\varepsilon}$ after transferring w from Y_ε into $Y_{-\varepsilon}$ by the mapping $(x,y,z) \mapsto (x,iy,iz)$.

4°. Whitney composed intersections.

Let c be a composed cycle including the halves of two Whitney spheres w_1 and w_2. The cycle $-c+w_1+w_2$ is a composed cycle built up for the same division of $\mathbb{R}^2$ by X_0 but with the opposite orientation of the real x-axis. So $c^2=(-c+w_1+w_2)^2$. By an obvious symmetry $(c,w_1)=(c,w_2)$ we conclude that $(c,w_1)=(c,w_2)=(w_1^2+w_2^2)/4$.

5°. Morse composed intersections.

Let m and c be the cycles associated to the sets U and ℓ of the division of $\mathbb{R}^2$. If closures of U and ℓ have only one common point then the intersection of m and c is inherited from the intersection of m with a corresponding Whitney sphere.

Now consider an intersection of m^+ and c^+ when ℓ^+ is in the closure of U^+. Then m^+ and c^+ have a common segment. Let U^+ lie in a half-plane $y>0$. Consider on Y_ε a flow of a vector field $\chi = -(2\tilde{h} + y\tilde{h}_y)\partial_x + y\tilde{h}_x\partial_y$, where $\tilde{h}_y = \partial\tilde{h}/\partial y$. One can easily check that in pure imaginary time $i\tau$, $\tau>0$, this flow takes m^+ to the cycle not intersecting c^+.

To obtain the same result for the set U^+ in a half-plane $y<0$ we must take $\tau<0$.

Working on $Y_{-\varepsilon}$ we similarly show that $(c^-, m^-)=0$.

6°. Neighbouring composed cycles.

Consider the intersection of cycles c^+ and c^- in a neighbourhood of a D_∞-point where $\tilde{f} = xy^2 + z^2$. We take these cycles in the local forms introduced in section 4. Then they have a common quarter of the Whitney sphere. So we must move one of the cycles.

Consider a diffeomorphism of Y_ε

$$d_t : (x,y,z) \mapsto \left(\varepsilon x/(t^2x + 2txy+\varepsilon),\, y + t\,,\, z\sqrt{\varepsilon}/\sqrt{t^2x + 2txy + \varepsilon}\right).$$

We take $t = \alpha(-1+i)$ with a real-valued non-negative smooth function α on $\mathbb{C}^{n+2}$. We define α to be identically a small positive in a neighbourhood of our D_∞-point containing the Whitney sphere and zero outside of some larger neighbourhood. One can check by elementary computations that the cycle c^- moved by diffeomorphism d_t meets c^+ exactly in one point. This point belongs to the real (x,y)-plane and has coordinates (x_0,y_0) such that $x_0y_0^2 = \varepsilon$, $y_0 > 0$, $|y_0-t| = 4\sqrt{\varepsilon}$.

Some not too complicated manipulations with the orientations show that $(c^-,c^+) = (-1)^{n(n+1)/2}$.

The case of local $\tilde{f} = -xy^2+z^2$ is treated in the same way.

7°. Selfintersection of a composed cycle.

We consider c^+ (the c^-case is isomorphic to this one). Move c^+ by the flow of the vector field χ from 5°. In a pure imaginary time $i\tau$, $\tau > 0$, we get a cycle $\tilde{c}$ which has common points with c^+ only on the corresponding Whitney hemispheres.

Let us consider a neighbourhood of the "left" hemisphere, i.e. where $\tilde{f}$ is written as $xy^2 + z^2$. Here the displacement made by χ is equivalent to a turn $(x,y,z) \mapsto (x.\kappa^2,y.\kappa^{-1},z)$, where the complex number κ is close to 1, has a small positive imaginary part and $|\kappa| = 1$.

Now we move $\tilde{c}$ in our neighbourhood by the diffeomorphism d_t from 6° with $t = i\alpha$. Some computations show that we get a cycle which meets c^+ exactly in one point (in the neighbourhood we consider). This point comes from the Whitney hemisphere of $\tilde{c}$, lies on the real (x,y)-plane and its coordinates (x_0, y_0) are defined by relations $x_0y_0^2 = \varepsilon$, $y_0 < 0$, $|y_0-t| = \sqrt[4]{\varepsilon}$. This is one of two points of intersection of c^+ with the whole Whitney sphere w moved by d_t. As for n odd

$(c^+, w) = 2\cdot(-1)^{(n+1)/2}$ we see that our neighbourhood gives $(-1)^{(n+1)/2}$ to the index of intersection of c^+ and $\tilde{c}$.

In the same way we consider a neighbourhood of the other D_∞-point and produce that $(c^+)^2 = (c^+, \tilde{c}) = 2\cdot(-1)^{(n+1)/2}$.

Thus we have showed that the intersection form on $H_{n+1}(Y_\varepsilon, \mathbb{Z})$ in our distinguished basis of vanishing cycles coincides with the formal bilinear form q_n constructed by the division of $\mathbb{R}^2$ by the curve $y^2\tilde{h} = 0$.

D. Siersma informed that intersection matrices for certain isolated line singularities were also computed by C. Cox (Utrecht; unpublished). He used almost the same basis, but with an ad-hoc definition.

References

[1] V.I. Arnol'd, Critical points of functions on a manifold with boundary, the simple Lie groups B_k, C_k, F_4 and singularities of evolutes, *Russian Math. Surveys*, 33 (1978), 99-116.

[2] A.M. Gabrielov, Intersection matrices for certain singularities, *Functional Analysis and Applications*, 7 (1973), 182-193.

[3] S.M. Gusein-Zade, Intersection matrices for certain singularities of functions of two variables, *Functional Analysis and Applications*, 8 (1974), 10-13.

[4] H. Hamm, Lokale topologische Eigenschaften komplexer Räume, *Math. Ann.*, 191 (1971), 235-252.

[5] R. Pellikaan, Hypersurface singularities and resolutions of Jacobi modules, Thesis, Rijksuniversiteit Utrecht, 1985.

[6] D. Siersma, Isolated line singularities, Preprint 217, Utrecht University, Dept. of Math., 1981.

[7] D. Siersma, Singularities with critical locus a 1-dimensional icis and transveral type A_1, *Topology and its Applications*, 27 (1987), 51-73.

Moscow Aviation Institute
Moscow
USSR

On the Degree of an Equivariant Map

S.M. Gusein-Zade

Let $\mathbb{R}^n$ be a vector space with a representation T of a finite group G on it. Let $F:(\mathbb{R}^n,0)\to(\mathbb{R}^n,0)$ be a germ of a finite analytic map which is G-equivariant with respect to the action T of the group G: $F(T_g(x)) = T_g(F(x))$ for $x\in\mathbb{R}^n$, $g\in G$. The complexification $F_{\mathbb{C}}: (\mathbb{C}^n,0)\to(\mathbb{C}^n,0)$ of the map F is G-equivariant with respect to the complexification of the representation T.

If the group G is trivial then there is defined the degree (or the index) of the map F. One of its definitions is the following. There always exists a perturbation $\tilde{F}$ of the germ F (defined in a neighbourhood of the origin) such that the origin is not a critical value of the map $\tilde{F}$ (it is possible to take $\tilde{F} = F - c$, where $c\in\mathbb{R}^n$ is a suitable constant vector). This means that each preimage p_i of the origin is isolated and the value $J_{\tilde{F}}(p_i)$ of the Jacobian $J_{\tilde{F}}$ of the map $\tilde{F}$ is not equal to zero. In this case

$$\deg F = \sum \operatorname{sign}(J_{\tilde{F}}(p_i))$$

where the sum is taken over all preimages p_i of the origin.

Theorem 1. There exists a G-equivariant real analytic perturbation $\tilde{F}_{\mathbb{C}}$ of the germ $F_{\mathbb{C}}$ (defined in a neighbourhood of the origin) for which the origin is not a critical value.

The proof follows from the following

Proposition. For any $x_0\in\mathbb{C}^n$ there exists such G-invariant real analytic function $\psi: \mathbb{C}^n\to\mathbb{C}$, that $\psi(x_0)=0$, $d\psi(x_0)=0$, $d^2\psi(x_0)$ is a nondegenerate quadratic form;

for (with ψ as in the proposition with respect to $x_0=0$) by adding $\varepsilon.\mathrm{grad}\psi$ to F, the singular point at 0 splits into a regular zero at 0, and other zeros of lower multiplicity. Now repeat this procedure at each of the degenerate zeros of $F+\varepsilon.\mathrm{grad}\psi$, and continue until all zeros are regular.

Let $\mathcal{O}$ be the ring of germs of analytic functions $(\mathbb{R}^n, 0)\to\mathbb{R}$ and let $Q=\mathcal{O}/(F)$ be the local ring of the germ F, where $(F)=(f_1,f_2,\dots f_n)$ is the ideal generated by the components f_i of the map F. The ideal (F) is invariant with respect to the natural action of the group G on the ring $\mathcal{O}$. Consequently there is a natural action of the group G on the local ring Q.

Let $J=\det(\partial f_i/\partial x_j)$ be the Jacobian of the map F. It is known that $J\neq 0$ in the ring Q (see, for example, [EL]). The Jacobian J is invariant with respect to the action of the group G. Consequently there is a G-invariant linear function $\varphi \in Q^*$ for which $\varphi(J)>0$. Let $<,>_\varphi$ be the non-degenerate G-invariant bilinear form on Q determined by the equality $<\alpha,\beta>_\varphi=\varphi(\alpha\cdot\beta)$. By the theorem of Eisenbud, Levine and Khimshiashvili ([EL], [Kh]) the signature $\mathrm{sgn}<,>_\varphi$ of the bilinear form $<,>_\varphi$ (or rather of the corresponding quadratic form) is equal to the degree of the map-germ F.

Let RG be the ring of real representations of the group G. If E is a G-module then its class in the ring RG we shall denote by [E]; [1] is the class of the trivial one-dimensional representation of any group. If H is a subgroup of the group G, and W a representation of the group H, then by $i_H^G W$ we shall denote the induced representation of the group G. For a G-invariant quadratic form Q on a G-module E there can be defined a G-equivariant signature in the ring RG as follows. There exists a decomposition $E=E_+\oplus E_0\oplus E_-$ of the module E into G-submodules E_+, E_0 and E_- which are orthogonal to each other with respect to the bilinear form Q: $Q=0$ on E_0, $Q>0$ on E_+ and $Q<0$ on E_-. Then $\mathrm{sgn}_G Q=[E_+]-[E_-]$.

<u>Definition.</u> The Eisenbud-Levine-Khimshiashvili G-equivariant degree (or G-degree) of the (equivariant) map F, which we shall denote by $\deg_G(F)$, is the G-signature of the (G-invariant) quadratic form $<,>_\varphi$.

In [D] J.Damon offered a definition of the G-equivariant degree as an element of the ring of modular representations of the group G (which is isomorphic to the ring of characters determined only for elements of G of odd order). He used it for a description of the permutation representation of the action of the group G on branches of a reduced curve singularity. It follows from [D] that the G-degree considered there is equal to the image of the degree $\deg_G$ under the natural homomorphism of the ring RG into the ring of modular representations.

Let $\tilde{F}_{\mathbb{C}}$ be a perturbation of the germ $F_{\mathbb{C}}$ described in theorem 1. The set $\tilde{F}_{\mathbb{C}}^{-1}(0)\subset\mathbb{C}^n$ of pre-images of the origin consists of the G-orbits of the following three types: 1) real orbits P_i, i.e. orbits which consist of real points; 2) non-real orbits P_j' which are mapped into themselves by complex conjugation; 3) pairs of different complex conjugate orbits. For a real orbit P_i let H_i be the isotropy subgroup of one of the points $p_i \in P_i$ (i.e. $H_i = \{g \in G: T_g(p_i) = p_i\}$; isotropy subgroups of other points from the orbit P_i are conjugate to the subgroup H_i in the group G). Let P_j' be a non-real orbit of the group G which is invariant under complex conjugation. The group G acts both on the orbit P_j' and on the quotient of this orbit modulo the action of complex conjugation. Let H_j' be the isotropy subgroup of one of the points $p_j' \in P_j'$ and let $\tilde{H}_j'$ be the isotropy subgroup of the corresponding point of the factor of the orbit P_j' modulo the action of the complex conjugation ($H_j' \subset \tilde{H}_j'$, $\tilde{H}_j'/H_j' \simeq \mathbb{Z}_2$). Let us denote by $-1 = -1_{\tilde{H}_j'}$ the one-dimensional representation of the group $\tilde{H}_j'$ which is trivial on the subgroup H_j' and which coincides with the multiplication by (-1) for elements from $\tilde{H}_j' \setminus H_j'$. Let $\varepsilon_i = \mathrm{sign}(J_{\tilde{F}_{\mathbb{C}}}(p_i))$, $\varepsilon_j' = \mathrm{sign}(J_{\tilde{F}_{\mathbb{C}}}(p_j'))$.

<u>Theorem 2.</u> $$\deg_G(F) = \sum \varepsilon_i \cdot i_{H_i}{}^G[1] + \sum \varepsilon_j' \cdot \{i_{\tilde{H}_j'}{}^G[1] - i_{\tilde{H}_j'}{}^G[-1]\}.$$

<u>Proof</u>: This theorem can be obtained by a suitable modification of the arguments in [EL]. Almost all cunstructions in [EL] are compatible with the group action. In particular, using the notation of [EL], we have $\mathrm{sgn}_G <,>_\varphi = \mathrm{sgn}_G <,>_{T_y}$. In this equality the bilinear form $<,>_{T_y}$ is defined on an algebra Q_y, which (when G is trivial) is the algebra of functions on the preimage $F^{-1}(y)$ for a regular value y of the map F. If G is not trivial the deformation $F_y(x) = F(x) - y$ may not be G-equivariant; instead we appeal to Theorem 1 and replace Q_y by the algebra Q of real functions (i.e. invariant under complex conjugation) on $\tilde{F}_{\mathbb{C}}{}^{-1}(0)$. We define the linear form T which determines the bilinear form $<,>_T$ (via $<a,b>_T = T(a.b)$) by $T(f) = \mathrm{Tr}(f/J_{\tilde{F}_{\mathbb{C}}})$. The algebra Q with the representation of G is the direct product of the algebras of real functions on the self-conjugate orbits and pairs of different complex conjugate orbits of the action of the group G on $\tilde{F}_{\mathbb{C}}{}^{-1}(0)$. The algebra corresponding to a pair of complex conjugate orbits is the complexification of a real algebra with a G-representation. In accordance with the analogue of Corollary 3.7 of [EL], the G-signature of the corresponding form on this summand is zero. The space of functions on a non-real orbit which is invariant under complex conjugation is the direct sum of subspaces of functions with real and purely imaginary values respectively. These subspaces are invariant under the action of G, and representations of the subgroup H_j on them are entire multiples of [1] and [-1] respectively. The Jacobian of $\tilde{F}_{\mathbb{C}}$ is real (and constant) on such an orbit and the form $<,>_T$ is positive definite on one of these spaces and negative definite on the other. Further arguments offer no difficulties.

In [G-Z2] there is the following definition of a notion of G-equivariant index $i_G f$ for a G-invariant germ of an analytic function $f:(\mathbb{R}^n,0)\to(\mathbb{R},0)$, with G finite. Let $V_z=\{x\in\mathbb{C}^n: f_{\mathbb{C}}(x)=z, \|x\|\leq\rho\}$ be the Milnor fibre of the complexification $f_{\mathbb{C}}$ of the germ f $(0<\|z\|\ll\rho$, ρ small enough). The representation T determines the action of the group G on the manifold V_z and consequently on its homology groups $H_{n-1}(V_z)=H_{n-1}(V_z;\mathbb{R})$ and $H_{n-1}(V_z,\partial V_z)$. There is the operator of variation Var: $H_{n-1}(V_z, \partial V_z)\to H_{n-1}(V_z)$ associated with the germ $f_{\mathbb{C}}$ (see, for example, [AVG]) which is an isomorphism and which commutes with the representations of the group G in the homology groups $H_{n-1}(V_z)$ and $H_{n-1}(V_z, \partial V_z)$. If z is real and positive then complex conjugation preserves the Milnor fibre V_z and defines the operator $\sigma_+:H_{n-1}(V_z)\to H_{n-1}(V_z)$. The operator $D=\mathrm{Var}^{-1}\sigma_+: H_{n-1}(V_z)\to H_{n-1}(V_z,\partial V_z)$ acts from a linear space into its dual and consequently determines a (non-degenerate) G-invariant bilinear form on the space $H_{n-1}(V_z)$. It can be shown that this form is symmetric. The G-equivariant index $i_G f$ of the germ f was defined as $(-1)^{n(n+1)/2}\mathrm{sgn}_G D$. Earlier ([G-Z1]) it was shown that for the trivial group G this expression coincides with the ordinary index of the gradient vector field of the germ f. To the germ of a G-invariant function $f:(\mathbb{R}^n,0)\to(\mathbb{R},0)$ there corresponds the G-equivariant map $\mathrm{grad}\, f:(\mathbb{R}^n,0)\to(\mathbb{R}^n,0)$ $(\mathrm{grad}\, f(x)= =(\partial f/\partial x_1(x), \ldots,\partial f/\partial x_n(x)))$. Let [det] be the one-dimensional representation of the group G for which the action of an element $g\in G$ coincides with the multiplication by $\det T_g$ ($T_g\in GL(\mathbb{R}^n,\mathbb{R}^n)$ - the operator corresponding to the element g).

<u>Theorem 3</u>. $i_G f=\deg_G(\mathrm{grad}\, f)\otimes[\det]$.

<u>Proof</u>: This follows from a direct comparison of contributions of G-orbits of $(\mathrm{grad}\, \tilde{f}_{\mathbb{C}})^{-1}(0)$ to $\deg_G(\mathrm{grad}\, f)$ and $i_G f$. The contributions to $\deg_G(\mathrm{grad}\, f)$ are described in theorem 2 and those to $i_G f$ are described in [G-Z2].

The difference between $i_G f$ and $\deg_G(\mathrm{grad}\, f)$ is similar to the difference between the representations of the finite group G in the local ring and in the homology group of the Milnor fiber for a G-invariant singularity of a function ([W]).

<u>Remark</u>. In [V] in particular it was shown that (for the trivial group G) the index of the gradient vector field of a germ f can be determined in the following way. Let us suppose that the number of variables n is odd, n=2k+1, let $(\cdot,\cdot)$ be the intersection form on the

homology group $H_{n-1}(V_z)$ of the Milnor fibre. Let us consider the quadratic form $(\sigma_+\cdot,\cdot)$ and its restriction $(\sigma_+\cdot,\cdot)_{\neq 1}$ to the subspace of the space $H_{n-1}(V_z)$ corresponding to eigenvalues of the classical monodromy operator of the singularity f different from $+1$. Then $\operatorname{ind} f=(-1)^k \operatorname{sgn}(\sigma_+\cdot,\cdot)_{\neq 1}$. As was communicated to the author by A.N.Varchenko his proof of this proposition is compatible with an action of a finite group G and so the analogue of theorem 3 with the left part defined as $(-1)^k \operatorname{sgn}_G(\sigma_+\cdot,\cdot)_{\neq 1}$ is an immediate consequence of his proof.

The ordinary degree (for trivial G) of a germ $F:(\mathbb{R}^n,0)\to(\mathbb{R}^n,0)$ can be expressed in terms of invariants of real preimages of the origin for a general perturbation $\tilde{F}$ of it. From Theorem 2 it can be seen that it is not so for a non-trivial G (it can be seen (and it was shown in [D]) that this property is preserved for G of odd order). Nevertheless it is possible to describe the degree $\deg_G F$ in terms of invariants of real pre-images of the origin for a suitable map.

Let $M=\mathbb{C}^n/G$ be the quotient space of $\mathbb{C}^n$ by the action of the group G. M is a (usually singular) algebraic variety (a subvariety of the space $\mathbb{C}^N$ where N is the number of generators for the ring of G-invariant polynomials on $\mathbb{C}^n$). If G is a direct product $G_1\times G_2\times\ldots\times G_r$ of irreducible Weil groups and $\mathbb{R}^{k_i}$ $(i=1,2,\ldots,r)$ is the vector space of the standard representation of the group G_i, and the action of the group G on the space $\mathbb{R}^{k_1}\times\mathbb{R}^{k_2}\times\ldots\times\mathbb{R}^{k_r}\times\mathbb{R}^m$ $(k_1+k_2+\ldots+k_r+m=n)$ is the direct product of the standard actions on the first r factors and the trivial action on the last one $(\mathbb{R}^m)$ then $M=\mathbb{C}^n/G$ is isomorphic to n-dimentional complex linear space. The image of the union of singular orbits is isomorphic to the bifurcation diagram of the corresponding (multi-germ) singularity. Complex conjugation on $\mathbb{C}^n$ induces a complex conjugation on the variety M. The set $M_{\mathbb{R}}$ of fixed points of this conjugation is the real part of M (it contains the factor space $\mathbb{R}^n/G$ of the real part $\mathbb{R}^n$ of the space $\mathbb{C}^n$ but does not coincide with it for non-trivial groups G). The maps $F_{\mathbb{C}}$ and $\tilde{F}_{\mathbb{C}}$ determine maps $\underline{F}_{\mathbb{C}}$ and $\underline{\tilde{F}}_{\mathbb{C}}$ from M into itself (determined in a neighbourhood of the point $0\in M$ which corresponds to the origin $0\in\mathbb{C}^n$). It can be easily seen that all the points that take part in the description of the degree $\deg_G F$ in theorem 2 correspond to points from the real part $M_{\mathbb{R}}$ of the variety M. So the degree $\deg_G F$ has been expressed in terms of invariants of real pre-images of the origin $0\in M$ for the map $\underline{\tilde{F}}_{\mathbb{C}}$.

Examples.

1) Let G be the group $\mathbb{Z}_2$ of two elements. It has two one-dimensional representations: the trivial one $[1]$ and the non-trivial one $[-1]$. In this case $M_{\mathbb{R}}$ consists of points of three types: 1) images of real orbits on which the action of the group $\mathbb{Z}_2$ is free; 2) images of real fixed points; 3) images of non-real orbits which are preserved by complex conjugation. In accordance with theorem 2 their contributions to the degree $\deg_G F$ are equal to $[1]+[-1]$, $[1]$ and $[1]-[-1]$ respectively.

Let the group $\mathbb{Z}_2$ act on $\mathbb{C}^n$ with coordinates $x_1,x_2,\dots,x_n$ by reflection in the hyperplane $\{x_1=0\}$: $(x_1,x_2,\dots,x_n)\to(-x_1,x_2,\dots,x_n)$. The factor space $\mathbb{C}^n/Z_2$ is isomorphic to the complex vector space $\underline{\mathbb{C}}^n$ with the real part $\underline{\mathbb{R}}^n$ (the isomorphism can be determined by the transformation $(x_1,x_2,\dots,x_n)\to(\underline{x}_1,\underline{x}_2,\dots,\underline{x}_n)=(x_1^2,x_2,\dots,x_n)$). Points of the upper half-space $\underline{\mathbb{R}}^n_+=\{(\underline{x}_1,\underline{x}_2,\dots,\underline{x}_n): \underline{x}_1>0\}$ correspond to real orbits with free action of the group $\mathbb{Z}_2$. Points of the hyperplane $\underline{\mathbb{R}}^{n-1}=\{(\underline{x}_1,\underline{x}_2,\dots,\underline{x}_n): \underline{x}_1=0\}$ correspond to real fixed points. Points of the lower half-space $\underline{\mathbb{R}}^n_-= =\{(\underline{x}_1,\underline{x}_2,\dots,\underline{x}_n): \underline{x}_1<0\}$ correspond to non-real orbits which are preserved by complex conjugation.

If the group $\mathbb{Z}_2$ is acting on the space $\mathbb{C}^n$ by reflection in the plane $\{x_1=x_2=0\}$ of codimention 2: $(x_1,x_2,x_3,\dots,x_n)\to(-x_1,-x_2,x_3,\dots,x_n)$, then the quotient space $\mathbb{C}^n/\mathbb{Z}_2$ is isomorphic to the direct product of the quadratic cone $\{(u,v,w)\in\mathbb{C}^3: uv=w^2\}$ and of the complex linear space with coordinates $x_3,\dots,x_n$ (u, v, and w correspond to the invariant functions x_1^2, x_2^2 and x_1x_2). The described partitioning of its real part is the following: $\{uv=w^2, u+v>0\}$, $\{u=v=0\}$, $\{uv=w^2, u+v<0\}$.

2) Let G be the Weil group of type A_2 (which is isomorphic to the group S_3 of permutations of three elements) acting on $\mathbb{R}^n=\mathbb{R}^2\times\mathbb{R}^{n-2}$ as the direct product of its standard action on the space $\mathbb{R}^2$ and trivial action on the space $\mathbb{R}^{n-2}$. The group S_3 has three irreducible real representations. Two of them are one-dimensional: the trivial one $[1]$ and the non-trivial one which we shall denote by $[-1]$. The third is the standard two-dimensional representation $[r]$. We have, $i_{\{1\}}^{S_3}[1]=[1]+[-1]+2[r]$, $i_{\mathbb{Z}_2}^{S_3}[1]=[1]+[r]$, $i_{\mathbb{Z}_2}^{S_3}[-1]=[-1]+[r]$. The factor space $\mathbb{C}^n/G$ is isomorphic to the complex linear space $\underline{\mathbb{C}}^n$ (in a certain system of coordinates the isomorphism can be determined by $(x_1,x_2,x_3,\dots,x_n)\to(u,v,x_3,\dots,x_n)$, where $u=x^2+y^2$, $v=y^3-3x^2y$). The image of the union of singular orbits coincides with the direct product of the semicubical parabola $\{u^3=v^2\}$ in the plane $\underline{\mathbb{R}}^2$ and the linear space $\underline{\mathbb{R}}^{n-2}$. The quotient $\mathbb{C}^n\to\underline{\mathbb{C}}^n$ maps the real subspace

$\mathbb{R}^n\subset\mathbb{C}^n$ into the part $\{u^3\geq v^2\}$ of the real subspace $\underline{\mathbb{R}}^n\subset\underline{\mathbb{C}}^n$. The real subspace $\underline{\mathbb{R}}^n$ is partitioned into the following strata which correspond to orbits of the G-action of different types: $\{u^3>v^2\}$, $\{u^3<v^2\}$, $\{u^3=v^2\neq 0\}$, $\{u=v=0\}$. In accordance with theorem 2 the contributions of points from these strata into the degree $\deg_G F$ (up to signs of the Jacobian) are equal to $[1]+[-1]+2[r]$, $[1]-[-1]$, $[1]+[r]$, and $[1]$ respectively.

3) Let the group $G=\mathbb{Z}_2\times\mathbb{Z}_2$ act on the space $\mathbb{R}^n=\mathbb{R}^2\times\mathbb{R}^{n-2}$ in accordance with the formula $(\varepsilon_1,\varepsilon_2)(x_1, x_2, x_3, ..., x_n)=(\varepsilon_1 x_1, \varepsilon_2 x_2, x_3, ..., x_n)$, where $\varepsilon_i=\pm 1\in\mathbb{Z}_2$, $(\varepsilon_1,\varepsilon_2)\in\mathbb{Z}_2\times\mathbb{Z}_2$ (germs of $\mathbb{Z}_2\times\mathbb{Z}_2$-invariant functions on the space $(\mathbb{R}^n,0)$ can be regarded as germs of functions on a manifold with a corner). The group $\mathbb{Z}_2\times\mathbb{Z}_2$ has four irreducible real (one-dimensional) representations. We shall denote them by 1_{++}, 1_{+-}, 1_{-+} and 1_{--} where $1_{\delta_1\delta_2}$ (δ_i=+ or -) is such a representation that its restriction to the subgroup $\mathbb{Z}_2\times\{1\}$ (respectively $\{1\}\times\mathbb{Z}_2$) is equal to $\delta_1 1$ (i.e. +1 or −1) (respectively $\delta_2 1$). The factor space $\mathbb{C}^n/\mathbb{Z}_2\times\mathbb{Z}_2$ is isomorphic to the n-dimensional complex linear space $\underline{\mathbb{C}}^n$ (the isomorphism can be determined by the transformation $(x_1, x_2, x_3,..., x_n)\to(\underline{x}_1,\underline{x}_2, \underline{x}_3,..., \underline{x}_n)=(x_1^2,x_2^2, x_3,..., x_n)$). The image of the union of singular orbits coincides with the union of coordinate hyperplanes $\{\underline{x}_1=0\}\cup\{\underline{x}_2=0\}$. The factorisation $\mathbb{C}^n\to\underline{\mathbb{C}}^n$ transforms the real subspace $\mathbb{R}^n\subset\mathbb{C}^n$ into the set of points from $\underline{\mathbb{R}}^n$ with non-negative coordinates $\underline{x}_1$ and $\underline{x}_2$. The partitioning of the real subspace $\underline{\mathbb{R}}^n$ into strata corresponding to orbits of different types and contributions of points from these strata to the degree $\deg_G F$ are the following:

1) $\{\underline{x}_1>0,\ \underline{x}_2>0\}$, $[1_{++}]+[1_{+-}]+[1_{-+}]+[1_{--}]$;
2) $\{\underline{x}_1<0,\ \underline{x}_2>0\}$, $[1_{++}]+[1_{+-}]-[1_{-+}]-[1_{--}]$;
3) $\{\underline{x}_1>0,\ \underline{x}_2<0\}$, $[1_{++}]-[1_{+-}]+[1_{-+}]-[1_{--}]$;
4) $\{\underline{x}_1<0,\ \underline{x}_2<0\}$, $[1_{++}]-[1_{+-}]-[1_{-+}]+[1_{--}]$;
5) $\{\underline{x}_1>0,\ \underline{x}_2=0\}$, $[1_{++}]+[1_{-+}]$;
6) $\{\underline{x}_1<0,\ \underline{x}_2=0\}$, $[1_{++}]-[1_{-+}]$;
7) $\{\underline{x}_1=0,\ \underline{x}_2>0\}$, $[1_{++}]+[1_{+-}]$;
8) $\{\underline{x}_1=0,\ \underline{x}_2<0\}$, $[1_{++}]-[1_{+-}]$;
9) $\{\underline{x}_1=\underline{x}_2=0\}$, $[1_{++}]$.

Let us regard the germ of the $(\mathbb{Z}_2\times\mathbb{Z}_2)$-invariant map $F:\mathbb{R}^2\to\mathbb{R}^2$ (m=0) defined by $(x,y)\to(x^3-xy^2, x^2y+y^5)$ ($x=x_1$, $y=x_2$ in previous notation). The multiplicity of this map at the origin is equal to 11. As generators of the local ring $Q=\mathcal{O}/(F)$ we can take monomials x^iy^j with $0\leq i\leq 2$, $0\leq j\leq 2$ or $i=0, j=3,4$. All the monomials x^iy^j except the chosen generators, x^3, x^4, y^5 and y^6 lie in the ideal (F); there are following equalities in the local ring Q: $x^3=xy^2$, $x^4=x^2y^2$, $y^5=-x^2y$ and $y^6=-x^2y^2$. The Jacobian J of the map F is equal to $3x^4+3x^2y^2+15x^2y^4-5y^6$, consequently $J=11x^2y^2$ in Q. As the $(\mathbb{Z}_2\times\mathbb{Z}_2)$-invariant linear function $\varphi:Q\to\mathbb{R}$ we can take the function which is equal to zero on all the chosen monomial generators of the local ring Q except x^2y^2, $\varphi(x^2y^2)=1$.

The subspace of the local ring Q corresponding to the representation $[1_{++}]$ is five-dimensional and is generated by the monomials $1, x^2, y^2, x^2y^2$ and y^4. The matrix of the form $<,>_\varphi$ on this subspace is equal to

$$\begin{bmatrix} 0 & 0 & 0 & 1 & 0 \\ 0 & 1 & 1 & 0 & 0 \\ 0 & 1 & 0 & 0 & -1 \\ 1 & 0 & 0 & 0 & 0 \\ 0 & 0 & -1 & 0 & 0 \end{bmatrix}$$

Its signature is equal to 1. The subspace of the local ring Q corresponding to the representation $[1_{+-}]$ is generated by the monomials y, x^2y and y^3. The matrix of the form $<,>_\varphi$ on this subspace is equal to

$$\begin{bmatrix} 0 & 1 & 0 \\ 1 & 0 & 0 \\ 0 & 0 & -1 \end{bmatrix}$$

Its signature is equal to (-1). The subspace corresponding to the representation $[1_{-+}]$ is generated by the monomials x and xy^2. We have $<x,x>_\varphi=<xy^2,xy^2>_\varphi=0$, $<x,xy^2>_\varphi=1$. The signature of the form $<,>_\varphi$ on this subspace is equal to 0. The subspace corresponding to the representation $[1_{--}]$ is generated by xy, $<xy,xy>_\varphi=1$. The signature of the form $<,>_\varphi$ on this subspace is equal to 1. So

$$\deg_{\mathbb{Z}_2\times\mathbb{Z}_2}F=[1_{++}]+[1_{--}]-[1_{+-}].$$

As a ($\mathbb{Z}_2\times\mathbb{Z}_2$-equivariant) perturbation of the map F (for which pre-images of the origin are not multiple) we can take $\tilde{F}(x,y)=(x^3-xy^2-\varepsilon x,\ x^2y+y^5+\varepsilon y)$, $\varepsilon>0$. The map $\underline{\tilde{F}}:\underline{\mathbb{C}}^2\to\underline{\mathbb{C}}^2$ has five pre-images of the origin (in a neighbourhood of the origin): three real points q_1 $(x=0, y=0)$, q_2 $(x=\varepsilon, y=0)$, and

$$q_3\ \left(\underline{x}=-\frac{1}{2}-\varepsilon+\sqrt{\frac{1}{4}-2\varepsilon}\ ,\ \underline{y}=-\frac{1}{2}+\sqrt{\frac{1}{4}-2\varepsilon}\ \right)$$

(this point came from a non-real orbit under the factorisation $\mathbb{C}^2\to\underline{\mathbb{C}}^2$), and two non-real points $x=0, y=\pm i\sqrt{\varepsilon}$. In accordance with theorem 2,

$$\deg_{\mathbb{Z}_2\times\mathbb{Z}_2}F=\operatorname{sign}(J_{\tilde{F}}(p_1))\cdot[1_{++}]+\operatorname{sign}(J_{\tilde{F}}(p_2))\cdot([1_{++}]+[1_{-+}])+$$
$$+\operatorname{sign}(J_{\tilde{F}}(p_3))\cdot([1_{++}]-[1_{+-}]-[1_{-+}]+[1_{--}])$$

(here p_i are points from the orbits P_i which are pre-images of the points q_i under the factorisation $\mathbb{C}^2 \to \underline{\mathbb{C}}^2$). It can be easily calculated that $\mathrm{sign}(J_{\tilde{F}}(p_1))=-1$, $\mathrm{sign}(J_{\tilde{F}}(p_2))=+1$. It follows that $\mathrm{sign}(J_{\tilde{F}}(p_3))=+1$ and

$$\deg_{\mathbb{Z}_2\times\mathbb{Z}_2} F = -[1_{++}] + ([1_{++}] + [1_{-+}]) + ([1_{++}] - [1_{+-}] - [1_{-+}] + [1_{--}])$$
$$= [1_{++}] + [1_{--}] - [1_{+-}].$$

Acknowledgements This paper was stimulated by the paper [D] of James Damon. The most part of this research was done while the author was at the University of Warwick on the occasion of the Symposium on Singularity Theory and its Applications. I want to express my deep gratitude to the University and to the organizers of the Symposium for support and hospitality.

References

[AVG] Arnol'd V.I., Varchenko A.N., Gusein-Zade S.M. Singularities of differentiable maps. Vol.2. Birkhäuser, 1988.

[D] Damon J. G-signature, G-degree, and symmetries of the real branches of curve singularities. Preprint, University of Warwick (1989).

[EL] Eisenbud D., Levine H.I. An algebraic formula for the degree of a C^∞ map germ. Annals of Math., **106** (1977), 19-44.

[G-Z1] Gusein-Zade S.M. Index of a singular point of a gradient vector field. Functional Analysis and its Applications, **18** (1984), 6-10.

[G-Z2] Gusein-Zade S.M. An equivariant analogue of the index of a gradient vector field. Lecture Notes in Mathematics, **1214**, 196-210; Springer-Verlag, 1986.

[Kh] Khimshiashvili G.M. On the local degree of a smooth map. Soobshch.Akad.Nauk GruzSSR, **85**, no 2 (1977), 309-311 (in Russian).

[V] Varchenko A.N. On the local residue and the intersection form on the vanishing cohomology. Math.USSR Izvestiya, **26**:1 (1986), 31-52.

[W] Wall C.T.C. A note on symmetry of singularities. Bull. London Math. Soc. **12** (1980), 169-175.

Faculty of Geography
Moscow State University
119899 Moscow
U.S.S.R.

AUTOMORPHISMS OF DIRECT PRODUCTS OF ALGEBROID SPACES

Herwig Hauser and Gerd Müller

1. Introduction

Consider complex algebroid spaces Z defined by some formal power series and factorizations $Z = Z_1 \times \ldots \times Z_p$ of Z into a direct product of spaces Z_i. We propose to study the connection between the automorphism group $\operatorname{Aut} Z$ of Z and the automorphism groups $\operatorname{Aut} Z_i$ of its factors. Clearly the product group $\operatorname{Aut} Z_1 \times \ldots \times \operatorname{Aut} Z_p$ is a subgroup of $\operatorname{Aut} Z$, but equality need not hold in general: Whenever two factors Z_i coincide their permutation gives an element of $\operatorname{Aut} Z$ not belonging to the product of the $\operatorname{Aut} Z_i$'s. And there may be even quite different types of automorphisms of Z which are not a product. Take for instance X defined in affine space $\mathbf{A}^m$ by homogeneous polynomials and an arbitrary Y in $\mathbf{A}^n$. Denoting by x and y the corresponding coordinates, the automorphism

$$\phi(x,y) = ((1+y_1)\cdot x_1, .., (1+y_1)\cdot x_m, y_1, ..., y_n)$$

of $X \times Y$ is not a product.

Let (R, m) be the local ring of the algebroid space Z. An abstract subgroup G of $\operatorname{Aut} Z$ equipped with the structure of an algebraic group is called an *algebraic subgroup* of $\operatorname{Aut} Z$ if the induced representations on the finite dimensional vector spaces R/m^{k+1} are rational.

The purpose of this paper is to show that if an automorphism of Z belongs to a reductive algebraic subgroup of $\operatorname{Aut} Z$ then it is already a combination of a permutation and a product of automorphisms of the factors of Z.

Let us make this statement more precise. First note that by [H-M 2, Theorem 3] one knows that an algebroid space Z has a factorization

$$Z \simeq Z_o^{n_o} \times Z_1^{n_1} \times \ldots \times Z_p^{n_p}$$

into indecomposable, pairwise not isomorphic factors Z_i, which is unique up to isomorphism and permutation. An algebroid space is called indecomposable if it is not a direct product of two factors both different from the reduced point. Allowing possibly $n_o = 0$ we may collect in $Z_o^{n_o}$ all smooth factors of Z, i.e., $Z_o = \mathbf{A}$ the one-dimensional affine space with local ring $\mathbf{C}[[x]]$. This factor $Z_o^{n_o} = \mathbf{A}^{n_o}$ will play a special rôle.

Let S_n be the symmetric group on n elements acting naturally on the n-fold direct product of the group $\operatorname{Aut} X$ for any X, so that we can form the semi-direct product $S_n \ltimes (\operatorname{Aut} X)^n$. Clearly this is a subgroup of $\operatorname{Aut} X^n$. Consider now the subgroup $\operatorname{Auto} Z$ of $\operatorname{Aut} Z$ of trivial automorphisms of Z, i.e., combinations of permutations of identical factors and product automorphisms:

$$\operatorname{Auto} Z := \operatorname{Aut} \mathbf{A}^{n_o} \times (S_{n_1} \ltimes (\operatorname{Aut} Z_1)^{n_1}) \times \ldots \times (S_{n_p} \ltimes (\operatorname{Aut} Z_p)^{n_p}).$$

The way that Auto Z sits inside Aut Z depends on the projections $Z \to Z_i$ making Z a direct product. Hence as a subgroup of Aut Z it is only defined up to conjugation.

We can now formulate the main result of this paper:

Theorem. Any reductive algebraic subgroup of Aut Z is up to conjugation contained in the group Auto Z of trivial automorphisms of Z.

In particular, any automorphism of Z of finite order is trivial. Observe that a reductive subgroup of Aut $\mathbf{A}^m$ need not be contained (even up to conjugation) in $S_m \ltimes (\mathrm{Aut}\,\mathbf{A})^m$. Take for instance $\mathrm{GL}_m(\mathbf{C}) \subset \mathrm{Aut}\,\mathbf{A}^m$.

One would like to have a similar theorem for analytic space germs instead of algebroid spaces. We do not know how to prove this in general. The case of reduced space germs is solved and treated in [M 1].

2. The Ephraim subspace of an algebroid space

In order to prove the theorem let us first study the situation for automorphisms of direct products $Z = \mathbf{A}^m \times X$ with m maximal, i.e., such that the second factor X of Z does not involve another smooth factor. We shall need the description of the product structure of a morphism given in [H-M 1, Theorem 2 and Remark (b) following it]: For any morphism $\tau : X \to S$ of algebroid spaces with section $\sigma : S \to X$ there exists a unique algebroid subspace T of S with the following universal property:

For any base change $S' \to S$ the induced morphism $\tau' : X' = X \times_S S' \to S'$ is trivial along the induced section $\sigma' : S' \to X'$ if and only if $S' \to S$ factors through T. The space T is called the *trivial locus of the pair* (τ, σ). (We say that τ' is trivial along σ' if there is an isomorphism $X' \simeq S' \times X_o$ over S' mapping $\sigma'(S')$ onto $S' \times 0$, where X_o is the special fiber of τ'.)

In analogy with the isosingular locus of an analytic space germ studied in [E, definition 0.1 and Theorem 0.2] we have in the algebroid category:

Proposition 1. For an algebroid space Z let $\tau : Z \times Z \to Z$ be the projection on the first factor with section the diagonal map $\sigma : Z \to Z \times Z$.

(a) The trivial locus E of the pair (τ, σ) is a smooth algebroid subspace of Z and $Z \simeq E \times X$ for some algebroid space X.

(b) Every automorphism of Z stabilizes E.

(c) The dimension of E is the maximum dimension of smooth spaces E' such that $Z \simeq E' \times X$ for some X.

We shall call E the *Ephraim subspace* of Z. Heuristically speaking, E is the maximal subspace of Z along which Z is trivial.

Proof. Assertion (a) is proved in [H-M 1, sec. 4] and (b) follows from the universal property of E. To see (c) assume $Z = E' \times X$. Consider the cartesian square

$$\begin{array}{ccc} E' \times (E' \times X) & \hookrightarrow & Z \times Z \\ \downarrow \tau' & & \downarrow \tau \\ E' & \hookrightarrow & Z \end{array}.$$

with induced section $\sigma' : E' \to E' \times (E' \times X)$. Since E' is smooth there is an automorphism of $E' \times E'$ mapping the diagonal onto $E' \times 0$. Combining with the identity on X we get an automorphism of $E' \times (E' \times X)$ trivializing τ' along σ'. The universal property of E then implies $E' \subset E$.

Proposition 2. Let Z be an algebroid space, E its Ephraim subspace and G a reductive algebraic subgroup of Aut Z.

(a) There is a G-stable subspace Y of Z such that $Z \simeq E \times Y$.

(b) The group G is conjugate to a subgroup of $\mathrm{Aut}\, E \times \mathrm{Aut}\, Y$ in Aut Z.

Proof. Write $Z = E \times X$ for some X and choose a minimal embedding $X \subset \mathbf{A}^n$. By [M 2, Satz 6 i)] the action of G on Z can be extended to an action on $E \times \mathbf{A}^n$. Choose coordinates v on E and x on $\mathbf{A}^n$. For $g \in G \subset \mathrm{Aut}(E \times \mathbf{A}^n)$ let g' denote the linear part of g w.r.t. (v, x). This gives a linear action of G on $E \times \mathbf{A}^n$. Every automorphism of Z stabilizes E. As E is a linear subspace of $E \times \mathbf{A}^n$ w.r.t. the coordinates (v, x) the linear action will stabilize E. Since every rational representation of a reductive algebraic group is completely reducible there exists a linear subspace of $E \times \mathbf{A}^n$ complementary to E and stabilized by the action of G.

By [M 2, Satz 6 ii)] there is an automorphism ϕ of $E \times \mathbf{A}^n$ tangent to the identity such that $\phi \circ g' = g \circ \phi$ for all $g \in G$. We thus obtain a smooth algebroid subspace V of $E \times \mathbf{A}^n$ transverse to E of dimension n which is stable under the original action of G. Therefore the intersection $Y = Z \cap V$ is G-stable and $Z \simeq E \times Y$, cf. [E, p. 359]. This proves (a). Note that in general G does not stabilize the original subspace $X = 0 \times X$ of Z.

To see assertion (b) observe that restriction to E and Y induces a group homomorphism $\alpha : G \to \mathrm{Aut}\, E \times \mathrm{Aut}\, Y \subset \mathrm{Aut}\, Z$. Any g in G has the same effect on the tangent space of Z as its image $\alpha(g)$. By [M 2, Satz 2], G and $\alpha(G)$ are conjugate.

3. Direct products without smooth factor

By what we have seen before we may assume for the proof of the theorem that Z has no smooth factor, i.e., trivial Ephraim subspace $E = 0$. We claim that there is a group homomorphism $\alpha : G \to \mathrm{Auto}\, Z \subset \mathrm{Aut}\, Z$ such that for every g in G both g and $\alpha(g)$ have the same effect on the tangent space of Z. Applying again [M 2, Satz 2], the theorem will be established. Our claim follows from:

Proposition 3. Let $Z_1, ..., Z_p$ be indecomposable algebroid spaces not isomorphic to $\mathbf{A}$ and let ϕ be an automorphism of $Z = Z_1 \times ... \times Z_p$. There exists a permutation

$\rho \in S_p$ such that, identifying Z_i with $0 \times Z_i \times 0 \subset Z$, we have $\phi(Z_i) = Z_{\rho(i)} \subset Z$ for all i, and hence $\phi|_{Z_i}$ is an isomorphism between Z_i and $Z_{\rho(i)}$.

Proof. Let $Z' = Z_2 \times ... \times Z_p$ and $\pi_1 : Z \to Z_1$, $\pi' : Z \to Z'$ be the projections. In the proof of the unique factorization property in [H-M 2, sec. 3] it was shown that after composition with a suitable permutation both $\pi_1 \circ \phi|_{Z_1} : Z_1 \to Z_1$ and $\pi' \circ \phi|_{Z'} : Z' \to Z'$ are isomorphisms. We shall prove that $\pi_1 \circ \phi|_{Z'} = 0$ and $\pi' \circ \phi|_{Z_1} = 0$ and can then proceed by induction on the number of factors of Z. Consider the cartesian square

$$\begin{array}{ccc} Z_1 \times Z' & \to & Z' \times Z' \\ \downarrow \tau' & & \downarrow \tau \\ Z_1 & \to & Z' \end{array}$$

where τ denotes projection on the first factor with section the diagonal map, and the map $Z_1 \to Z'$ is given by $\pi' \circ \phi|_{Z_1}$. The induced section of τ' is given by $\sigma' = (\mathrm{id}_{Z_1}, \pi' \circ \phi|_{Z_1})$. Then $(\pi_1, \pi' \circ \phi)$ is an automorphism of $Z_1 \times Z'$ and its inverse trivializes τ' along σ'. As Z' has zero Ephraim subspace Proposition 1 implies that $\pi' \circ \phi|_{Z_1} = 0$. Symmetrically $\pi_1 \circ \phi|_{Z'} = 0$. This concludes the proof of the proposition and of the theorem.

4. When is an automorphism of Z a product?

In general it seems hard to decide whether a given automorphism of infinite order is contained in a reductive algebraic subgroup of Aut Z. We shall give a sufficient condition for this. Let us call $\phi \in \mathrm{Aut}\, Z$ semi-simple if ϕ is semi-simple on all R/m^{k+1}. If γ_k denotes the natural representation of Aut Z on R/m^{k+1} the image $H_k = \gamma_k(\mathrm{Aut}\, Z)$ is an algebraic subgroup of $\mathrm{Aut}(R/m^{k+1})$, cf. [M 2, Satz 3].

Proposition 4. Let $\phi \in \mathrm{Aut}\, Z$ be semi-simple and such that $\gamma_1(\phi)$ belongs to the unity component H_1^0 of H_1. Then ϕ is contained in a reductive algebraic subgroup of Aut Z, in fact contained in a torus.

Proof. Every semi-simple element of a connected algebraic group is contained in a subtorus. Choose a torus $T_1 \subset H_1$ containing $\gamma_1(\phi)$. We claim that there are tori $T_k \subset H_k$ containing $\gamma_k(\phi)$ such that T_{k+1} is mapped isomorphically onto T_k by the natural map $H_{k+1} \to H_k$. Taking the projective limit one obtains a torus in Aut Z containing ϕ.

Obviously the homomorphism $\pi : H_{k+1} \to H_k$ is surjective. By Cartan's Uniqueness Theorem [M 2, Satz 1], it is injective on every reductive subgroup of H_{k+1}. In particular, Ker π is unipotent hence solvable and connected. Therefore $S = \pi^{-1}(T_k)$ is solvable and connected. In addition Ker π is the normal subgroup of unipotent elements of S. Consequently $S = \mathrm{Ker}\, \pi \rtimes T_{k+1}$ with a subtorus $T_{k+1} \subset S$ containing the semi-simple element $\gamma_{k+1}(\phi)$ of S. Clearly T_{k+1} is mapped isomorphically on T_k.

Remark. It should be noted that the preceding argument only works for algebroid spaces. We do not know how to prove a similar result in the category of analytic space germs.

References

[E] Ephraim, R.: Isosingular loci and the cartesian product structure of complex analytic singularities. Trans. Amer. Math. Soc. **241**, 357-371 (1978).

[H-M 1] Hauser, H., Müller, G.: The trivial locus of an analytic map germ. Ann. Inst. Fourier **39**, 831-844 (1989).

[H-M 2] Hauser, H., Müller, G.: The cancellation property for direct products of analytic space germs. Math. Ann. **286**, 209-223 (1990).

[M 1] Müller, G.: Endliche Automorphismengruppen von direkten Produkten komplexer Raumkeime. Arch. Math. **45**, 42-46 (1985).

[M 2] Müller, G.: Reduktive Automorphismengruppen analytischer C-Algebren. J. Reine Angew. Math. **364**, 26-34 (1986).

Herwig Hauser
Institut für Mathematik
Universität Innsbruck
A-6020 Austria.

Gerd Müller
Fachbereich Mathematik
Universität Mainz
D-6500 Germany.

Disentanglements.

by T. de Jong and D. van Straten.

Universität Kaiserslautern
Fachbereich Mathematik
Erwin-Schrödinger-Straße, Geb. 48
W-6750 Kaiserslautern
Germany.

Introduction.

Consider a hypersurface germ $X \subset \mathbb{C}^{n+1}$, defined by an equation $f = 0$, $f \in \mathcal{O} := \mathbb{C}\{x_0, x_1, \ldots, x_n\}$ and let Σ be a subscheme of the singular locus Sing(X) (with structure ring $\mathcal{O}/(f, J_f)$, J_f the Jacobian ideal). In [J-S1] we introduced the functor Def(Σ,X) of *admissible deformations* of the pair (Σ,X). An admissible deformation (Σ_S, X_S) over a base S consists of flat deformations Σ_S and X_S over S, such that Σ_S is contained in the critical locus of the map $X_S \longrightarrow S$. This notion of deformation was first considered by R. Pellikaan ([Pe1], [Pe2]) and leads under the condition that the space of first order deformations

$$T^1(\Sigma,X) = \mathrm{Def}(\Sigma,X)(\mathbb{C}[\epsilon]/\epsilon^2)$$

is *finite dimensional* to the existence of a semi-universal admissible deformation. We will give a short sketch of its construction in §1. (See also [J-S1] or [J-S2] for the formal case.)

An interesting situation arizes when we consider a map $\varphi: \tilde{X} \longrightarrow \mathbb{C}^{n+1}$, where $\tilde{X}$ is an n-dimensional Cohen-Macaulay (multi-) germ with (say) isolated singular points. As an example one could have in mind the situation where $X \subset \mathbb{C}^N$ and φ is induced by a generic linear projection $L : \mathbb{C}^N \longrightarrow \mathbb{C}^{n+1}$. The image $X = \varphi(X)$ then is a hypersurface with a singular locus Σ of codimension 2 in $\mathbb{C}^{n+1}$, the double locus of φ in the target. The map $\overline{\varphi}: \tilde{X} \longrightarrow X$ can be identified with the *normalization map* of X. The deformation theory of this situation is related to that of admissible deformations in the following way:

Theorem:
Assume that the conductor $\mathcal{C} = \mathcal{H}om\,(\mathcal{O}_{\tilde{X}}, \mathcal{O}_X)$ is *reduced* and let $\Sigma \subset X$ be defined by $\mathcal{C}$. Then we have natural equivalences:

$$\mathrm{Def}(\tilde{X} \longrightarrow \mathbb{C}^{n+1}) \xrightarrow{\approx} \mathrm{Def}(\tilde{X} \longrightarrow X) \xrightarrow{\approx} \mathrm{Def}(\Sigma,X)$$

Furthermore, the natural forgetful transformation

$$\mathrm{Def}(\tilde{X} \longrightarrow \mathbb{C}^{n+1}) \longrightarrow \mathrm{Def}(\tilde{X}) \quad \text{is smooth.}$$

Here the first two functors refer to deformations of the *diagram* (see [Bu]). The first map is induced by forming the *image* of φ, the second by forming the *conductor*. The first and the second statement together imply that the functor $Def(\Sigma,X)$ is as complicated as $Def(\tilde{X})$. For proofs of these statements we refer to [J-S1], §4 and the forthcoming paper [J-S3].

Let $\tilde{\mathfrak{X}} \longrightarrow B$ be the semi-universal deformation of $\tilde{X}$. An irreducible component of the base space B is called a *smoothing component* if the fibre $\tilde{X}_s$ over a general point s of this component is a smooth space. The corresponding notion for the functor $Def(\Sigma, X)$ is that of what we call a *disentanglement component*. These are components of the base space of the semi-universal admissible deformation for which the fibre X_s over a general point s of the component has *smooth* normalization $\tilde{X}_s$ and the mapping from $\tilde{X}_s$ to X_s is *stable*. For the dimension of smoothing components there is a formula conjectured by J. Wahl [Wa] and proved by G.-M. Greuel and E. Looijenga [G-L]. In §2 we apply their ideas to find similar results for the functor $Def(\Sigma,X)$. In the theory of hypersurface singularities one has to distinguish between deformations of the *hypersurface* X and deformations of a *function* f that defines X. It is useful to have a similar distinction for admissible deformations. This leads to a functor $Def(\Sigma,f)$ (which maps smoothly onto $Def(\Sigma,X)$) for which the result is more natural. In §3 we concentrate on the case that X is a weakly normal surface singulary in $\mathbb{C}^3$. We prove that the difference in dimension of two disentangelement components is even. This implies the same statement for smoothing components of normal surface singularities, a fact first discovered by J. Wahl [Wa]. In §4 we give a proof of a conjecture of D. Mond, first formulated as a question in [Mo2], on the $\mathcal{A}_e$ - codimension of a map germ $\varphi: \mathbb{C}^2 \longrightarrow \mathbb{C}^3$. (For a different proof see the paper of D. Mond [Mo3] in these proceedings.)

Acknowledgement.

We would like to thank the organizers of the Symposium on Singularity Theory and its Applications held at the University of Warwick in 1989 not only for inviting us with full financial support, but also for creating an exciting atmosphere which we truly enjoyed. This work was done while the first author was supported by a stipendium of the European Community, and the second was supported with a stipendium by the Deutsche Forschungs Gemeinschaft.

§ 1 The Semi-universal Admissible Deformation.

As in [J-S1] and [J-S2], we consider a pair of germs of analytic spaces $\Sigma \subset X$, where $\Sigma \subset \mathrm{Sing}(X)$. The singular locus is defined by the Fitting ideal of Ω^1_X, as usual. Our strategy to construct a semi-universal deformation for the functor $\mathrm{Def}(\Sigma,X)$ is very near to the one used by H.Hauser [Ha] to construct one for isolated singularities. The idea is to construct first a very big object in the Banach analytic category and to come down to a finite dimensional space by putting in the extra geometrical conditions. The following five steps outline this procedure.

Step 1: First embed Σ and X in $\mathbb{C}^N$. Let $I_\Sigma = (g_1, \dots, g_r)$ and $I_X = (f_1, \dots, f_m)$ be the ideals of Σ and X. Consider the map

$$F: \mathbb{C}^N \longrightarrow \mathbb{C}^r \times \mathbb{C}^m \; ; \; x \longmapsto (g_1(x), \dots, g_r(x), f_1(x), \dots, f_m(x))$$

and the projections $p_\Sigma : \mathbb{C}^r \times \mathbb{C}^m \longrightarrow \mathbb{C}^r$ and $p_X: \mathbb{C}^r \times \mathbb{C}^m \longrightarrow \mathbb{C}^m$.
Note that $(p_X F)^{-1}(0) = X$ and $(p_\Sigma F)^{-1}(0) = \Sigma$.

Step 2: Construct the semi-universal unfolding of the map F, with groups of coordinate transformations at the right which respect the projections p_Σ and p_X. Let the base space be $\mathcal{B}$, a Banach analytic space.

Step 3: Form the families $(p_X F_{\mathcal{B}})^{-1}(0) =: X_{\mathcal{B}}$ and $(p_\Sigma F_{\mathcal{B}})^{-1}(0) =: \Sigma_{\mathcal{B}}$ over the space $\mathcal{B}$. Use a *flatifier* to get the subspace $\mathcal{F} \subset \mathcal{B}$ such that the induced families $\Sigma_{\mathcal{F}}$ and $X_{\mathcal{F}}$ over $\mathcal{F}$ are flat.

Step 4: Over $\mathcal{F}$ we can form the critical space C of $X_{\mathcal{F}} \longrightarrow \mathcal{F}$. Analoguous to the flatifier there is a notion of *containifier*. We use this to restrict our families to the sub-space B of $\mathcal{F}$ such that over B we have $\Sigma_B \subset C_B$. We now have an admissible family (Σ_B, X_B) over B.

Step 5: If the space $T^1(\Sigma,X)$ is *finite dimensional*, then B is an analytic space, having $T^1(\Sigma,X)$ as Zariski tangent space. The family $\xi_B = ((\Sigma_B, X_B) \longrightarrow B) \in \mathrm{Def}(\Sigma,X)(B)$ is versal in the following sense: Given any admissible deformation $\xi_A \in \mathrm{Def}(\Sigma,X)(A)$ over A, induced by $\alpha: A \longrightarrow B$, and any admissible deformation $\xi_C \in \mathrm{Def}(\Sigma_A, X_A)(C)$ over $C \supset A$, there exists a map $\gamma: C \longrightarrow B$, extending α and inducing ξ_C from ξ_B. Further more, the principle of openness of versality holds.

We want to stress however that the results in §3 and §4 are *independent* of this construction because in those cases $\mathrm{Def}(\Sigma,X)$ can be related to other functors for which the convergence of the semi-universal deformation and openness of versality is already known.

§ 2 The Relative T^1 - sequences.

We consider a hypersurface X, with an equation $f = 0$, $f \in \mathcal{O}$. Let Σ be defined by an ideal $I \subset \mathcal{O}$. The condition that $\Sigma \subset \mathrm{Sing}(X)$ is that we have $(f, J_f) \subset I$. (Or, $f \in \int I$). Here $J_f = (\partial f/\partial x_0, \ldots, \partial f/\partial x_n)$ is the Jacobian ideal of f. For reasons of simplicity and because of the applications we have in mind we assume:

1) Σ is a reduced Cohen-Macaulay germ.

2) $\dim(\mathrm{supp}(I/(f, J_f))) < \dim(\mathrm{Sing}(X))$.

3) $\dim T^1(\Sigma, X) < \infty$.

Under these circumstances $\Sigma = \mathrm{Sing}(X)_{red}$, so Σ is completely determined by X alone (and $\mathrm{Def}(\Sigma, X)$ becomes a sub-functor of $\mathrm{Def}(X)$, see [J-S1] and [J-S2]).Transverse to a generic point of Σ the hypersurface X has an A_1 - singularity (cf. [Pe 1]).

There is an exact sequence computing the space $T^1(\Sigma, X)$ of first order admissible deformations:

$$0 \longrightarrow \Theta_X \longrightarrow \Theta_{\mathbb{C}^{n+1}} \otimes \mathcal{O}_X \longrightarrow P_X(\mathcal{A}) \longrightarrow T^1(\Sigma, X) \longrightarrow 0 \qquad (1)$$

Here $P_X(\mathcal{A})$ is called the ideal of *admissible functions*. A precise definition of $P_X(\mathcal{A})$ can be found in [J-S1] and [J-S2]. The important properties that we will use here are that $P_X(\mathcal{A})$ is an *ideal* and that it occurs in the exact sequense (1).

As in [G-L], we study next what happens in a one parameter family. Let $\xi_\Delta = ((\Sigma_\Delta, X_\Delta) \longrightarrow \Delta) \in \mathrm{Def}(\Sigma, X)(\Delta)$ be an admissible deformation over a small disc Δ. Then analogous to (1) we have a *relative* sequence:

$$0 \longrightarrow \Theta_{X_\Delta/\Delta} \longrightarrow \Theta_{\mathbb{C}^{n+1} \times \Delta/\Delta} \longrightarrow P_{X_\Delta}(\mathcal{A}_\Delta) \qquad (2)$$

The cokernel of the last map we denote by $T^1(\Sigma_\Delta, X_\Delta)_{rel}$. It is naturally an $\mathcal{O}_\Delta$-module.

Proposition (2.1):

The elements of $T^1(\Sigma_\Delta, X_\Delta)_{rel}$ are in 1-1 correspondence with isomorphism classes of admissible deformations of (Σ, X) over $\Delta \times \mathrm{Spec}(\mathbb{C}[\varepsilon]/\varepsilon^2)$ which restrict to the given $\xi_\Delta \in \mathrm{Def}(\Sigma, X)(\Delta)$

proof: This is a matter of definition reading and is similar to the proof of (1) in [J-S1]. (A more systematic approach to relative groups will appear in [J-S2].) ⊠

Now, as in [G-L], there is a commutative diagram:

$$\begin{array}{ccccccc}
0 \longrightarrow & \Theta_{\mathbb{C}^{n+1} \times \Delta/\Delta} \otimes \mathcal{O}_{X_\Delta} & \xrightarrow{t} & \Theta_{\mathbb{C}^{n+1} \times \Delta/\Delta} \otimes \mathcal{O}_{X_\Delta} & \longrightarrow & \Theta_{\mathbb{C}^{n+1}} \otimes \mathcal{O}_X & \longrightarrow 0 \\
& \downarrow & & \downarrow & & \downarrow & \\
0 \longrightarrow & P_{X_\Delta}(\mathcal{A}_\Delta) & \xrightarrow{t} & P_{X_\Delta}(\mathcal{A}_\Delta) & \longrightarrow & P_X(\mathcal{A}) &
\end{array}$$

with exact rows, induced by multiplication by t, a local parameter on Δ. Hence, by the snake lemma, we deduce a six-term exact sequence:

$$0 \longrightarrow \Theta_{X_\Delta/\Delta} \xrightarrow{t} \Theta_{X_\Delta/\Delta} \longrightarrow \Theta_X \longrightarrow T^1(\Sigma_\Delta, X_\Delta)_{rel} \xrightarrow{t} T^1(\Sigma_\Delta, X_\Delta)_{rel} \longrightarrow T^1(\Sigma, X) \qquad (3)$$

(In fact, one can define higher T^i 's to prolong the sequence to the right.)

Definition (2.2): (With the notation as above)

An *admissible deformation* of (Σ, f) over a base S is a pair (Σ_S, f_S) where Σ_S is a flat deformation of Σ over S, f_S a deformation of f over S (i.e. a function parametrized by S) such that $(\Sigma_S, X_S := f_S^{-1}(0)) \in \mathrm{Def}(\Sigma, X)(S)$. The functor $S \longmapsto \{$Isomorphism classes of admissisble deformations of Σ, f over S $\}$ is denoted by $\mathrm{Def}(\Sigma, f)$. Here isomorphism is defined in the obvious way. (See also [J-S2].)

The functor $\mathrm{Def}(\Sigma, f)$ is closely related to $\mathrm{Def}(\Sigma, X)$ and one has:

Proposition (2.3):

1) The forgetful transformation $\mathrm{Def}(\Sigma, f) \longrightarrow \mathrm{Def}(\Sigma, X)$ is *smooth*.
2) If X is quasi-homogeneous, then one has an isomorphism of vector spaces $T^1(\Sigma, f) \longrightarrow T^1(\Sigma, X)$.

Analoguous to the exact sequence (1) one has an exact sequence

$$0 \longrightarrow \Theta_f \longrightarrow \Theta_{\mathbb{C}^{n+1}} \longrightarrow P(\mathcal{A}) \longrightarrow T^1(\Sigma, f) \longrightarrow 0 \qquad (4)$$

Here $\Theta_f := \{\vartheta \in \Theta_{\mathbb{C}^{n+1}} \mid \vartheta(f) = 0\}$ is the module of vector fields killing f and $P(\mathcal{A})$ is again the ideal of admissible functions (but now it is an ideal in $\mathcal{O}$ instead of $\mathcal{O}_X$). In the same way as we derived the exact sequence (3) from (1), we can derive from (4) a six-term exact sequence associated with an element $(\Sigma_\Delta, f_\Delta)$ of $\mathrm{Def}(\Sigma, f)(\Delta)$:

$$0 \longrightarrow \Theta_{f_\Delta/\Delta} \xrightarrow{t} \Theta_{f_\Delta/\Delta} \longrightarrow \Theta_f \longrightarrow T^1(\Sigma_\Delta, f_\Delta)_{rel} \xrightarrow{t} T^1(\Sigma_\Delta, f_\Delta)_{rel} \longrightarrow T^1(\Sigma, f) \qquad (5)$$

Here the relative group $T^1(\Sigma_\Delta, f_\Delta)_{rel}$ has an interpretation similar to the one in proposition (2.1). We leave it to the reader to spell it out.

Now let $\xi_B = ((\Sigma_B, X_B) \longrightarrow B) \in \mathrm{Def}(\Sigma, X)(B)$ be the semi-universal admissible deformation of (Σ, X). (See also §1.) By versality, our given family $(\Sigma_\Delta, X_\Delta) \longrightarrow \Delta$ is induced via a map $\alpha: \Delta \longrightarrow B$ from ξ_B and as in [G-L] we see that the dimension of the image of $T^1(\Sigma_\Delta, X_\Delta)_{rel}$ in $T^1(\Sigma, X)$ is equal to the dimension of the Zariski tangent space to B at a general point of the image of α. Of course, similar statements hold for $\mathrm{Def}(\Sigma, f)$ and hence by the exactness of sequences (3) and (5) we get:

Proposition (2.4):

The dimension of the Zariski tangent space to the base space of the semi-universal admissible deformation at a general point of the image of α is equal to:

A. For $Def(\Sigma,X)$: $rank_{\mathcal{O}_\Delta}(T^1(\Sigma_\Delta,X_\Delta)_{rel}) + dim_{\mathbb{C}}(Coker(\Theta_{X_\Delta/\Delta} \longrightarrow \Theta_X))$

B. For $Def(\Sigma,f)$: $rank_{\mathcal{O}_\Delta}(T^1(\Sigma_\Delta,f_\Delta)_{rel}) + dim_{\mathbb{C}}(Coker(\Theta_{f_\Delta/\Delta} \longrightarrow \Theta_f))$

Corollary (2.5):

A. Suppose we have a deformation (Σ_Δ,X_Δ) over Δ such that at a generic point of Δ the fibre has only rigid singularities (for the functor $Def(\Sigma,X)$ of course). Then the dimension of the component to which α maps is equal to $dim(Coker(\Theta_{X_\Delta/\Delta} \longrightarrow \Theta_X))$.

B. Suppose we have an admissible deformation (Σ_Δ,f_Δ) over Δ such that for a generic point of Δ f_Δ has only rigid singularities in the zero fibre and some A_1 - points outside the zero fibre. Then the dimension of the component to which α maps is equal to $\# A_1 + dim(Coker(\Theta_{f_\Delta/\Delta} \longrightarrow \Theta_f))$.

The corollary follows, because the rank terms of proposition (2.4) are zero in case A. and $\#A_1$ in case B. By openness of versality it follows that the components in question are generically reduced, so the dimension to the Zariski tangent space at a generic point is equal to its dimension.

Lemma (2.6):

With the notations as above one has :

$$Coker(\Theta_{f_\Delta/\Delta} \longrightarrow \Theta_f) = Coker(H_1(\mathcal{O}_\Delta,\{\partial f_\Delta/\partial x_i\}) \longrightarrow H_1(\mathcal{O},\{\partial f/\partial x_i\})$$

Here $H_\bullet(R,\{f_i\})$ denotes *Koszul homology* of the elements f_i on R.

proof: An element of Θ is a vector field $\vartheta = \Sigma_{i=0}^{n} a_i\partial/\partial x_i$ such that $\vartheta(f) = \Sigma_{i=0}^{n} a_i\, \partial f/\partial x_i = 0$. This means exactly that $(a_0,\dots,a_n)$ is in the kernel of the first Koszul differential The image of the second Koszul differential then corresponds to the span of the 'trivial vector fields' $\partial f/\partial x_j\, .\partial/\partial x_i - \partial f/\partial x_i\, .\partial/\partial x_j$. These can be lifted for trivial reasons. ⊠

Proposition (2.7):

Let $J = (f_0, f_1, \dots, f_n) \subset \mathcal{O}$ be an ideal defining a variety of codimension m. Then one has:

$$H_{n+1-m}(\mathcal{O},\{f_i\}) \approx Ext_{\mathcal{O}}^{m}(\mathcal{O}/J,\mathcal{O}).$$

This should be 'well-known'. For a discussion and proof see [Pe 3].

Something very interesting happens in case $\mathbf{dim(\Sigma) = 1}$:

Corollary (2.8):

Let $(\Sigma_\Delta, f_\Delta)$ be an admissible deformation of (Σ, f) over a disc Δ. If $\dim(\Sigma) = 1$, then

$$\mathrm{Coker}(\Theta_{f_\Delta/\Delta} \longrightarrow \Theta_f) = 0 .$$

proof: Of course, we apply (2.7) with $f_i = \partial f/\partial x_i$ and $m = n$. Because by assumption $\dim_{\mathbb{C}}(I/(fJ_f)) < \infty$ it follows that $H_1(\mathcal{O}, \{\partial f/\partial x_i\}) = \mathrm{Ext}^n_{\mathcal{O}}(\mathcal{O}/J_f, \mathcal{O}) = \mathrm{Ext}^n_{\mathcal{O}}(\mathcal{O}/I, \mathcal{O}) \approx \omega_\Sigma$, the *dualizing module of* Σ. But in a flat family one has : $\omega_{\Sigma_\Delta/\Delta} \otimes \mathcal{O}_\Sigma = \omega_\Sigma$, as one easily checks. The assertion then follows from (2.6). ⊠

Corollary (2.9):

If $\dim(\Sigma) = 1$, then the dimension of the component of the base space of $\mathrm{Def}(\Sigma, f)$ to which α maps is equal to the number of A - points that split off. ⊠

This is very similar to the case of an *isolated* hypersurface singularity.

Question (2.10):

Is is true in general (under the stated conditions) for an admissible deformation that $\mathrm{Coker}(\Theta_{f_\Delta/\Delta} \longrightarrow \Theta_f) = 0$? This sounds rather implausible, but it would be extremely interesting to know the answer, especially for Σ of codimension 2.

§3 Applications to Surface Singularities.

From now on we will restrict further to the case X is a hypersurface germ in $\mathbb{C}^3$. Then the conditions of §2 are equivalent to X being *weakly normal*, i.e. X having a singular locus Σ, which is an ordinary double curve away from the point 0. The normalization X will be a (multi-) germ of a normal surface singularity. As was mentioned in the introduction, one has an equivalence of functors between $\mathrm{Def}(\tilde{X} \longrightarrow X)$ and $\mathrm{Def}(\Sigma, X)$, whereas $\mathrm{Def}(\tilde{X} \longrightarrow X) \longrightarrow \mathrm{Def}(\tilde{X})$ is smooth. So there is in this case a 1-1 correspondence between components of the base space of $\tilde{X}$ and components of the base space of (Σ, X). We now spell out the notions corresponding to *smoothing* and *smoothing component*.

Definition (3.1):

A. Let $X \subset \mathbb{C}^3$ be a weakly normal surface singularity, with $\Sigma = \mathrm{Sing}(X)$. A *disentanglement* of (Σ, X) over Δ is an admissible deformation (Σ, X) over Δ such that for a general $t \in \Delta$ the *disentanglement fibre* X_t has only the following types of singularities: ordinary double curve (type A_∞), ordinary pinch point (type D_∞), ordinary triple point (type $T_{\infty,\infty,\infty}$).

B. Let $f \in \mathcal{O} = \mathbb{C}\{x,y,z\}$ such that $X := f^{-1}(0)$ is a weakly normal surface singularity with singular locus Σ. A *disentanglement* of (Σ,f) over Δ is an admissible deformation $(\Sigma_\Delta, f_\Delta)$ over Δ such that $(\Sigma, X := f^{-1}(0))$ is a disentanglement in the above sense and such that for a general $t \in \Delta$ the *disentanglement function* f_t has at most A_1 - points away from the zero fibre.

C. An irreducible component of the base space of the semi-universal admissible deformation is called a *disentanglement component* when over it disentanglement occurs. On each such component the number of pinch points and triple points of the disentanglement fibre (and the number of A_1 - points of the disentanglement function) is constant and will be denoted by $\#D_\infty$, $\#T$ (and $\#A_1$) respectively. Note that corollary (2.5) and (2.9) can be applied to these components.

Remark (3.2):

There exist weakly normal surfaces X that:

* have no disentanglement at all.
* have several disentanglement components.
* have components in their base space which are not disentanglement components.

This follows from the equivalence of functors and the fact that there exist normal surface singularities $\tilde{X}$ with the corresponding properties.

However, in the case that the function f is an element of $I^2 \subset \int I$ there is a *special* disentanglement component in the base space of $\mathrm{Def}(\Sigma,X)$ and $\mathrm{Def}(\Sigma,f)$. This component can be described as follows: (see also [Pe2], Ex.2.3) Write $f = \sum_{i,j=1}^{r} h_{ij}\, \Delta_i\, \Delta_j$, where $I = (\Delta_1,\ldots,\Delta_r)$. Choose representatives $g_1, g_2, \ldots, g_p$ for a basis of the vector space $I^2/I^2 \cap J_f$ and write these as $g_k = \sum_{i,j=1} \varphi_{kij} \cdot \Delta_i \cdot \Delta_j$. Let S be the (smooth) base space of the semi-universal deformation of the curve Σ and let $\Delta_i(s)$ be generators for the ideal of the curve Σ_s, $s \in S$. Consider the function

$$F: \mathbb{C}^3 \times \mathbb{C}^p \times S \longrightarrow \mathbb{C}$$

$$F(x,y,z,t_1,t_2,\ldots,t_p,s) = \sum_{i,j=1}^{r} \Big(h_{ij} + \sum_{k=1}^{p} t_k \cdot \varphi_{kij}\Big) \Delta_i(s) . \Delta_j(s)$$

Then F is a disentanglement function over $\mathbb{C}^p \times S$. For general $s \in S$ the curve Σ_s is smooth, so in this disentanglement no triple points occur. It is not obvious at all that this really is a *component* of the base space of $\mathrm{Def}(\Sigma,f)$. For this one has to prove that no element of $\int I/I^2$ can be lifted over this deformation, a fact that ultimately depends on $T_2(\Sigma) = T^2(\Sigma) = 0$ for a space curve. For details we refer to [J-S2].

Example (3.3): **The Pinkham - Pellikaan example.**

Let $F(x,y,z;a,b,c,\mu) := X^2 + Y^2 + Z^2 + 2.\lambda(XY + YZ + ZX) + 2\mu xyz$, where

$X := (y-b)(z+c) + 4bc$; $Y := (z-c)(x+a) + 4ac$; $Z := (x-a)(y+b) + 4ab$

and where λ is a fixed complex number, $\lambda^2 \neq 1$.

Let $X(a,b,c,\mu) := \{(x,y,z) | F(x,y,z;a,b,c,\mu) = 0\}$.

The surface $X := X(0,0,0,0)$ is just the cone over a three-nodal quartic in $\mathbb{P}^2$, with singular locus defined by the ideal $I = (yz, zx, xy)$. Hence its normalization $\tilde{X}$ is the cone over the rational normal curve of degree 4 in $\mathbb{P}^4$. This singularity has two different smoothing components, as H. Pinkham discovered [Pi]. The surface X has two different disentanglement components, a fact discovered by R. Pellikaan [Pe1], [Pe2],Ex.2.4. The surfaces $X(a,b,c,0)$ are fibres over the big component, $X(0,0,0,\mu)$ over the small component. Below a graphical impression of the real part of these surfaces is given. ($\lambda < -1$.)

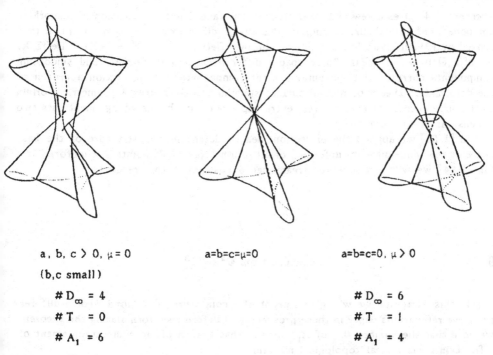

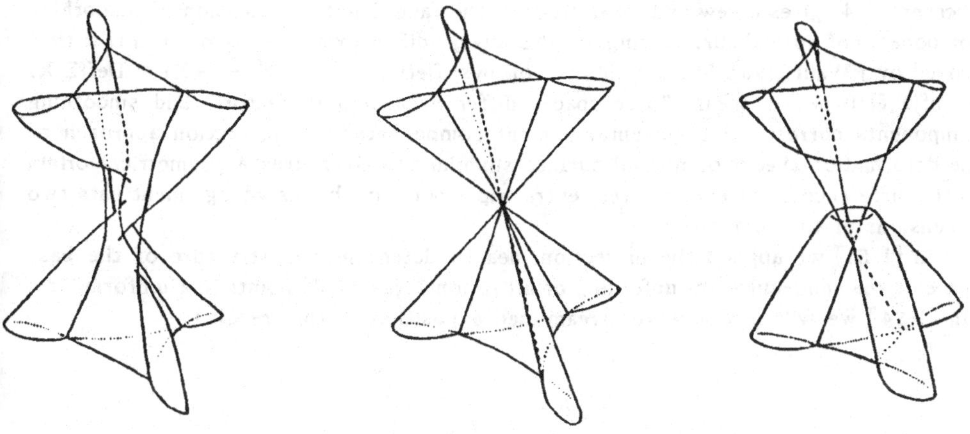

$a, b, c > 0, \mu = 0$ (b,c small) $\qquad$ $a=b=c=\mu=0$ $\qquad$ $a=b=c=0, \mu > 0$

	$a, b, c > 0, \mu = 0$ (b,c small)	$a=b=c=0, \mu > 0$
$\# D_\infty$	= 4	= 6
$\# T$	= 0	= 1
$\# A_1$	= 6	= 4

Probably these pictures should be considered as an artists impression; we challenge computer graphicians to provide better ones! We remark that the A_1 - points cannot be real all at the same time.

Theorem (3.4):

Let X be a germ of a weakly normal surface singularity in $\mathbb{C}^3$, with singular locus Σ, defined by a function $f \in \mathbb{C}\{x,y,z\}$. Then dimensions of disentanglement components differ by even numbers.

proof : As $\mathrm{Def}(\Sigma,f) \longrightarrow \mathrm{Def}(\Sigma,X)$ is smooth, it suffices to consider disentanglement components of f. For those of $\mathrm{Def}(\Sigma,f)$ we have by (2.9) that the dimension is equal to $\#A_1$, the number of A_1 - points that split off. We have the following formulae:

- $j(f)\ (:= \dim(I/(J_f)) = \#A_1 + \#D_\infty$ (see [Pe2])
- $VD_\infty(f) = \#D_\infty - 2.\#T$ (see [Jo])

Here $VD_\infty(f)$ is the so-called 'virtual number of D_∞- points of f as introduced in [Jo]. The left hand sides are invariants of f and do not refer to any deformation of f. Hence: $\#A_1 = (j(f) - VD_\infty(f)) - 2.\#T$, and so $\#A_1$ is a mod 2 invariant of f. ⊠

Remark (3.5):

Theorem (3.4) gives a new and local proof of the fact that the dimension of smoothing components of normal surface singularities always differ by an even number, a fact first proved by J. Wahl [Wa]. We see this as follows: $\mathrm{Def}(\tilde{X}) \sim \mathrm{Def}(\tilde{X} \longrightarrow X) \approx \mathrm{Def}(\Sigma,X) \sim \mathrm{Def}(\Sigma,f)$ (where ~ means:"base spaces differ by a smooth factor") and smoothing components correspond to disentanglement components. Our projection approach to the deformation theory of normal surface singularities thus gives a geometrical origin to the difference in dimension: every extra triple point in the disentanglement eats two dimensions of the component.

In [J-S1] we applied the projection idea to determine the structure of the base space of the semi-universal unfolding of all rational quadruple points in a uniform way. (In [J-S4] we will give a more streamlined exposition of this result.)

§4 Mappings from $\mathbb{C}^2$ to $\mathbb{C}^3$

In this paragraph we will give a proof of a conjecture of D. Mond. (For a different proof we refer to his paper in these proceedings.) Before even formulating the theorem, we note that the number $\#A_1$ of A_1 - points that branch off in a disentanglement of a function f has a clear topological meaning :

Lemma (4.1):

Consider a disentanglement $(\Sigma_\Delta, f_\Delta) \longrightarrow \Delta$ of function $f \in \mathbb{C}\{x,y,z\}$ defining a weakly normal surface X with double locus Σ, over a disc Δ. Let $X_t = f_t^{-1}(0)$, $t \neq 0$, be the disentanglement fibre, Σ_t its singular locus and $\tilde{X}_t$ its normalization. Then we have:

1) $\chi(X_t) - 1 = \#A_1$

2) $\chi(\Sigma_t) - 1 = 2.\#T - \mu(\Sigma)$

3) $\chi(\tilde{X}_t) = \chi(X_t) + \chi(\Sigma_t) - \#D_\infty + \#T$

where χ denotes the topological Euler characteristic.
(Of course, for these statements to make sense, one needs to take appropriate representatives. For simplicity of statement, we simply ignore this.)

Sketch of proof : 1) and 2) are "jump formulae" computing the jump in topology in terms of local data. 1) is just a very special case of a general result for functions. (We refer to the paper of D. Siersma in these proceedings [Si]. In fact, X_t has the homotopy type of a wedge of $\#A_1$ 2-spheres, see also [Mo3].) We only have to remark that during the disentanglement the fibration at the boundary of the Milnor sphere does not change, essentially because outside 0 the surface X has only A_∞ - singularities, which are rigid for admissible deformations. Formula 2) is just the jump property of the milnor number $\mu(\Sigma)$ of a curve singularity (see [B-G]). Formula 3) is an easy exercise in topology. ⊠

Now consider a map-germ $\varphi : (\mathbb{C}^2, 0) \longrightarrow (\mathbb{C}^3, 0)$. The space of first order deformations of this diagram, $T^1(\mathbb{C}^2 \xrightarrow{\varphi} \mathbb{C}^3)$, is the same as the space of first order deformations of φ , modulo left-right equivalence:

$$T^1(\mathbb{C}^2 \xrightarrow{\varphi} \mathbb{C}^3) = \varphi^*\Theta_{\mathbb{C}^3} / \left(d\varphi.\Theta_{\mathbb{C}^2} + \varphi^{-1}\Theta_{\mathbb{C}^3}\right)$$

The dimension of this vector space is called the $\mathcal{A}_e$ - codimension of φ, cod(φ), and if this number is finite, φ has a semi-universal unfolding with of course a *smooth* base space of this dimension. In [Mo1], D.Mond started to classify such φ with small $\mathcal{A}_e$-codimension. In [Mo2], he posed a question, which is equivalent to the following:

Conjecture of D. Mond (4.2):

Let $\varphi : (\mathbb{C}^2, 0) \longrightarrow (\mathbb{C}^3, 0)$ a map-germ with $\text{cod}(\varphi) < \infty$. Let X be an appropriate representative of the image-germ $\varphi(\mathbb{C}^2, 0)$, and put $\tilde{X} = \varphi^{-1}(X)$. (So $\tilde{X}$ is just a small neighbourhood of 0 in $\mathbb{C}^2$.) Let φ_t be a generic perturbation of φ, with $t \in \Delta$, a small disc . Then one has:

$$\text{cod}(\varphi) \leq \chi(\varphi_t(\tilde{X})) - 1$$

with equality in case that φ is *quasi-homogeneous*.

proof : Because cod(φ) $<\infty$, the surface X is weakly normal, with double locus Σ. Let f=0 be an equation for X. The map $\varphi : \tilde{X} \longrightarrow X$ can be identified with the normalization map of X. We have: Def(Σ,f) $\sim$ Def(Σ,X) = Def($\tilde{X} \longrightarrow X$), so Def($\Sigma$,f) and Def($\Sigma$,X) have *smooth* base spaces. On the other hand, $X_t = \varphi_t(\tilde{X})$ can be seen as a disentanglement fibre, so by (4.1) , (2.9) and (2.3):

$$\chi(X_t) - 1 = \#A_1 = \dim T^1(\Sigma,f) \geq \dim T^1(\Sigma,X) = \mathrm{cod}(\varphi)$$

Equality holds when f or, what is easily seen to be equivalent, φ is quasi-homogeneous.

⊠

Remark (4.3):

In the mean time D. Mond generalized his question or conjecture. It is the same as (4.2), only now for map-germs $\varphi : \mathbb{C}^n \longrightarrow \mathbb{C}^{n+1}$. We remark that our proof would generalize to this situation *if we had a positive answer to question (2.10).*

References

[Bu]: R.-O. Buchweitz; *Thesis*, Universite Paris VII, (1981).

[B-G]: R.-O. Buchweitz & G.-M. Greuel; *The Milnor Number and Deformations of Complex Curve Singularities*, Inv. Math. 58, (1980), 241-281.

[G-L]: G.-M. Greuel and E. Looijenga; *The Dimension of Smoothing Component*, Duke Math. J. 52, (1985), 263-272.

[Ha]: H. Hauser; ' *Sur la construction de la deformation semi-universelle d'une singularite isolee* ', Thesis, Universite de Paris Sud,(1980).

[Jo]: T. de Jong; *The Virtual Number of D_∞ points*, Topology 29, 175-184 ,(1990).

[J-S1]: T. de Jong and D. van Straten; *Deformations of Non-Isolated Singularities*, Preprint Utrecht, also part of the thesis of T. de Jong, Nijmegen, 1988.

[J-S2]: T. de Jong and D. van Straten; *A Deformation Theory for Non-Isolated Singularities*, To appear in Abh. Math. Sem. Hamburg.

[J-S3]: T. de Jong and D. van Straten; *Deformations of the Normalization of Hypersurface Singularities*, To appear in Math. Ann.

[J-S4]: T. de Jong and D. van Straten; *On the Base Space of a Semi-universal Deformation of Rational Quadruple Points*, Submitted to Ann. of Math.

[Mo1]: D. Mond; *On the Classification of Germs of Maps from $\mathbb{R}^2$ to $\mathbb{R}^3$*, Proc. Lond. Math. Soc.(3),50(1985), 333-369.

[Mo2]: D. Mond; *Some Remarks on the Geometry and Classification of Germs of Maps from Surfaces to 3-space*, Topology 26, (1987), 361-383.

[Mo3]: D. Mond; *Vanishing Cycles for Analytic Maps*, In these Proceedings.

[Pe 1]: R. Pellikaan; *Hypersurface Singularities and Resolutions of Jacobi Modules*, Thesis, Rijksuniversiteit Utrecht, (1985).

[Pe 2]: R. Pellikaan; *Deformations of Hypersurfaces with a One-Dimensional Singular Locus*, Preprint Vrije Universiteit Amsterdam, (1987).

[Pe 3]: R. Pellikaan, *Projective Resolutions of the Quotient of Two Ideals*, Indag. Math. 50, (1988), 65-84.

[Pi]: H. Pinkham; *Deformations of Algebraic Varieties with G_m - action*, Asterisque 20, (1974).

[Si]: D. Siersma; *Vanishing Cycles and Special Fibres*, In these Proceedings.

[Wa]: J. Wahl, *Smoothings of Normal Surface Singularities*, Topology 20, (1981), 219-246.

——— 0 ———

Theo de Jong * Duco van Straten **

Universität Kaiserslautern
Fachbereich Mathematik
Erwin Schrödinger straße, Geb. 48
D - 6750 Kaiserslautern
West-Germany.

*) supported by a stipendium of the European Community.
**) supported with a stipendium by the Deutsche Forschungs Gemeinschaft.

The Euler characteristic of the disentanglement of the image of a corank 1 map germ.

W. L. Marar[1]

Introduction

Let $f_0 : (\mathbb{C}^n,0) \to (\mathbb{C}^p,0)$, $2 \le n < p$ be a finitely $\mathcal{A}$-determined map germ with (n,p) in the range of nice dimensions in the sense of J. Mather. In this paper we express the Euler characteristic of the image of a stabilization f_t of f_0 (see definition below) in terms of the Milnor number of the multiple point schemes of f_0 ([4]), in the case where f_0 is of corank 1.

1. Stabilization and disentanglement

Let $f_0 : (\mathbb{C}^n,0) \to (\mathbb{C}^p,0)$, $2 \le n < p$ be as above. Let $F: (\mathbb{C}^{n+d},0) \to (\mathbb{C}^{p+d},0)$, $F(x,t)=(f_t(x),t)$ be an $\mathcal{A}_e$-versal unfolding of f_0, let $\pi : \mathbb{C}^{p+d} \to \mathbb{C}^d$ be the natural projection and $B \subsetneq \mathbb{C}^d$ the bifurcation set of F.

(1.1) Theorem: There exists an $\varepsilon > 0$, neighbourhoods U of the origin of $\mathbb{C}^{n+d}$ and T of the origin of $\mathbb{C}^d$ and a proper, finite to one representative of the unfolding $F:U \to B_\varepsilon \times T$ (good representative) such that, if $X = F(\overline{U})$ then the stable type stratified mapping $F: \overline{U} \cap F^{-1}(\overline{B}_\varepsilon \times (T-B)) \to X \cap (\overline{B}_\varepsilon \times (T-B))$ is locally topologically trivial over T-B with respect to the stratified submersion $\pi : \overline{B}_\varepsilon \times (T-B) \to T-B$.

Proof: The proof uses standard techniques in stratification theory, in particular Thom's Second Isotopy Lemma (see [5]). The stratification of the image X of F by stable types (see [1]) is locally analytically trivial, and hence certainly Whitney regular over T-B (see [3])□.

Thus, we obtain :

(i) a 'fibration' of the mapping $F:\overline{U} \to X$ whose 'fibre' over a parameter $t \in T-B$ is the stable mapping $f_t : U_t \to X_t$, where $U_t=\{x \in \mathbb{C}^n : (x,t) \in \overline{U}\}$ is contractible and $X_t=X \cap (\overline{B}_\varepsilon \times \{t\})$.

---

[1] partially supported by CNPq and FAPESP

(ii) a topological fibration of the image X of F, locally trivial over T-B, whose fibre over $t \in T-B$ is the image X_t of the stable mapping f_t.

Definition: f_t as above is called a *stabilization* of $f_0 : (\mathbb{C}^n,0) \to (\mathbb{C}^p,0)$ and its image X_t the *disentanglement* of the image of f_0.

2. Multiple point schemes.

In [4] the *multiple point schemes* of a mapping $g : \mathbb{C}^n \to \mathbb{C}^p$, $2 \le n < p$ are introduced. If g is of corank at most 1, then the k-tuple point scheme embed in $\mathbb{C}^{n-1} \times \mathbb{C}^k$.

If $\gamma(k)=(r_1,\dots,r_m)$ is an ordered partition of an integer k, i.e., $\Sigma\, r_i=k$ and $r_i \ge r_{i+1}$ then we denote by $\tilde{D}^k(g,\gamma(k)) \subseteq \mathbb{C}^{n-1} \times \mathbb{C}^k$ the k-tuple point scheme of g associated to the partition $\gamma(k)$.

If g is a corank 1 map then a generic point y of $\tilde{D}^k(g,\gamma(k))$ is of the form $(x,y_1,\dots,y_1,\dots,y_m,\dots,y_m)$, with $x \in \mathbb{C}^{n-1}$, $y_i \in \mathbb{C}$, y_i repeated r_i times, $y_i \neq y_j$ for $i \neq j$, $g(x,y_1)=\dots=g(x,y_m)$ and the local algebra of g at (x,y_i) isomorphic to $\dfrac{\mathbb{C}[z]}{(z^{r_i})}$.

(2.1) Theorem: Let $f : (\mathbb{C}^n,0) \to (\mathbb{C}^p,0)$, $2 \le n < p$ be a corank 1 map germ. Then, f is finitely $\mathcal{A}$-determined (respectively stable) if and only if $\tilde{D}^k(f,\gamma(k))$ is an ICIS (respectively smooth) of dimension $p-k(p-n+1)+m$, if not empty □.

For the proof see [4].

Let $f_0 : (\mathbb{C}^n,0) \to (\mathbb{C}^p,0)$, $2 \le n < p$ be a finitely $\mathcal{A}$-determined corank 1 map germ. Let $F : (\mathbb{C}^{n+d},0) \to (\mathbb{C}^{p+d},0)$, $F(x,t)=(f_t(x),t)$ be an $\mathcal{A}_e$-versal unfolding of f_0. Thus $\left(\tilde{D}^k(F,\gamma(k)),0\right)$ is an ICIS. Let $\pi_{\gamma(k)}$ be the restriction to $\tilde{D}^k(F,\gamma(k))$ of the projection from $\mathbb{C}^{n-1} \times \mathbb{C}^k \times \mathbb{C}^d$ to $\mathbb{C}^d$. According to theorem (2.1), the bifurcation set B of F contains the disciminant set of all mappings $\pi_{\gamma(k)} : \tilde{D}^k(F,\gamma(k)) \to T$. Therefore, the projection $\tilde{D}^k(F,\gamma(k)) \cap \pi^{-1}_{\gamma(k)}(T-B) \to T-B$ defines a Milnor fibration with typical fibre $\tilde{D}^k(f_t,\gamma(k))$ and critical fibre $\tilde{D}^k(f_0,\gamma(k))$ (see [3]).

So, over a parameter $t \in T-B$ we have the following diagram:

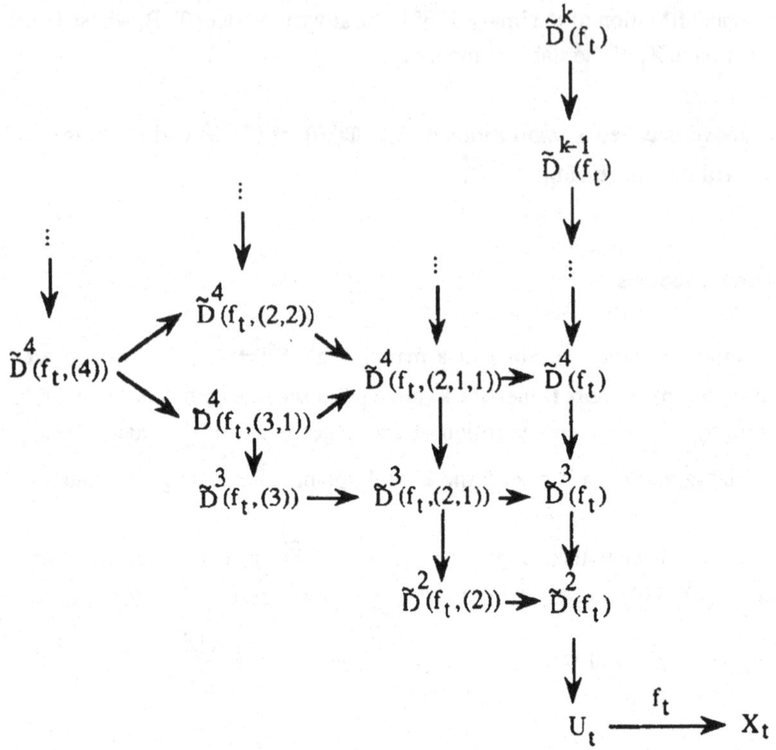

Notation: (1^k) stands for the partition $(1,\ldots,1)$ of k and $\tilde D^k(f_t,(1^k)) = \tilde D^k(f_t)$.

Let $\rho_{\gamma(k)}$ be the mapping obtained as composition of the inclusion $\tilde D^k(f_t,\gamma(k)) \hookrightarrow \tilde D^k(f_t)$, the projection $\tilde D^k(f_t) \to U_t$ and f_t.

3. Statement of the theorem

The mappings $\rho_{\gamma(k)}$ and f_t are (by construction) proper and finite to one. Moreover, f_t is branched over the points of the image of $\rho_{(1^2)}$ (and hence over the points of the image of all $\rho_{\gamma(k)}$) in X_t, elsewhere it is one to one. $\rho_{\gamma(k)}$ is branched over the points of the image of all $\rho_{\gamma'(r)}$ with $r \geq k$ and such that $\gamma(k) < \gamma'(r)$. Here the symbol $\gamma(k) < \gamma'(r)$ means that $\tilde D^r(f_t,\gamma''(r)) \supsetneqq \tilde D^r(f_t,\gamma'(r))$, where $\gamma''(r)=(\gamma(k),1,\ldots,1)$ is a partition of r.

We recall that since $\tilde D^k(f_t,\gamma(k))$ is a typical Milnor fibre of $\tilde D^k(F,\gamma(k))$ and $\tilde D^k(f_0,\gamma(k))$ is the critical fibre, we have: $\chi(\tilde D^k(f_t,\gamma(k)))=1+(-1)^s\mu(\tilde D^k(f_0,\gamma(k)))$, where s is the complex dimension of the ICIS $\tilde D^k(f_0,\gamma(k))$ and $\mu(\tilde D^k(f_0,\gamma(k)))$ its Milnor number.

So, to express the Euler characteristic of the disentanglement X_t in terms of the Milnor number of the multiple points schemes $\tilde{D}^k(f_0,\gamma(k))$ is equivalent to finding numbers $\beta_{\gamma(k)}$ such that:

$$\chi(X_t)=\beta_0\,\chi(U_t)+\sum_{k\geq 2}\sum_{\gamma(k)}\beta_{\gamma(k)}\,\chi(\tilde{D}^k(f_t,\gamma(k))), \qquad (*)$$

where $\gamma(k)$ runs through the set of all ordered partitions of k.

The properties of the mappings $\rho_{\gamma(k)}$ and f_t allow us to show, by combinatorial methods the following:

(3.1) Theorem : If $\gamma(k)=(r_1,\ldots,r_m)$, $r_i \geq r_{i+1}$, $\Sigma\, r_i=k$ and $\alpha_i=\#\{j : r_j=i\}$ then the coefficients in $(*)$ above are $\beta_0=1$, $\beta_{\gamma(k)}=\dfrac{-(-1)^{\Sigma\alpha_i}}{\prod_{i\geq 1} i^{\alpha_i}\alpha_i!}$ if the correspondent multiple point scheme $\tilde{D}^k(f_t,\gamma(k))$ is non-empty, and zero otherwise.

The proof goes as follows (details are given in section **4** below). Firstly we triangulate X_t in such way that the pull-back of this triangulation provides triangulations for U_t and all $\tilde{D}^k(f_t,\gamma(k))$. Those triangulations are such that the problem of proving the equality $(*)$ is reduced to proving that

$$C_0^{X_t}=\beta_0\,C_0^{U_t}+\sum_{k\geq 2}\sum_{\gamma(k)}\beta_{\gamma(k)}\,C_0^{\gamma(k)}, \qquad (**)$$

where $C_0^{X_t}$, $C_0^{U_t}$ and $C_0^{\gamma(k)}$ are respectively the number of zero-cells in the triangulation of X_t, U_t and $\tilde{D}^k(f_t,\gamma(k))$. Secondly, in studying the degrees of the mappings $\rho_{\gamma(k)}$ and f_t, we shall determine the number of zero cells in U_t and $\tilde{D}^k(f_t,\gamma(k))$ that come from the pull-back of each zero-cell of X_t. In this way, $(**)$ is equivalent to the system of equations:

$$\begin{bmatrix} 1 \\ 1 \\ 1 \\ 1 \\ \vdots \\ 1 \end{bmatrix} = M \begin{bmatrix} \beta_0 \\ \beta_{(1^2)} \\ \beta_{(2)} \\ \beta_{(1^3)} \\ \vdots \\ \vdots \end{bmatrix} \qquad (***)$$

where M is the square matrix whose entries are precisely the degrees of the mappings $\rho_{\gamma(k)}$

and f_t. M is constructed in the following way: its first column contains the degrees of f_t at a generic point of the image of f_t, $\rho_{(1^2)}, \rho_{(2)}, \rho_{(1^3)}, \rho_{(2,1)}, \rho_{(3)}, \rho_{(1^4)}, \ldots$ and so on for all partitions $\gamma(k)$, following the order $<$ on the partitions. The second column of M contains the degrees of the mapping $\rho_{(1^2)}$, the third column those of $\rho_{(2)}$, and so on, following the order $<$ on the partitions. Thus M is a non-singular lower triangular matrix and hence the system of equations above has unique solution.

Finally the solution will follow from the following:

(3.2) Lemma : Let $e^{(k)}$ be the kth elementary symmetric function on the variables $x_1,\ldots,x_q$, i.e., $e^{(k)} = \sum x_{i_1} \cdot \ldots \cdot x_{i_k}$, summed over $1 \le i_1 < \ldots < i_k \le q$. Let $\gamma(k)=(a_1,\ldots,a_h)$, with $a_i \ge a_{i+1}$, be a partition of k and $\alpha_i= \#\{j : a_j=i\}$. Then

$$e^{(k)} = \sum_{\gamma(k)} \frac{(-1)^{k-\Sigma\alpha_i}}{\prod_{i\ge1} i^{\alpha_i}\alpha_i!} \prod_{i\ge1}(x_1^i+\ldots+x_q^i)^{\alpha_i},$$

where $\gamma(k)$ runs through the set of all partitions of k.

For the proof see [2] p.17.

4. Proof of the theorem.

We start by triangulating the stable type stratified space X_t so that the interior of each simplex lies in a single stratum of X_t. Then we lift this triangulation to obtain triangulations for U_t and all $\tilde{D}^k(f_t,\gamma(k))$. Let $C_i^{X_t}$, $C_i^{U_t}$ and $C_i^{\gamma(k)}$ be respectively the number of i-cells in the triangulations of X_t, U_t and $\tilde{D}^k(f_t,\gamma(k))$. Then, the equality (*) above is equivalent to:

$$\sum_i(-1)^i C_i^{X_t} = \beta_0 \sum_i(-1)^i C_i^{U_t} + \sum_{k\ge2}\sum_{\gamma(k)} \beta_{\gamma(k)} \sum_i(-1)^i C_i^{\gamma(k)}.$$

Thus, if we find coefficients β_0 and $\beta_{\gamma(k)}$ (for all $\gamma(k)$) independent of i, such that

$$C_i^{X_t}=\beta_0 C_i^{U_t} + \sum_{k\ge2}\sum_{\gamma(k)} \beta_{\gamma(k)} C_i^{\gamma(k)}$$

for all i, $0 \le i \le 2n$, then these coefficients will be the solution of equation (*). So, let us concentrate on solving

$$C_0^{X_t}=\beta_0 C_0^{U_t} +\sum_{k\ge2}\sum_{\gamma(k)} \beta_{\gamma(k)} C_0^{\gamma(k)} \qquad (**)$$

The numbers $C_0^{U_t}$ and $C_0^{\gamma(k)}$ are obtained from $C_0^{X_t}$ according to the following:

Claim: If x is a generic point of U_t, (i.e., x is not on the image of the projection of $\tilde{D}^2(f_t)$ over U_t) then $f_t(f_t^{-1}(x)) = 1$.

If **y** is a generic point of $\tilde{D}^k(f_t,\gamma(k))$, where $\gamma(k)=(a_1,...,a_h)$, with $a_i \geq a_{i+1}$, then $\# f_t^{-1}(\rho_{\gamma(k)}(\mathbf{y})) = h$.

If **y** is a generic point of $\tilde{D}^k(f_t,\gamma'(k))$, where $\gamma'(k)=(b_1,...,b_q)$, with $b_i \geq b_{i+1}$ and $\gamma(k) < \gamma'(k)$ then $\#\rho_{\gamma(k)}^{-1}(\rho_{\gamma'(k)}(\mathbf{y}))$ is the coefficient of the monomial $x_1^{b_1}\cdot ... \cdot x_q^{b_q}$ in the polynomial $\prod_{i\geq 1}(x_1^{a_i}+...+x_q^{a_i})$.

If **y** is a generic point of $\tilde{D}^r(f_t,\gamma'(r))$, where $r > k$, $\gamma'(r)=(b_1,...,b_q)$, with $b_i \geq b_{i+1}$ and $\gamma(k) < \gamma'(r)$ then $\#\rho_{\gamma(k)}^{-1}(\rho_{\gamma'(r)}(\mathbf{y}))$ is the sum of the coefficients of the monomial $x_1^{c_1}\cdot ... \cdot x_q^{c_q}$ in the polynomial $\prod_{i\geq 1}(x_1^{a_i}+...+x_q^{a_i})$, with $(c_1,...,c_q) \in \mathbb{N}_0^q$ and such that $c_i \leq b_i$ and $\Sigma c_i=k$.

In fact, a generic point $\mathbf{y} \in \tilde{D}^r(f_t,\gamma'(r)) \subseteq \mathbb{C}^{n-1}\times \mathbb{C}^r$ is of the form $\mathbf{y}=(x,y_1,...,y_1,...,y_q,...,y_q)$, with $x \in \mathbb{C}^{n-1}$, $y \in \mathbb{C}$, $y_i \neq y_j$ for $i \neq j$ and y_i repeated b_i times. On the other hand the points of the fibre $\rho_{\gamma(k)}^{-1}(\rho_{\gamma'(r)}(\mathbf{y}))$ are the (generic and non-generic) points of $\tilde{D}^k(f_t,\gamma(k))$ whose coordinates are chosen out of the coordinates of **y** (the way to choose is such that the resulting point belongs to $\tilde{D}^k(f_t,\gamma(k))$). Thus, $\#\rho_{\gamma(k)}^{-1}(\rho_{\gamma'(r)}(\mathbf{y}))$ is the number of all such possible choices. Finally, we consider the following one to one correspondence between the generic points $\mathbf{y}=(x,y_1,...,y_1,...,y_q,...,y_q)$ of $\tilde{D}^r(f_t,\gamma'(r))$ and the monomials $x_1^{b_1},..., x_q^{b_q}$: namely, to the sequence of coordinates $y_i,...,y_i$ we associate $x_i^{b_i}$. So, the claim follows by expanding the product $\prod_{i\geq 1}(x_1^{a_i}+...+x_q^{a_i})$.

Now, if we consider the matrix M (constructed in section 3 above) whose entries are the degrees of the mappings f_t and $\rho_{\gamma(k)}$ then the equality (**) is equivalent to the system (***) whose equations are :

$$1=\beta_0\cdot 1$$
$$1=\beta_0\cdot 2+\beta_{(1^2)}\cdot 2$$
$$1=\beta_0\cdot 1+\beta_{(1^2)}\cdot 1+\beta_{(2)}\cdot 1$$
$$1=\beta_0\cdot 3+\beta_{(1^2)}\cdot 6+\beta_{(2)}\cdot 0+\beta_{(1^3)}\cdot 6$$

.

.

.

$$1=\beta_0\cdot q+\sum_{k=2}^{r}\sum_{\gamma(k)}\beta_{\gamma(k)}\cdot \#\rho^{-1}_{\gamma(k)}(\rho_{\gamma'(r)}(y)),$$

where $\gamma'(r)=(b_1,\dots,b_q)$, with $b_i \geq b_{i+1}$, is a partition of $s \geq 2$. However, if $\gamma(k)=(a_1,\dots,a_h)$, with $a_i \geq a_{i+1}$, and $\alpha_i=\#\{j : a_j = i\}$ then it follows from lemma (3.2) that

$$\sum_{\gamma(k)}\frac{(-1)^{\Sigma\alpha_i}}{\prod_{i\geq 1} i^{\alpha_i}.\alpha_i!}\ \#\rho^{-1}_{\gamma(k)}\ (\rho_{\gamma'(r)}(y)) = (-1)^{k-1}\binom{q}{k}.$$

Therefore, the coefficient $\beta_{\gamma(k)}$ must be equal to $\dfrac{-(-1)^{\Sigma\alpha_i}}{\prod_{i\geq 1} i^{\alpha_i}.\alpha_i!}$ (if the corresponding multiple point scheme $\tilde{D}^k(f_t,\gamma(k))$ is non-empty).

5. Some symmetries

The k-tuple point scheme $\tilde{D}^k(f_t)$ is embedded in $\mathbb{C}^{n-1}\times\mathbb{C}^k$ where it lies invariant under the action of the symmetric group S_k, which permutes the coordinates in $\mathbb{C}^k$ ([4]). Now, considering the finite mapping $\rho^{(k)}$: $\tilde{D}^k(f_t) \to \tilde{D}^k(f_t)/S_k$ we can repeat the combinatorial procedure described in sections 3 and 4 above, to obtain:

(5.1) Proposition : For each $k \geq 2$, $\chi(\tilde{D}^k(f_t)/S_k) = \sum_{\gamma(k)} \upsilon_{\gamma(k)}\,\chi(\tilde{D}^k(f_t,\gamma(k)))$,

where $\gamma(k) = (r_1,\dots,r_m)$, with $r_i \geq r_{i+1}$, runs through the set of all partitions of k,

$\alpha_i=\#\{j : r_j = i\}$ and $\upsilon_{\gamma(k)}=\dfrac{1}{\prod_{i\geq 1} i^{\alpha_i}.\alpha_i!}$ (or zero if the corresponding multiple point scheme $\tilde{D}^k(f_t,\gamma(k))$ is empty).

Proof: This is analogous to the proof in section 4 above and is a consequence of the following:

(5.2) Lemma: Let h_r be the rth complete symmetric function in the variables $x_1,...,x_q$, i.e., h_r is the sum of all monomials of degree r in the variables $x_1,...,x_q$. Then

$$h_r = \sum_{\gamma(k)} \frac{1}{\prod_{i\geq 1} i^{\alpha_i}\alpha_i!} \prod_{i\geq 1} (x_1^i + ... + x_q^i)^{\alpha_i},$$

where $\gamma(k)$ runs through the set of all ordered partitions of k.

For the proof see [2] p.16.

Using (5.1) to replace $\chi(\tilde{D}^k(f_t))$ by $\chi(\tilde{D}^k(f_t)/S_k)$ in (*) and using the fact that $\tilde{D}^k(f_t,\gamma(k))$ and $\tilde{D}^k(f_t)/S_k$ are Milnor fibres, we obtain:

(5.3) Theorem: Let X_t be the disentanglement of the image of a finitely $\mathcal{A}$-determined corank 1 map germ $f_0 : (\mathbb{C}^n,0) \to (\mathbb{C}^p,0)$, $2 \leq n < p$. Then, if all the multiple point schemes of f_0 are non-empty,

$$\chi(X_t) = 1 + \sum_{k=2}^{p} [(-1)^{k+1} + \sum_{\gamma(k)} \frac{(-1)^k - (-1)^{\Sigma\alpha_i}}{\prod_{i\geq 1} i^{\alpha_i}\alpha_i!}] +$$

$$+ \sum_{k=2}^{p} (-1)^{p-k(p-n+1)+1} [\mu(\tilde{D}^k(f_0)/S_k) +$$

$$+ \sum_{\gamma(k)} \frac{1-(-1)^{k+1+\Sigma\alpha_i}}{\prod_{i\geq 1} i^{\alpha_i}\alpha_i!} \mu(\tilde{D}^k(f_0,\gamma(k)))],$$

where $\gamma(k)=(r_1,...,r_m) \neq (1^k)$, with $r_i \geq r_{i+1}$, runs through the set of all partitions of k, and $\alpha_i = \#\{j : r_j = i\}$.

Remark: If p=n+1 then $\chi(X_t)$ is semicontinuous with respect to the parameter t since $(-1)^{p-k(p-n+1)+1}$ is independent of k. Futhermore, D. Mond ([6]) proved that, in this case, the disentanglement X_t has the same homotopy type of a wedge of n-spheres. Hence, the formula above express the number of spheres in that wedge.

Acknowlegment: I am very grateful to David Mond for his constant help. Thanks are also due to T. Gaffney anf F. Kouwenhover for helpful conversations.

References

[1] T. Gaffney, Polar multiplicities and equisingularity of map germs, Topology (to appear)

[2] I. Macdonald, Symmetric functions and Hall polynomials, Oxford math. monographs, Oxford University Press, 1977.

[3] W. L. Marar, Mapping fibrations and multiple point schemes, Ph.D. Thesis, University of Warwick,1989.

[4] W. L. Marar and D. Mond, Multiple point schemes for corank 1 maps, J. London Math. Soc. (2) 39 (1989), 553-567.

[5] J. N. Mather, Stratifications and mappings, in Dynamical Systems, M. Peixoto (ed.), Academic Press, 1973, 195-232.

[6] D. Mond, Vanishing cycles for analytic maps. These Proceedings.

Address:

ICMSC-USP
Caixa Postal 668
13560-São Carlos (SP)
Brazil

Vanishing cycles for analytic maps

David Mond

In this paper we discuss the topology of the image of a stable perturbation of a finitely determined map-germ $f_0:(\mathbb{C}^n,0)\to(\mathbb{C}^{n+1},0)$, and its relation to the number of parameters necessary for a versal unfolding of f_0, $\mathrm{cod}(\mathcal{A}_e,f_0)$. The results are the following:

Theorem 1 If $f_0:(\mathbb{C}^n,0)\to(\mathbb{C}^{n+1},0)$ is finitely $\mathcal{A}$-determined, with (n,n+1) nice dimensions (i.e. $n\le 6$), and $f_t:B_\varepsilon\to\mathbb{C}^{n+1}$ is a good representative of a stable perturbation of f_0, then the image X_t of f_t has the homotopy type of a wedge of n-spheres.

The term "good representative" of a stable perturbation, used in Theorem 1, will be explained in Section 1 below. Theorem 1 may be read as "the vanishing image of a map-germ $(\mathbb{C}^n,0)\to(\mathbb{C}^{n+1},0)$ has the homotopy type of a wedge of n-spheres". The cycles carried by the spheres in the wedge, are the vanishing cycles of the title of this paper.

Theorem 2 If n=2, then the number σ of spheres in the wedge referred to in Theorem 1 satisfies

i) $\mathrm{cod}(\mathcal{A}_e,f_0)\le\sigma$

ii) If f is quasihomogeneous, then $\mathrm{cod}(\mathcal{A}_e,f_0)=\sigma$.

The number σ of vanishing spheres may be calculated from other geometrical invariants of the map-germ f_0; from 3.3 and 3.4 of [16] it follows that

$$\chi(X_t) = 1/2\{\mu(D^2_f)+C-4T+1\}$$

where C is the number of pinch-points, and T the number of triple points, in X_t, and D^2_f is the curve germ in $(\mathbb{C}^2,0)$ consisting of points x for which $f_0^{-1}(f_0(x)) \supsetneq \{x\}$. Generalisations of this formula, for map-germs $(\mathbb{C}^n,0)\to(\mathbb{C}^p,0)$ of corank 1, when p>n>2, are given in [15]. A related formula is given in [10].

Theorem 2 was first observed empirically (in the case of the simple singularities), merely as a relation between the vanishing Euler characteristic and the $\mathcal{A}_e$-codimension, in [19] and [16] and in this form was proved first by Theo de Jong and Ruud Pellikaan, (unpublished), and then by Theo de Jong and Duco van Straten ([11], these proceedings).

The proof given here was found jointly with Theo de Jong and Duco van Straten, and I am grateful for their agreement to use it here.

The corresponding result for germs of maps $(\mathbb{C},0) \to (\mathbb{C}^2,0)$ is easily verified; in consequence, we conjecture that the statement of Theorem 2 holds for all n. Other questions remain: what can be said of the image of a stable perturbation of a map-germ $f_0:(\mathbb{C}^n,0) \to (\mathbb{C}^{n+k},0)$ for $k\geq 2$? Examples seem to show that here there is no simple relation between the vanishing Euler characteristic of the image, and the $\mathcal{A}_e$-codimension of f_0, even when f_0 is quasihomogeneous.

In Theorems 1 and 2 consider, instead of germs $(\mathbb{C}^n,0) \to (\mathbb{C}^{n+1},0)$, germs of maps $(\mathbb{C}^{n+k},0) \to (\mathbb{C}^k,0)$. If we replace "image" by "fibre", and $\mathcal{A}$ by $\mathcal{K}$, we obtain statements of the well known theorems of Hamm [9], that the Milnor fibre has the homotopy type of a wedge of spheres, and of Greuel [8] and Looijenga and Steenbrink [14] relating the Tjurina and Milnor numbers of an isolated complete intersection singularity. The statements obtained by replacing "image" by "discriminant" (and retaining $\mathcal{A}$) are also true (though now the spheres making up the wedge are of dimension p-1): these are proved in [5]. The resemblance between all these results is striking; presumably there is a more general theorem, of which they are all special cases.

Conversations with Theo de Jong, Lê Dung Trang, James Montaldi, Ruud Pellikaan and Duco van Straten, on the topics discussed in this article, have been very helpful, and I thank them all. I also thank V.I. Arnold for indicating the need for a theory of vanishing cycles for analytic maps.

Illustration (on following page) The lower drawing shows an exploded view of the image X_t of a real stable perturbation f_t of the singularity H_2 (defined by $f_0(x,y) = (x,y^3,xy+y^5)$), exhibiting two pinch points (at α and β) and one triple point (at $\square$). This space is, by inspection, homotopy equivalent to a wedge of two 2-spheres. The upper figure shows a topological disc U_t in $\mathbb{R}^2$, whose image under f_t is X_t. The curve Σ of double points in X_t has been drawn with double thickness, as has $f_t^{-1}(\Sigma)$ in U_t, and $f_t(\partial U_t)$ has been drawn with triple thickness.

While, by Theorems 1 and 2, we expect the image of a complex stable perturbation of H_k to have the homotopy type of a wedge of k 2-spheres (for k is the $\mathcal{A}_e$-codimension), it is a curious fact that the singularities of the series H_k have *real* stable perturbations with a full complement of triple and pinch points, and having the homotopy type of a wedge of k 2-spheres. (See [19] for a description of the H_k, and [16] for the construction of the real stable perturbations.)

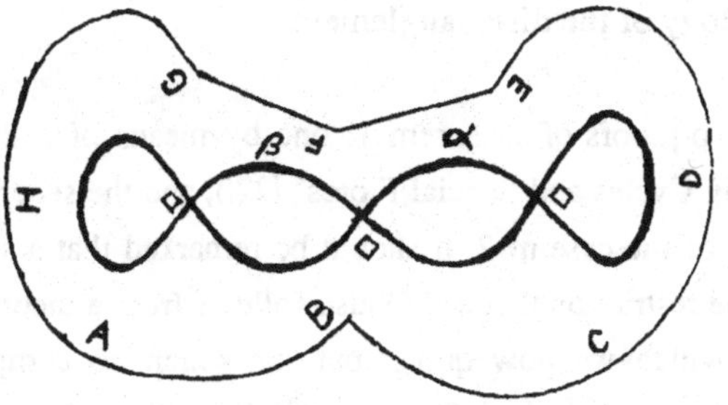
G
E
β
F
α
H
D
A
B
C

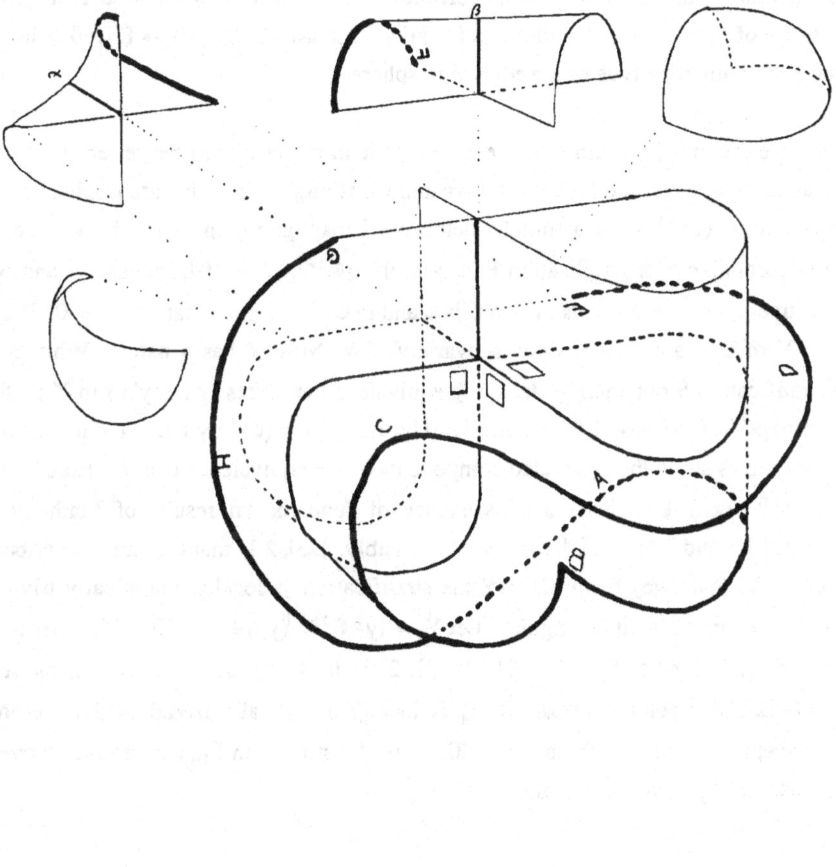
α
β
E
G
E
D
C
H
A
B

Section 1 The topology of the disentanglement

In this section we give two proofs of Theorem 1: one by means of a theorem of Dirk Siersma (2.1 in 'Vanishing Cycles and Special Fibres', [22]), and the second, by means of an elementary argument, for the case n=2. It should be remarked that a stronger version of Theorem 1 (without the restriction that $n \leq 6$) also follows from a theorem of Lê Dung Trang (see 4.4 of [13]) which we now quote, but for which no complete proof has appeared in print. The argument by which Theorem 1 follows from 1.1 is effectively the same as the proof of a similar theorem concerning the discriminant of a topological stabilisation of a finitely $\mathcal{A}$-determined map-germ $(\mathbb{C}^n,0) \to (\mathbb{C}^p,0)$ with $n \geq p$, which is given in § 5 of [4].

1.1 Theorem Let (X,x) be the germ of a complete intersection of dimension n+1 (not necessarily with isolated singularity) and let $h:(X,x) \to \mathbb{C}$ be the germ of a function having isolated singularity, in the sense of [13], Definition 2.7. Then if $h:B_\varepsilon \cap X \to \mathbb{C}$ is a good representative of h, and $\eta \ll \varepsilon$, h induces a fibration $B_\varepsilon \cap X \cap h^{-1}(D_\eta - 0) \to D_\eta - 0$ whose fibres have the homotopy type of a wedge of n-spheres. □

Before proving Theorem 1, we need to state it in rather more precise terms which make clear the notions of good representative and vanishing cycle which are implicit in it. So let $f_0:(\mathbb{C}^n,0) \to (\mathbb{C}^{n+1},0)$ be a finitely-determined map-germ, and let $F:U \to V$ be a proper representative of a stabilisation $F:(\mathbb{C}^n \times \mathbb{C},0) \to (\mathbb{C}^{n+1} \times \mathbb{C},0)$ of f_0, where U and V are, respectively, neighbourhoods of 0 in $\mathbb{C}^n \times \mathbb{C}$ and in $\mathbb{C}^{n+1} \times \mathbb{C}$, such that $F^{-1}(0) = 0$. Then the image Y of U is a closed analytic subvariety of V. Now Y has a natural Whitney-regular stratification $\mathcal{S}$ obtained by declaring equivalent the points (y,t), (y',t') in Y if the map-germs $f_t:(U_t, f_t^{-1}(y)) \to (\mathbb{C}^{n+1},y)$ and $f_{t'}:(U_{t'}, f_{t'}^{-1}(y')) \to (\mathbb{C}^{n+1},y')$ are $\mathcal{A}$-equivalent, and then taking as strata the connected components of the equivalence classes. Here U_t is the set $\{x \in \mathbb{C}^n \mid (x,t) \in U\}$. It is a consequence of fundamental results of Mather (in particular [17] 1.6 and 2.1) and [5], (or, more accessibly, [23], 2.1) that the strata described are smooth, and that away from $(0,0) \in Y$ the stratification is locally analytically trivial, and therefore *a fortiori* Whitney regular. Let $Y_t = \{y \in \mathbb{C}^{n+1} \mid (y,t) \in Y\}$. Then Y_t inherits a stratification $\mathcal{S}_t$, defined by $\mathcal{S}_t = \mathcal{S} \cap \mathbb{C}^{n+1} \times \{t\}$. $\mathcal{S}_t$ is itself regular: since f_t is stable for $t \neq 0$, and f_0 is stable for away from 0, $\mathcal{S}_t$ is locally analytically trivial (and therefore regular), except at 0 when t = 0; however, {0} is itself a stratum in $\mathcal{S}_0$, and regularity over a 0-dimensional stratum is automatic.

Now choose a *Milnor radius* for Y_0; that is, choose $\varepsilon > 0$ such that for all ε' with $0 < \varepsilon' \leq \varepsilon$, Y_0 is stratified transverse to the sphere $S^{2n+1}_{\varepsilon'}$. Then by the "conic structure lemma" (3.2 of [3]) $Y_0 \cap B_\varepsilon$ is a cone on its boundary $Y_0 \cap S^{2n+1}_\varepsilon$. It follows also that there exists $\eta > 0$ such that for all η' with $0 < \eta' \leq \eta$, Y_t is stratified transverse to S^{2n+1}_ε. Then one has

1.2 Theorem The projections $\pi_T: \mathbb{C}^{n+1} \times \mathbb{C} \to \mathbb{C}$ and $\pi_S: \mathbb{C}^n \times \mathbb{C} \to \mathbb{C}$ induce locally trivial fibre bundles $Y \cap (B_\varepsilon \times (D_\eta - 0)) \to D_\eta - 0$ and $F^{-1}(Y \cap (B_\varepsilon \times (D_\eta - 0))) \to D_\eta - 0$, and the family of mappings

$$\begin{array}{ccc} F^{-1}(Y \cap (B_\varepsilon \times (D_\eta - 0))) & \to & Y \cap (B_\varepsilon \times (D_\eta - 0)) \\ & \searrow \quad \swarrow & \\ & D_\eta - 0 & \end{array}$$

is locally trivial over $D_\eta - 0$. Moreover, the fibres of $\pi_S: F^{-1}(Y \cap (B_\varepsilon \times (D_\eta - 0))) \to D_\eta - 0$ are contractible. □

For the proof of this, and for more details on the construction described before 1.2, we refer the reader to Chapter 1 of [15]. A similar, slightly weaker result is proved by Lê in [12].

Following de Jong and van Straten in [10] we shall refer to the fibre of $\pi_T: Y \cap (B_\varepsilon \times (D_\eta - 0)) \to D_\eta - 0$ as a *disentanglement* of the germ of the image of f_0, and we shall denote it by X_t.

1.3 Lemma The topology of the disentanglement is independent of the choice of stabilisation.

Proof As f_0 is finitely determined, it has a versal unfolding $F: (\mathbb{C}^n \times \mathbb{C}^d, 0) \to (\mathbb{C}^{n+1} \times \mathbb{C}^d, 0)$. Since f_0 has stabilisations, the bifurcation set β is a proper analytic subset of the base $(\mathbb{C}^d, 0)$ of F, and thus does not separate it. By choosing appropriate representatives, one constructs a fibration π of the image $\mathcal{Y}$ of F over $\mathbb{C}^d - \beta$, cf. [15]. By definition of versality, any stabilisation of f_0 is induced from F by base change, and it follows that the corresponding disentanglement of f_0 is homeomorphic to the fibre of $\mathcal{Y} - \pi^{-1}(\beta) \to \mathbb{C}^d - \beta$. □

The existence of stabilisations is not guaranteed when (n,n+1) are not nice dimensions i.e. when $n > 6$. However, one can prove results similar to 1.2 and 1.3 for *topological* stabilisations, which exist even outside the nice dimensions, by replacing the stratification by local analytic type used in 1.2, by the pull-back via the jet extension map of the canonical stratification of the jet bundle.

1.4 Theorem Let $f_0:(\mathbb{C}^n,0) \to (\mathbb{C}^{n+1},0)$ be a finitely $\mathcal{A}$-determined map-germ, with $n \le 6$, and let X_t be a disentanglement of the image of f_0. Then X_t has the homotopy type of a wedge of n-spheres.

Proof This is an immediate consequence of proposition 2.3 of [22]. It is necessary only to show

(i) that all the fibres over $\mathbb{C}-0$ of an appropriately chosen defining equation $g_t:B_\varepsilon \to \mathbb{C}$ of the disentanglement X_t are either smooth, or have isolated singularities, and

(ii) that over the Milnor sphere S_ε^{2n+1}, g_t is a topologically trivial deformation of g_0.

The first condition requires a little work, although in the case n=2 it follows directly from a theorem of Ruud Pellikaan (quoted as 2.6 below). The proof in the general case follows from the following three facts:

1) $\mathrm{cod}(\mathcal{A}_e,f_0) = \mathrm{cod}(\mathcal{K}_{V,e}, j_0)$, where V is the image of a stable unfolding $F_S:(\mathbb{C}^m,0) \to (\mathbb{C}^{m+1},0)$ of f_0, $j_0: (\mathbb{C}^{n+1},0) \to (\mathbb{C}^{m+1},0)$ is an immersion which recovers f_0 from F_S as a fibre product (i.e. by base change), and $\mathcal{K}_V$-equivalence is just contact-equivalence in which the diffeomorphism of $(\mathbb{C}^{n+1}\times\mathbb{C}^{m+1},0)$ preserves $\mathbb{C}^{n+1}\times V$;

2) if $\mathrm{cod}(\mathcal{K}_{V,e}, j_0) < \infty$, then $\mathrm{cod}(\mathcal{K}_{H,e}, j_0) < \infty$, where now H is a defining equation for (V,0) and $\mathcal{K}_H$-equivalence is contact-equivalence in which the diffeomorphism of $(\mathbb{C}^{n+1}\times\mathbb{C}^{m+1},0)$ preserves not just $\mathbb{C}^{n+1}\times V$ but *all* fibres of H, and

3) let j_t be a deformation of j_0 and let g_t be the equation for the image of f_t obtained as $g_t = H\circ j_t$; if $j_t(y) \notin V$ then

$$\mathrm{cod}(\mathcal{K}_{H,e}, j_t)_y = \dim_{\mathbb{C}} \mathcal{O}_{\mathbb{C}^{n+1},y} / J_{g_t}\mathcal{O}_{\mathbb{C}^{n+1},y}.$$

Proofs of (1) and (3) may be found in [4], Theorem 1 and [5], Proposition 6.6 respectively. The proof of (2) is as follows: if $\mathrm{cod}(\mathcal{K}_{H,e}, j_0)$ is infinite, then there is a curve germ $(\Sigma,0) \subseteq (\mathbb{C}^{n+1},0)$ such that at each point y of $j_0(\Sigma)$,

$$tj_0(\theta_{\mathbb{C}^{m+1},y}) + j_0^*\mathrm{Derlog}(H)_y \subsetneq j_0^*(\theta_{\mathbb{C}^{m+1}})_y$$

and thus

$$dj_0(T_y\mathbb{C}^{m+1}) + \mathrm{Derlog}(H)_y(j_0(y)) \subsetneq T_{j_0(y)}\mathbb{C}^{m+1},$$

where Derlog(H) is the sheaf of germs of vector fields annihilating H. Off V, the fibres of

H are smooth, so $\mathrm{Derlog}(H)_y(j_0(y))$ is the tangent space to the fibre of H through $j_0(y)$, and by composing j_0 with the normalisation of Σ, we see that in fact H must be constant along $j_0(\Sigma)$. Hence $j_0(\Sigma) \subseteq H^{-1}(0) = V$. Now since $\mathrm{cod}(\mathcal{K}_{V,e}, j_0) < \infty$, at each point y of $\Sigma - 0$, j_0 is $\mathcal{K}_V$-stable; since (n,n+1) are in the nice dimensions, it follows from 3.23 of [5] that j_0 is also $\mathcal{K}_H$-stable, i.e. $tj_0(\theta_{\mathbb{C}^{m+1},y}) + j_0^*\mathrm{Derlog}(H)_y = j_0^*(\theta_{\mathbb{C}^{m+1}})_y$.

The proof of (i) is now as follows: first, from the upper semicontinuity of codimension in a deformation, we have $\sum \mathrm{cod}(\mathcal{K}_{H,e}, j_t)_y \leq \mathrm{cod}(\mathcal{K}_{H,e}, j_0)$, where the sum is taken over all points y in $\mathbb{C}^{n+1}$ where $\mathrm{cod}(\mathcal{K}_{H,e}, j_t)_y$ is non-zero, which move away from 0 during the deformation. As f_0 is finitely determined, then by (1) and (2), the right hand side of this inequality is finite, and then the finiteness of the Milnor number of each singularity of g_t occuring off $g_t^{-1}(0)$ follows from (3).

Now we prove (ii). Let $F:U \to V$ be a proper representative of the stabilisation F, as in the discussion preceding the statement of 1.2, and let $G:V \to \mathbb{C}$ be a defining equation of $Y = F(U)$. Denote by g_t the restriction of G to V_t. Then since $g_0^{-1}(0)$ is transverse to S_ε^{2n+1} and all remaining fibres of g_t have at most isolated singularities, it follows that there exist $\delta > 0$ and $\eta_1 > 0$ such that for $|t| \leq \eta_1$ and $|s| \leq \delta$, $g_t^{-1}(s)$ is transverse to S_ε^{2n+1}. It now follows by a standard argument that $\{g_t\}_{t \leq \eta_1}$ is a topologically trivial family on S_ε^{2n+1}. □

We now give a more geometrical proof of 1.4 in the case where n=2. It is based on the idea of using pinch points (cross-caps, or Whitney umbrellas) to cancel the obstructions to lifting, back to the source, a loop in the image of a map.

1.5 Proposition Suppose that U is a smooth complex surface, and that $p:U \to \mathbb{C}^3$ is a proper, locally stable map with the property that on each irreducible component of the curve $\Sigma \subseteq p(U)$ of singular points of p(U) there is at least one pinch-point. Then the induced homomorphism $p_*:\pi_1(U,u) \to \pi_1(p(U),p(u))$ is an epimorphism.

Proof Base points play no part in the argument, so we ignore them. Suppose that $\gamma:S^1 \to p(U)$ is a continuous path, representing an element of $\pi_1(p(U))$. Then γ can be approximated within the same homotopy class by a smooth, regular path which meets Σ at smooth points only in one of the two ways, whose real analogues are depicted in Figure 1 below: either it remains on the same sheet as it crosses Σ, as in Figure 1(a) or it changes sheet, as in Figure 1(b). Replace γ by such a path. The obstructions to a continuous lifting of γ are now presented only by points of the second kind.

Figure 1 (a)

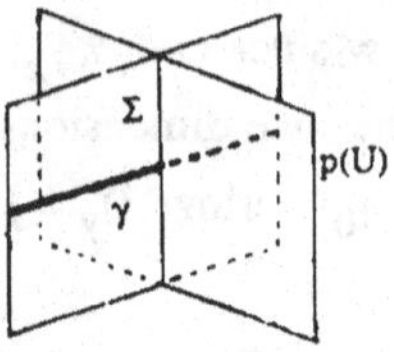

(b)

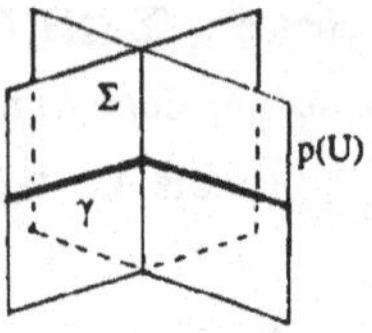

If on each irreducible component of Σ there is a pinch-point, we can perform a homotopy of γ as indicated in Figure 2: we pull the point q where γ crosses Σ, along Σ until it coincides with a pinch-point.

Figure 2 (a)

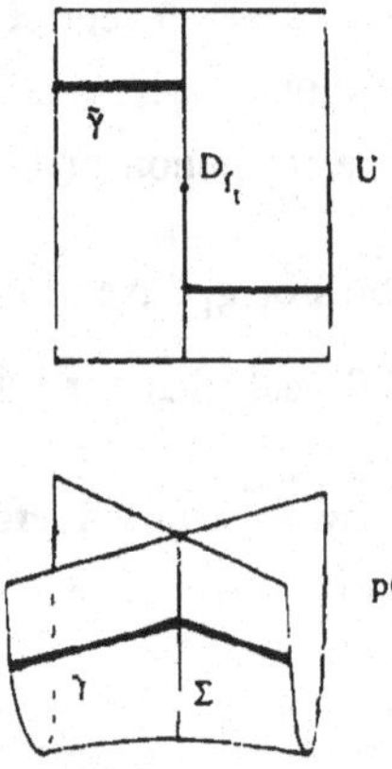

(b)

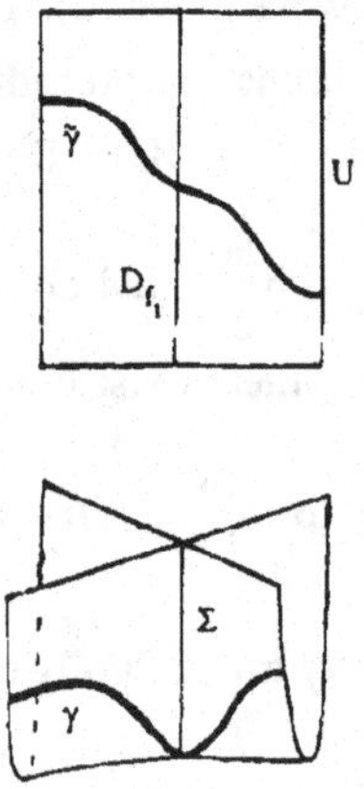

Evidently, the discontinuous lift $\tilde{\gamma}$ of γ shown in Figure 2(a), becomes continuous as soon as q reaches the pinch-point, as in 2(b). It follows that each element of $\pi_1(p(U))$ can be represented by a path in p(U) which has a continuous lift to a path in U. □

1.6 Corollary The disentanglement X_t of a finitely determined map-germ $f_0:(\mathbb{C}^2,0) \to (\mathbb{C}^3,0)$ is simply connected.

Proof Let Σ_t denote the set of double points in X_t, let $D_{f_t} = f_t^{-1}(\Sigma_t)$, and let $\tilde{D}_{f_t}$ be the closure in $U_t \times U_t$ of the set $\{(x_1,x_2) \mid f_t(x_1) = f_t(x_2), x_1 \neq x_2\}$. Similarly, define $\tilde{D}_F$ to be the closure in $U \times U$ of the set $\{((x_1,t_1),(x_2,t_2)) \mid F(x_1, t_1) = F(x_2, t_2), (x_1, t_1) \neq (x_2, t_2)\}$ Then for $t \neq 0$, $\tilde{D}_{f_t}$ is smooth; this is clear, since the only singularities of D_{f_t} are nodes, corresponding to the triple points in X_t , and these are resolved in $\tilde{D}_{f_t}$. In fact $\tilde{D}_F \cap ((B_\varepsilon \times D_\eta) \times (B_\varepsilon \times D_\eta))$ is a normal surface, since it is singular only at ((0,0),(0,0)) and thus regular in codimension 1, and is Cohen-Macaulay (see [16], Section 3). Moreover, theprojection $\tilde{D}_F \cap ((B_\varepsilon \times D_\eta) \times (B_\varepsilon \times D_\eta)) \to \mathbb{C}$ (where $\mathbb{C}$ is the parameter space of the stabilisation) exhibits $\tilde{D}_{f_t} \cap (B_\varepsilon \times B_\varepsilon)$ as a Milnor fibre of the isolated curve singularity $(\tilde{D}_{f_0},(0,0))$. It follows by Theorem 4.2.2 of [1] that $\tilde{D}_{f_t} \cap (B_\varepsilon \times B_\varepsilon)$ is *connected.* Thus, Σ_t as the image of $\tilde{D}_{f_t} \cap (B_\varepsilon \times B_\varepsilon)$ under projection to one copy of U, followed by f_t, is irreducible. Now Σ_t must contain at least one pinch-point, (except in the trivial case where f_0 is an immersion, when Σ_t is in fact empty and the corollary is trivially true); this is proved in 2.9 below. It follows that 1.5 applies, and, since X_t is the image of a simply connected set (one of the fibre of π_S in 1.2), it is simply connected itself □

1.7 Alternative proof of Theorem 1 for the case n=2 By general properties of affine complex analytic varieties, cf. [7] Part II Chapter 1,we know that X_t has the homotopy type of a CW complex of dimension ≤ 2. Since D_{f_t} is the fibre product of Σ_t and $f_t^{-1}(X_t)$, there is a Barratt-Whitehead exact sequence beginning $0 \to H_2(X_t, \mathbb{Z}) \to H_1(D_{f_t}, \mathbb{Z})$, and since $H_1(D_{f_t}, \mathbb{Z})$ is a free abelian group, $H_2(X_t, \mathbb{Z})$ is also free. By wedging together maps $S^2 \to X_t$ representing the free generators of $H_2(X_t, \mathbb{Z})$, we obtain a map which induces isomorphisms on all homotopy groups, and therefore is the desired homotopy equivalence between X_t and a wedge of 2-spheres. □

1.8 Remark The argument of 1.6 shows also that if f_t is a *real* stable perturbation of the map-germ $f_0:(\mathbb{R}^2,0) \to (\mathbb{R}^3,0)$ and if there is a non-immersive point on each connected and ireducible component of the curve Σ_t, then the image X_t of f_t is simply connected. This applies for example in the case of H_k.

Section 2 **The relation between the topology of the disentanglement and the $\mathcal{A}_e$-codimension**

The theorem of Siersma used in Section 1 establishes more than just that the disentanglement is homotopy-equivalent to a wedge of n-spheres. It also tells us how to count these spheres: their number is equal to the sum of the Milnor numbers of the isolated singularities in the fibres of g_t distinct from $g_t^{-1}(0)$, where g_t is a defining equation of X_t as in 1.5. This, together with a theorem of Ragni Piene ([21]), provides the key to the proof of Theorem 2 that we now give.

2.1 Proposition Let g_0 be a defining equation of the image $(X_0,0)$ of the finitely determined map-germ $f_0:(\mathbb{C}^n,0) \to (\mathbb{C}^{n+1},0)$. Then evaluation on g_0 defines an isomorphism of $\mathcal{O}_{\mathbb{C}^{n+1},0}$-modules

$$\frac{f_0^*(\theta_{\mathbb{C}^{n+1}})_{,0}}{df_0(\theta_{\mathbb{C}^n,0})+f_0^{-1}(\theta_{\mathbb{C}^{n+1}})_0} \to \frac{J_{g_0}\mathcal{O}_{\mathbb{C}^n,0}}{J_{g_0}\mathcal{O}_{X_0,0}}$$

Proof The elements of the top line on the left hand side this expression should be thought of as vector fields on (but not necessarily tangent to) X_0, with coefficients in $\mathcal{O}_{\mathbb{C}^n,0}$. Given such a vector field $\alpha = \sum_{i=1}^{n+1} \alpha_i\partial/\partial Y_i$, with $\alpha_i \in \mathcal{O}_{\mathbb{C}^n,0}$, define $ev_{g_0}(\alpha) = \sum_{i=1}^{n+1} \alpha_i\, \partial g_0/\partial Y_i$.

Clearly $ev_{g_0}(\alpha) \in J_g\mathcal{O}_{\mathbb{C}^n,0}$. We must now show that if $\alpha \in df_0(\theta_{\mathbb{C}^n,0})+f_0^{-1}(\theta_{\mathbb{C}^{n+1},0})$ then $ev_{g_0}(\alpha) \in J_{g_0}\mathcal{O}_{X_0,0}$. Suppose first that $\alpha = df_0(\zeta)$ At regular points of the image X_0, the vector field α is tangent to X_0; since g_0 is constant on X, $ev_{g_0}(\alpha)$ is equal to 0 at such points, and since regular points are dense in X_0, it follows that $ev_{g_0}(\alpha) = 0$. On the other hand, if $\alpha = \beta\circ f$, then clearly $ev_{g_0}(\alpha) \in J_{g_0}\mathcal{O}_{X_0,0}$, since then $\alpha_i \in \mathcal{O}_{X_0,0}$.

Thus, we do indeed have a well defined morphism. It is evidently surjective, so it remains only to show that it is injective. For this, we make use of a theorem of Ragni Piene (Theorem 1, but see also Example 1, of [21]), which in this context simply states that the quotient $c = (-1)^i(\partial g_0/\partial Y_i)/\det(df_1,\ldots, df_{i-1}, df_{i+1},\ldots, df_{n+1})$ (which is evidently independent of i), is analytic on X_0,0, and is in fact an $\mathcal{O}_{\mathbb{C}^n,0}$ generator of the conductor ideal $\mathcal{C}$ of $\mathcal{O}_{\mathbb{C}^n,0}$ in $\mathcal{O}_{X_0,0}$ (here the f_i, for i = 1, ..., n+1, are the component functions of f_0). Suppose that $\alpha \in f_0^*(\theta_{\mathbb{C}^{n+1},0})$, and that $ev_{g_0}(\alpha) = \sum \alpha_i\, \partial g_0/\partial Y_i$ lies in $J_{g_0}\mathcal{O}_{X_0,0}$. That is, there exist $\tilde{\alpha}_i \in \mathcal{O}_{X_0,0}$ such that $\sum(\alpha_i - \tilde{\alpha}_i\circ f_0)\, \partial g_0/\partial Y_i = 0$. Now $\mathcal{O}_{\mathbb{C}^n,0}$ is a domain, so we obtain $\sum(-1)^i(\alpha_i - \tilde{\alpha}_i\circ f_0)\det(df_1,\ldots, df_{i-1}, df_{i+1},\ldots, df_{n+1}) = 0$. It follows that the matrix obtained by adding to the $n\times(n+1)$ matrix of df_0 the n+1'st column $(\alpha_1 - \tilde{\alpha}_1,\ldots,\alpha_{n+1} - \tilde{\alpha}_{n+1})^t$ has determinant identically 0. At points p where the rank of df_0 is equal to n, there exist unique $\zeta_i \in \mathcal{O}_{\mathbb{C}^n,p}$, i = 1,..., n, such that this n+1'st column is equal to the sum of the ζ_i times the remaining columns; that is, $(\alpha_1 - \tilde{\alpha}_1,\ldots,\alpha_{n+1} - \tilde{\alpha}_{n+1})^t = \sum\zeta_i\, \partial f_0/\partial x_i$, or, in other words, $\sum(\alpha_i-\tilde{\alpha}_i)\partial/\partial Y_i = df_0(\sum\zeta_i\partial/\partial x_i)$ in $f_0^*(\theta_{\mathbb{C}^{n+1}})_{,p}$. As f_0 is by assumption finitely determined, the set of non-immersive points is of codimension at least 2; it follows by Hartog's theorem that the ζ_i extend to elements of $\mathcal{O}_{\mathbb{C}^n,0}$, and hence that $\sum(\alpha_i - \tilde{\alpha}_i \circ f_0)\,\partial/\partial Y_i = df_0(\sum \zeta_i \partial/\partial x_i)$ in $f_0{}^*(\theta_{\mathbb{C}^{n+1}}),0$. From this we conclude that $\alpha = df_0(\zeta) + \tilde{\alpha} \circ f_0$, as desired. □

From 2.1, it is immediate that $\operatorname{cod}(\mathcal{A}_e, f_0) = \dim_{\mathbb{C}}(J_{g_0}\mathcal{O}_{\mathbb{C}^n,0}/J_{g_0}\mathcal{O}_{X,0})$, for the module in the left hand side of the isomorphism of 2.1 is just $\theta(f_0)/tf_0(\theta_{\mathbb{C}^n,0}) + \omega f_0(\theta_{\mathbb{C}^{n+1},0})$, whose dimension as a complex vector space is, by definition, $\operatorname{cod}(\mathcal{A}_e, f_0)$. Now in order to relate this to the topology of the disentanglement, we make use of the exact sequence

$$(2.2) \qquad 0 \to J_{g_0}\mathcal{O}_{\mathbb{C}^n,0}/J_{g_0}\mathcal{O}_{X,0} \to C/J_{g_0}\mathcal{O}_{X,0} \to C/J_{g_0}\mathcal{O}_{\mathbb{C}^n,0} \to 0$$

The second arrow is just an inclusion, since the $\partial g_0/\partial Y_i$ belong to C, and the third arrow is projection.

2.3 Proposition Let $\mathcal{R} \subseteq \mathcal{O}_{\mathbb{C}^n,0}$ be the *ramification ideal* of f_0, generated by the $n \times n$ minor determinants of the matrix of df_0 and let c be an $\mathcal{O}_{\mathbb{C}^n,0}$ generator of the conductor ideal C. Then multiplication by c induces an isomorphism $\mathcal{O}_{\mathbb{C}^n,0}/\mathcal{R} \to C/J_{g_0}\mathcal{O}_{\mathbb{C}^n,0}$.

Proof This is immediate from the theorem of Ragni Piene quoted in the proof of 2.3. □

When n=2, the quotient $\mathcal{O}_{\mathbb{C}^2,0}/\mathcal{R}$ is of particular significance to us:

2.4 Lemma $\dim_{\mathbb{C}}(\mathcal{O}_{\mathbb{C}^2,0}/\mathcal{R})$ is equal to the number of pinch points in the disentanglement X_t.

Proof Let F be a stabilisation of f_0 and let $\mathcal{R}_F$ be the ideal in $\mathcal{O}_{\mathbb{C}^3,0}$ generated by the 3×3 minors of the 3×4 matrix of dF. Since $V(\mathcal{R}_F)$ is the set of of non-immersive points of F and hence has codimension 2, by the Hilbert-Burch theorem the quotient $\mathcal{O}_{\mathbb{C}^3,0}/\mathcal{R}_F$ is a Cohen Macaulay ring of Krull dimension 1 ([2]). It follows by the Auslander Buchsbaum formula (see e.g [18] Theorem 19.1) that if $\pi:\mathbb{C}^3 \to \mathbb{C}$ is projection onto the parameter space of the stabilisation, then $\pi_*(\mathcal{O}_{\mathbb{C}^{n+1},0}/\mathcal{R}_F)$ is a free $\mathcal{O}_{\mathbb{C}}$ module. Thus we have

$$\dim_{\mathbb{C}}\left(\pi_*(\mathcal{O}_{\mathbb{C}^3,0}/\mathcal{R}_F),t)\otimes_{\mathcal{O}_{\mathbb{C},t}}\mathcal{O}_{\mathbb{C},t}/m_t\right) = \dim_{\mathbb{C}}\left(\pi_*(\mathcal{O}_{\mathbb{C}^3,0}/\mathcal{R}_F),0)\otimes_{\mathcal{O}_{\mathbb{C},0}}\mathcal{O}_{\mathbb{C},0}/m_0\right)$$

$$= \dim_{\mathbb{C}}(\mathcal{O}_{\mathbb{C}^2,0}/\mathcal{R}).$$

The left hand side of the preceding equality is just the sum of contributions from each of the non-immersive points of f_t. As f_t is stable for $t \neq 0$, each of its non-immersive points is a pinch point, contributing 1 to the sum. □

It follows from 2.1, 2.2 , 2.3 and 2.4 that

(2.5) $$\dim_{\mathbb{C}}(C/J_{g_0}\mathcal{O}_{X,0}) = \mathrm{cod}(\mathcal{A}_e, f_0) + \#\text{pinch points in } X_t$$

Now let $\mathfrak{C}$ be the preimage of C with respect to the projection $\mathcal{O}_{\mathbb{C}^3,0} \to \mathcal{O}_{X,0}$. We use the following theorem of Ruud Pellikaan (Proposition 7.20(iii) of [20]):

2.6 Proposition $\dim_{\mathbb{C}}(\mathfrak{C}/J_{g_0}\mathcal{O}_{\mathbb{C}^3,0}) = \#$ pinch points in $X_t + \sum \mu(g_t; y_t)$
where the sum is taken over all isolated singular points in the fibres of the deformation g_t of g_0. □

The proof of Theorem 2 is now straightforward:

2.7 Proof of Theorem 2 From 2.5 and 2.6 we have

$$\sum \mu(g_t; y_t) = \mathrm{cod}(\mathcal{A}_e, f_0) + \dim_{\mathbb{C}}(\mathfrak{C}/J_{g_0}\mathcal{O}_{\mathbb{C}^3,0}) - \dim_{\mathbb{C}}(C/J_{g_0}\mathcal{O}_{X,0}).$$

By 2.1 of [22], the left hand side is equal to the number σ of 2-spheres making up the wedge to which X_t is homotopy-equivalent. On the other hand, we have

$$C/J_{g_0}\mathcal{O}_{X,0} = \mathfrak{C}/\left(J_{g_0}\mathcal{O}_{\mathbb{C}^3,0} + (g_0)\right)$$

and thus $\mathrm{cod}(\mathcal{A}_e, f_0) \le \sigma$. Moreover, if f_0 is quasihomogeneous, then g_0 is quasihomogeneous also, and hence lies in its Jacobian ideal, and it follows that $\mathrm{cod}(\mathcal{A}_e, f_0) = \sigma$ □

Bibliography

[1] R.O. Buchweitz and G.-M. Greuel, The Milnor number and deformations of complex curve singularities, Invent. Math. **58** (1980), 241-281

[2] L. Burch, On ideals of finite homological dimension in local rings, Math. Proc. Camb. Phil. Soc. **64** (1968), 941-946

[3] J. Damon, $\mathcal{A}$-equivalence and the equivalence of sections of images and discriminants, *these proceedings*

[4] J. Damon and D.Mond, $\mathcal{A}$-codimension and the vanishing topology of the discriminant, Preprint, University of Warwick, 1989

[5] T. Gaffney, Properties of finitely determined germs, Ph.D. thesis, Brandeis, 1976

[6] C.G. Gibson, A. A. du Plessis, E. J. N. Looijenga, K. Wirthmuller, Topological stability of smooth mappings, Lecture Notes in Math. 552, Springer Verlag, Berlin, 1977

[7] M. Goresky, R. MacPherson, Stratified Morse theory, Ergebnisse der Mathematik und ihrer Grenzgebiete, Springer Verlag, Berlin, 1988

[8] G.-M. Greuel, Dualitat in der lokalen kohomologie isolierter Singularitäten, Math. Ann. **250** (1980,) 157-173

[9] H. Hamm, Lokale topologische Eigenschaften komplexer Räume, Math. Annalen **191** (1971), 235-252.

[10] T. de Jong, D. van Straten, Deformations of surface singularities, preprint, University of Kaiserslautern, 1988

[11] T. de Jong, D. van Straten, Disentanglements, *these proceedings*

[12] Lê D. T., Some remarks on relative monodromy, *in* Real and complex singularities, Per Holm (ed.), Proceedings of Nordic Summer School, Oslo 1976, Sijthoff and Noordhoff, Alphen aan den Rijn, 1977.

[13] Lê D. T., Le concept de singularité isolée de fonction analytique, Advanced Studies in Pure Math. **8** (1986), Complex Analytic Singularities, 215-227

[14] E. J. N. Looijenga and J. H. M. Steenbrink, Milnor numbers and Tjurina numbers of complete intersections, Math. Annalen **271** (1985), 121-124

[15] W.L. Marar, The Euler characteristic of the disentanglement of the image of a corank 1 map germ, *these proceedings*

[16] W. L. Marar, D. Mond, Multiple point schemes for corank 1 maps, J. London Math. Soc. (2) **39** (1989), 553-567

[17] J. N. Mather, Stability of C^{∞} mappings IV: Classification of stable germs by R-algebras, Publ. Math. I. H. E. S. **37** (1969), 223-248

[18] H. Matsumura, Commutative ring theory, Cambridge University Press, Cambridge, 1986

[19] D. Mond, Some remarks on the geometry and classification of germs of maps from surfaces to 3-space, Topology **26**, 1987, 361-383

[20] G. R. Pellikaan, Hypersurface singularities and resolutions of Jacobi modules, Thesis, Rijksuniversiteit Utrecht, 1985

[21] R. Piene, Ideals associated to a desingularisation, Proc. Summer meeting Copenhagen 1978, Lecture Notes in Math. 732, Springer Verlag, Berlin, 1979, 503-517

[22] D. Siersma, Vanishing cycles and special fibres, *these proceedings*

[23] C. T. C. Wall, Finite determinacy of smooth map-germs, Bull. London Math. Soc. **13** (1981), 481-539

Mathematics Institute
University of Warwick
Coventry CV4 7AL

ON COMPLETE CONDITIONS IN ENUMERATIVE GEOMETRY

Ruud Pellikaan

§0 Introduction.

In enumerative geometry one deals with geometrical figures on which one imposes conditions. If the set of geometrical figures is a variety then one says that a condition is r-fold if the variety of figures which satisfy this condition has codimension r. In particular if one imposes an n-fold condition on geometrical figures of dimension n then one expects a finite number of solutions and one seeks to compute this number. Moreover varying this condition continuously, the "principle of continuity" also called the "principle of conservation of number", says that this number counted with appropriate multiplicities stays constant.

The first example of the above is Bézout's theorem [2]. Geometers like Chasles [4], Halphen [12] gave an abundance of examples culminating in Schubert's book "Kalkül der abzählenden Geometrie" [31], see also Zeuthen and Pieri [45], [46].

Hilbert poses as his 15th problem the question how the principles used by the enumerative geometers could be justified and whether the numbers obtained by them were correct, see [18].

After the foundational work of Severi [34], Van der Waerden [41], Weil [44] and Grothendieck [11] algebraic geometry got a rigorous basis. Algebraic varieties were defined and the language of schemes was developed to take account of non-reduced structures. In analogy with the cohomology ring in algebraic topology [40], the Chow ring was developed and Schubert calculus could be justified in any characteristic by doing the calculations in the Chow ring. Intersection theory was strongly developed by Fulton, Kleiman and MacPherson [8], [21]. Although the basic notions are well defined, it was still not clear whether all the principles underlying enumerative geometry were justified. This was clearly stressed by Kleiman [18]. For instance it is not obvious that the number one computes is obtained by taking the condition figures in general position, nor whether the "principle of conservation of number" is valid. Kleiman justified these for the class of enumerative problems where an algebraic group acts transitively on the geometrical figures [17]. The paradigm of enumerative geometry where a naive approach fails, are the plane conics. They are parametrized by $\mathbb{P}^5$ and the condition for a conic to be tangent to a given conic defines a hypersurface of degree 6 in $\mathbb{P}^5$. Hence one concludes, like Steiner did, that there are 6^5 conics tangent to 5 given conics, by Bézout's theorem. But the double lines are tangent to any given conic, hence one always has infinitely many solutions for any choice of the 5 given conics. Remark that the group of projective transformations does not act transitively on the plane conics, in fact one has three orbits: non-singular conics, the union of two lines and double lines.

One can proceed in two different ways. One can try to compute residual or excess intersections, that is an intersection theory where the intersections are not proper and one seeks to compute those solutions which are outside a base locus of degenerate solutions. Classical geometers like Severi developed a dynamic intersection theory. A modern treatment is given by Fulton and MacPherson [8] by their method of deforming to the normal cone and by Vogel and Stückrad [36], [38]. The link between them is given by Van Gastel [9]; he also gives a historical account of excess intersection theory. Up to now this approach only works if the geometrical figures are lying in projective n-space. Classically one remedied the naive approach by considering the collection of so called "complete conics", that is to say by considering a conic together with its dual. Chasles obtained the correct number 3264 of conics tangent to 5 given conics in case the characteristic is not 2. There are a lot of examples of complete geometrical figures, see (2.1), but no general definition seems to be known.

The primary aim of this paper is not to compute the numbers of a specific enumerative problem but to see which properties of a condition imply that the numbers one computes make sense, that is to say they are obtained by taking a generic choice of the condition figures, and the intersection of two conditions have again this property.

In Section 1 we define the notion of a condition and construct the sum and the intersection of two conditions and the pull back under a morphism of a condition and pose the question whether they are again conditions. In Section 2 we consider the construction of complete conics. In Section 3 we define the class of proper conditions which satisfies the sum, intersection and pull back property. In Section 4 we introduce the notion of a flat condition and show that it satisfies the intersection and pull back property. In Section 5 we prove that a condition has a flattening. In Section 6 we define Cohen-Macaulay conditions and show that they satisfy the intersection and pull back property and the principle of conservation of number. In Section 7 we give examples of conditions and consider their properties. In Section 8 we sketch a Schubert calculus on singular varieties.

§1 r-fold conditions.

Let k be an algebraically closed field. All schemes considered will be of finite type over k.

(1.1) **Definition.** Let X be a scheme of finite type over k. An *r-fold condition* on X is a triple (X,Γ,Y), where Y is a scheme of finite type over k and Γ is a closed subscheme of $X\times Y$, such that all its irreducible components have codimension r.

(1.2) **Some terminology.** Let $f:V\to W$ be a morphism of schemes over k. Let w be a point of W and $k(w)$ the residue field at w and let $\operatorname{Spec} k(w)\to W$ be the natural morphism. Then we define the fibre $f^{-1}(w)$ of f over the point w to be the scheme

$$f^{-1}(w)=V\times_W \operatorname{Spec} k(w),$$

see [13] II.3.3.

Let (X,Γ,Y) be an r-fold condition on X. We have two projections $\phi:X\times Y\to X$ and $\psi:X\times Y\to Y$. If Γ is a subscheme of $X\times Y$ then we denote the restrictions of ϕ and ψ to Γ by ϕ_Γ and ψ_Γ respectively. If y is a point of Y then we denote the fibre $\psi_\Gamma^{-1}(y)$ of ψ_Γ by Γ_y. By abuse of notation we denote $(X \times Y)_y$ by X_y. Now Γ_y is a subscheme of X_y. Similarly, if x is a point of X then define Γ^x to be the fibre $\phi_\Gamma^{-1}(x)$ and denote $(X \times Y)^x$ by Y^x. In enumerative problems X and Y will parametrize geometrical figures in some scheme Z. For instance, take for Z projective n-space and for the geometrical figures subschemes of a given dimension and degree or with a given Hilbert polynomial and take for Γ a relation between these figures like incidence or tangency. We abstract from this and forget that X and Y are parameter schemes of certain geometrical figures in some Z. We call X the scheme of geometrical figures and Y the scheme of condition figures and Γ a condition on X imposed by Y. For a point y of Y we call Γ_y the specialization of the condition Γ at y.

(1.3) **Proposition.** Let $f : V \to W$ be a flat morphism of schemes of finite type over k, and assume that W is irreducible. Then the following conditions are equivalent:

(i) Every irreducible component of V has dimension $\dim W + n$.

(ii) For any point w of W every irreducible component of the fibre $f^{-1}(w)$ has dimension n.

(iii) There exists an open dense subset U of W such that for any closed point w of U, every irreducible component of the fibre $f^{-1}(w)$ has dimension n.

Proof. See [13] III Corollary 9.6. and [11] IV_2 6.9.1.

(1.4) **Proposition.** Let X and Y be schemes of finite type over k. Suppose X is equidimensional and Y is integral. If (X, Γ, Y) is an r-fold condition on X then there exists an open dense subset V of Y such that Γ_y is empty for all points y of V or ψ_Γ is flat above V and all irreducible components of Γ_y have codimension r in X_y for all points y of V.

Proof. This follows from [11] IV_2 6.9.1 and Proposition (1.3).

(1.5) **Remark.** Let V be the open dense subset of Y mentioned in Proposition (1.4) and suppose $\psi_\Gamma^{-1}(V)$ is not empty. Define similarly the corresponding open dense subset U of X in case x is integral, after interchanging the roles of X and Y and suppose $\phi_\Gamma^{-1}(U)$ is not empty. Let $a = \dim X$, $b = a - r$, $c = \dim Y$ and $d = c - r$. Then $b = \dim \Gamma_y$ for all y of V and $d = \dim \Gamma^x$ for all x of U. Thus

$$\dim \Gamma = a + d = c + b.$$

This is called: *Prinzip der Konstantenzählung*, see [43].

(1.6) **Definition.** Let (X, Γ, Y) and (X, Λ, Z) be r-fold, respectively s-fold, conditions on X. Let

$$\tau : Y \times X \times Z \to X \times Y \times Z$$

be the isomorphism which interchanges the factors X and Y. Define the *sum* $(X, \Gamma + \Lambda, Y \times Z)$ of (X, Γ, Y) and (X, Λ, Z) by

$$\Gamma + \Lambda = (\Gamma \times Z) \cup \tau(Y \times \Lambda)).$$

Define the *product* or *intersection* $(X, \Gamma \circ \Lambda, Y \times Z)$ of (X, Γ, Y) and (X, Λ, Z) by

$$\Gamma \circ \Lambda = (\Gamma \times Z) \cap \tau(Y \times \Lambda)).$$

Define by induction

$$1\Gamma = \Gamma \text{ and } (n+1)\Gamma = (n\Gamma) + \Gamma$$

$$\Gamma^1 = \Gamma \text{ and } \Gamma^{n+1} = \Gamma^n \circ \Gamma.$$

(1.7) **Proposition.** Let (X, Γ, Y) and (X, Λ, Z) be r-fold, respectively s-fold, conditions on X and let x, y and z be closed points of X, Y and Z respectively. Then

(i) $$(\Gamma \circ \Lambda)_{(y,z)} = \Gamma_y \cap \Lambda_z$$

(ii) $$(\Gamma \circ \Lambda)^x \cong \Gamma^x \times \Lambda^x$$

where the schemes are considered as subschemes of X and Y, after identifying X_y and X_z with X and Y^x with Y.

Proof. (i) We have that $(\Gamma \circ \Lambda) = (\Gamma \times Z) \cap \tau(Y \times \Lambda)$.
Hence

$$\begin{aligned}(\Gamma \circ \Lambda)_{(y,z)} &= (\Gamma \times Z) \cap \tau(Y \times \Lambda) \cap X \times \{(y,z)\}\\ &= \Gamma_y \times \{(y,z)\} \cap \Lambda_z \times \{(y,z)\}\\ &= (\Gamma_y \cap \Lambda_z) \times \{(y,z)\}.\end{aligned}$$

We get the desired equality after the above mentioned identification.

(ii) $$\begin{aligned}(\Gamma \circ \Lambda)^x &= (\Gamma \times Z) \cap \tau(Y \times \Lambda) \cap \{x\} \times Y \times Z\\ &\cong (\Gamma^x \times Z) \cap \tau(Y \times \Lambda^x)\\ &= \Gamma^x \times \Lambda^x.\end{aligned}$$

(1.8) **Remark.** The underlying sets of $(\Gamma + \Lambda)_{(y,z)}$ and $\Gamma_y \cup \Lambda_z$ are the same for all closed points y and z of Y and Z respectively, but it is not true in general that $(\Gamma + \Lambda) = \Gamma_y \cup \Lambda_z$ as schemes, see Example (7.1). The underlying sets of $(\Gamma + \Lambda)^x$ and $(\Gamma^x \times Z) \cup (Y \times \Lambda^x)$ are the same for all closed points of X.

(1.9) **Proposition.** If (X, Γ, Y) and (X, Λ, Z) are both r-fold conditions on X then $(X, \Gamma + \Lambda, Y \times Z)$ is an r-fold condition on X.

Proof. All irreducible components of Γ have codimension r in $X \times Y$, hence all irreducible components of $\Gamma \times Z$ have codimension r in $X \times Y \times Z$, the same holds for $\tau(Y \times \Lambda)$, since τ is an isomorphism. Therefore all irreducible components of $(\Gamma \times Z) \cup \tau(Y \times \Lambda))$ have codimension r in $X \times Y \times Z$. Thus $(X, \Gamma + \Lambda, Y \times Z)$ is an r-fold condition on X. This proves the proposition.

(1.10) **Remark.** If (X, Γ, Y) and (X, Λ, Z) are r-fold, respectively s-fold, conditions on X then it is not in general true that $(X, \Gamma \circ \Lambda, Y \times Z)$ is an $(r+s)$-fold condition. Take for example a subvariety V of X of codimension $r > 0$. Let $\Gamma = V \times Y$. Then (X, Γ, Y) and $(X, \Gamma \circ \Gamma, Y \times Y)$ are both r-fold conditions on X, but $(X, \Gamma \circ \Gamma, Y \times Y)$ is not an $(r+r)$-fold condition, see (1.15) and Propositions (3.5), (4.7) and (6.6).

(1.11) **Definition.** Let (X, Γ, Y) be an r-fold condition on X. Let $f: X' \to X$ be a morphism of schemes of finite type over k. Define

$$f^{-1}(\Gamma) = (f \times \mathrm{id}_Y)^{-1}(\Gamma).$$

Then we call $(X', f^{-1}(\Gamma), Y)$ the *pull back* of (X, Γ, Y) under f.

(1.12) **Definition.** Let Z be a closed subscheme of a scheme X of finite type over k. If V is an irreducible component of Z then $O_{V,Z}$ is an Artinian local ring. Define the multiplicity of Z at V to be the length $O_{V,Z}$

$$m(V,Z) = \text{length } O_{V,Z}.$$

Define the cycle $[Z]$ associated to Z by

$$[Z] = \Sigma\, m(V,Z)\, V,$$

where the summation runs over all irreducible components V of Z, see [8] I.1.5. If Z is a zero dimensional subscheme of X then define

$$\int Z = \Sigma\, m(P,Z),$$

where the summation runs over all closed points P of Z. If (X, Γ, Y) is an n-fold condition on a scheme X of dimension n then Γ_y is called the *solutions* of the condition Γ on X at the point y of Y, in case Γ_y is a zero dimensional subscheme of X_y. If moreover Y is irreducible then $\int \Gamma_Y$ is called the *generic number of solutions*.

(1.13) **Remark.** In case X and Y are smooth varieties over k, and (X, Γ, Y) is an n-fold condition on X, and X has dimension n, we could associate to the points of Γ_y the intersection multiplicities, according to Severi [33], [34], [35] which was made rigorous by Van der Waerden [39], [42] and Weil [44]. They proved the principle of conservation of number with their assignment of multiplicities. The disadvantage of this way is that one has to look at the number of solutions which emerge from a special solution to the generic solutions. So one has saved the principle of conservation of number, but one has to know the generic solutions in order to compute the multiplicities, whereas the classical geometers used the principle to get the number of generic solutions by specializing, where it is easier to compute the number of solutions. Another approach is to assign multiplicities according to Serre's Tor formula [32], of the intersection $\Gamma_y = \Gamma \cap X_y$ in $X \times Y$ for every closed point y in Y. Then the number of solutions would be constant for varying y in Y in case Y is irreducible, but this will not justify the methods used in the enumerative geometers, since this was not their way of assigning multiplicities to solutions. So one can say that our assignment of multiplicities is the naive one.

(1.14) **Proposition.** Let X and Y be varieties over k and let $n = \dim X$. Let (X, Γ, Y) be an n-fold condition X. If $\mathrm{char}(k) = 0$ and Γ is reduced then there exists an open dense subset V of Y such that all solutions of Γ_y have multiplicity 1 for all y in V.

Proof. Γ is reduced, hence there exists a closed subscheme Λ of Γ such that $\Gamma\backslash\Lambda$ is smooth and open dense in Γ. Let Z be the closure of $\psi_\Gamma(\Lambda)$ in Y. Then $Y\backslash Z$ is open dense in Y and there exists an open dense subset V of $Y\backslash Z$ such that all the fibres of ψ_Γ over V are smooth, by [11] IV_3.9.9.4, since $\mathrm{char}(k) = 0$. Hence all the points of Γ_y have multiplicity 1 for all y in V. This proves the proposition.

(1.15) **Definition.** Let X be a scheme of finite type over k. Define

$$C^r(X) = \{(X,\Gamma,Y) \mid (X,\Gamma,Y) \text{ is an } r\text{-fold condition on } X\}$$

$$C(X) = \bigcup \{C^r(X) \mid 0 \le r \le n\}, \text{ where } n = \dim X.$$

If (X,Γ,Y) is an r-fold condition then define

$$V(\Gamma) = \{y \in Y \mid \text{all irreducible components of } \Gamma_y \text{ have codimension } r \text{ in } X_Y\}.$$

If S is a subset of $C(X)$ then define $S^r(X) = S \cap C^r(X)$.
In this paper we are concerned with finding subsets S of $C(X)$ which have one or more of the following properties:

(i) **Sum property.**
If (X,Γ,Y), $(X,\Lambda,Z) \in S^r$ then $(X,\Gamma+\Lambda,Y\times Z) \in S^r$.

(ii) **Intersection property.**
If $(X,\Gamma,Y) \in S^r$ and $(X,\Lambda,Z) \in S^s$ then $(X,\Gamma\circ\Lambda,Y\times Z) \in S^{r+s}$.

(iii) **Principle of conservation of number.**
If $(X,\Gamma,Y) \in S^n$ and $n = \dim X$ then the number of solutions of Γ at y stays constant for all y in Y as long as it is finite, that is to say $\int \Gamma_y$ is the same for all y in $V(\Gamma)$.

If moreover $f : X' \to X$ is a morphism of schemes of finite type over k and T is a subset of $C(X')$ then we are interested in the following property of S and T:

(iv) **Pull back property.**
If $(X,\Gamma,Y) \in S^r$ then $(X',f^{-1}(\Gamma),Y) \in T^r$.

(1.16) **Remark.** $C(X)$ has the sum property by Proposition (1.9) but not the intersection property, by (1.10) nor does it satisfy the principle of conservation of number, by (7.1).

§2 Complete conics.

(2.1) **Example.** A suggestion to an answer of the above questions is given by the paradigm of enumerative geometry: plane conics, see Kleiman [20] for a historical survey of Chasles's work and Casas and Xambó [3] for an account of Halphen's work. Assume char(k) is not 2. The plane conics are parametrized by $\mathbb{P}^5$ and the lines in $\mathbb{P}^2$ are parametrized by $\check{\mathbb{P}}^2$. Consider the 1-fold condition $(\mathbb{P}^5,\Lambda,\check{\mathbb{P}}^2)$, where Λ is the subscheme of $\mathbb{P}^5 \times \check{\mathbb{P}}^2$ defining the tangency between a conic and a line. So if l is a line then Λ_l is as a set the collection of all conics q which are tangent to 1. If q is a double line then q is tangent to every line. Let D be the subscheme of $\mathbb{P}^5$ parametrizing all double lines. Then Λ^4 contains the subscheme $D\times(\check{\mathbb{P}}^2)^4)$, which has dimension 10. But for a 4-fold condition $(\mathbb{P}^5,\Gamma,(\check{\mathbb{P}}^2)^4)$ irreducible components of Γ have dimension 9. Thus Λ^4 is not a 4-fold condition on $\mathbb{P}^5$. The problem is that the double lines are too often a solution of the imposed condition Λ, that is to say

$$\dim \Lambda_q = \dim \phi_\Lambda^{-1}(q) = \begin{cases} 1 \text{ in case } q \in \mathbb{P}^5 \backslash D \\ 2 \text{ in case } q \in D. \end{cases}$$

To remedy this one classically considers the variety of complete conics. Let q be a (3×3)-symmetric matrix and define $\hat{q}$ by $\hat{q}_{ij} = (-)^{i+j}$ determinant of the submatrix of q obtained by deleting the ith row and the jth column. Now if we consider q as a point of $\mathbb{P}^5$ then $\hat{q} = 0$ if and only if q is an element of D. We have a map

$$f : \mathbb{P}^5 \backslash D \to \mathbb{P}^5,$$

sending q to $\hat{q}$. Define

$$X = \text{closure of the graph of } f \text{ in } \mathbb{P}^5 \times \mathbb{P}^5.$$

Geometrically X consists of 4 kinds of complete conics:

(i) non-singular conics

(ii) union of two different lines

(iii) a double line with two different points on it

(iv) a double line with a double point on it.

Let Λ be the subscheme of $X \times \mathbb{P}^2$ defining the tangency correspondence between complete conics and lines. In the first two cases we already know what tangency means. If $\hat{q}$ is a double line with two points p_1 and p_2 on it then the line 1 is tangent to $\hat{q}$ if and only if p_1 or p_2 belongs to 1. We now have that all the fibres of the map $\phi_\Lambda : \Lambda \to X$ have dimension 1. For a conic being tangent to a line is dual to going through a point and there are no degenerate conics with respect to this last condition, see (3.1). But the condition for a conic to be tangent to a given conic is self dual and one can complete the conics in the above way in order to get a sensible answer.

(2.2) **Remark.** In a lot of enumerative problems one has to complete the scheme of geometrical figures in order to get a sensible answer. One considers for instance complete quadrics, see (7.7), complete collineations, complete correlations, see Laksov [23], [24], [26] and the references given there, complete triangles [30], complete twisted cubics [1], [27], [28] and complete symmetric varieties [5], [6]. It seems that the name giving "completeness" in this context is rather ad hoc. Of course there is the notion of a complete scheme [13] II.4.10, but this is not the only property of complete geometrical figures, since $\mathbb{P}^5$ parametrizing the plane conics is a complete scheme. In our opinion completeness is a property of the condition (X,Γ,Y) rather than of the scheme of geometrical figures, and sometimes one has to replace the degenerate figures, see (3.1), by new figures to get a scheme X' and a condition (X',Γ',Y) such that Γ' is "complete", see (5.1). Halphen pointed out that no matter how one completes the non-degenerate conics, there is always a condition whose number of solutions makes no sense for the completed object, [12], [3] 14.8. In the sequel we will investigate proper, flat and Cohen-Macaulay conditions and whether the classical enumerative conditions have these properties. We will not propose a definition for a complete condition and leave it as a classical terminology, but we think that a flat condition and a flattening of a condition are most close to what should be a complete condition and a completion of a condition.

§3 Proper conditions.

(3.1) **Definition.** Let X be a scheme of finite type over k and (X,Γ,Y) an r-fold condition on X. Define the *degeneracy locus* $D(\Gamma)$ or D for short, of the condition by

$$D(\Gamma)=\{x\in X \mid \Gamma^x \text{ is empty or has not codimension } r \text{ in } Y^x\}.$$

We call the condition *non-degenerate* or *proper* if the degeneracy locus is empty, that is, for all x in X the subscheme Γ^x of Y^x is not empty and has codimension r.

(3.2) **Example.** Let X be a variety and G an algebraic group over k acting transitively on the closed points of X. Let Y be a scheme of finite type over k such that G also acts on Y. Let (X,Γ,Y) be an r-fold condition on X such that the action of G on the product $X\times Y$ leaves Γ invariant. Then (X,Γ,Y) is a proper condition, see [17] and see (7.6).

(3.3) **Example.** Let $C(r,d,n)$ be the Chow variety of effective cycles of dimension r and degree d in $\mathbb{P}^n$ and Γ the subscheme of $C(r,d,n)\times\mathbb{P}^n$ defined by

$$(c,p)\in\Gamma \text{ if and only if } p \text{ is in the support of } c\ ,$$

where c and p are closed points of $C(r,d,n)$ and $\mathbb{P}^n$ respectively. Then $(C(r,d,n),\Gamma,\mathbb{P}^n)$ is a proper $(n-r)$-fold condition.

(3.4) **Remark.** For the definition of a proper morphism and a complete variety we refer to [13] II.4.6 and II.4.10 respectively.

(3.5) **Proposition.** Let X,Y and Z be smooth varieties over k. Suppose Y and Z are complete. If (X,Γ,Y) and (X,Λ,Z) are r-fold, respectively s-fold, proper conditions on X, then $(X,\Gamma\circ\Lambda,Y\times Z)$ is an $(r+s)$-fold proper condition on X. If $r=s$ then $(X,\Gamma+\Lambda,Y\times Z)$ is a proper r-fold condition on X.

(3.6) **Remark.** Let X be a smooth variety over k. Define

$$P^r(X)=\{(X,\Gamma,Y)\mid (X,\Gamma,Y) \text{ is a proper } r\text{-fold and } Y \text{ is a smooth complete variety}\}$$

$$P(X)=\bigcup\{P^r(X)\mid 0\le r\le n\},\ \text{ where } n=\dim X.$$

Then $P(X)$ satisfies the sum and the intersection property, by Proposition (3.5) and since the product of two complete smooth varieties over k is again a complete smooth variety over k, by [11] IV_2 6.8.5.

Proof of (3.5). The irreducible components of $\Gamma\times Z$ and $\tau(Y\times\Lambda)$ have codimension r and s respectively in $X\times Y\times Z$. Hence all irreducible components of $(\Gamma\times Z)\cap\tau(Y\times\Lambda)$ have codimension at most $r+s$, by [32] since $X \times Y \times Z$ is smooth. Let Σ be an irreducible component of $\Gamma \circ \Lambda$. Then

$$\dim \Sigma \geq l+m+n-(r+s),$$

where $l=\dim Z$, $m=\dim Y$ and $n=\dim X$. Let x be a closed point of X. Then $(\Gamma \circ \Lambda)^x = \Gamma^x \times \Lambda^x$, by (1.7), and $\dim \Gamma^x = m-r$ and $\dim \Lambda^x = l-s$, since Γ and Λ are both proper conditions. Hence $(\Gamma \circ \Lambda)^x$ has dimension $l+m-(r+s)$ for all closed points x of X. Y and Z are complete, so $Y \times Z$ is complete. Σ is a closed subscheme of $X \times Y \times Z$, hence $\phi(\Sigma)$ is closed in X. Therefore

$$\dim \Sigma \leq l+m-(r+s)+\dim \phi(\Sigma) \leq l+m+n-(r+s).$$

Hence equality holds and $\phi(\Sigma)=X$, since X is irreducible and X and $\phi(\Sigma)$ have the same dimension. Furthermore $\dim \phi_\Sigma^{-1}(x) = l+m-(r+s)$ for all closed points x of X, so $(\Gamma \circ \Lambda)^x$ has codimension $r+s$ for all closed points x of X. Thus $(X, \Gamma \circ \Lambda, Y \times Z)$ is a proper $(r+s)$-fold condition on X.

In case $r=s$ we know already that $(X, \Gamma+\Lambda, Y \times Z)$ is an r-fold condition. The fibre $(\Gamma+\Lambda)^x$ has the same underlying set as $(\Gamma^x \times Z) \cup (Y \times \Lambda^x)$, see Remark (1.8), which has constant codimension r in $Y \times Z$ for all closed points x in X. Hence the condition is proper. This proves the proposition.

(3.7) **Proposition.** Let X and Y be smooth varieties over k. Suppose Y is complete. Let (X, Γ, Y) be a proper r-fold condition on X and Z a subvariety of X. Then $(Z, \Gamma \cap (Z \times Y), Y)$ is a proper r-fold condition on Z.

Proof. All irreducible components of Γ have codimension r in $X \times Y$ and $Z \times Y$ has codimension s in $X \times Y$, if Z has codimension s in X. Further $X \times Y$ is smooth, hence all irreducible components of $\Gamma \cap (Z \times Y)$ have codimension at most $r+s$ in $X \times Y$, by [32]. Let Σ be such an irreducible component, then

$$\dim \Sigma \geq m+n-(r+s),$$

where $\dim X = n$ and $\dim Y = m$. Then $\dim \phi_\Sigma^{-1}(x) \leq m-r$, since $\dim \phi_\Gamma^{-1}(x) = m-r$ for all closed points in $\phi(\Sigma)$. Σ is closed in $X \times Y$ and Y is complete, hence $\phi(\Sigma)$ is closed in X and contained in Z. Hence

$$\dim \Sigma \leq m-r+\dim \phi(\Sigma) \leq m-r+n-s.$$

Hence equality holds. Z and $\phi(\Sigma)$ have the same dimension, Z is irreducible and $\phi(\Sigma)$ is a closed subvariety, hence $\phi(\Sigma)=Z$. Furthermore $\dim \phi_\Sigma^{-1}(x) = m-r$ for all closed points in Z. Thus $(Z, \Gamma \cap (Z \times Y), Y)$ is a proper r-fold condition on Z. This proves the proposition.

(3.8) **Definition.** A scheme is called *pure dimensional* if all its components have the same dimension and it has no embedded components. Let V and W be two subschemes of a smooth variety X of pure codimension r and s respectively. Then the intersection $V \cap W$ is called *proper* if it is empty or of pure codimension $r+s$ in X.

(3.9) **Corollary.** Under the same assumptions as in Proposition (3.7) we have that there exists an open dense subset U of Y such that $\Gamma_y \cap Z$ is a proper intersection for all closed points y in U.

Proof. This follows from the Propositions (1.4) and (3.7).

(3.10) **Corollary.** Let X and Y be smooth varieties over k. Suppose Y is complete and $n = \dim X$. Let (X, Γ, Y) be a proper n-fold condition on X and Z a subvariety of X, not equal to X. Then there exists an open dense subset U of Y such that the solutions of Γ at y are disjoint from Z, for all closed points y of U.

Proof. This follows from Corollary (3.9), since $\Gamma \cap (Z \times Y)$ is an n-fold condition on Z which has dimension less than n.

§4 Flat conditions.

(4.1) **Definition.** An r-fold condition (X, Γ, Y) is called *flat* if the map $\phi_\Gamma : \Gamma \to X$ is flat.

(4.2) **Remark.** The idea of considering flat conditions and the flattening of a condition stems from Piene and Schlessinger [28] where they consider flat specializations of non-degenerate twisted cubics and the Hilbert scheme to make a completion.

(4.3) **Example.** (3.2) is also an example of a flat condition, since the map ϕ_Γ is flat over an open dense subset U of X, by Proposition (1.4) after interchanging the rôles of X and Y, and the group G acts transitively on the closed points of X, [17].

(4.4) **Proposition.** Let X and Y be schemes of finite type over k. Suppose X and Y are irreducible. If (X, Γ, Y) is a flat r-fold condition then all irreducible components of Γ^x have codimension r in Y^x for all points x of X, in particular (X, Γ, Y) is a proper condition.

Proof. Γ and X are schemes of finite type over k. The morphism ϕ_Γ is flat and X is irreducible and all irreducible components of Γ have the same codimension r in $X \times Y$, hence all irreducible components of Γ^x have codimension r in Y^x for all points x in X, by Proposition (1.3). Hence (X, Γ, Y) is a proper condition. This proves the proposition.

(4.5) **Remark.** We give in (7.2) an example of a condition which is proper but not flat.

(4.6) **Remark.** Let Y be a projective scheme over k and X a scheme of finite type over k. If (X, Γ, Y) be a flat r-fold condition on X and X is irreducible then the Hilbert polynomial of Γ^x in Y^x is the same for all points x of X, by [11] III.2.2.1. Conversely, if (X, Γ, Y) is an r-fold condition on X and the Hilbert polynomial of Γ^x in Y^x is the same for all x of X then (X, Γ, Y) is a flat condition, by [14]. Let P be a polynomial in one variable with rational coefficients. Let $\mathrm{Hilb}^P(Y)$ be the Hilbert scheme of subschemes of Y with Hilbert polynomial P, see [10] exposé 221. The Hilbert scheme has the following universal property. Let $\Lambda(P)$ be the subscheme of $\mathrm{Hilb}^P(Y) \times Y$ such that Λ^h is the subscheme of Y^h corresponding to the point h of $\mathrm{Hilb}^P(Y)$. Then $\phi_{\Lambda(P)}$ is flat. Moreover, for every flat condition (X, Γ, Y) such that Γ^x has Hilbert polynomial P in Y^x for all points x of X, there exists a unique map $f : X \to \mathrm{Hilb}^P(Y)$ such that $\Gamma^x = \Lambda(P)^{f(x)}$ for all x of X.

(4.7) **Proposition.** Let X, Y and Z be schemes of finite type over k. Suppose X, Y and Z are irreducible. Let (X, Γ, Y) and (X, Λ, Z) be flat r-fold, respectively s-fold, conditions on X. Then

$(X, \Gamma \circ \Lambda, Y \times Z)$ is a flat $(r+s)$-fold condition on X.

(4.8) **Remark.** Let X be an irreducible scheme of finite type over k. Define

$$F^r(X) = \{(X, \Gamma, Y) \mid (X, \Gamma, Y) \text{ is a flat } r\text{-fold condition on } X \text{ and } Y \text{ is irreducible}\}$$

$$F(x) = \bigcup \{F^r(X) \mid 0 \leq r \leq n\}, \quad \text{where } n = \dim X.$$

Then F(X) satisfies the intersection property by Proposition (4.7).
Before we proof the proposition we need a lemma.

Proof of (4.7). Consider the following commutative diagrams of schemes of finite type over k

$$\begin{array}{ccccc} \Gamma \circ \Lambda & \to & Y \times \Lambda & \to & \Lambda \\ \downarrow & & \downarrow & & \downarrow \\ \Gamma \times Z & \to & X \times Y \times Z & \to & X \times Z \\ \downarrow & & \downarrow & & \downarrow \\ \Gamma & \to & X \times Y & \to & X \end{array} \qquad \begin{array}{cccccc} & \Gamma \circ \Lambda & \xrightarrow{g} & \Lambda & \\ f & \downarrow & & \downarrow & \phi_\Lambda \\ & \Gamma & \xrightarrow{\phi_\Gamma} & X & \end{array}$$

, where the morphisms in the left diagram are the obvious inclusions or projections, g is the composition of the two morphisms in the top row and f is the composition of the two morphisms in the left column. The four squares in the left diagram are fibred, hence the right diagram is fibred by diagram chasing.. Now ϕ_Λ is flat, hence f is flat by the base change property of flatness [13] III Proposition 9.2.b. Further ϕ_Γ and f are flat, so the composition $\phi_{\Gamma \circ \Lambda}$ of ϕ_Γ and f is flat, by transitivity [13] III Proposition 9.2.c. Let x be a closed point of X, then

$$(\Gamma \circ \Lambda)^x \cong \Gamma^x \times \Lambda^x,$$

by (1.7.i). The irreducible components of Γ^x have codimension r in Y and the irreducible components of Λ^x have codimension s in Z, by Proposition (4.4). Hence all irreducible components of $(\Gamma \circ \Lambda)^x$ have codimension $r+s$ in $Y \times Z$. Hence all irreducible components of $\Gamma \circ \Lambda$ have codimension $r+s$ in $X \times Y \times Z$, by Proposition (1.3). Thus $(X, \Gamma \circ \Lambda, Y \times Z)$ is a flat $(r+s)$-fold condition on X. This proves the proposition.

(4.9) **Proposition.** Let Y and X' be irreducible schemes of finite type over k and $f : X' \to X$ a morphism of schemes over k. If (X, Γ, Y) is a flat r-fold condition on X then $(X', f^{-1}(\Gamma), Y)$ is a flat r-fold condition on X'.

(4.10) **Remark.** In this case F(X) and F(X') have the pull back property.

Proof of (4.9). Let $\Lambda = f^{-1}(\Gamma)$. The morphism ϕ_Γ is flat. The following diagram is a fibered square

$$\begin{array}{ccccc} & \Lambda & \to & \Gamma & \\ \phi_\Lambda & \downarrow & & \downarrow & \phi_\Gamma \\ & X' & \to & X & \end{array}$$

Hence ϕ_Λ is a flat morphism, by the base change property [13] III Proposition 9.2.b. The fibre $\Lambda^{x'}$ is isomorphic to $\Gamma^{f(x')}$ for all x' in X'. All irreducible components of Γ^x have the same dimension for all x in X, by Proposition (4.4). Hence all irreducible components of $\Lambda^{x'}$ have the same dimension for all x' in X'. Furthermore the map ϕ_Λ is flat and X' is irreducible. Thus all irreducible components of $f^{-1}(\Gamma)$ have codimension r in $X'\times Y$, by Proposition (1.3). Thus $(X', f^{-1}(\Gamma), Y)$ is a flat r-fold condition on X'. This proves the proposition.

§5 Flattening of a condition.

(5.1) **Definition.** Let (X,Γ,Y) be an r-fold condition on X. A flat r-fold condition (X',Γ',Y) on X' together with a morphism $\pi : X' \to X$ is called a *flattening* of (X,Γ,Y) if there exists an open dense subset U of X such that

$$\pi : \pi^{-1}(U) \to U \quad \text{and}$$

$$\pi\times \mathrm{id}_Y : \Gamma' \cap (\pi^{-1}(U)\times Y) \to \Gamma \cap (U\times Y)$$

are isomorphisms.

A flattening (X',Γ',Y) with a morphism $\pi : X' \to X$ is called *universal* if for every flattening (X'',Γ'',Y) together with a morphism $\pi' : X'' \to X$ there exists a unique morphism $f : X'' \to X'$ such that $\pi' = \pi \circ f$ and Γ'' is the pull back of Γ' under f.

(5.2) **Remark.** It is clear that the universal flattening is unique up to canonical isomorphisms, if it exists.

(5.3) **Proposition.** Let X be a scheme of finite type over k. Let (X,Γ,Y) be an r-fold condition on X. Then there exists a flattening of (X,Γ,Y).

Proof. The existence of a flattening is a result of Raynaud and Gruson [29]. We give a Hilbert scheme proof of the existence under the assumption that X is integral and Y is projective. The existence proof is in the same spirit as the way one constructs a completion of geometrical figures, see (2.1) and (7.5). If no irreducible component of Γ dominates X then we take $X' = X$ and $\Gamma' = \emptyset$. Otherwise there exists an open dense subset U of X such that ϕ_Γ is flat over U and the irreducible components of Γ^x have codimension r in Y^x for all x in U, by Proposition (1.4). The Hilbert polynomial P of Γ^x in Y^x is constant for all x in U, by Remark (4.5). Let $H = \mathrm{Hilb}^P(Y)$. Then there exists a map $g : U \to H$, where $g(x)$ is the point in H corresponding to the subscheme Γ^x of Y^x for all x in U. Consider the graph Γ_g in $U\times H$ of the map g. Let X' be the closure of Γ_g in $X\times H$. Let $\pi : X' \to X$ be the restriction to X' of the projection $X\times H \to X$ and let $g' : X' \to H$ be the restriction to X' of the projection $X\times H \to H$. Then $\pi : \pi^{-1}(U) \to U$ is an isomorphism and g' and $g\circ\pi$ are the same on $\pi^{-1}(U)$. Let $\Lambda(P)$ be the subscheme of $H\times Y$ as defined in Remark (4.6). Let Γ' be the pull back of $\Lambda(P)$ via g'. Then $\phi_{\Gamma'}$ is flat, by Proposition (4.9) and using the fact that $\phi_{\Lambda(P)}$ is flat, by Remark (4.5). Further

$$\pi\times \mathrm{id}_y : \Gamma' \cap (\pi^{-1}(U)\times Y) \to \Gamma \cap (U\times Y)$$

is an isomorphism, since $\Gamma^x = \Lambda(P)^{g(x)}$ for all x in U, and g' and $g\circ\pi$ are the same on $\pi^{-1}(U)$, hence $\Gamma^{x'} = \Gamma^{\pi(x')}$ for all x' in $\pi^{-1}(U)$. All irreducible components of $\Gamma^{x'}$ have codimension r in $Y^{x'}$ for all points x' in $\pi^{-1}(U)$, since this holds for $\Gamma^{\pi(x')}$. Furthermore $\pi^{-1}(U)$ is an open dense subset of X' and $\phi_{\Gamma'}$ is flat. Hence all irreducible components of Γ' have codimension r in $X \times Y$, by Proposition (1.3). Thus (X',Γ',Y) is a flat r-fold condition on X', together with the map π it gives a completion of (X,Γ,Y). This proves the proposition.

(5.4) **Remark.** If (X,Γ_i,Y_i), $i = 1,\dots,r$ is a finite collection of conditions on X then there exists a scheme X' and a morphism $\pi : X' \to X$ and conditions (X',Γ_i',Y_i), $i = 1,\dots,r$ such that they are flattenings of the original conditions. One can do this inductively by first making a flattening $(X^{(1)},\Gamma_1^{(1)},Y_1)$ of the first condition and pulling back conditions Γ_i, for $i = 2,\dots,r$ to $X^{(1)}$ and then making a flattening of the second condition on $X^{(1)}$, etc. Another way is doing it in one step by taking the closure of the map $g = g_1 \times \cdots g_r : U \to H_1 \times \cdots \times H_r$, where $H_i = \mathrm{Hilb}^{P_i}(Y_i)$ and $g_i : U \to H_i$ is defined by $g_i(x)$ is the point in H_i corresponding to the subscheme Γ_i^x in Y_i^x.

(5.5) **Proposition.** Let X and Y be smooth varieties over k and let $n = \dim X$. Suppose Y is complete. Let (X,Γ,Y) be an r-fold condition and (X',Γ',Y) together with a morphism $\pi : X' \to X$ a flattening such that $\pi : \pi^{-1}(U) \to U$ is an isomorphism for an open dense subset U of X. Then there exists an open dense subset V of Y such that all the solutions of Γ_y' lie in $\pi^{-1}(U)$.

Proof. This follows from Corollary (3.10) by taking $Z = X' \setminus \pi^{-1}(U)$, since the condition Γ' is proper, by Proposition (4.4).

(5.6) **Corollary.** The generic number of solutions of a flattening is independent of the chosen flattening.

§6 Cohen-Macaulay conditions.

(6.1) **Definition.** An r-fold condition (X,Γ,Y) is called *Cohen-Macaulay*, or *CM* for short, if the condition is proper and Γ is a Cohen-Macaulay scheme.

(6.2) **Proposition.** Let $f : V \to W$ be a morphism of irreducible schemes of finite type over k.

(i) If V is *CM* and W is smooth then the map f is flat if and only if the fibres of $f^{-1}(w)$ have constant dimension $\dim V - \dim W$ for all (closed) points w of W.

(ii) If the map is flat then V is *CM* if and only if W is *CM* and the fibres $f^{-1}(w)$ are *CM* for all (closed) points w of W.

Proof. See [26] 5.1 and 23.1 for (i) and the corollary of 23.3 for (ii).

(6.3) **Lemma.** Let X and Y be irreducible smooth schemes of finite type over k. Suppose X is smooth. Then the following are equivalent:

(i) (X,Γ,Y) is an r-fold *CM* condition.

(ii) (X,Γ,Y) is a flat r-fold condition and Γ^x is a *CM* scheme for all points x of X.

(iii) (X,Γ,Y) is a flat r-fold condition and Γ^x is a *CM* scheme for all closed points x of X.

Proof. (i) $\Rightarrow$ (ii). If (X,Γ,Y) is an r-fold *CM* condition then Γ is *CM* and the condition is proper hence the fibre Γ^x of ϕ_Γ at x has constant dimension $\dim\Gamma - \dim X$ for all closed points x of X and X is smooth, thus the map ϕ_Γ is flat, by Proposition (6.2.i), and (X,Γ,Y) is a flat r-fold condition. Furthermore Γ^x is *CM* for all points x of X, by Proposition (6.2.ii).
(ii) $\Rightarrow$ (iii) is trivial.
(iii) $\Rightarrow$ (i). If (X,Γ,Y) is a flat r-fold condition then it is a proper r-fold condition, by Proposition (4.4) since X is irreducible. Furthermore X is *CM* and the fibres Γ^x of ϕ_Γ at x are *CM* for all closed points x of X by assumption, and the map ϕ_Γ is flat, hence Γ is *CM*, by Proposition (6.2.ii). Thus (X,Γ,Y) is an r-fold *CM* condition. This proves the lemma.

(6.4) **Remark.** In (7.4) we give an example of a flat condition which is not *CM*.

(6.5) **Proposition.** Let X and Y be smooth varieties and Γ a hypersurface in $X\times Y$. If (X,Γ,Y) is a proper 1-fold condition then it is *CM*.

Proof. This follows immediately, since a hypersurface in a smooth scheme is *CM*.

(6.6) **Proposition.** Let X and Y be irreducible smooth schemes of finite type over k. Suppose X is smooth. If (X,Γ,Y) and (X,Λ,Z) are r-fold, respectively s-fold, *CM* conditions, then $(X,\Gamma\circ\Lambda,Y\times Z)$ is an $(r+s)$-fold *CM* condition.

(6.7) **Remark.** Let X and Y be irreducible smooth schemes of finite type over k. Suppose X is smooth. Define

$$\mathrm{CM}^r(X) = \{(X,\Gamma,Y) \mid (X,\Gamma,Y) \text{ is an } r\text{-fold } CM \text{ condition on } X\}$$

$$\mathrm{CM}(X) = \bigcup \{CM^r(X) \mid 0 \le r \le n\}, \quad \text{where } n = \dim X.$$

Then $\mathbf{CM}(X)$ satisfies the intersection property by Proposition (6.6). In general $\mathbf{CM}(X)$ does not satisfy the sum property, see example (7.3).

Proof of (6.6). (X,Γ,Y) and (X,Λ,Z) are flat r-fold respectively s-fold conditions, by Lemma (6.3). Hence $(X,\Gamma\circ\Lambda,Y\times Z)$ is a flat $(r+s)$-fold condition, by Proposition (4.7), and $\phi_{\Gamma\circ\Lambda}$ is a flat map. The fibers Γ^x and Λ^x of ϕ_Γ and ϕ_Λ respectively are *CM* for all closed points x of X. The fibre $(\Gamma\circ\Lambda)^x$ of $\phi_{(\Gamma\circ\Lambda)}$ is isomorphic with the product $\Gamma^x\times\Lambda^x$, by (1.7.ii) and is therefore also *CM*, for all closed points x of X. Thus $(X,\Gamma\circ\Lambda,Y\times Z)$ is an $(r+s)$-fold *CM* condition on X, by Lemma (6.3) since we assumed X to be irreducible and smooth. This proves the proposition.

(6.8) **Proposition.** Let X,Y and X' be irreducible schemes of finite type over k and $f: X'\to X$ a morphism. Suppose X and X' are smooth. If (X,Γ,Y) is an r-fold *CM* condition on X then $(X',f^{-1}(\Gamma),Y)$ is an r-fold *CM* condition on X'.

(6.9) **Remark.** In this case $\mathbf{CM}(X)$ and $\mathbf{CM}(X')$ have the pull back property.

Proof. If (X,Γ,Y) is a r-fold *CM* condition then this condition is flat and Γ^x is *CM* for all closed points of X, by Lemma (6.3). Hence $(X,f^{-1}(\Gamma),Y)$ is flat, by Proposition (4.10) and $f^{-1}(\Gamma)^{x'} = \Gamma^{f(x')}$ is *CM* for all closed points x' of X'. Thus $(X',f^{-1}(\Gamma),Y)$ is an r-fold *CM* condition, by Lemma (6.3) since X' is irreducible and smooth. This proves the proposition.

(6.10) **Proposition.** Let X be a complete variety over k of dimension n and Y an irreducible smooth variety over k. If (X,Γ,Y) is an n-fold CM condition then the principle of conservation of number holds.

Proof. Γ is a closed subscheme of $X \times Y$ and X is complete, hence the map ψ is closed and ψ_Γ is a proper map. We define $V(\Gamma) = \{y \in Y \mid \Gamma_y$ is zero dimensional$\}$ in (1.15). Let $U = \psi_\Gamma^{-1}(V(\Gamma))$. Then $\psi_\Gamma : U \to V(\Gamma)$ is a finite map. Moreover U is CM since it is an open subscheme of Γ which is CM by assumption. The sheaf $\psi_{\Gamma *}(O_U)$ is a coherent on $V(\Gamma)$ and depth $\psi_{\Gamma *}(O_U)_p = \dim_p Y$ for all points p of Y. Furthermore Y is smooth, hence

$$\text{depth } M_p + \text{projdim } M_p = \dim_p Y ,$$

for all points p of Y, by [32]. Thus $\psi_{\Gamma *}(O_U)$ is a locally free sheaf on $V(\Gamma)$ of constant rank, say N, since Y is irreducible. Thus for every y in $V(\Gamma)$ one has that

$$N = \Sigma \text{ length } O_{p,U} ,$$

where p runs through all the points in the fibre $\psi_\Gamma^{-1}(y)$, which is equal to $\int \Gamma_y$. Hence the number of solutions $\int \Gamma_y$ is constant and equal to N for all points y of $V(\Gamma)$. Thus the principle of conservation of number holds. This proves the proposition.

§7 Examples.

(7.1) **Example.** Let $(\mathbb{P}^2, \Gamma, \check{\mathbb{P}}^2)$ be the condition defining the incidence between points and lines in $\mathbb{P}^2$, where the lines of $\mathbb{P}^2$ are parametrized by $\check{\mathbb{P}}^2$. Γ is defined by the bihomogeneous ideal

$$(l_0 x_0 + l_1 x_1 + l_2 x_2) \quad \text{in} \quad k[x_0, x_1, x_2, l_0, l_1, l_2] ,$$

the bihomogeneous coordinate ring of $\mathbb{P}^2 \times \check{\mathbb{P}}^2$. The condition Γ^2 is defined by the ideal (lx, mx) in the trihomogeneous coordinate ring $k[x,l,m]$ of $\mathbb{P}^2 \times (\check{\mathbb{P}}^2)^2$, where we denote (x_0, x_1, x_2) by x and $l_0 x_0 + l_1 x_1 + l_2 x_2$ by lx. The condition $\Gamma^2 + \Gamma^2$ is defined by the ideal

$$(lx, mx) \cap (l'x, m'x) = ((lx)(l'x), (lx)(m'x), (mx)(l'x), (mx)(m'x)) ,$$

in the 5-homogeneous coordinate ring $k[x,l,m,l',m']$ of $\mathbb{P}^2(\check{\mathbb{P}}^2)^4$. Let l and m be two different lines in $\mathbb{P}^2$ which intersect at P and let $y = (l,m)$ then the scheme $(\Gamma^2 + \Gamma^2)_{(y,y)}$ is supported at P and has multiplicity 3, whereas $\Gamma^2_y \cup \Gamma^2_y$ is equal to Γ^2_y and is supported at P with multiplicity 1. Thus

$$(\Gamma^2 + \Gamma^2)_{(y,y)} \neq \Gamma^2_y \cup \Gamma^2_y ,$$

and the principle of conservation does not hold for $\Gamma^2 + \Gamma^2$.

(7.2) **Example.** Let Γ be the closed subscheme of $\mathbb{P} \times \mathbb{P}^3$ defined by the bihomogeneous ideal

$$(y_0, y_1) \cap (y_0, y_2) \cap (x_0 y_0 + x_1 y_1 + x_2 y_3 , y_1 + y_2)$$

in the bihomogeneous coordinate ring $k[x_0, x_1, x_2, y_0, y_1, y_2, y_3]$ of $\mathbb{P}^2 \times \mathbb{P}^3$. Then $(\mathbb{P}^2, \Gamma, \mathbb{P}^3)$ is a proper 2-fold condition, since Γ^* consists of three lines in $\mathbb{P}^3$ for all closed points x in $\mathbb{P}^2$. If $x=(0:1:0)$ then Γ^x is defined by the ideal $(y_0,y_1)\cap(y_0,y_2)\cap(y_1,y_2)$, so it is the union of three lines going through $(0:0:0:1)$ and not lying in a plane, it has Hilbert polynomial $3t+1$. If $x=(1:0:0)$ then Γ^x is defined by the ideal $(y_0,y_1)\cap(y_0,y_2)\cap(y_0,y_1+y_2)$, so it is the union of three lines in a plane going through one point and it has Hilbert polynomial $3t$. So the Hilbert polynomial of Γ^x is not constant, hence the map ϕ_Γ is not flat, by Remark (4.6), and the condition is not flat.

(7.3) **Example.** Let $G(2,4)$ be the Grassmann variety of planes in $\mathbb{P}^4$ and $(G(2,4),\Gamma,\mathbb{P}^4)$ the condition such that $(v,p)\in\Gamma$ if and only if p is a point of v, for all v and p closed points of $G(2,4)$ and $\mathbb{P}^4$ respectively. Then Γ is a 2-fold CM condition, see (7.6) and $\Gamma\circ\Gamma$ is a 4-fold CM condition, by Proposition (6.6). We have the following exact sequence of local rings

$$0\to O_{p,\Gamma+\Gamma}\to O_{p,\Gamma\times\mathbb{P}^4}\oplus O_{p,\mathbb{P}^4\times\Gamma}\to O_{p,\Gamma\circ\Gamma}\to 0,$$

for every closed point p of $\Gamma\circ\Gamma$. The middle term is a direct sum of two CM local rings of dimension 12, hence it has depth 12. The third term is a CM local ring of dimension 10, so of depth 10. Thus the first term has depth 11, whereas it has dimension 12. Therefore $\Gamma+\Gamma$ is not CM.

(7.4) **Example. Twisted cubics.** Let $P=3t+1$. Then $\mathrm{Hilb}^P(\mathbb{P}^3)$ consists of two smooth irreducible components H and H' of dimensions 12 and 15 respectively and the intersection is smooth of dimension 11. A point of $H_0=H\setminus H'$ corresponds to a non-degenerate twisted cubic, a point of $H'\setminus H$ corresponds to a plane cubic curve with a point outside the plane and a point of $H\cap H'$ corresponds to a singular plane cubic curve with an embedded point at a singular point. See Piene and Schlessinger [28]. Let Λ be the pull back of $\Lambda(P)$ under the inclusion of H in $\mathrm{Hilb}^P(\mathbb{P}^3)$. Then $(H,\Lambda,\mathbb{P}^3)$ is a flat 2-fold condition, since the Hilbert polynomial of Γ^x is constant $3t+1$. The scheme Γ^x has an embedded component for all points x of $H\cap H'$ and therefore is not CM. Thus the condition is not CM, by Lemma (6.3).

(7.5) **Example. Complete twisted cubics** according to Piene [27]. Suppose char(k) is not 2 or 3. Consider the conditions $(H,\Gamma,G(1,3))$ and $(H,\Lambda^\vee,\mathbb{P}^3)$, where for $c\in H_0$ and $l\in G(1,3)$, the Grassmann variety of lines in $\mathbb{P}^3$, we have that

$(c,l)\in\Gamma$ if and only if the line l is tangent the curve c,

$(c,h)\in\Lambda^\vee$ if and only if h is an osculating plane of the curve c.

If $c\in H_0$ then Γ^c is the tangent curve c^* of c in $G(1,3)$, which is a rational normal curve of degree 4 and has Hilbert polynomial $4t+1$, and $(\Lambda^\vee)^c$ is the dual curve $c^\vee$ of osculating planes in $\mathbb{P}^3$, which is again a twisted cubic with Hilbert polynomial $3t+1$. Let $G=\mathrm{Hilb}^{4t+1}(G(1,3))$. Thus we have morphisms

$$g: H_0\to G \text{ defined by } g(c)=c^*,$$

$$f : H_0 \to H \text{ defined by } f(c) = c^\vee .$$

The closure T of the graph of $g \times f$ in $H \times G \times H$ is called the scheme of complete twisted cubics by Piene [27]. The restrictions to T of the projections to the second and third factor we denote by g' and f' respectively. The pull backs of $\Lambda(4t+1)$ and $\Lambda(3t+1)$ under the morphisms g' and f' we denote by Γ' and $\Lambda^{\vee\prime}$ respectively, they are flattenings of Γ and $\Lambda^\vee$, by the proof of Proposition (5.3) and Remark (5.4).

(7.6) **Example. Schubert conditions on Grassmannians.** Let $G(r,n)$ be the Grassmannian of all r-planes in $\mathbb{P}^n$, it is a smooth variety of dimension $(r+1)(n-d)$. Let $\mathbf{a} = (a_0, \ldots, a_r)$ be a sequence of integers such that $0 \le a_0 < \cdots < a_r \le n$. Let $F(\mathbf{a},n)$ be the flag variety of all flags $\mathbf{A}$ in $\mathbb{P}^n$, where $\mathbf{A} = (A_0, \ldots, A_r)$ and A_i is an a_i-dimensional linear subspace in $\mathbb{P}^n$ and $A_i \subset A_{i+1}$ for all $i = 0, \ldots, r-1$. Let $(G(r,n), \Omega, F(\mathbf{a},n))$ be the condition where

$$(B, \mathbf{A}) \in \Omega \text{ if and only if } \dim(B \cap A_i) \ge i \text{ for all } i = 0, \ldots, r ,$$

for closed points B in $G(r,n)$ and $\mathbf{A}$ in $F(\mathbf{a},n)$. Then $\Omega_{\mathbf{A}}$ is called a Schubert variety and has codimension $\Sigma(a_i - i)$. For every two flags $\mathbf{A}$ and $\mathbf{B}$ in $F(\mathbf{a},n)$ there exists an invertible projective transformation ϕ of $\mathbb{P}^n$ which induces an isomorphism of $G(r,n)$ and carries $\Omega_{\mathbf{A}}$ into $\Omega_{\mathbf{B}}$, see [22]. Hence we are in the situation of Example (3.2) and (4.3) and the condition is flat $\Sigma(a_i - i)$-fold. Moreover the Schubert varieties $\Omega_{\mathbf{A}}$ are CM, by [15], [16] thus the condition is even CM, by Lemma (6.3).

(7.7) **Example. Complete quadrics,** see Laksov [24] and the references given there. Suppose $\mathrm{char}(k) \ne 2$. Let V be a vector space of dimension $n+1$ over k, with coordinates $x_0, \ldots, x_n$. We denote the projectivization of V by $\mathbb{P}(V)$. A quadric q in $\mathbb{P}(V)$ is given by the zero locus of a quadratic form

$$\Sigma q_{ij} x_i x_j ,$$

where $q = (q_{ij})$ is a non-zero symmetric $(r+1) \times (r+1)$-matrix. Thus quadrics in $\mathbb{P}(V)$ are parametrized by $\mathbb{P}(S^2 V)$, where $S^2 V$ is the vector space of symmetric maps $V \to V^*$. We denote a quadric in $\mathbb{P}^n$, its symmetric matrix and the point in $\mathbb{P}(S^2 V)$ representing it, by the same q. The Grassmann variety $G(r,n)$ of r-planes in $\mathbb{P}^n$ can be embedded in $\mathbb{P}(\Lambda^{r+1} V)$ with Plücker coordinates (x_I), where $I = (i_0, \ldots, i_r)$ and $0 \le i_0 < \cdots < i_r \le n$. The 1-fold condition $(\mathbb{P}(S^2 V), \Gamma(r), G(r,n))$ describing the tangency between a quadric q and an r-plane x in $\mathbb{P}^n$ is defined by the zero locus of the quadratic form

$$\Sigma(\Lambda^{r+1} q)_{I,J}\, x_I x_J ,$$

where I and J are multi-indices $I = (i_0, \ldots, i_r)$, $J = (j_0, \ldots, j_r)$ such that $0 \le i_0 < \cdots < i_r \le n$, $0 \le j_0 < \cdots < j_r \le n$ and $(\Lambda^{r+1} q)_{I,J}$ is the determinant of the $(r+1) \times (r+1)$-submatrix of q consisting of the rows $i_0, \ldots, i_r$ and columns $j_0, \ldots, j_r$. One can view $\Lambda^{r+1} q$ as an element of $\mathbb{P}(S^2 \Lambda^{r+1} V)$, the latter we will denote by M_r. So the condition $\Gamma(r)$ is the pull back of $\Lambda(r)$, where $(M_r, \Lambda(r), G(r,n))$ is the 1-fold condition defined by the zero locus of the quadratic form

$$\Sigma Q_{I,J} x_I x_J ,$$

where $(Q_{I,J}) \in \mathbb{P}(S^2 \Lambda^{r+1} V) = M_r$.

Let U_r be the open dense set of quadrics which have rank at least $r+1$. Then we have a morphism

$$\lambda_r : U_r \to M_r ,$$

defined by $\lambda_r(q) = \Lambda^{r+1} q$. Let $U = U_n$ and let

$$\lambda : U \to M_1 \times \cdots \times M_{n-1} ,$$

be the morphism defined by $\lambda = \lambda_1 \times \cdots \times \lambda_{n-1}$. Define the variety B of complete quadrics to be the closure of the graph of λ in $M_0 \times M_1 \times \cdots \times M_{n-1}$. Let π_r be the projection of B to M_r and $\Gamma(r)'$ the pull back to B of $\Lambda(r)$ under π_r. The variety B can be obtained by a sequence of *blowing ups* with smooth centers starting with M_0 which is smooth, hence B is smooth, see Vainsencher [37].

A complete n-quadric is some k-tuple $\mathbf{q} = (q_1, \ldots, q_k)$, where q_1 is a quadric in $\mathbb{P}^n$ of rank r_1, and q_i is a quadric in the singular locus of q_{i-1} of rank r_i, for all $i = 2, \ldots, k$ and $r_1 + \cdots + r_k = n+1$. The closed points of B are in one to one correspondence with complete n-quadrics. Thus if $\mathbf{q} = (q_1, \ldots, q_k)$ is a complete n-quadric then $\mathbf{q}' = (q_2, \ldots, q_k)$ is a complete $(n+1-r_1)$-quadric in the singular locus of q_1, see Finat [7]. The condition $\Gamma(r)'$ can be expressed inductively as follows. Let $\mathbf{q}$ be a complete n-quadric with ranks $(r_1, \ldots, r_k)$ and x an r-plane in $\mathbb{P}^n$. Then $(\mathbf{q}, x) \in \Gamma(r)'$ if and only if

(i) x is tangent to q_1 in case $r < r_1$

(ii) x intersects $\mathrm{Sing}(q_1)$ non-transversally or x intersects $\mathrm{Sing}(q_1)$ transversally and $(\mathbf{q}', x \cap \mathrm{Sing}(q_1)) \in \Gamma(r - r_1)'$ in case $r \geq r_1$.

From this description it follows by induction that all the fibres $\Gamma(r)_{\mathbf{q}}'$ at the closed point $\mathbf{q}$ of B have codimension 1 in $G(r,n)$. Hence $(B, \Gamma(r)', G(r,n))$ is a proper 1-fold condition. Moreover the condition $\Gamma(r)'$ is a hypersurface in $B \times G(r,n)$ and B and $G(r,n)$ are smooth, hence the condition is even *CM*, by Proposition (6.5).

Note that the complete quadrics are not obtained by the use of Hilbert scheme flattening as in the proof of Proposition (5.3), although it is very similar to it. It would be interesting to know whether these two completions are isomorphic, see also Kleiman's question [21] page 362.

§8 Schubert calculus.

We sketch how to get an intersection theory on singular varieties.

Let X be an irreducible scheme of finite type over k of dimension n. Now $\mathbf{F}^r(X)$ is the collection of flat r-fold conditions (X, Γ, Y) such that Y is irreducible. Then $\mathbf{F}^r(X)$ satisfies the intersection property, by Proposition (4.7). Let $F^r(X)$ be the free abelian group generated by $\mathbf{F}^r(X)$ and

$$F^*(X) = \oplus \{ F^r(X) \mid r = 0, \ldots, n \}.$$

Define the map

$$\int : F^n(X) \to \mathbb{Z} ,$$

by $\int(X,\Gamma,Y) = \int \Gamma_Y$, the generic number of solutions, see (1.15), on generators of $F^n(X)$ and extend by linearity. The map $\int$ is a morphism of groups.
Define an intersection

$$\circ : F^r(X) \otimes F^s(X) \to F^{r+s}(X)$$

by

$$(X,\Gamma,Y) \otimes (X,\Lambda,Z) \to (X,\Gamma \circ \Lambda, Y \times Z)$$

on generators of $F^r(X)$ and $F^s(X)$, and extend by linearity.
Define the following numerical equivalence relation $\sim$ on $F^r(X)$ by

$$\Sigma a_i(X,\Gamma_i,Y_i) \sim \Sigma b_j(X,\Lambda_j,Z_j)$$

if and only if

$$\Sigma a_i \int (Z,f^{-1}(\Gamma_i),Y_i) = \Sigma b_j \int (Z,f^{-1}(\Lambda_j),Z_j)$$

for all closed embeddings $f : Z \to X$ such that Z is irreducible of dimension r. This is well-defined by the pull back property, see Proposition (4.9). Let $N^r(X)$ be the subgroup of $F^r(X)$ of elements numerically equivalent to zero and let

$$N^*(X) = \oplus \{N^r(X) \mid r = 0, \ldots, n\}.$$

Define

$$S^r(X) = F^r(X) / N^r(X)$$

and

$$S^*(X) = \oplus \{S^r(X) \mid r = 0, \ldots, n\}.$$

Then the intersection $\circ$ is well-defined on $S^*(X)$ and this product is distributive with respect to +, associative, commutative and has unit the class of $(X, X \times Y, Y)$, where $Y = \mathrm{Spec}(k)$. Hence $S^*(X)$ is a commutative ring with a unit. We call $S^*(X)$ the *Schubert ring* of X.
If (X,Γ,Y), (X,Λ,Z) and $(X,\Gamma+\Lambda,Y \times Z)$ are elements of $F^r(X)$ then

$$(X,\Gamma,Y) + (X,\Lambda,Z) \equiv (X,\Gamma+\Lambda,Y \times Z) \bmod N^r(X).$$

So if we denote (X,Γ,Y) by Γ then the two meanings of $\Gamma + \Lambda$ are equal modulo $N^r(X)$.
In case X is a smooth quasi projective variety we can associate to every condition (X,Γ,Y) the cycle $[\Gamma_y]$, with y some element of $V(\Gamma)$, which gives a well-defined cycle class modulo algebraic equivalence, for every choice y in $V(\Gamma)$. An element of $F^*(X)$ which is numerically equivalent to zero gives a cycle class in the Chow group $A^*_{alg}(X)$ modulo algebraic equivalence, which is numerically equivalent to zero. Thus we have a well defined morphism of rings

$$S^*(X) \to A^*_{num}(X).$$

It would be interesting to know what its image is.
For every r-fold condition (X,Γ,Y) and for every embedding $f: Z \to X$ of a variety Z of dimension r we have that

$$[Z] \cdot [\Gamma_y] = [f^{-1}(\Gamma)_y]$$

in $A^0_{alg}(X)$, for all $y \in Y$ such that $Z \cap \Gamma_y$ consists of finitely many points. Hence the morphism $S^*(X) \to A^*_{num}(X)$ is injective.
We call a class in $S^r(X)$ effective if it has a representative $\Sigma a_i(X,\Gamma_i,Y_i)$ such that $a_i \geq 0$ for all i. Let $S^r_{eff}(X)$ be the set of elements of $S^r(X)$ which are effective. It follows from the definition of the product that if $a \in S^r_{eff}(X)$ and $b \in S^s_{eff}(X)$ then $a \circ b \in S^{r+s}_{eff}(X)$.
Consequently, if X is the blow up of $\mathbb{P}^2$ at a point and E is the exceptional divisor then E has self intersection -1, so its class in $A^1_{num}(X)$ does not lie in the image of the map $S^1(X) \to A^1_{num}(X)$.
It is not always possible to define a ring structure on the Chow group $A^*(X)$ in case X is singular, see [13] appendix A, (1.1.2).
Thus we have defined a Schubert ring for every variety X over k, even in the case X is singular. This is of some importance in enumerative geometry, since after flattening (completion) of geometrical figures one may end up with a singular variety. It is for instance not known whether the variety T of complete twisted cubics, example (7.5), is smooth.

Acknowledgement. We like to thank the organizers of the Symposium on Singularity Theory and its Applications. Part of this work was financed by an SERC research fellowship.

References

[1] A.R. Alguneid, Analytical degenerations of complete twisted cubics, Proc. Cambridge Phil. Soc. **52** (1962), 202-208.

[2] E. Bézout, Théorie générale des équations algébriques, Pierres, Paris, 1779.

[3] E. Casas-Alvero and S. Xambó-Descamps, Halphen's enumerative theory of conics, Lect. Notes Math. **1196**, Springer-Verlag, Berlin Heidelberg New York, 1986.

[4] M. Chasles, Détermination du nombre de sections coniques qui doivent toucher cinq courbes données d'ordre quelconque, ou satisfaire à diverses autres conditions, C.R.Ac.Sc. **58** (1864), 297-308.

[5] C. De Concini and C. Procesi, Complete symmetric varieties, Invariant Theory (Proc. Conf., Montecatini, 1982), F. Gheradelli editor, Lect. Notes in Math. **996**, Springer-Verlag, Berlin Heidelberg New York, 1983, 1-44.

[6] C. De Concini and C. Procesi, Complete symmetric varieties, II, Intersection theory, Alg. Groups and Related Topics (Kyoto/Nagoya, 1983), Adv. Stud. Pure Math., vol. 6, North-Holland, 1985, 481-513.

[7] J. Finat, A combinatorial presentation of the variety of complete quadrics, Conf. La Rabida 1984, Hermann.

[8] W. Fulton, Intersection theory, Ergeb. Math. Grenzgeb., 3 Folge, Band 2, Springer-Verlag, Berlin Heidelberg New York, 1984.

[9] L. van Gastel, Excess intersections, Thesis Rijksuniversiteit Utrecht, 1989.

[10] A. Grothendieck, Fondements de la géometrie algébrique, Extraits du séminaire Bourbaki, 1957-1962, Sécretariat Math., Paris, 1962.

[11] A. Grothendieck and J. Dieudonné, Eléments de géométrie algébrique, I, II, III, IV_1, IV_2, IV_3, IV_4, Publ. Math. IHES, **4** (1960), **8** (1961), **11** (1961), **17** (1964), **20** (1964), **24** (1965), **28** (1966), **32** (1967).

[12] G.H. Halphen, Sur le nombre des coniques qui dans un plan satisfont à cinq conditions projectives et independentes entre elles, Proc. London Math. Soc. **10** (1978-79), 76-91. Oeuvres II, Gauthier-Villars, Paris (1918), 275-289.

[13] R. Hartshorne, Algebraic geometry, Graduate Texts in Math. **52**, Springer-Verlag, Berlin Heidelberg New York, 1977.

[14] R. Hartshorne, Connectedness of the Hilbert scheme, Publ. Math. IHES, **29** (1966), 5-48.

[15] M. Hochster, Grassmannians and their Schubert subvarieties are arithmetically Cohen-Macaulay, J. Algebra **25** (1973), 40-57.

[16] M. Hochster and J.A. Eagon, Cohen-Macaulay rings, invariant theory, and the generic perfection of determinantal loci, Amer. J. Math. **93** (1971), 1020-1059.

[17] S.L. Kleiman, The transversality of a general translate, Compositio Math. **28** (1974), 287-297.

[18] S.L. Kleiman, Problem 15. Rigorous foundations of Schubert's enumerative calculus, Mathematical Developments arising from Hilbert Problems, Proc. Sympos. Pure Math. vol. **28**, Amer. Math. Soc., Providence R.I., 1976, 445-482.

[19] S.L. Kleiman, An introduction to the reprint edition of Schubert's Kalkül der abzählenden Geometrie, Springer-Verlag, Berlin Heidelberg New York, 1979.

[20] S.L. Kleiman, Chasles's enumerative theory of conics: A historical introduction, Studies in algebraic geometry, A. Seidenberg, editor, MAA Stud. Math. **20**, Math. Assoc. America, Washington D.C., 1980, 117-138.

[21] S.L. Kleiman, Intersection theory and enumerative geometry: a decade in review, Algebraic geometry Bowdoin 1985, Proc. Sympos. Pure Math. **46** (1987), 321-370.

[22] S.L. Kleiman and D. Laksov, Schubert calculus, Amer. Math. Monthly **79** (1972), 1061-1082.

[23] D. Laksov, Notes on the evolution of complete correlations, Enumerative and classical algebraic geometry (Proc., Nice, 1981) P. Le Barz and Y. Hervier editors, Progress in Math. vol. **24**, Birkhäuser, 1982, 107-132.

[24] D. Laksov, Completed quadrics and linear maps, Algebraic geometry Bowdoin 1985, Proc. Sympos. Pure Math. **46** (1987), 371-387.

[25] D. Laksov, The geometry of complete linear maps, Arkiv för Mathematik **26** (1988), 231-263.

[26] H. Matsumura, Commutative ring theory, Cambridge Stud. Adv. Math. **8**, Cambridge Un. Press, Cambridge, 1986.

[27] R. Piene, Degenerations of complete twisted cubics, Enumerative and classical algebraic geometry (Proc., Nice, 1981) P. Le Barz and Y. Hervier editors, Progress in Math. vol. **24**, Birkhäuser, 1982, 37-50.

[28] R. Piene and M. Schlessinger, On the Hilbert scheme compactification of the space of twisted cubics, Amer. J. Math. **107** (1985), 761-774.

[29] M. Raynaud and L. Gruson, Critère de platitude et de projectivité, Invent. Math. **13** (1971), 1-89.

[30] J. Roberts and R. Speiser, Schubert's enumerative geometry of triangle's from a modem viewpoint, Algebraic geometry, (Proc. Conf. Chicago Circle 1980) Lect. Notes in Math. **862**, Springer-Verlag, Berlin Heidelberg New York, 1981, 272-281.

[31] H. Schubert, Kalkül der abzählenden Geometrie, Teubner, Leibzig 1879, reprint edition, Springer-Verlag, Berlin Heidelberg New York, 1979.

[32] J-P. Serre, Algèbre locales, Multiplicités, Lect. Notes Math. **11**, Springer-Verlag, Berlin Heidelberg New York, 1965.

[33] F. Severi, Sul principio della conservazione del numero, Rendiconti del Circolo Mathematico di Palermo, **33** (1912), 313-327.

[34] F. Severi, Über die Grundlagen der algebraischen Geometrie, Abh. Math. Sem. Hamburg Univ. **13** (1939), 101-112.

[35] F. Severi, I fundamenti della geometria numerativa, Annali di Mat. 4, **19** (1940), 153-242.

[36] J. Stückrad and W. Vogel, An algebraic approach to the intersection theory, in: Curves Seminar at Queens University, vol. II, Queens Papers Pure Appl. Math., **61**, Kingston Ontario, 1982, 1-32.

[37] I. Vainsencher, Schubert calculus for complete quadrics, Enumerative and classical algebraic geometry, (Proc. Nice, 1981), P. Le Barz and Y. Hervier editors, Progress in Math., **24**, Birkhäuser, 199-235.

[38] W. Vogel, Results on Bézout's theorem, Tata Lect. Notes **74**, Springer-Verlag, Berlin Heidelberg New York, 1984.

[39] B.L. van der Waerden, Der Multiplizitätsbegriff in der algebraischen Geometrie, Math. Annalen **97** (1927), 736-774.

[40] B.L. van der Waerden, Topologische Begründung des Kalküls der abzählenden Geometrie, Math. Annalen **102** (1929), 337-362.

[41] B.L. van der Waerden, Zur algebraischen Geometrie, I-XV, Math. Ann. (1933-1938).

[42] B.L. van der Waerden, The foundation of algebraic geometry from Severi to André Weil, Arch. Hist. Exact. Sci. 7, (1970-1971), 171-179.

[43] B.L. van der Waerden, Einführung in die algebraischen Geometrie, Springer-Verlag, Berlin 1939.

[44] A. Weil, Foundations of algebraic geometry, 1946, Revised and enlarged edition, Amer. Math. Soc. Colloq. Publ. **29**, 1962.

[45] H. Zeuthen, Lehrbuch der abzählenden Geometrie, Teubner, Leibzig, 1914.

[46] H. Zeuthen and M. Pieri, Geometrie enumerative, Encyclopedie des Science Mathematiques III.2, Gauthier-Villars, Paris, 1915, 260-331.

Author's address:
University of Technology Eindhoven
Department of Mathematics and Computing Science
P.O. Box 513
5600 MB Eindhoven
The Netherlands

Right-Symmetry of Mappings

Andrew du Plessis, Leslie Charles Wilson[1]

§ 1. Introduction

In this paper we study symmetries in the domain of C^∞ or analytic map-germs. In this we were inspired by the earlier work of Jänich [J] and Wall [Wa] and by the possibility of applying the present results to global equivalence problems (see [GW1], [GW2] and [Wi]).

The main theorem of Jänich was:

(1.1) Theorem. *Let* $f : (\mathbb{R}^n,0) \to (\mathbb{R},0)$ *be a* C^∞ *finitely determined germ. Then any compact group* G *of diffeomorphism-germs of* $(\mathbb{R}^n,0)$ *preserving* f *is contained in a maximal such group, and any two such maximal compact subgroups are conjugate by a diffeomorphism-germ preserving* f.

Wall considered C^∞ map-germs $f : (\mathbb{R}^n,0) \to (\mathbb{R}^p,0)$ which were finitely determined with respect to any of the groups $\mathcal{A}$, $\mathcal{K}$, $\mathcal{R}$, $\mathcal{C}$ or $\mathcal{L}$ of singularity theory, and proved the analogous results to (1.1). He also considered the complex analytic case, in which compact groups are replaced by reductive groups and the analogous results hold.

In this paper, we begin by considering the action of the group $\mathcal{R}$ of source diffeomorphisms on map-germs $f: (\mathbb{R}^n,0) \to (\mathbb{R}^p,0)$. If $p > 1$, f cannot be finitely $\mathcal{R}$-determined, so the arguments of [J] and [Wa] must be altered somewhat. Nevertheless, we have not only succeeded in proving a result like (1.1), but in calculating exactly what the maximal compact subgroups are. Actually, our result is for compact subgroups of a certain subgroup $\mathcal{R}_f$ of $\mathcal{R}$. Proposition (1.3) shows that this is not a drastic restriction, in that $\mathrm{Iso}(\mathcal{R}_f,f) = \mathrm{Iso}(\mathcal{R},f)$ for a very large class of map-germs. Our theorem holds under a very mild hypothesis on f (namely, that its Jacobian ideal be closed). Under this hypothesis, we can replace the usual $\mathcal{R}$-determinacy results by the theory of $\mathcal{R}_f$-determinacy developed in [duPW]. Before we state the main results, we will present the necessary definitions and notations.

Let k denote either $\mathbb{R}$ or $\mathbb{C}$, as context determines. Let E_n denote the germs at 0 in k^n of real C^∞ or k-analytic k-valued functions (which we will refer to in each case as *smooth* functions). Let m_n denote the unique maximal ideal of E_n. Let $E_n^{\times p}$ denote p-tuples of elements of E_n (often interpreted as map-germs). Let $\mathcal{R}$ denote the group of diffeomorphism-germs (or bianalytic-germs) of $(k^n,0)$ to itself.

A subgroup G of $\mathcal{R}$ is a *compact* (resp. *reductive*) *Lie subgroup* if there exists a compact (resp. reductive) Lie group G′ and a group isomorphism φ from G′ onto G induced by a smooth germ $\Phi : G' \times (k^n,0) \to (k^n,0)$ such that: for every representative

[1] Leslie Wilson was partially supported by the National Science Foundation under Grant no. MCS 81-00779, and was a guest in Aarhus while most of this work was done.

$\tilde{\Phi}: W \to k^n$ of Φ, there exist neighbourhoods $U \supset V$ of 0 such that $W \supset G' \times U$ and the induced map $\tilde{\varphi}: G' \to C^\infty(U,k^n)$ satisfies, for all $g_1, g_2 \in G'$, $\tilde{\varphi}(g_1) \cdot \tilde{\varphi}(g_2) \mid V$ is defined and equals $\tilde{\varphi}(g_1,g_2) \mid V$. Note that if $r \in \mathcal{R}$ and G is a compact Lie subgroup of $\mathcal{R}$, then so is $G^r = \{r^{-1}gr : g \in G\}$.

Let J denote an ideal in E_n and let $\mathcal{R}_J = (1 + J^{\times n}) \cap \mathcal{R}$ (where 1 is the identity in $\mathcal{R}$). For $f \in E_n^{\times p}$, let J(f) denote the ideal generated by the p by p minors of the Jacobian matrix of f, and let $\mathcal{R}_f = \mathcal{R}_{J(f)}$. Let rk(f) denote the rank of the Jacobian matrix of f at 0. For G a subroup of $\mathcal{R}$, let $\mathrm{Iso}(G,f) = \{h \in G : f \circ h = f\}$.

We will say that a subset of $C^\infty(U,\mathbb{R}^p)$ is *closed* if it is closed in the compact-open C^∞ topology (e.g.: in the analytic case, all submodules are closed). We will say J(f) is closed if it has a representative $\tilde{f} \in C^\infty(U,\mathbb{R}^p)$ for which $J(\tilde{f})$ is closed in $C^\infty(U,\mathbb{R})$. In this case, $1 + J(\tilde{f})^{\times n}$ is closed in $C^\infty(U,\mathbb{R}^n)$; it is also convex.

(1.2) Theorem. *Assume* $f \in E_n^{\times p}$ *has* J(f) *closed.*

(a) *Every compact (reductive) Lie subgroup of* $\mathrm{Iso}(\mathcal{R}_f,f)$ *is contained in a maximal compact (reductive) Lie subgroup* G_f *of* $\mathrm{Iso}(\mathcal{R}_f)$, *which is unique up to conjugation by elements of* $\mathrm{Iso}(\mathcal{R}_f)$

(b) $n < p : \mathrm{Iso}(\mathcal{R}_f,f) = \{1\}$.

(c) $n \geq p$: (i) $\mathrm{rk}f = p : G_f \simeq O(n-p)$ *(the orthogonal group in* k^{n-p}).

(ii) $\mathrm{rk}f = p-1$: *let* r *be the rank and* i *the index of the second intrinsic derivative of f; then* $G_f \simeq O(i) \times O(r-i)$ *if* $k = \mathbb{R}, G_f \simeq O(r)$ *if* $k = \mathbb{C}$.

(iii) $\mathrm{rk}f \leq p-2 : G_f \simeq \{1\}$. (The assumption that J(f) is closed is unnecessary for this part.)

Call f a *critical simplification* (CS for short) if J(f) = I(C(f)) (= the ideal of all smooth functions vanishing on the critical set C(f) of f) and f | C(f) is generically one-to-one. It is proved in [GduPW] that this is equivalent to certain transversality conditions holding generically. In particular, all stable map-germs, all finitely $\mathcal{A}$-determined map-germs with $p \geq 2$ and $C(f) \neq \{0\}$, and all analytic topologically stable map-germs are critical simplifications. If f is a CS and $h \in \mathrm{Iso}(\mathcal{R},f)$, then h must leave C(f) pointwise fixed, hence $h \in \mathcal{R}_f$. Thus we have:

(1.3) Proposition. *If* f *is a* CS, *then* $\mathrm{Iso}(\mathcal{R},f) = \mathrm{Iso}(\mathcal{R}_f,f)$.

In §3, we will prove the following theorem, which is useful for the global study of equivalence of mappings.

(1.4) Theorem. $\mathrm{Iso}(\mathcal{R}_f,f)/G_f$ *is contractible in the sense of Jänich (see [J] or §3).*

In §4 we study the relationship between $\mathcal{R}$ symmetries, $\mathcal{A}$ symmetries, and symmetries of the discriminant D = f(C(f)). If f is finitely $\mathcal{A}$-determined, it follows from Wall's result that there exists a MCS G_a of $\mathrm{Iso}(\mathcal{A},f)$. Let G_r denote the MCS of $\mathrm{Iso}(\mathcal{R}_f,f)$. Let Inv(D) denote the group of diffeo-germs on $(\mathbb{R}^p,0)$ such that $h(D) = D$. Let $\mathrm{Inv}^o(D)$ consist of those elements of Inv(D) which preserve the orientation of cok(df). Let $\mathrm{Inv}^{\sim}(D) = \mathrm{Inv}^o(D)$ if df has rank p−1 and $\mathrm{rk}(d^2f) \neq 2\cdot \mathrm{index}(d^2f)$, where d^2f is the second intrinsic derivative of f, and $\mathrm{Inv}^{\sim}(D) = \mathrm{Inv}(D)$ otherwise; in the reductive case, $\mathrm{Inv}^{\sim}(D) = \mathrm{Inv}(D)$.. We will call a compact (resp. reductive) Lie subgroup of $\mathrm{Inv}^{\sim}(D)$ *linearizable* if it can be conjugated to a subgroup of GL(p) by an element h of $\mathrm{Inv}^{\sim}(D)$ having $j^1h(0) = 1$.

In §4 we prove:

(1.5) Theorem. *If f is finitely $\mathcal{A}$-determined and a critical normalization, then, among all linearizable compact (resp. reductive) Lie subgroups of* $\mathrm{Inv}^{\sim}(D)$, *there exists an MCS* G_ℓ, *unique up to conjugation by elements of* $\mathrm{Inv}^{\sim}(D)$, *such that the natural sequence*

$$0 \to G_r \to G_a \to G_\ell \to 0$$

is split exact.

The notion of critical normalization is defined and studied in [GW1] and [GduPW]; a map-germ $f : (\mathbb{R}^n,0) \to (\mathbb{R}^p,0)$ is a critical normalization if it is a critical simplification, and if $f|C(f) : C(f) \to fC(f)$ is a C^∞ normalization in the sense of [GW1]. It is proved in [GduPW] that this is equivalent to certain transversality conditions holding generically. In particular, all stable map-germs, all finitely-$\mathcal{A}$-determined map-germs with $p \geq 3$ and $C(f) \neq \{0\}$, and all analytic, topologically stable map-germs are critical normalizations.

§2. Proof of Theorem (1.2)

We will concentrate on the compact - C^∞ case. At the end of this section, we will discuss the reductive - holomorphic case.

If $r \in \mathcal{R}$, let $r^* \in GL(n)$ denote the Jacobian matrix of r at 0. If G is a subgroup of $\mathcal{R}$, let G^* denote the subgroup of GL(n) of all r^*'s. Observe that if G is a compact Lie subgroup, then G^* is a compact Lie subgroup of GL(n). The next theorem is essentially the Bochner Linearization Theorem (see e.g. [Boc]).

(2.1) Theorem. *Suppose* G *is a compact Lie subgroup of* $\mathcal{R}$. *Suppose there are* G', U, V, φ and $\tilde{\varphi}$ *as in the definition of compact Lie subgroup and* $\tilde{N}$ *a closed, convex subset of* $C^\infty(U,\mathbb{R}^n)$ *such that* $\tilde{\varphi}(g)^*\tilde{\varphi}(g^{-1})$ *is contained in* $\tilde{N}$ *for all* g *in* G. *Denote* $\tilde{N}_0$ *by* N. *Then there exists* $\alpha \in N$ *such that* $\alpha^* = 1$ *and* $G^\alpha = G^*$.

Proof. Let $\tilde{\alpha} = \int_G \tilde{\varphi}(g)^*\tilde{\varphi}(g^{-1})dg$; $\tilde{\alpha}$ is C^∞ and defined on U. Let $\alpha = \tilde{\alpha}_0$. As in the usual proof of Bochner linearization (see [Boc]), α is a diffeomorphism-germ, $G^\alpha = G^*$ and $\varphi^* = 1$. For all g, the above integrand is in $\tilde{N}$. Since $\tilde{\alpha}$ lies in the convex closure of the integrands, it lies in $\tilde{N}$; hence $\alpha \in N$. □

Assume $n < p$. Then $J(f) = 0$, so $\mathcal{R}_f = \{1\}$. Theorem 1.1 is trivial in this case.

From now on in this section, we will assume $n \geq p$. Suppose $\mathrm{rk} f \leq p-2$. By (1.2.ii) of [duPW], $J(f) \subset m_n^2$. Suppose G is a compact Lie subgroup of $\mathrm{Iso}(\mathcal{R}_f, f)$. By (2.1), $G \simeq G^* = \{1\}$, so the Theorem is proven.

The case $\mathrm{rk} f = p$ is essentially in [Wa], but we will reprove it simultaneously with the case $\mathrm{rk} f = p-1$. In both cases, the method is to replace the action of $\mathcal{R}_f$ on $E_n{}^{\times p}$ by the Lie group $\mathcal{R}_f{}^*$ acting algebraically on a finite dimensional vector space $F^{\#}$. One then applies to $G = \mathrm{Iso}(\mathcal{R}_f{}^*, f^{\#})$ the following:

(2.2) Proposition. (See p.180 of [H1]). *Let G be a Lie group with finitely many connected components. Then any compact subgroup is contained in a maximal compact subgroup, any two MCS's are conjugate, and the quotient of G by a MCS is contractible.*

One calculates the MCS G_f in $\mathrm{Iso}(\mathcal{R}_f{}^*, f^{\#})$ and then uses a finite determinacy result to prove G_f has the same properties in $\mathrm{Iso}(\mathcal{R}_f, f)$.

Let $F = f + J(f)\tau f(E_n{}^{\times n})$, let $F_1 = f + (m_n^2 \cap J(f))\tau f(E_n{}^{\times n})$, and let $F^{\#} = F/F_1$. If $g \in F$, $g^{\#}$ denotes its projection into $F^{\#}$.

In case $\mathrm{rk} f = p$, $J(f) = E_n$, so $\mathcal{R}_f = \mathcal{R}$, $\mathcal{R}_f{}^* = GL(n)$ and $F^{\#} = A(n,p)$, the affine maps from $\mathbb{R}^n$ to $\mathbb{R}^p$. The action of $\mathcal{R}$ on $E_n{}^{\times p}$ by composition induces the action of $GL(n)$ on $A(n,p)$ by composition.

In case $\mathrm{rk} f = p-1$, it follows from Proposition 0.4 and Theorem 0.7 of [duPW] that the action of $\mathcal{R}_f$ on F induces a well-defined algebraic action on $F^{\#}$ depending algebraically on the 2-jet of f.

We will show that the groups G_f of Theorem 1.2 are MCS's in $\mathrm{Iso}(\mathcal{R}_f{}^*, f^{\#})$, but first we want to show that it suffices to replace f by a "normal form".

(2.3) Proposition. *Suppose $g = \ell \circ f \circ r$ for diffeo-germs ℓ and r. Conjugation by r maps $\mathcal{R}_f$ isomorphically onto $\mathcal{R}_g$, $\mathrm{Iso}(\mathcal{R}_f, f)$ iso-morphically onto $\mathrm{Iso}(\mathcal{R}_g, g)$, taking compact Lie subgroups to compact Lie subgroups. Two compact Lie subgroups G_1 and G_2 of $\mathrm{Iso}(\mathcal{R}_f, f)$ are conjugate by a $\beta \in \mathrm{Iso}(\mathcal{R}_f, f)$ iff $G_1{}^r$ and $G_2{}^r$ are conjugate by $\beta^r \in \mathrm{Iso}(R_g, g)$.*

Proof. In this proof, all citations of results will be to Proposition 1.3 of [duPW]. From (A.a.iii) of that proposition, $J(fr) = J(f)r$; it is similar and easy to see that $J(\ell f) = J(f)$. Thus, $\mathcal{R}_{j(f)}{}^r = r^{-1}(\mathcal{R}_{J(f)} \cdot 1)r + \mathcal{R}_{J(f)r} \cdot 1$ (by (A.a.i) and (B.b.i)) $= \mathcal{R}_{J(g)}$. Also, if $f\alpha = f$, then $g = \ell f r = \ell f \alpha r = \ell f r \alpha^r = g\alpha^r$; thus, conjugation by r does map $\mathrm{Iso}(\mathcal{R}_f, f)$ to $\mathrm{Iso}(\mathcal{R}_g, g)$. The remaining claims are routine to check. □

We need three lemmas before we calculate a MCS of Iso($\mathcal{R}_f$,f). The first lemma is just the fact that every unipotent representation of a compact Lie group is trivial:

(2.4) Lemma. *Suppose* G *is a compact Lie group. Suppose a continuous representation* $\mu : G \to GL(n)$ *satisfies: for each* $g \in G$ *there is a matrix* A *so that*

$$\mu(g) = \begin{vmatrix} I & 0 \\ A & I \end{vmatrix}.$$

Then each of these A's = 0. *(The same conclusion holds for* A's *above the diagonal rather than below).*

Proof. By [H1], every continuous representation of G is semisimple, so every invariant subspace of $\mu(G)$ has an invariant complement. Since the image of the matrix $| 0 \ I |^t$ is invariant, the invariant complement is the image of a matrix $| \ I \ B \ |^t$. Since

$$\begin{vmatrix} I & 0 \\ A & I \end{vmatrix} \cdot \begin{vmatrix} I \\ B \end{vmatrix} = \begin{vmatrix} I \\ A+B \end{vmatrix},$$

A = 0. □

(2.5) Lemma. *Each* MCS *of* GL(r) *is conjugate to* O(r).

(2.6) Lemma. *Each* MCS *of* Iso(GL(n),Q), *where*

$$Q = x_1^2 + \dots + x_i^2 - x_{i+1}^2 - \dots - x_r^2$$

is conjugate to O(i)×O(r−i).

Proof. Let G be such a MCS. By (2.5), G is conjugate to a subgroup G′ of O(n). Of course, this conjugation transforms Q into another quadratic form Q′ of rank r and index r−i. By the Principal Axes Theorem, there is an orthogonal transformation which transforms Q′ into $Q'' = \Sigma\, a_j^2 x_j^2 - \Sigma a_k^2 x_k^2$, $1 \le j \le i < k \le r$. This orthogonal transformation conjugates G′ to a G″ in O(n) which preserves Q″. Choose $h \in G''$. The proof will be done if we can show $h \in O(i) \times O(r-i)$.

We may assume $a_{i+1} \le a_{i+2} \le \dots \le a_r$. Since h is orthogonal, it preserves $a_r(\Sigma x_j^2 + \Sigma x_k^2) = Q'''$. Thus h preserves Q‴ + Q″, which is positive-semidefinite. The kernel of this form is $x_1 = \dots = x_s = 0$, where $x_{s+1},\dots,x_r$ are the variables having coefficient a_r in Q″. This kernel is preserved by h, as is its complement. So h preserves $Q'' \mid \langle x_1,\dots,x_s\rangle$. Repeat the argument until we have that h preserves $\langle x_1,\dots,x_i\rangle$ and $\langle x_{i+1},\dots,x_r\rangle$. Thus $h \in O(i) \times O(r-i)$. □

Now we calculate the MCS of Iso($\mathcal{R}^*$,f#).
Suppose rkf = p. Then without loss of generality we may assume

(2.7) $$f(x_1,...,x_n) = (x_1,...,x_p).$$

Note that Iso(GL(n),f) = $1_p \times$ GL(n-p). Let $G_f = 1_p \times O(n-p)$.

(2.8) Claim. G_f *is a* MCS *of* Iso(GL(n),f).

Proof. Assume G is a compact subgroup of Iso(GL(n),f) containing G_f. Since G preserves f-fibres, each element of G is represented in the given coordinates by a matrix of the form

$$g = \begin{vmatrix} I & 0 \\ A & B \end{vmatrix}$$

where A and B depend on g. The projection $g \to B(g)$ is a continuous homomorphism, so its image G' is a compact Lie subgroup of GL(n-p) and, by assumption, contains O(n-p). By Lemma 2.5, G' = O(n-p). Thus, if g is as above, B $\in$ O(n-p) and so

$$\begin{vmatrix} I & 0 \\ 0 & B^{-1} \end{vmatrix} \in G, \text{ so } \begin{vmatrix} I & 0 \\ A & I \end{vmatrix} \in G. \text{ The map } g \to \begin{vmatrix} I & 0 \\ A & I \end{vmatrix}$$

is a continuous representation of G. By (2.4), A = 0. Thus, G_f = G. □

Now suppose rkf = p-1. By (3.1) of [duPW] and by (2.3), we may assume f is a map of the form

(2.9) $$f(u_1,...,u_a, v_1,...,v_b, x_1,...,x_b, y_1,...,y_c, z_1,...,z_d) = (u_1,...,u_a, v_1,...,v_b, v_1x_1 + ... + v_bx_b + Q(\underline{y}) + F(\underline{u},\underline{x},\underline{z}))$$

where $Q = -y_1^2 - ... - y_i^2 + y_{i+1}^2 + ... + y_c^2$ and $F \in m_n^3$ (of course, a + 2b + c + d = n and a + b + 1 = p). Let G_f denote $O(i) \times O(c-i) \times 1_{n-c}$, where the orthogonal groups are acting on $\langle \underline{y} \rangle$. Note that G_f is contained in Iso($\mathcal{R}_f$,f) and hence in Iso($\mathcal{R}_f^*$, f#) (see (2.12) below).

(2.10) Claim. G_f *is a* MCS *of* Iso($\mathcal{R}_f^*$,f#).

Proof. Suppose G is a compact subgroup of Iso($\mathcal{R}_f^*$,f#) containing G_f. Since G is contained in $\mathcal{R}_f^* = \{1 + \langle \underline{y}, \underline{v} \rangle^{\times n}\} \cap GL(n)$, each g $\in$ G is expressed in the given coordinates as a matrix of the form (rearranging the variables in the indicated order)

$$\begin{array}{c|cc|} & uxz & yv \\ \hline uxz & 1 & M_1 \\ yv & 0 & M_2 \end{array}.$$

Since g also preserves u,v levels,

$$g = \begin{array}{c|ccccc|} & u & x & z & y & v \\ u & 1 & 0 & 0 & 0 & 0 \\ x & 0 & 1 & 0 & A & B \\ z & 0 & 0 & 1 & C & D \\ y & 0 & 0 & 0 & E & F \\ v & 0 & 0 & 0 & 0 & 1 \end{array}.$$

The projection $g \to E$ is a continuous homomorphism, and so its image G' is a compact Lie subgroup which, by assumption, contains $O(i) \times O(c-i)$. Furthermore, G' lies in $\mathrm{Iso}(GL(c),Q)$. By Lemma 2.6, $G' = O(i) \times O(c-i)$. Thus,

$$\begin{vmatrix} 1 & 0 & 0 & 0 & 0 \\ 0 & 1 & 0 & 0 & 0 \\ 0 & 0 & 1 & 0 & 0 \\ 0 & 0 & 0 & E^{-1} & 0 \\ 0 & 0 & 0 & 0 & 1 \end{vmatrix} \in G, \text{ hence } \mu(g) = \begin{vmatrix} 1 & 0 & 0 & 0 & 0 \\ 0 & 1 & 0 & A & B \\ 0 & 0 & 1 & C & D \\ 0 & 0 & 0 & 1 & F \\ 0 & 0 & 0 & 0 & 1 \end{vmatrix} \in G$$

The map μ is a continuous representation of G. By (2.4), A, B, C, D and F = 0. Thus $G = G_f$. □

We need some determinacy results to pass from MCS's in $\mathrm{Iso}(\mathcal{R}_f{}^*, f^\#)$ to MCS's in $\mathrm{Iso}(\mathcal{R}_f, f)$.

(2.11) Proposition. *If* $f,g \in E_n{}^{\times p}$ *satisfy* $g-f \in J\tau f(E_n{}^{\times n})$, *then* f *and* g *are* $\mathcal{R}_J$ *equivalent, where* $J = J(f) \cap m_n{}^2$.

Proof. This is a consequence of (2.7) of [duPW], with the J in (2.7) the same as in the statement of this Proposition. □

(2.12) Corollary. $\mathrm{Iso}(\mathcal{R}_f{}^*, f^\#) = (\mathrm{Iso}(\mathcal{R}_f, f))^*$.

Proof. $\supset$: If $\varphi \in \mathrm{Iso}(\mathcal{R}_f, f)$, then $f = f\varphi$, so $f^\# = (f\varphi)^\# = f^\# \varphi^*$.

$\subset$: Pick $\alpha \in \mathrm{Iso}(\mathcal{R}_f{}^*, f^\#)$. Then $\alpha = \varphi^*$ for some $\varphi \in \mathcal{R}_f$ and $(f\varphi)^\# = f^\# \varphi^* = f^\#$. By (2.11), there is a $\Phi \in \mathcal{R}_f$ such that $\Phi^* = 1$ and $f\varphi\Phi = f$. Thus $\varphi\Phi \in \mathrm{Iso}(\mathcal{R}_f, f)$ and $(\varphi\Phi)^* = \varphi^* = \alpha$. □

Next is an equivariant version of (2.11).

(2.13) Proposition. *Suppose* f *and* g *are as in* (2.11) *and there is a compact Lie subgroup* L *of* $\mathcal{R}_f$ *leaving both* f *and* g *invariant. Then there is a* $k \in \mathcal{R}_f$ *such that* $k^* = 1$, k *is* L*-equivariant and* $gk = f$.

Proof. In the proof of (2.7) of [duPW], one constructs a vector field φ such that $F(x,t) := f(x) + t(g(x) - f(x)) = tf\cdot\varphi$, for $t \in [0,1]$. One solves φ to get $h(x,t)$ such that $h(x,1)$ is the desired k. To produce such a φ, one looks for a germ φ at each $(0,t)$ which satisfies

(a) $$(\partial F/\partial x) \cdot \varphi + (\partial F/\partial t) = 0;$$

also, φ should satisfy

(b) $$\varphi \in M := (m_n{}^2 \cap J(f))E_{n+1}{}^{\times n}$$

(to simplify the notation, we have identified germs at (0,t) with germs at (0,0)).

L acts on vector fields by $(\ell,\varphi) \to (d\ell/dx)^{-1}\varphi\ell$. Since f and g are invariant under L, $(\partial F/\partial t)\ell = \partial F/\partial t$ and $F\ell = F$; differentiating the latter equation, $((\partial F/\partial x)\ell)(d\ell/dx) = \partial F/\partial x$. Composing (a) with ℓ yields

$$((\partial F/\partial x)\ell) \cdot (d\ell/dx) \cdot (d1/dx)^{-1} \cdot (\varphi\ell) + (\partial F/\partial t)\ell = 0,$$

and hence

$$(\partial F/\partial x) \cdot (\ell \cdot \varphi) + (\partial F/\partial t) = 0.$$

In other words, the set of φ satisfying (a) is invariant under L. Since L is in $\mathcal{R}_f$, (b) is also invariant under L. Furthermore, the set of φ satisfying (a) and (b) is linear and closed. Thus, the average φ^- of these φ over L satisfies (a) and (b) and is L-invariant.

It is not hard to check that the corresponding solution $h^-(x,t)$ is invariant under the action $(\ell,h) \to \ell^{-1}h\ell$. Then $k^-(x) = h^-(x,1)$ is the desired k of this Proposition. □

We are ready to complete the proof of Theorem (1.2). Assume rkf = p or p−1. Assume f is the appropriate normal form: (2.7) or (2.9). Let H be a compact Lie subgroup of $\mathrm{Iso}(\mathcal{R}_f, f)$. Then H^* is a compact Lie subgroup of $(\mathrm{Iso}(\mathcal{R}_f, f))^* = \mathrm{Iso}(\mathcal{R}_f{}^*, f^\#)$ (by (2.12)). Let G_f be the MCS of $\mathrm{Iso}(\mathcal{R}_f{}^*, f^\#)$ computed in (2.8) or (2.10).

By Proposition 2.2, there is a $\Phi \in \mathrm{Iso}(\mathcal{R}_f{}^*, f^\#)$ such that H^Φ is contained in G_f.

Furthermore, $\Phi = \varphi^*$ for some $\varphi \in \mathrm{Iso}(\mathcal{R}_f,f)$. Let $K = H^{\varphi}$; K is in $\mathrm{Iso}(\mathcal{R}_f,f)$ and K^* is in G_f, so is also in $\mathrm{Iso}(\mathcal{R}_f,f)$. By (2.1), there is an $\alpha \in \mathcal{R}_f$ with $\alpha^* = 1$ and $K^{\alpha} = K^*$. So K^* also leaves $f\alpha$ invariant. Since $f^{\#} = f^{\#}\alpha^* = (f\alpha)^{\#}$, (2.13) implies there is a K^* equivariant $\beta \in \mathcal{R}_f$ such that $\beta^* = 1$ and $f\alpha\beta = f$. Thus $\delta = \varphi\alpha\beta \in \mathrm{Iso}(\mathcal{R}_f,f)$ and $H^{\delta} = K^{\alpha\beta} = (K^*)^{\beta} = K^*$ is contained in G_f. □

Holomorphic, reductive case.

The proofs in this section hold for this case with only a few changes (for example, we no longer need (2.6)). One place in which we used the compactness of our group was in the proof of (2.4) : we used that every continuous representation μ is semisimple, so every invariant subspace of $\mu(G)$ has an invariant complement. However, this holds for reductive Lie groups as well. We also used compactness to allow us to average over G (in (2.1) and (2.13)). See [Wa] for a discussion of how to carry out these arguments in the complex reductive case.

§3. Contractibility of the quotient.

For our proof of Theorem 1.4, we need an extension of (2.9) to maps dependent on parameters. For this we will need to make some minor modifications to the argument in §2 of [duPW]. In this section, all maps and manifolds will be C^{∞}.

Let N, P and Q be manifolds, $x_0 \in N$, $y_0 \in P$ and A a closed subset of Q. Let

$$F : (N \times Q \times \mathbb{R}, x_0 \times A \times [0,1]) \to (P \times Q \times \mathbb{R}, y_0 \times A \times [0,1])$$

be the germ of a $Q \times \mathbb{R}$ level preserving mapping. Let $J \subset C = C^{\infty}((N \times Q \times \mathbb{R}, x_0 \times A \times [0,1]),\mathbb{R})$ be an ideal contained in $m_0 = \{\alpha \in C : \alpha = 0 \text{ on } x_0 \times A \times [0,1]\}$.

(3 . 1) Definition. F is $\mathcal{R}_J$-*trivial* (*on* A) if there is a germ of a $Q \times \mathbb{R}$ level preserving diffeo-germ H of $(N \times Q \times \mathbb{R}, x_0 \times A \times \mathbb{R})$ which is a J-approximation to the identity such that $FH(x,q,t) = (F_0(x,q),t)$ (as germs), where

$$F_0 : (N \times Q, x_0 \times A) \to (P \times Q, y_0 \times A)$$

is defined by $F_0(x,q) = \pi_{P \times Q} F(x,q,t_0)$.

Let $J_{q,t}$ be the ideal in $C^{\infty}((N \times Q \times \mathbb{R}, x_0 \times q \times t),\mathbb{R})$ of germs of elements of J at $x_0 \times q \times t$. Let θ (resp. θ') be the space of germs at $x_0 \times q \times t$ (resp. at $x_0 \times A \times [0,1]$) of vector fields on $N \times Q \times \mathbb{R}$ with zero $TQ \times T\mathbb{R}$-component.

(3.2) Proposition. F *(as in* (3.1)) *is* $\mathcal{R}_J$-*trivial if*

$$(\partial/\partial t)F \in J_{q,t}\tau F(\theta) \text{ } \textit{for each } q \in A \textit{ and } t \in [0,1].$$

Proof. For each $q \in A$ and $t \in [0,1]$, there is a $\varphi_{q,t} \in J_{q,t}\theta$ such that $(\partial/\partial t)F = \tau F(\varphi_{q,t})$. Since $J_{q,t}\theta$ is convex, we can apply a partition of unity on $Q \times E$ to get a $\varphi \in J\theta'$ such that $(\partial/\partial t)F = \tau F(\varphi)$. Let $\tilde{\varphi}$ be a representative of φ. Because $\tilde{\varphi} = 0$ on $x_0 \times A \times [0,1]$ in $N \times Q \times \mathbb{R}$, there is a neighbourhood W of $x_0 \times A \times [0,1]$ on which the flow Φ of $\tilde{\varphi}$ is defined and $\Phi((x,q,t),s)$ stays in the domain of $\tilde{\varphi}$ for all $s \in [0,1]$ and $(x,q,t) \in W$. Then the rest of the proof of (2.2) of [duPW] goes through with only minor changes to yield the desired result. □

The next lemma follows from (3.2) and the proof of (2.6) of [duPW] with only minor notational changes.

(3.3) Lemma. *Let* $f,g : (\mathbb{R}^n \times Q, 0 \times A) \to (\mathbb{R}^p \times Q, 0 \times A)$ *be Q-level-preserving map-germs and let* $J(f)$ *denote the ideal in* $C = C^\infty((\mathbb{R}^n \times Q, 0 \times A),\mathbb{R})$, *generated by the* $p \times p$ *minors of that Jacobian matrix of* f *in which we take only partials of* $f^1,\dots,f^p$ *with respect to* $x_1,\dots,x_n$. *Let* J *be an ideal in* $m_0 = \{\alpha \in C : \alpha = 0 \text{ on } 0 \times Q\}$. *Assume* g *is a* $J\tau F(C^{\times n})$ *approximation to* f. *Suppose* $\tau(f-g) = \tau f\{\alpha_{ij}\}$ *for some* $\alpha_{ij} \in C$ *such that* $\{\alpha_{ij}(0,q)\}$ *is a matrix with all eigenvalues* 0 *for all* $q \in A$. *Then* f *and* g *are* $\mathcal{R}_J$*-equivalent, i.e. there exists a Q-level-preserving diffeo-germ* h *of* $(\mathbb{R}^n \times Q, 0 \times A)$, $h \in 1 + J^{\times n}$, *such that* $f = gh$.

The next proposition follows from the proof of (2.7) of [duPW] with only minor notational changes (the main point being that we still have the canonical form for f used in the proof of (2.7) of [duPW]).

(3.4) Proposition. *Let* $f : (N \times Q, x_0 \times A) \to (P \times Q, y_0 \times A)$ *be a map-germ of the form* $f_0 \times 1_Q$. *Let* $J \subset J(f)$ *(*$J(f)$ *as in (3.3)) be a proper ideal in* $C^\infty((N \times Q, x_0 \times A),\mathbb{R})$. *Then* f *is*

$$J\tau f(\theta) \cap \{J(f) \cap m_0^2)\tau f(\theta) - \mathcal{R}_J\text{-determined.}$$

(Here θ *is the space of germs at* $x_0 \times A$ *of vector fields on* $N \times Q$ *with zero* TQ*-component;* $m_0 = \{\alpha \in C^\infty((N \times Q, x_0 \times A),\mathbb{R}) : \alpha = 0 \text{ on } x_0 \times Q\}$; $\mathcal{R}_J$ *is as in (3.3)).*

(3.5) Corollary. *Let* $f, g : (N \times Q, x_0 \times A) \to (P \times Q, y_0 \times A)$ *be Q-level-preserving map-germs,* $f_0 = f_{q_0}$ *for some* $q_0 \in Q$, g_q *and* f_q $\mathcal{R}_{f_0}$*-equivalent to* f_0 *for all* $q \in A$. *Furthermore, suppose* $g_q = f_q$ *for all* $q \in B \subset A$. *Then* $f = gh$, *where* h *is a Q-level-preserving diffeo-germ of* $(N \times Q, x_0 \times A)$, *is an* $\mathcal{R}_{f_0}$*-approximation to the identity and satisfies* $h_q = 1_N$ *for all* $q \in B$.

Proof. By (1.3.B.a.ii) of [duPW], $J(f_q) = J(g_q) = J(f_0)$ for all $q \in A$, so $J(f) = J(g) = J(f_0 \times 1_Q)$. Let $J = J(f) \cap I(B)$, where $I(B) = \{\alpha \in C^\infty((N \times Q, x_0 \times A), \mathbb{R}) : \alpha(x,q) = 0$ for $q \in B\}$. Both f and g are $J\tau F(\theta)$-approximations to $F = f_0 \times 1_Q$. By (3.4), they are $\mathcal{R}_J$-equivalent to F and hence to each other. □

Proof of Theorem 1.4.

In case $n < p$, $\mathcal{R}_f = 1$ and the result is trivial. So assume $n \geq p$. The theorem is known for the case of our jet model; we will apply (3.5) with A = M, a smooth manifold, and B the boundary of M to lift this result to the smooth case.

First let's restate what we are trying to prove. We are given a smooth family of diffeo-germs $h_u \in \mathrm{Iso}(\mathcal{R}_f, f)$, $u \in B$. Let G_f be a MCS of $\mathrm{Iso}(\mathcal{R}_f, f)$. We want a smooth family $k_u \in \mathrm{Iso}(\mathcal{R}_f, f)$, $u \in A$, and $\ell_u \in G_f$, $u \in B$, such that $h_u \ell_u = k_u \mid B$.

If $g = \ell f r$ and if we have proved the theorem for f, then it will be proved also for g. To see this, use the correspondence $h_u \in \mathrm{I\,so}\,(\mathcal{R}_f, f)$ iff $r^{-1} h_u r \in \mathrm{Iso}(\mathcal{R}_g, g)$ and $r^{-1} G_f r = G_g$ given in (2.3), and note that $h_u \ell_u = k_u \mid B$ iff $r^{-1} h_u r r^{-1} \ell_u r = r^{-1} k_u r \mid B$. Therefore we assume f and G_f are given as in (2.7) and (2.8) if rkf = p or as in (2.9) and (2.10) if rkf = p−1. Define $f^\#$ and $\mathcal{R}_f{}^*$ as before. Recall that if $\mathrm{rkf} \leq p-2$, then $\mathcal{R}_f{}^* = 1$.

$\mathrm{Iso}(\mathcal{R}_f{}^*, f^\#)$ is an algebraic group, so, by the contractibility of $\mathrm{Iso}(\mathcal{R}_f{}^*, f^\#)/G_f$ (see (2.2)), there is a $k_u' \in \mathrm{Iso}(\mathcal{R}_f{}^*, f^\#)$ depending smoothly on $u \in A$ and an $\ell_u \in G_f$ on $u \in B$ such that $h_u{}^* \ell_u = k_u' \mid B$.

Now we find $k_u'' \in \mathcal{R}_f$, $u \in A$, such that $(k_u'')^* = k_u'$ for all $u \in A$: in case $\mathrm{rkf} \neq p-1$, $\mathcal{R}_f{}^*$ is contained in $\mathcal{R}_f$, so we need only consider the case rkf = p−1. Then J(f) is generated by $v_i'' = v_i - (\partial f/\partial x_i)$ $(1 \leq i \leq b)$, y_i $(1 \leq i \leq c)$ and $\partial f/\partial z_i$ $(1 \leq i \leq d)$; $J(f)^* = \langle v_i, y_i \rangle$, so the j-th component function of $k_u' \in \mathcal{R}_f{}^*$ is of the form (j-th variable) + (linear combination of v_i's and y_i's). Replacing v_i by v_i'' yields a k_u'' such that $(k_u'')^* = k_u'$.

Next we find $k_u''' \in \mathcal{R}_f$ such that $(k_u''')^* = k_u'$ for all $u \in A$ and $h_u \ell_u = k_u''' \mid B$. Note that, for each fixed u, the collection of diffeo-germs which satisfy these conditions is convex. Thus, if we cover A by coordinate charts and construct local solutions, we can piece these together with a partition of unity to get the desired k_u'''. On a coordinate neighbourhood in the interior of A, we can let $k_u''' = k_u''$. So now suppose we have a coordinate chart U of a boundary point together with a smooth retraction π of U to $U \cap B$. Then

$$k_u''' = h_{\pi(u)} \ell_{\pi(u)}{}^{-1} - k_{\pi(u)}'' + k_u'', \text{ for all } u \in U,$$

is the desired diffeo-germ (note that $\ell_u{}^{-1} \in G_f \subset \mathcal{R}_f$).

Now we apply (3.5) to A = M, B = bndry(M), Q = double of M, $f = f \times 1_Q$, and

$g = fk_u'''$ to get $h_u' \in \mathcal{R}_f$ depending smoothly on $u \in A$ such that $h_u' = 1$ for all $u \in B$ and $f = fk_u'''j_u'$. Then $k_u = k_u'''h_u'$ is the family of diffeo-germs required for our theorem. □

§4. Maximal compact subgroups of $\mathcal{A}$, $\mathcal{R}$ and Inv(D).

Our main goal in this section is to prove (1.5). You should review the terminology of the paragraph before that proposition. We will use CN as an abbreviation for "critical normalization". In this section, we will concentrate on the compact - C^∞ case. In §5, we will describe some properties of reductive groups which will show that the arguments of this section still work in the holomorphic-reductive case.

(4.1) Lemma. *Assume that* f *is a* CN. *Then the sequence*

$$1 \to \mathrm{Iso}(\mathcal{R},f) \to \mathrm{Iso}(\mathcal{A},f) \xrightarrow{\ p\ } \mathrm{Inv}^{\sim}(D) \to 1$$

is well-defined and exact, where $i(r) = (r,1)$ *and* $p(r,\ell) = \ell$.

Proof. The only parts of this which are not completely obvious are that p maps into and onto $\mathrm{Inv}^{\sim}(D)$.

"Into": Suppose $\ell fr = f$. Then $r(C(f)) = C(f)$ and so $\ell(D) = D$. If $\mathrm{rk} f = p-1$ and $d\ell$ reverses the orientation of cok(df), then since $d\ell(d^2f)(dr,dr) = d^2f$, therefore $\mathrm{index}(d^2f) = \mathrm{rk}(d^2f) - \mathrm{index}(d^2f)$.

"Onto": Suppose $\ell \in \mathrm{Inv}^{\sim}(D)$. Let $g = \ell f$. Then g is a CN, $D(g) = D(f)$ and, if $\mathrm{rk} f = p-1$, d^2f and d^2g have the same rank and index. By Theorem 2.1 of [GW1] (which was only stated there, but which is an easy consequence of (1.12) of [GW2] and (0.14) of [duPW]), f and g are $\mathcal{R}$-equivalent; that is, there is an r such that $(r,\ell) \in \mathrm{Iso}(\mathcal{A},f)$. □

If f is finitely $\mathcal{A}$-determined, (1.3) of [Wa] guarantees that G_a exists, and is unique up to conjugation by elements of $\mathrm{Iso}(\mathcal{A},f)$. Furthermore, f is $\mathcal{A}$-equivalent to a polynomial p, and J(p) is a closed ideal, hence so is J(f). Thus we can apply Theorem 1.2 to see that a MCS G_r of $\mathrm{Iso}(\mathcal{R}_f,f)$ exists, and is unique up to conjugation. If f is a CN, then a fortiori it is a CS, so $\mathrm{Iso}(\mathcal{R}_f,f) = \mathrm{Iso}(\mathcal{R},f)$ by (1.3).

Assume f is both finitely $\mathcal{A}$-determined and a CN. Let G_r be a MCS of $\mathrm{Iso}(\mathcal{R},f)$. Then $i(G_r)$ is a compact Lie subgroup of $\mathrm{Iso}(\mathcal{A},f)$, so is contained in a MCS G_a of $\mathrm{Iso}(\mathcal{A},f)$. Now $G' = \ker(p) \cap G_a = i(\mathcal{R}) \cap G_a$ is a compact subgroup of $\mathrm{Iso}(\mathcal{A},f)$; $i^{-1}(G')$ is a compact Lie subgroup of $\mathrm{Iso}(\mathcal{R},f)$ and contains G_r, hence equals G_r. Thus the sequence

$$1 \to G_r \xrightarrow{\ i\ } G_a \xrightarrow{\ p\ } p(G_a) \to 1$$

is exact. Clearly $p(G_a)$ is a compact Lie subgroup of $\mathrm{Inv}^{\sim}(D)$. It only remains to prove that $p(G_a)$ is a MCS among all linearizable compact Lie subgroups, unique up to conjugation in $\mathrm{Inv}^{\sim}(D)$, and that the sequence splits.

If $g = \ell f r$, then $\ell(D(f)) = D(g)$ and ℓ conjugates $\mathrm{Inv}^{\sim}(D(f))$ onto $\mathrm{Inv}^{\sim}(D(g))$ and conjugates compact Lie subgroups to compact Lie subgroups, preserving conjugates and preserving linearizability. Also (ℓ,r) conjugates compact Lie subgroups of $\mathrm{Iso}(\mathcal{A},f)$ to those of $\mathrm{Iso}(\mathcal{A},g)$. Thus it is enough to prove the result for any representative of the $\mathcal{A}$ equivalence class of f.

In Wall's proof of the Maximal Compact Subgroup Theorem for $\mathcal{A}$ equivalence (p. 348 of [Wal]), he shows that f can be taken to be a polynomial having a MCS G of $\mathrm{Iso}(\mathcal{A},f)$ which acts linearly (i.e., $G \subset GL(n) \times GL(p)$). Actually, there is a gap in Wall's argument. One starts with a MCS G of $\mathrm{Iso}(j^k\mathcal{A}, j^k f)$. Wall says there is a diffeomorphism φ with G^{φ} acting linearly. But G is in $j^k\mathcal{A}$, not in $\mathcal{A}$, so is not given as an action on $(\mathbb{R}^n \times \mathbb{R}P, 0)$, so we cannot directly apply Bochner linearization. However, as already pointed out in §2 of [J], Bochner's argument shows there is an α in $j^k\mathcal{A}$ with $G^{\alpha} = j^k(G^*)$ (where G^* is as in our §2, with $\mathcal{R}$ replaced by $\mathcal{A}$). Then if φ is any element of $\mathcal{A}$ with $j^k\varphi = \alpha$, Walls's argument works for this φ.

Now we show p(G) is a MCS of $\mathrm{Inv}^{\sim}(D)$, unique up to conjugation by members of $\mathrm{Inv}^{\sim}(D)$. Let H be any linearizable compact Lie subgroup of $\mathrm{Inv}^{\sim}(D)$. It suffices to show H can be conjugated in $\mathrm{Inv}^{\sim}(D)$ to a subgroup of p(G).

First we will paraphrase Theorem XV.3.7 of [H1]:

(4.2) Theorem. *Suppose* G *is a Lie group with finitely many connected components,* N *is a closed connected normal subgroup and* G/N *is compact. If* L *is a* MCS *of* G, *then* G = LN.

(4.3) Corollary. *If* N *has finitely many connected components, the Theorem remains true.*

Proof. Let N_1 denote the connected component of 1 of N. Then G/N_1 is a finite sheeted covering of G/N, hence is compact. (4.2) implies that $G = LN_1$; a fortiori, G = LN. □

(4.4) Corollary. *Suppose* $p : H \to H'$ *is an algebraic map between real algebraic groups. Then every compact subgroup of* H' *is the image under* p *of a compact subgroup of* H.

Proof. Suppose K is a compact subgroup of H'. Let G denote $p^{-1}(K)$. Then G is a Lie group with only finitely many connected components. Let N be the kernel of $p \mid G$; N is a closed, connected, normal subgroup of G, and has only finitely many connected components. Thus by (4.3) p(L) = K, where L is any MCS of G. □

Let $\mathcal{A}_s = \{(\ell,r) \in \mathcal{A}: j^s\ell(0) = j^s\mathrm{Id} \text{ and } j^s r(0) = j^s\mathrm{Id}\}$. If f is finitely $\mathcal{A}$-det., then it is also finitely $\mathcal{A}_s$-det. for any s (see Proposition 2.8 of [duPl]). Let $\pi^k{}_s$ denote the projection from k-jets to s-jets.

(4.5) Lemma. *Suppose* f *is* k-$\mathcal{A}_s$-*determined. Then*

$$\pi^k{}_s \mathrm{Iso}(j^k\mathcal{A}, j^k f(0)) = j^s(\mathrm{Iso}(\mathcal{A}, f)).$$

Proof. Pick $\alpha^k \in \mathrm{Iso}(j^k\mathcal{A}, j^k f(0))$; $\alpha^k = j^k\varphi(0)$ for some $\varphi \in \mathcal{A}$. $j^k(\varphi \cdot f) = \alpha^k \cdot (j^k f)(0) = j^k f(0)$. There is a $\Phi \in \mathcal{A}_s$ with $f = \Phi \cdot (\varphi \cdot f)$. Thus $\Phi \cdot \varphi \in \mathrm{Iso}(\mathcal{A}, f)$ and $j^s(\Phi \cdot \varphi)(0) = j^s\varphi(0) = \pi^k{}_s\alpha^k$. □

Now choose k such that f is k-$\mathcal{A}_1$-determined.

$j^k p: \mathrm{Iso}(j^k\mathcal{A}, j^k f(0)) \to j^k p(\mathrm{Iso}(j^k\mathcal{A}, j^k f(0)))$ is an algebraic map between algebraic groups. (4.4) implies that there is a compact group G^k in $\mathrm{Iso}(j^k\mathcal{A}, j^k f(0))$ such that $j^k(p)(G^k) = j^k H$. By (2.2), α^k conjugates G^k into G. By (4.5), there is a $\varphi \in \mathrm{Iso}(\mathcal{A}, f)$ such that $j^1\varphi = \pi^k{}_1\alpha^k$. By (4.1), $p(\varphi) = \ell \in \mathrm{Inv}^{\sim}(D)$; also H^{ℓ} is a compact Lie subgroup of $\mathrm{Inv}^{\sim}(D)$ such that $j^1(H^{\ell}) = K \subset p(G)$. By hypothesis, there is a $\Phi \in \mathrm{Inv}^{\sim}(D)$ such that $(H^{\ell})^{\Phi} = K$. (Since $\mathrm{Inv}^{\sim}(D)$ is not convex, we can't apply (2.1)). Thus $\Phi\ell \in \mathrm{Inv}^{\sim}(D)$ and conjugates H into K.

Finally, we show that the sequence

$$1 \to G_r \xrightarrow{\ i\ } G_a \xrightarrow{\ p\ } G_{\ell} \to 1$$

splits. This is trivial unless rkf = p-1 or p, for otherwise $G_r = \{1\}$; if rkf = p then clearly $G_{\ell} = O(p)$. Assume rkf = p-1, f is of the normal form (2.9) and $G_r = O(i) \times O(r-i) \times 1$. Then $(G_a)^*$ leaves the kernel $Y = \langle y_1, \ldots, y_r \rangle$ of $d^2 f$ invariant. For each $(r, \ell) \in G_a$,

$$r' = r^* \mid Y \in O(i) \times O(r-i).$$

Let $r'' = r' \times 1 \in G_r$. Since $(r'', 1) \in G_a$, $(R, \ell) \in G_a$, where $R = (r'')^{-1} r$ has $R^* \mid Y = 1_Y$.

(4.5) Claim. *For each* $(r, \ell) \in G_a$, *there is a unique* R *such that* $R^* \mid Y = 1_Y$ *and* $(R, \ell) \in G_a$, *and the map* $\ell \to (R, \ell)$ *is a splitting of the above exact sequence.*

Proof. We have already shown that R exists. Suppose (R, ℓ) and (R', ℓ) both satisfy the hypotheses. Then $(R^{-1}R', 1) \in G_a$. Hence $R^{-1}R' \in G_r$. But $(R^{-1}R')^* \mid Y = 1_Y$. Thus $R^{-1}R' = 1$, proving uniqueness.

Now we show $\ell \to R$ is a homomorphism. Given ℓ and ℓ' and their associated R and R', $(R, \ell) \in G_a$ and $(R', \ell') \in G_a$ imply $(RR', \ell\ell') \in G_a$ and $R^* \mid Y = 1_Y$ and $R'^* \mid Y = 1_Y$ imply $(RR')^* \mid Y = 1_Y$. Hence RR' is the source diffeo-germ associated to $\ell\ell'$. □

§5. Holomorphic, reductive case.

We now discuss some properties of reductive groups, which show that the arguments of Section 4 still work in the holomorphic, reductive case.

What is needed are "reductive" versions of (4.3) and (4.4). These are provided by (5.7) and (5.8), when combined with (5.10).

Let G be an affine algebraic group, R a subgroup. R is a *linearly reductive* subgroup of G if the restriction to R of any polynomial finite dimensional representation of G is semi-simple. G is a linearly reductive (affine algebraic) group if it is a linearly reductive subgroup of itself.

We will assume characteristic zero, unless we specifically state otherwise. The following structure theorem makes discussion of linear reductiveness fairly straightforward in characteristic zero.

Let G_u denote the unipotent radical of G, i.e. the unique maximal normal subgroup of G amongst subgroups, the restrictions to which all finite dimensional polynomial representations of G are unipotent.

(5 . 1) Theorem. *Let* R *be a maximal linearly reductive subgroup of* G, *an affine algebraic group. Then* G *is isomorphic to the semi-direct product of* G_u *with* R.

Proof. See 4.3, p. 117, of [H2]. □

(5 . 2) Corollary. *If* R *is a maximal linearly reductive subgroup of* G, *then* R *is a linearly reductive group.*

Proof. By the theorem, and the remarks on p. 70 of [H2], inclusion induces $R \cong G/G_u$ as algebraic groups; combining this with the projection $G \to G/G_u$ yields an algebraic epimorphism $\pi : G \to R$ which is the identity on R. It follows that any polynomial representation ρ of R extends to a polynomial representation $\rho \circ \pi$ of G. So, since R is a linearly reductive subgroup of G, it is linearly reductive. □

Linear reductiveness is preserved on taking normal subgroups:

(5.3) Lemma. (Arbitrary characteristic). *If* G *is a linearly reductive group, and* $R < G$, *then* R *is a linearly reductive subgroup of* G.

Proof. This is a trivial fact of representation theory; see p. 212 of [H1] or p.67 of [H2]. □

(5 . 4) Lemma. *If* G *is a linearly reductive group, and* $R < G$, *then* R *is a linearly reductive group.*

Let us first observe a very useful corollary of (5.1):

(5.5) Corollary. (Characteristic zero). *An affine algebraic group* G *is linearly reductive if, and only if,* $G_u = \{1\}$.

Proof of (5.4). Since $R < G$, so is its Zariski closure. So it is enough to treat the case where R is an affine algebraic subgroup of G.

Since $R < G$, $gR_ug^{-1} \subset R$ for all $g \in G$, and is a subgroup of G. We claim that it is a unipotent subgroup of G. For if ρ is a finite dimensional polynomial representation of G, and $k \in R_u$, then, for $n \in \mathbb{Z}$,

$$\rho(gkg^{-1})^n = \rho(g)\rho(k)^n\rho(g)^{-1};$$

since $\rho(k)$ is unipotent, $\rho(k)^n = 1$ and hence $\rho(gkg^{-1})^n = 1$ for sufficiently large n.

Moreover, $gR_ug^{-1} < R$; for if $k \in R$, then $g^{-1}k\,g \in R$ (since $R < G$), so $g^{-1}kg\, R_u(g^{-1}kg)^{-1} \subset R_u$ (since $R_u < R$), i.e. $k(gR_ug^{-1})k^{-1} \subset gR_ug^{-1}$. Thus gR_ug^{-1} is a normal, unipotent subgroup of R, hence is in R_u.

Since this holds for all $g \in G$, $R_u < G$. Since it is a unipotent subgroup, $R_u \subset G_u$. But G is linearly reductive, so $G_u = \{1\}$, so $R_u = \{1\}$, so R is linearly reductive. □

Linear reductiveness is also preserved by images of algebraic homomorphisms:

(5.6) Lemma. *If* $\varphi : G \to H$ *is a surjection of affine algebraic groups and* G *is reductive, then so is* H.

Proof. Since φ is onto, it follows by (4.40), p. 120, of [H2], that $\varphi(G_u) = H_u$. Since G is linearly reductive, $G_u = \{1\}$, hence so is H_u, so H is linearly reductive. □

The structure theorem also has other consequences:

(5.7) Corollary. *Let* G *be an affine algebraic group. Let* N *be a closed normal subgroup such that* G/N *is linearly reductive. If* H *is a maximal linearly reductive subgroup of* G, *then* $G = HN$.

Proof. G/N is linearly reductive, so $(G/N)_u = \{[N]\}$. Now unipotent elements map to unipotent elements, and the image of a normal subgroup by an epimorphism is normal in the target, so G_u projects to $[N]$; i.e. $G_u \subset N$. Thus the result follows by (5.1). □

(5.8) Corollary. *Let* $\varphi : G \to H$ *be an algebraic epimorphism. Let* S *be an algebraic subgroup of* H *which is linearly reductive. Then there exists an algebraic subgroup* R *of* G *which is linearly reductive such that* $\varphi(R) = S$.

Proof. Let $M = \varphi^{-1}(S)$ and let $N = \ker \varphi$. Then $S \cong M/N$ and is linearly reductive. By (5.7), if R is a maximal linearly reductive subgroup of M, then we have $M = RN$. So $S = \varphi(M) = \varphi(RN) = \varphi(R)$. □

Reductive versus linearly reductive

(5 . 9) Lemma. *A linearly reductive algebraic group over $\mathbb{C}$ is a complex-analytic reductive group.*

Proof. It is well-known (see e.g. (4.3) of [Hr]) that a linearly reductive affine algebraic group G is reductive as an algebraic group, which means that the radical of G (i.e. its unique maximal normal solvable subgroup) is a $\mathbb{C}$-toroid, that is up to isomorphism a product of copies of $\mathbb{C}^*$. It is also well-known (see e.g. p. 245 of [Boul) that a finite dimensional complex-analytic representation ρ of a complex-analytic Lie group is semi-simple if, and only if, its restriction to the radical of the group is semi-simple. This will be the case if the radical is a product of copies of $\mathbb{C}^*$, since $(\mathbb{C}^*)^k$ is reductive as a complex-analytic Lie group (e.g. because it is clearly the universal complexification of $(S^1)^k$, so we can apply p.208 of [H1]). Thus all complex-analytic representations of G are semi-simple. Since G has a faithful finite-dimensional representation (since all affine algebraic groups do ((1.10), p. 101, of [Bor])), we see that indeed G is reductive as a complex-analytic Lie group. □

(5 . 10) Corollary. *Let* G *be an affine algebraic group over* $\mathbb{C}$. *A subgroup of* G *is a maximal linearly reductive subgroup if, and only if, it is a maximal* $\mathbb{C}$*-analytic reductive subgroup.*

Proof. If H is a $\mathbb{C}$-analytic reductive subgroup of G, then all its finite dimensional $\mathbb{C}$-analytic representations are semi-simple, so a fortiori all its finite dimensional polynomial representations are. So H is a linearly reductive subgroup of G.

Thus a maximal $\mathbb{C}$-analytic reductive subgroup of G is contained in a maximal linearly reductive subgroup of G. Conversely, if H is a maximal linearly reductive subgroup of G, then it is a linearly reductive group (Corollary 5.2)), so by Lemma (5.8), it is reductive as a $\mathbb{C}$-analytic group. □

References

[Boc] Bochner, S., Compact groups of differentiable transformations. Ann. Math. **46**, 372-381 (1945).

[Bor] Borel, A., Linear algebraic groups. Benjamin, 1969.

[Boul] Bourbaki, N., Groupes et algèbres de Lie, Chaps. 2 & 3. Hermann, Paris, 1972.

[GW1] Gaffney, T., Wilson, L., Equivalence of generic mappings and C^∞ normalization. Compositio Mathematica **49**, 291-308 (1983).

[GW2] Gaffney, T., Wilson, L., Equivalence theorems in global singularity theory. AMS Symposium on Singularities, Arcata, Proceedings of Symposia in Pure Math.**40**, part 1, 439-447 (1983).

[GduPW] Gaffney, T., du Plessis, A., Wilson, L., On map-germs determined by their discriminants. In preparation.

[H1] Hochschild, G., The structure of Lie groups. San Francisco, Holden-Day 1965.

[H2] Hochschild, G., Basic theory of algebraic groups and Lie algebras. Springer-Verlag, 1981.

[Hr] Hochster, M., Invariant theory of commutative rings. Group actions on rings, Contemporary Math. AMS. 43, 161-180.

[J] Jänich, K., Symmetry properties of singularities of C^∞-functions. Math. Ann. **238**, 147-156 (1978).

[duP] du Plessis, A., On the determinacy of smooth map-germs. Invent. Math.**58**, 107-160 (1980).

[duPW] du Plessis, A., Wilson, L., On right-equivalence. Math. Z. **190**, 163-205 (1985).

[Wa] Wall, C.T.C., A second note on symmetry of singularities. Bull. London Math. Soc. **12**, 347-354 (1980).

[Wi] Wilson, L., Global singularity theory. Institute of Mathematics, University of Aarhus (Denmark), Seminar Notes No. 1, 129-137 (1982).

Matematisk Institut
Aarhus Universitet
Ny Munkegade
DK-8000 Aarhus C
Denmark

Department of Mathematics
University of Hawaii at Manoa
2505 The Mall
Honolulu,
HI 96822, USA

Deformations and the Milnor Number of Non-Isolated Plane Curve Singularities

Rob Schrauwen

Abstract

We consider deformations of non-isolated plane curve singularities and compute the number of special points occuring in such a deformation. These numerical data are used to give formulae for the Milnor number of the non-isolated singularity.

1 Introduction

We consider non-zero holomorphic function germs $f : (\mathbf{C}^2, 0) \to (\mathbf{C}, 0)$ and certain deformations. The programme follows the established path laid by work of R. Pellikaan and T. de Jong. In his thesis [Pe1], Pellikaan developed the deformation theory for application in the case of singularities of arbitrary dimensions with a one-dimensional singular locus and transversal type A_1. De Jong [Jo] considered the case that the singular locus is a smooth curve but with more complicated transversal types. In later versions, Pellikaan stated his results more generally ([Pe2], [Pe3]), and we can obtain our key results by easy and straightforward application of his theorems.

Our study is, however, in a sense transverse to that of Pellikaan, since we consider arbitrary transversal types but only in the plane curve case.

We start by defining the Jacobi number $j_I(f) = \dim_{\mathbf{C}} I/J_f$ (where I is the ideal defining the singular locus Σ and J_f the Jacobi ideal) and prove that finite Jacobi number is equivalent to finite I-codimension and to f having prescribed transversal singularities along the branches of the singular locus. For f with finite Jacobi number we consider deformations and count the number of special points in such a deformation. We prove that $j_I(f)$ in fact equals the Milnor number of the associated reduced singularity f_R.

We carry on by following Siersma [Si2], in order to express the Milnor number $\mu(f)$ in the number of special points. This generalizes results of Siersma and De Jong (in the plane curve case). The answer is the following: Let Σ^k be the reduced curve whose branches are the branches of the singular locus of f where f has transversal type A_{k-1}. Let $\#D[p,q]$ be the number of points in a deformation f_t of f which makes each Σ^k smooth, where, in local coordinates, the singularity of f_t is $x^p y^q$. Let d be the number of connected components of the Milnor fibre. Then:

$$\mu(f) = \sum_{p<q} (p+q-1) \cdot \#D[p,q] + \#D[1,1] + \sum_k (k-1)(\mu(\Sigma^k) - 1) + d - 1.$$

This work answers of a question of Dirk Siersma, who asked how the Jacobi and Milnor numbers of an arbitrary plane curve singularity could be expressed in the number of "p,q-points".

2 Invariants

2.1 We use the following notations: $\mathcal{O} = \mathbf{C}\{x,y\}$, $\mathfrak{m}$ is its maximal ideal. We call the elements of $\mathcal{O}$ *(plane) curve singularities.* For $n \in \mathbf{N}$ we put $\underline{n} = \{1,\ldots,n\}$.

We denote by $\mathrm{Sing}(f)$ the singular set of an analytic function germ $f \in \mathcal{O}$. If $C \subset \mathrm{Sing}(f)$ is one-dimensional, then f is *of transversal type A_{m-1} along C* if for all $c \in C$ we can find local coordinates u, v in a neighbourhood of c such that $f(u,v) = v^m$. The name comes from the fact that on a transversal slice X the zero-dimensional singularity $f : (X,c) \to \mathbf{C}$ is of type A_{m-1}.

2.2 Definition Let $p \geq 0, q \geq 1$ be integers. A germ of an analytic function $f : (\mathbf{C}^2,0) \to (\mathbf{C},0)$ is said to be *of type $D[p,q]$* if there are local coordinates x, y such that $f(x,y) = x^p y^q$. A function germ of type $D[p,p]$ is also called of type $A[p]$. We will also use Siersma's notations, such as $A_\infty = D[0,2]$ and $D_\infty = D[1,2]$. Note that $D[p,q] = D[q,p]$.

2.3 Let $I \subset \mathcal{O} = \mathbf{C}\{x,y\}$ be an ideal. Then we define the *primitive ideal* $\int I = \{f \in \mathcal{O} \mid (f) + J_f \subset I\}$. Here J_f is the Jacobi ideal of f, generated by the partial derivatives. This definition is due to Pellikaan.

Suppose $I = (g')\mathcal{O}$ and let $g' = g_1^{m_1-1}\cdots g_r^{m_r-1}$, where $m_i \geq 2$, be the decomposition of g in irreducible factors. Then it is easy to see that $\int I = (g_1^{m_1}\cdots g_r^{m_r})$, cf. [Pel] 1.7.

Furthermore let $\mathcal{D}$ be the group of local analytic isomorphisms $\psi : (\mathbf{C}^2,0) \to (\mathbf{C}^2,0)$. For an ideal I we define: $\mathcal{D}_I = \{\psi \in \mathcal{D} \mid \psi^*(I) = I\}$.

2.4 In the sequel we will always have the following situation:

(a) $I = (g_1^{m_1-1}\cdots g_r^{m_r-1})\mathcal{O}$ with g_i irreducible, $m_i \geq 2$ $(i \in \underline{r})$, and g_i and g_j having no common factor $(i \neq j)$.

(b) $\int I = g\mathcal{O} = (g_1^{m_1}\cdots g_r^{m_r})\mathcal{O}$.

(c) $\Sigma_i = Z(g_i)$, $\Sigma = \Sigma_1 \cup \cdots \cup \Sigma_r$.

If $f \in \int I$ then $\Sigma \subset \mathrm{Sing}(f)$. We are looking for conditions on f such that $\mathrm{Sing}(f) = \Sigma$ and f has transversal type precisely A_{m_i-1}. It will prove useful to give $(\Sigma,0)$ the (possibly non-reduced) analytic structure of I.

Observe that in this case $\mathcal{D}_I = \mathcal{D}_{\int I}$.

2.5 We have a right action of $\mathcal{D}_I$ on $\int I$ and hence for all $f \in \int I$ the orbit $\mathrm{Orb}_I(f) \subset \int I$. Consider the tangent space $T\mathcal{D}_I$ of $\mathcal{D}_I$ at the identity. We have $T\mathcal{D}_I \subset T\mathcal{D}$, the tangent space of $\mathcal{D}$ at the identity. We identify $T\mathcal{D}$ with $\mathfrak{m}\Theta$, the germs of vector fields vanishing at the origin.

According to [Pel], p.19 we have $T\mathcal{D}_I = \{\xi \in \mathfrak{m}\Theta \mid \xi(I) \subset I\}$.

2.6 Definition Let $f \in \int I$. Then we define $c_I(f) = \dim_{\mathbf{C}} \int I / T\mathcal{D}_I(f)$, where $T\mathcal{D}_I(f) = \{\xi(f) \mid \xi \in T\mathcal{D}_I\}$. $c_I(f)$ is called the *I-codimension* of f.

Suppose $I = (y^{m-1})$. Then one easily sees that $c_I(f) = 0$ if and only if $\int I = (f)$, i.e. f has transversal A_{m-1} singularities along $\mathrm{Sing}(f)$; and $c_I(f) = 1$ if and only if f is of type $D[1,m]$.

2.7 We now state some standard finite determinacy results (cf. [Sil]). If $\mathfrak{a}$ is an ideal in $\mathcal{O}$ and $k \in \mathbf{N}$, then $f \in \mathfrak{a}$ is called *k-determined in $\mathfrak{a}$* if $f + \mathfrak{m}^k\mathfrak{a} \subset \mathrm{Orb}_{\mathfrak{a}}(f)$.

Theorem *Let $f \in \int I$. Then:*

(a) If f is k-determined in $\int I$ then $\int I \cdot \mathfrak{m}^k \subset T\mathcal{D}_I(f) + \int I \cdot \mathfrak{m}^{k+1}$.

(b) *If* $\int I \cdot \mathfrak{m}^k \subset \mathfrak{m}T\mathcal{D}_I(f)$, *then* f *is* k*-determined in* $\int I$.

The proof is standard (cf. [Si1], [Pe2]). □

Corollary $c_I(f) < \infty$ *if and only if* f *is* k*-determined for some* $k \in \mathbf{N}$. *Furthermore, if* $c_I(f) < \infty$ *then* $\mathrm{Sing}(f) = \Sigma$.

Proof The first statement is obvious. Now suppose $c_I(f) < \infty$, so there is a k such that $\int I \cdot \mathfrak{m}^k \subset T\mathcal{D}_I(f)$. Because $T\mathcal{D}_I(f) \subset \mathfrak{m}J_f \cap \int I$, it follows that $Z(J_f) \cup \Sigma \subset \Sigma$, hence $\mathrm{Sing}(f) = \Sigma$. □

2.8 Definition Let I be as above. Then we define $j_I(f) = \dim_{\mathbf{C}} I/J_f$. We call $j_I(f)$ the *Jacobi number.*

The Jacobi number plays the same rôle as the Milnor number in the case of isolated singularities. Since $\dim_{\mathbf{C}}\mathcal{O}/J_f$ is infinite, we look at other quotients and it appears that I/J_f is the right choice.

Example If $I = (y^{m-1})$ and $f(x,y) = y^m$ then $j_I(f) = 0$. If f is of type $D[p,q]$, then choose coordinates on which $f(x,y) = x^p y^q$. Let $I = (x^{p-1}y^{q-1})$. Then $j_I(f) = 1$.

In the second section we will show that the Jacobi number equals the Milnor number of the reduced singularity associated to f. This fact seems to have been unnoticed before and it is definitely false in higher dimensions.

2.9 Proposition *Let* $f \in \int I$ *and suppose* $j_I(f) < \infty$ *and* $\mathrm{depth}(\mathcal{O}/I) > 0$. *Then:*

(a) $T\mathcal{D}_I(f) = \mathfrak{m}J_f \cap \int I$ $(\mathfrak{m}J_f = T\mathcal{D}(f))$,

(b) $c_I(f) < \infty$.

Proof This is proposition 5.3 of [Pe2]. Pellikaan also considers $c_{I,e}(f) = \dim_{\mathbf{C}} \int I/(J_f \cap \int I)$. It is clear that $c_{I,e}(f) \leq c_I(f)$. In proposition 5.3 of [Pe2] it is in fact proved that $c_{I,e}(f) \leq j_I(f)$. But $c_I(f)$ is finite if $c_{I,e}(f)$ is finite, because the quotient $M = (J_f \cap \int I)/(\mathfrak{m}J_f \cap \int I)$ is a finitely generated $\mathcal{O}$-module and $\mathfrak{m}(J_f \cap \int I) \subset T\mathcal{D}_I(f) = \mathfrak{m}J_f \cap \int I$, so $\mathfrak{m}M = (0)$ and $\dim_{\mathbf{C}} M < \infty$. □

This shows that $c_I(f)$ is an invariant of the right-equivalence class of f.

2.10 Theorem *Let* $f \in \int I$. *The following statements are equivalent:*

(i) $j_I(f) < \infty$,

(ii) $c_I(f) < \infty$,

(iii) f *is a singularity with singular locus* $\Sigma = \Sigma_1 \cup \cdots \cup \Sigma_r$ *and transversal type* A_{m_i-1} *along* $\Sigma_i \backslash \{0\}$ $(i \in \underline{r})$.

Proof (i) ⇒ (ii): Proposition 2.9.
(ii) ⇒ (iii): In corollary 2.7 it has been observed that $\mathrm{Sing}(f) = \Sigma$. We consider the sheaf $\mathcal{O}$ of analytic functions on a small neighbourhood V of the origin (for $a \in V$ the stalk at a is $\mathcal{O}_a$). We have ideal sheaves $\mathcal{I}$, $\int\mathcal{I}$ and $\mathcal{J}_f$ with the obvious meanings. Let $\mathcal{F} = \int\mathcal{I}/(\mathcal{J}_f \cap \int\mathcal{I})$. Then $\mathcal{F}$ is a coherent sheaf of $\mathcal{O}$-modules. Now because $c_I(f)$ is finite and $\dim_{\mathbf{C}}\mathcal{F}_0 = c_{I,e}(f)$, we can choose V such that $\dim_{\mathbf{C}}\mathcal{F}_a = 0$ for $a \in V\backslash\{0\}$ ($\mathcal{F}$ is concentrated in a finite set of points). Look at a point $a \in \Sigma_i\backslash\{0\}$. (Σ_i, a) is defined by $(g_i^{m_i-1})$ and g_i can be used as one of the local coordinates near a. From the remark following definition 2.6 it follows that f has only transversal A_{m_i-1} singularities along $\Sigma_i\backslash\{0\}$ $(i \in \underline{r})$.
(iii) ⇒ (i): According to example 2.8 the stalk of $\mathcal{I}/\mathcal{J}_f$ at $a \in V\backslash\{0\}$ is (0) if we choose V sufficiently small. So $\mathcal{I}/\mathcal{J}_f$ is concentrated in a finite set of points and therefore we obtain that $j_I(f)$, the dimension of the stalk at the origin, is finite. □

3 Deformations

3.1 Deformations of isolated singularities are thoroughly studied. They are used to split up the complicated isolated singularity into a number of simple singularities which are better understood, and in this case it is well-known that it splits up in A_1 or *Morse* singularities (in our notation: $D[1,1]$ singularities) and that their number is the Milnor number μ of the singularity.

In this section we will consider deformations of a non-isolated plane curve singularity f. In such a deformation, it is important to describe what happens to the singular set of f, because we want to recover various properties of f in the deformation (e.g. we would like that the Milnor fibrations of f and the deformed f_t are equivalent, cf. lemma 4.8). A theorem of Pellikaan is invoked to show that $j_I(f)$ is invariant under deformations.

We will consider two kinds of deformations in more detail and compute the number of special points (critical points) in such a deformation. In the next section we will use these two types to obtain two formulae for the Milnor number. The two special kinds are examples of deformations where the singularities of f split up into $D[p,q]$ singularities only. There are many other deformations possible, giving similar formulae for the Milnor number.

Our reference for this section is [Pe3].

3.2 Definition Let $I = (g_1^{m_1-1} \cdots g_r^{m_r-1})$ define $(\Sigma, 0)$ as before.

(a) We define $E_I = \{m_1, \ldots, m_r\}$.

(b) For $p \in E_I$, let Σ^p be the *reduced* curve defined by the product of all g_i such that $m_i = p$. We call Σ^p the *p-part* of Σ.

(c) Let $f \in \int I$ with $j_I(f) < \infty$. By theorem 2.10 we can write

$$f = \prod_{p \in E_I \cup \{1\}} f_{(p)}^p.$$

For $p, q \in E_I \cup \{1\}$, $p \neq q$, we define $d_{p,q}(f) = \dim_{\mathbf{C}} \mathcal{O}/(f_{(p)}, f_{(q)})$, and for other p, q we put $d_{p,q}(f) = 0$.

It will appear in the next section that for $p \neq q$ the number $d_{p,q}$ is the number of $D[p,q]$-points in a deformation. Notice that if $p, q > 1$, $d_{p,q} = \Sigma^p \cdot \Sigma^q$ (intersection number).

3.3 The following definitions come from [Pe3]. We think of Σ and I being as before, but for the definitions this is not important. Let $(\Sigma, 0)$ be a germ of an analytic space defined by the ideal I in $\mathcal{O}$. A *deformation* of $(\Sigma, 0)$ consists of a germ of an flat map $G : (\mathcal{X}, 0) \to (S, 0)$ of analytic spaces, together with an embedding $i : (\Sigma, 0) \to (\mathcal{X}, 0)$ such that $(i(\Sigma), 0) \cong (G^{-1}(0), 0)$ as analytic spaces. We can embed $(S, 0)$ in $(\mathbf{C}^\sigma, 0)$ and $(\mathcal{X}, 0)$ in $(\mathbf{C}^2 \times \mathbf{C}^\sigma, 0)$ such that the following diagram commutes:

$$\begin{array}{ccccc} (\Sigma,0) & \xrightarrow{i} & (\mathcal{X},0) & \xrightarrow{G} & (S,0) \\ \downarrow & & \downarrow & & \downarrow \\ (\mathbf{C}^2,0) & \xrightarrow{j} & (\mathbf{C}^2 \times \mathbf{C}^\sigma,0) & \xrightarrow{\pi} & (\mathbf{C}^\sigma,0) \end{array}$$

where $j(z) = (z, 0)$ and π is the projection on the second factor. Let $\tilde{\mathcal{O}}$ be the local ring of germs of analytic functions on $(\mathbf{C}^2 \times S, 0)$. Let $\tilde{I}$ be the ideal in $\tilde{\mathcal{O}}$ defining the germ $(\mathcal{X}, 0)$ considered as a subspace of $(\mathbf{C}^2 \times S, 0)$.

3.4 Definition Let $(\Sigma, 0)$ be a germ of an analytic space in $(\mathbf{C}^2, 0)$ defined by I and $G: \mathcal{X} \to S$ a deformation of Σ. Let $f: (\mathbf{C}^2, 0) \to (\mathbf{C}, 0)$ be a germ of an analytic function such that $f \in \int I$. Let $F: (\mathbf{C}^2 \times S, 0) \to (\mathbf{C}, 0)$ be a germ of an analytic map. Then (F, G) is called a *deformation* of $(f, \Sigma, 0)$ if $F(x, y, 0) = f(x, y)$ and

$$(F)\tilde{\mathcal{O}} + J_F \subset \tilde{I}$$

where $J_F = (\frac{\partial F}{\partial x}, \frac{\partial F}{\partial y})\tilde{\mathcal{O}}$ and x, y are local coordinates.

For example one could consider deformations of $(f, \Sigma, 0)$ with Σ fixed, as in [Pel] and [Si3]. We will not do this, because such a deformation only gives information about the $D[1,p]$-points.

If the context is clear, then we say that F is a deformation of $(f, \Sigma, 0)$, or even that F is a deformation of f.

3.5 Let $I = (g_1^{m_1-1} \cdots g_r^{m_r-1})$, $\int I = (g_1^{m_1} \cdots g_r^{m_r})$ and $f \in \int I$ as before. From now on we assume $j_I(f) < \infty$.

The first kind of deformation we consider, will be one with as much crossings as possible. It will be called a *network map* type of deformation[1].

Such a deformation arises as follows: First look at the *reduced* singularity f_R. This singularity can be deformed in such a way that it has only normal crossings and their number is δ, the virtual number of double points. An explicit construction, involving small translations of the branches in the resolution, was first given by A'Campo and by Gusein-Sade and can be found for instance in [AGV]. We now have a network map deformation $(f_R)_t$ of f_R, and get one for f by giving the branches of $(f_R)_t$ the correct multiplicities of f.

Write $f = g_1^{m_1} \cdots g_r^{m_r} h_{r+1} \cdots h_s$. Let the network map deformation of

$$f_R = g_1 \cdots g_r h_{r+1} \cdots h_s$$

be $F_R: (\mathbf{C}^2 \times \mathbf{C}^\sigma, 0) \to (\mathbf{C} \times \mathbf{C}^\sigma, 0)$. We may assume that we can write

$$F_R(z,t) = (G_1(z,t) \cdots G_r(z,t) H_{r+1}(z,t) \cdots H_s(z,t), t),$$

where the G_i and the H_i describe what happens to the branches of f_R (this is possible in the construction of A'Campo and Gusein-Sade), and $z = (x, y)$. Now let

$$F(z,t) = G_1(z,t)^{m_1} \cdots G_r(z,t)^{m_r} H_{r+1}(z,t) \cdots H_s(z,t)$$

and

$$G(z,t) = (G_1(z,t)^{m_1-1} \cdots G_r(z,t)^{m_r-1}, t).$$

Then (F, G) is a deformation of $(f, \Sigma, 0)$ and we say that (F, G) is a *network map* type deformation of f.

3.6 The second type of deformation will be a deformation of $(f, \Sigma, 0)$ which makes each of the p-parts Σ^p of the singular locus Σ smooth. Recall that we consider $(\Sigma^p, 0)$ $(p \in E = E_I)$ as a reduced curve. There exists a versal deformation $G_p: (\mathbf{C}^2 \times \mathbf{C}^{\sigma_p}, 0) \to (\mathbf{C} \times \mathbf{C}^{\sigma_p}, 0)$ of $(\Sigma^p, 0)$ with $G_p(z, t_p) = (G_{p1}(z, t_p), t_p)$. Let $\sigma = \sum_{p \in E} \sigma_p$. By $t = (t_p)_{p \in E}$ we denote local coordinates on $\mathbf{C}^\sigma$. Let

$$G(z,t) = (\prod_{p \in E_I} G_{p1}(z, t_p)^{p-1}, t),$$

[1] We specifically think of the London Underground network map designed by Harry Beck in 1933.

then G defines a deformation of Σ. Furthermore write $f = g_1^{m_1} \cdots g_r^{m_r} h$ and define $F : (\mathbf{C}^2 \times \mathbf{C}^\sigma \times \mathbf{C}^3, 0) \to (\mathbf{C}, 0)$ by

$$F(z,t,a,b,c) = \prod_{p \in E_I} G_{p1}(z,t)^p (h(z) + a + bx + cy).$$

Then (F, G) defines a deformation of $(f, \Sigma, 0)$, a *deformation which makes the p-parts smooth.*

3.7 The next theorem is an important result due to Pellikaan. In [Pe3] it is proved as a part of a larger theorem.

Theorem *Let I, $f \in \int I$, and Σ be as before and $j_I(f) < \infty$. Let $G : (\mathcal{X}, 0) \to (S, 0)$ be a deformation of $(\Sigma, 0)$ with non-singular base $(S, 0)$ and let (F, G) be a deformation of $(f, \Sigma, 0)$. Let $\pi : (\mathbf{C}^2 \times S, 0) \to (S, 0)$ be the projection. Then there exist representatives for all considered germs such that for all $t \in S$*

$$j_I(f) = \sum_{a \in \pi^{-1}(t)} j(f_t, a).$$

So the Jacobi number is invariant under deformations. □

The following theorems show that in our type of deformations, the number of special points is finite. We omit their straightforward but somewhat tedious proofs, which are analogous to the proofs of propositions 7.18 and 7.20 of [Pe1].

3.8 Theorem *Let I, $f \in \int I$, and Σ be as before and $j_I(f) < \infty$. Suppose $F : (\mathbf{C}^2 \times S, 0) \to (\mathbf{C}, 0)$ is a network map type deformation of $(f, \Sigma, 0)$, where $S = \mathbf{C}^\sigma$. Then there is a dense subset $V \subset S$ and an open neighbourhood U of $0 \in \mathbf{C}^2$ such that for all $t \in V$ sufficiently small:*

(a) $f_t^{-1}(0)$ has only normal crossings, and their number equals $\delta(f_R)$, the virtual number of double points of the reduced germ f_R,

(b) f_t has only A_1 singularities in $U \setminus \Sigma_t$,

(c) For $0 < p < q$, f_t has only $D[p,q]$ singularities on $\Sigma_t^q \cap U$, and their number is $d_{p,q}(f)$,

(d) f_t has only $D[0,q]$ and a finite number of $D[q,q]$ singularities on the rest of $\Sigma_t^q \cap U$. □

3.9 Corollary $j_I(f) = \mu(f_R)$, *i.e. the Jacobi number $j_I(f)$ equals the Milnor number $\mu(f_R)$ of the reduced singularity f_R.*

Proof A network map type of deformation of f arises from a deformation of of the same kind. It is well-known that $\mu(f_R)$ equals $\delta + \#A_1$, where δ is the number of crossing points and $\#A_1$ the number of A_1 singularities outside the zero-locus. By construction, the deformation of f has δ $D[p,q]$ crossing points. But it is also not difficult to see that there are as many A_1-points outside the zero locus as there are outside the zero locus in the deformation of f_R.

By theorem 3.7 it now follows that $j_I(f) = \mu(f_R)$, for we know that the Jacobi number of all $D[p,q]$ singularities is 1. □

Remark The Jacobi number being equal to the Milnor number of the reduced singularity gives interesting interpretations to several of the formulae to be found in Pellikaan's work. For instance, from [Pe1], 5.14, it follows that for an isolated plane curve singularity, the difference of the Milnor and Tjurina numbers $\mu(f) - \tau(f)$ equals the extended codimension $c_{I,e}(f^2)$.

3.10 Theorem *We use the same notations, $j_I(f) < \infty$. Suppose F is a deformation of $(f, \Sigma, 0)$ which makes the p-parts smooth. Then there exists a dense subset $V \subset S$ and an open neighbourhood U of $0 \in \mathbf{C}^2$, such that for all $t \in V$ sufficiently small:*

(a) Σ_t^p is smooth for each $p \in E_I$,

(b) f_t has only A_1 singularities in $U \setminus \Sigma_t$,

(c) For $0 < p < q$, f_t has only $D[p,q]$ singularities on $\Sigma_t^q \cap U$ and their number is $d_{p,q}(f)$,

(d) f_t has only $D[0,q]$ singularities on the rest of $\Sigma_t^q \cap U$. □

4 $D[p,q]$-points and the Milnor number

4.1 From now on, we will write F for the Milnor fibre of f, whereas f_t will denote a deformation of $(f, \Sigma, 0)$, with $\Sigma = \mathrm{Sing}(f)$, which is a network map deformation or a deformation which makes the p-parts smooth. The decomposition of f in irreducible factors is

$$f = f_1^{m_1} \cdots f_r^{m_r} f_{r+1} \cdots f_s,$$

where $r \geq 1$, $s \geq r$ and $m_i \geq 2$ ($i \in \underline{r}$). We put $m_{r+1} = \ldots = m_s = 1$. F will denote the Milnor fibre of f.

A *$D[p,q]$-point* of f_t is a point where f_t has a local singularity of type $D[p,q]$. We ignore $D[0,q]$-points. Denote the number of $D[p,q]$-points of f_t by $\#D[p,q]$, the number of $D[p,q]$-points on $f_t^{-1}(0)$ by $\#D^0[p,q]$. We assume that for $(p,q) \neq (1,1)$ all $D[p,q]$-points are in fact situated on $f_t^{-1}(0)$.

We will express the *Milnor number* μ, which is the dimension of $H_1(F;\mathbf{Z})$, in the number of $D[p,q]$-points of f_t. Put $d = \dim H_0(F;\mathbf{Z})$ (the number of connected components); d equals g.c.d.$(m_1, \ldots, m_s)$. In this section, the singular locus has its *reduced* structure, i.e. it is defined by $(f_1 \cdots f_r)$. This is important, as $\mu(\Sigma)$, the Milnor number of Σ, will come in.

4.2 Our formulae will generalize various known formulae for the Milnor number, which are, however, often valid for all dimensions. Some of them are outlined below.

(i) In the case of *isolated singularities*, where $m_1 = \ldots = m_s = 1$, we have the well-known

$$\mu = 2\delta - s + 1,$$

see [BG]. The number δ is the virtual number of double points and equals the maximum of $\#D^0[1,1]$ over all deformations of f. Another formula is

$$\mu = \#D[1,1],$$

as used for instance in the method of A'Campo and Gusein-Sade [AGV].

(ii) In the case of *transversal type* A_1, where $m_1 = \ldots = m_r = 2$ we have the formulae of Siersma:

$$\mu = 2\#D[1,2] + \#D[1,1] - \mu(\Sigma) + d - 2,$$

if Σ is deformed in such a way that it becomes *smooth*, see [Si2]; and:

$$\mu = 2\#D[1,2] + \#D[1,1] - 2\mu(\Sigma) - 1,$$

if $\#D[1,2] > 0$ and Σ stays *fixed* under the deformation ([Si3]).

(iii) In the case that Σ *is a non-singular curve and the transversal type is* A_{q-1} (e.g. $f = y^q f_2 \cdots f_s$):

$$\mu = q\#D[1,q] + \#D[1,1] - q + 1,$$

see De Jong [Jo].

Remark Here we only consider formulae of the Milnor number in terms of deformations. One can obtain expressions more directly from the topology. One of the ways to do this is described in [EN] and [Sch].

Below, we state two formulae for the Milnor number of f, one for each of the special types of deformations that we consider. The first of them, which may be regarded as known, we give for the sake of completeness. Then we give some examples of the computation of the Milnor number using our formulae. After that, we give the proofs. Recall that d is the number of connected components of the Milnor fibre F.

4.3 Theorem (Formula 1) *Let f_t be a network type deformation of f. Then:*

$$\mu(f) = \sum (p+q) \cdot \#D^0[p,q] - S + d$$

where the first sum runs over all $D[p,q]$-points on $f_t^{-1}(0)$ with $p \le q$, and $S = \sum_{i=1}^{s} m_i$, the number of branches counted with multiplicities.

4.4 Theorem (Formula 2) *Let f_t be a deformation which makes the p-parts Σ^p smooth. Then*

$$\mu(f) = \sum_{p<q} (p+q-1) \cdot \#D[p,q] + \#D[1,1] + \sum_k (k-1)(\mu(\Sigma^k) - 1) + d - 1,$$

the first summation over the $D[p,q]$-points with $p < q$, the second over all $k \in \{m_1, \ldots, m_r\}$.

4.5 Example We compute the Milnor number in four cases.

(i) $f(x,y) = x^p y^q$. Then $d = \text{g.c.d.}(p,q)$, $\mu(\Sigma^p) = \mu(\Sigma^q) = 0$, and $\#D[p,q] = 1$. Both formulae give $\mu = d$.

(ii) $f(x,y) = x^p y^q (x+y)$ with $1 \le p < q$. Then $d = 1$, $\mu(\Sigma^p) = \mu(\Sigma^q) = 0$ and $\#D[p,q] = \#D[1,p] = \#D[1,q] = \#D[1,1] = 1$. See figure 1. Both formulae give $\mu = p + q + 2$.

(iii) $f(x,y) = x^p y^p (x+y)$ with $p > 1$. Then $d = 1$ and $\mu(\Sigma) = 1$. See figure 1.

- Formula 1 only works with deformation 1 and gives: $\mu = 2p + 2$ (see (ii)).
- Formula 2 only works with deformation 2 and gives: $\mu = p\#D[1,p] + \#A_1 + (p-1)(\mu(\Sigma) - 1) = 2p + 2$.

(iv) $f(x,y) = (y^2 - x^3)^p (y^3 - x^2)^q$, with $p < q$. Then $d = \text{g.c.d.}(p,q)$, $\mu(\Sigma^p) = \mu(\Sigma^q) = 2$ and $\#D[p,q] = 4$. See figure 2.

- Formula 1 only works with deformation 1 and gives: $\mu = 5p + 5q + d$, because $\#D[p,p] = \#D[q,q] = 1$.
- Formula 2 only works with deformation 2 and gives: $\mu = 5p + 5q + d$, because $\#D[1,1] = 7$.

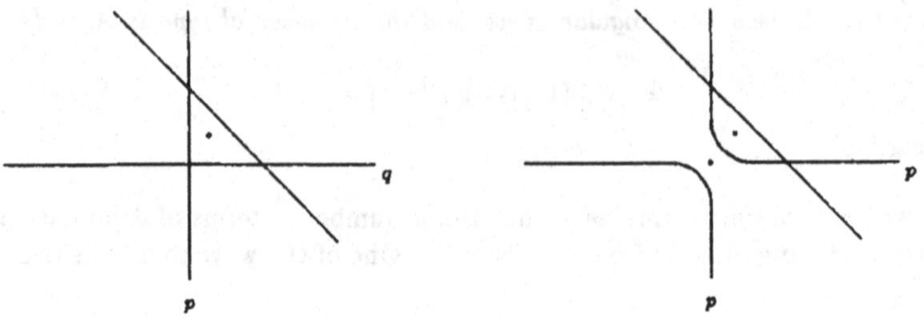

Figure 1: (ii), (iii) deformation 1; and (iii) deformation 2.

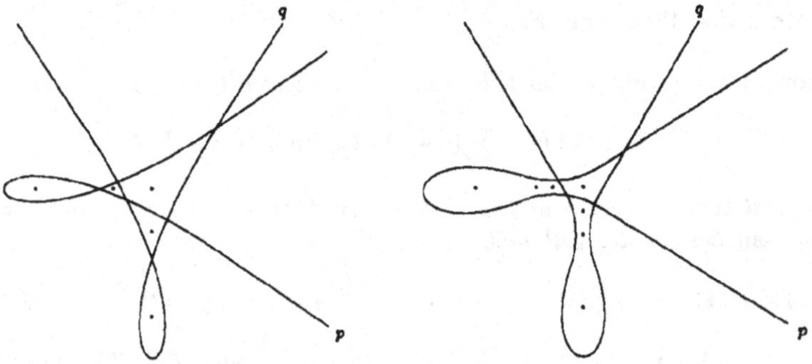

Figure 2: (iv) deformations 1 and 2.

4.6 Proof of theorem 4.3. Near Σ_i the Milnor fibre is a m_i-sheeted covering of the zero locus $X = f_t^{-1}(0)$, except in the multiple points. So start with S copies of the disc D^2, and cover the ith branch with m_i copies. If for each $D[p,q]$-point, we remove $p+q$ small discs and replace them by g.c.d.(p,q) small annuli (the local Milnor fibre of type $D[p,q]$), we obtain the Milnor fibre F. So the Euler-Poincaré characteristic of F is clearly

$$\chi(F) = S - \sum_{p,q}(p+q)\cdot \#D^0[p,q].$$

Since F has d connected components, we obtain

$$\mu(f) = \sum_{p,q}(p+q)\cdot \#D^0[p,q] - S + d.$$

This proves the theorem. □

4.7 The proof of theorem 4.4 (Formula 2) requires more work; we will follow Siersma [Si2]. We have to start with some definitions and lemmas. In the following, f_t will be a deformation of $(f,\Sigma,0)$ which makes the p-parts smooth and the notations are as in the theorem.

We write $X = f_t^{-1}(0)$. Let ε_0 be an admissible radius for the Milnor fibration, i.e. a positive number with the property that for all $\varepsilon \in [0,\varepsilon_0]$, $X \pitchfork \partial B_\varepsilon$ as stratified set. For each admissible

$\varepsilon > 0$ there exists a $\delta_\varepsilon > 0$ such that $f^{-1}(u) \pitchfork \partial B_\varepsilon$ for all $|u| < \delta_\varepsilon$. Fix $\varepsilon \leq \varepsilon_0$ and $\delta \leq \delta_\varepsilon$, and let D be the disc of radius δ. Put $X_D = f^{-1}(D) \cap B_\varepsilon$ and $X_{D,t} = f_t^{-1}(D) \cap B_\varepsilon$. Consider $f : X_D \to D$ and $f_t : X_{D,t} \to D$.

4.8 Lemma *For t, δ sufficiently small, we have:*

(a) $f_t^{-1}(u) \pitchfork B_\varepsilon$ for all $u \in D$.

(b) Over the boundary circle ∂D the fibrations induced by f and f_t are equivalent.

(c) X_D and $X_{D,t}$ are homeomorphic.

The proof is analogous to the one presented in [Si2]. □

4.9 We assume that $f_t : X_{D,t} \to D$ satisfies the conditions of the preceding lemma. Suppose $\mathrm{Sing}(f_t) = \Sigma_t \cup \{c_1, \ldots, c_\sigma\}$, where $c_1, \ldots, c_\sigma$ are the A_1-points (that is, $D[1,1]$-points) of f_t, with critical values $v_1, \ldots, v_\sigma$, respectively (so $\sigma = \#D[1,1]$). Let 0 be the critical value of all non-isolated singularities. We may assume that all critical values are distinct.

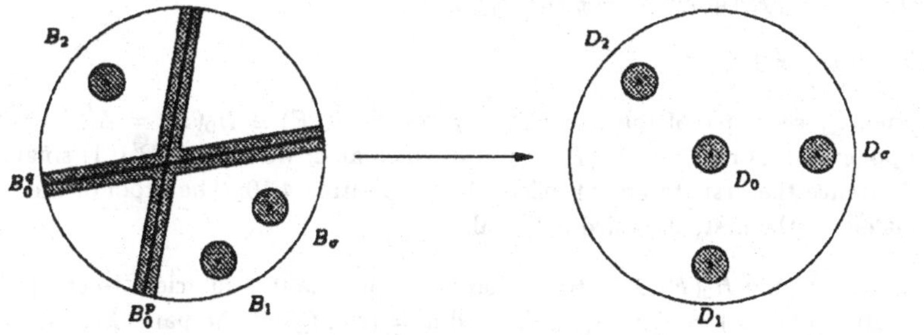

Figure 3: $f_t : B_\varepsilon \to D$

Now choose, as indicated in Fig. 1:

(a) Small disjoint balls B_i around c_i ($i \in \underline{\sigma}$);

(b) Small tubes B_0^p around Σ_t^p ($p \in \{m_1, \ldots, m_r\}$), all of the form $B_0^p = \Sigma_t^p \times D(\eta)$, where $D(\eta)$ is the disc of radius $\eta > 0$ (recall that Σ_t^p is smooth);

(c) Small disjoint discs $D_i \subset D$ around v_i ($i \in \underline{\sigma}$), and $D_0 \subset D$ around $v_0 = 0$, such that $f^{-1}(u) \pitchfork \partial B_i$ for all $u \in D_i$;

(d) Points $a_i \in \partial D_i$ and a point $a \in \partial D$.

Furthermore, (re)define:

$$
\begin{array}{lllllll}
B_0 & = & \cup_{p\in\{m_1,\ldots,m_r\}} B_0^p & & & & \\
\Sigma_t & = & \Sigma_t \cap B_0 & \Sigma_t^p & = & \Sigma_t^p \cap B_0 & \\
E & = & B \cap f^{-1}(D) & F & = & B \cap f^{-1}(a) & \\
E_i & = & B_i \cap f^{-1}(D_i) & F_i & = & B_i \cap f^{-1}(a_i) & (i \in \{0, \ldots, \sigma\}) \\
E_0^p & = & B_0^p \cap f^{-1}(D_0) & F_0^p & = & B_0^p \cap f^{-1}(0) & (p \in \{m_1, \ldots, m_r\})
\end{array}
$$

E is called the *Milnor ball*, F is still called the *Milnor fibre* and (E, F) the *Milnor pair*.

4.10 Proposition $\tilde{H}_{\bullet-1}(F) \cong H_\bullet(E,F) = \oplus_{i=0}^{\sigma} H_\bullet(E_i, F_i)$.

Proof The first isomorphism follows from the homology sequence of the pair (E,F), since E, being the Milnor ball, is contractible; for the second, see [Si2], (2.8). □

Unlike F, the Milnor pair (E,F) has homology that splits into a direct sum, hence (E,F) is easier to work with. We start by computing the homology of the Milnor pair of our basic singularity, the $D[p,q]$-point in the following easy lemma.

4.11 Lemma *The Milnor fibre of a $D[p,q]$-point is homeomorphic to e = g.c.d.(p,q) annuli. Therefore the Milnor pair of a $D[p,q]$-point has homology as follows:*

$$H_1(E_{D[p,q]}, F_{D[p,q]}) = \mathbf{Z}^{e-1}, \quad H_2(E_{D[p,q]}, F_{D[p,q]}) = \mathbf{Z}^{e}, \quad H_j(E_{D[p,q]}, F_{D[p,q]}) = 0 \text{ if } j \neq 1,2.$$

□

4.12 Proposition

(a) $H_1(E,F) = H_1(E_0,F_0) = \mathbf{Z}^{d-1}$,

(b) $H_2(E,F) = H_2(E_0,F_0) \oplus \mathbf{Z}^{\sigma}$ $(\sigma = \#D[1,1])$, *and*

(c) $H_j(E,F) = 0$ *if* $j \neq 1,2$.

Proof The homology sequence of the pair (E,F) gives $H_1(E,F) \cong \tilde{H}_0(F) = \mathbf{Z}^{d-1}$ (F has d connected components). For $i \in \underline{\sigma}$, $H_1(E_i,F_i) = 0$, since at c_i we have a $D[1,1]$ singularity, see lemma 4.11. Hence the first statement follows by proposition 4.10. The proof of the second statement is analogous the first, the third is trivial. □

4.13 It remains to compute $H_2(E_0,F_0)$. By the preceding lemma it is sufficient to compute the Euler characteristic $\chi(E_0,F_0) = \dim H_2(E_0,F_0) - \dim H_1(E_0,F_0)$ of the pair (E_0,F_0). Recall the following properties of the Euler characteristic:

(i) If (X,A) is a pair of topological spaces, then $\chi(X,A) = \chi(X) - \chi(A)$ ([Sp], p. 205);

(ii) If $\{X,Y\}$ is an excisive couple of spaces then $\chi(X \cup Y) = \chi(X) + \chi(Y) - \chi(X \cap Y)$ ([Sp], p. 205);

(iii) If $\pi : (\tilde{X}, \tilde{A}) \to B$ is a fibre bundle pair with fibre the pair (X,A), then $\chi(\tilde{X},\tilde{A}) = \chi(X,A) \cdot \chi(B)$ ([Sp], p. 481).

4.14 Recall $E_0^p = B_0^p \cap f^{-1}(D)$, where $B_0^p = \Sigma_t^p \times D(\eta)$ is a small tube around the smooth curve Σ_t^p. Let π_p be the projection onto the first factor. If η is chosen sufficiently small, then $\pi_p : (E_0^p, F_0^p) \to \Sigma^p$ is a fibre bundle pair, locally trivial outside the $D[p,q]$-points, and with general fibre equivalent to the Milnor pair $(\overline{E^p}, \overline{F^p})$ of the transversal A_{p-1} singularity. Observe that $\overline{F^p}$ consists of p points.

Definition For $Y \subset \Sigma_t^p$, define $E_Y = \pi_p^{-1}(Y) \cap E_0^p$, and $F_Y = \pi_p^{-1}(Y) \cap F_0^p$. The definition is extended in the obvious way to subsets $Y \subset \Sigma_t$ that are disjoint unions of real two dimensional manifolds with boundary, each of which is lying entirely in a Σ_t^p.

In each Σ_t^p $(p \in \{m_1,\ldots,m_r\})$ choose small discs $W_{p,q,i}$, $q \neq p$, $i \in \{1,\ldots,\#D[p,q]\}$ around the $D[p,q]$-points such that they do not meet each other. We may assume that $E_{W_{p,q,i}} = E_{W_{q,p,i}}$ and $F_{W_{p,q,i}} = F_{W_{q,p,i}}$. Let $W_p = \cup_{q,i} W_{p,q,i}$ and $M_p = \overline{\Sigma_t^p \setminus W_p}$.

4.15 Proposition *$\pi_p : (E_{M_p}, F_{M_p}) \to M_p$ is a trivial fibre bundle with fibres equivalent to the Milnor pair $(\overline{E}^p, \overline{F}^p)$ of the transversal A_{p-1} singularity.*

Proof Use [Si2] (4.7) in a somewhat more general setting. □

4.16 Let $p \in \{m_1, \ldots, m_r\}$. We have defined W_p as the (disjoint) union of all discs $W_{p,q,i}$ around the $D[p,q]$-points in Σ_t^p. Σ_t^p is a Riemann surface with holes and has a wegde of circles as deformation retract. Let B_p be the union of this wedge with $\#D[p,q]$ non-intersecting paths connecting the wedge point with the discs $W_{p,q,i}$, as in figure 2. Observe that $W_p \cap B_p$ consists of a finite set of points.

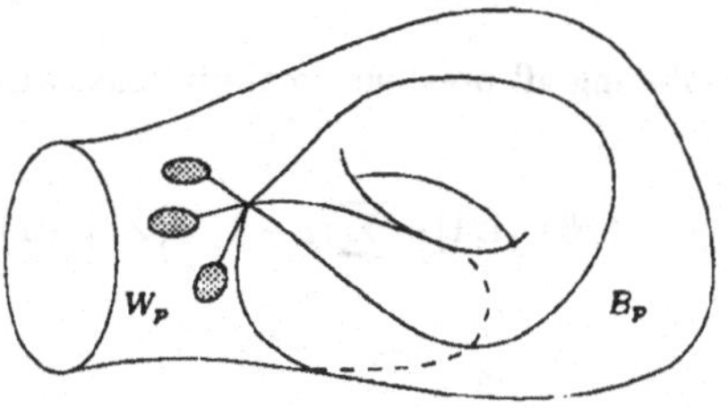

Figure 4: Σ_t^p

4.17 Proposition *Let $W = \cup_p W_p$, $B = \cup_p B_p$. Then:*

$$\chi(E_0, F_0) = \chi(E_{W\cup B}, F_{W\cup B}) = \chi(E_W, F_W) + \chi(E_B, F_B) - \chi(E_{W\cap B}, F_{W\cap B})$$

Proof The first equality follows from the fact that $(E_{W\cup B}, F_{W\cup B})$ is homotopy equivalent to (E_0, F_0). Indeed, for each p, $W^p \cup B^p$ is homotopy equivalent to Σ_t^p (rel. W_p), and therefore by the homotopy lifting property of the $\pi_p : (E_{M_p}, F_{M_p}) \to M_p$ ($p \in \{m_1, \ldots, m_r\}$), $(F_{W\cup B}, F_{W\cup B})$ is homotopy equivalent to (E_0, F_0).

The second equality follows from the fact that $(F_{W\cup B}; F_W, F_B) \subset (E_{W\cup B}; E_W, E_B)$ is an inclusion of excisive triads by the properties of the Euler characteristic 4.13. □

4.18 Lemma

(a) $\chi(E_W, F_W) = \sum_{p<q} \#D[p,q]$.

(b) $\chi(E_B, F_B) = \sum_p (p-1)(\mu(\Sigma^p) - 1)$.

(c) $\chi(E_{W\cap B}, F_{W\cap B}) = -\sum_{p<q}(p+q-2) \cdot \#D[p,q]$.

Proof

(a) W is the disjoint union of the $W_{p,q,i}$, $p < q$, $i \in \{1, \ldots, \#D[p,q]\}$. So $\chi(E_W, F_W)$ is the sum of the $\chi(E_{W_{p,q,i}}, F_{W_{p,q,i}})$ which are all equal to 1 (see lemma 4.11).

(b) B is the disjoint union of the B_p, so $\chi(E_B, F_B) = \sum_p \chi(E_{B_p}, F_{B_p})$. B_p is a wegde of $\mu(\Sigma^p)$ circles (Σ_t^p and Σ^p have the same homotopy type), so its Euler characteristic is $1 - \mu(\Sigma^p)$. (E_{B_p}, F_{B_p}) is a trivial fibre bundle pair over B_p with fibres $(\overline{E}^p, \overline{F}^p)$. Since $\overline{E}^p$ is a topological disc and $\overline{F}^p$ a set of p points, we have that $\chi(\overline{E}^p, \overline{F}^p) = 1 - p$. By 4.13 (iii) we obtain $\chi(E_{B_p}, F_{B_p}) = (p-1)(\mu(\Sigma^p) - 1)$.

(c) $W_p \cap B_p$ is a set of $\sum_{q \neq p} \#D[p,q]$ points. Above each point the fibre is equivalent to the Milnor pair of the transversal A_{p-1} singularity, which has Euler characteristic $1-p$ as we have seen in (b). So $\chi(E_{W_p \cap B_p}, F_{W_p \cap B_p}) = \sum_{q \neq p} (1-p) \#D[p,q]$. Hence $\chi(E_{W \cap B}, F_{W \cap B}) = \sum_p \sum_{q \neq p} (1-p) \#D[p,q] = -\sum_{p<q} (p+q-2) \#D[p,q]$.

□

4.19 Proof of theorem 4.4 By combining all previous computations, we obtain the desired formula

$$\mu(f) = \sum_{p<q} (p+q-1) \cdot \#D[p,q] + \#D[1,1] + \sum (p-1)(\mu(\Sigma^p) - 1) + d - 1.$$

□

5 Splicing of real morsifications

In this section, a deformation or morsification will be a *real* network map type of deformation f_t of a function germ f with real coefficients, as in [A'C], hypothèse **BR**. By considering parametrizations, it is not difficult to see that such real deformations exist. The necessary data are contained in the intersection of $\mathbf{R}^2$ and the level $f_t^{-1}(0)$, and we will even call this morsification diagram a morsification.

5.1 In the preceding section we showed a formula which expressed the Milnor number of a plane curve singularity in the number of special points of a deformation. It is not comparably easy to obtain from deformations more topological details, in particular the Waldhausen decomposition of the exterior of the link of the singularity. On the other hand, that decomposition may give us some results on deformations.

One such result will now be described briefly, merely as an illustration: an algorithmic way to obtain formally a morsification of an isolated plane curve singularity. This can be used to obtain a Dynkin diagram of the intersection form by the A'Campo–Gusein-Sade method ([A'C], [AGV]). Such algorithms are not new, cf. Schulze-Röbbecke [SR], who obtained a *Dynkindiagramm für jede Singularität*, where in that case "each singularity" meant "each irreducible isolated plane curve singularity."

We use the Waldhausen decomposition and splicing without detailed descriptions, the reader is referred to [EN].

5.2 Consider a plane curve singularity $(X,0)$. We write $X = \cup\, m_i X_i$, which means that the equation is of the form $f = f_1^{m_1} \cdots f_r^{m_r} = 0$. It is well-known that for $\varepsilon > 0$ small enough, $L = X \cap S_\varepsilon^3$ is a link, and that $f/|f| : S_\varepsilon^3 \setminus L \to S^1$ is the Milnor fibration (indeed equivalent to the descriptions we encountered before). As described in [EN], the link exterior $M = S_\varepsilon^3 \setminus N$ (where N is the union of open tubular neighbourhoods of the components of L) has a unique (upto isotopy) decomposition in Seifert manifolds, called the Waldhausen decomposition.

The construction of the Waldhausen decomposition is described in [EN]; an alternative way (using the polar filtration) is given in [LMW]. A Seifert manifold itself belongs to a, usually non-isolated, plane curve singularity. If we follow the construction given in [EN], Appendix to Chapter 1, we see that we may even assume that this singularity is $x^a y^b (y^p - x^q)^c$, where $a, b, c \geq 0$, $p, q \geq 1$ and $\gcd(p, q) = 1$. For such an uncomplicated singularity, a real network map deformation is readily found (for example by the A'Campo–Gusein-Sade method as also described earlier). In this way we construct real morsifications for our building blocks, the singularities belonging to the Seifert pieces.

5.3 The construction of the Waldhausen decomposition is by glueing the Seifert pieces together, which process is called *splicing*. We now have to describe how to splice two morsifications. Again, this should be seen primarily as a way of manipulating morsification diagrams.

The splicing takes place "along" an m_i-fold branch X_i of the one part, X, and an n_j-fold branch Y_j of the other part, Y. We have the following relationship (the splice condition): $n_j = X_i \cdot \bigcup_{k \neq i} m_k X_k$, the intersection number of X_i with the other branches counted with their multiplicities, and also $m_i = Y_j \cdot \bigcup_{l \neq j} n_l Y_l$. Perhaps a bit unusual is that the multiplicies may be 0, but this has been given a natural interpretation in [EN].

The first step consists of *multiplying* the branches X_i, Y_j by m_i resp. n_j. This means that one takes m_i resp. n_j parallel copies very near to each other, which are deformed slightly to take care of intersections between each of the copies. In figure 5 a doubled morsification of the cusp is depicted.

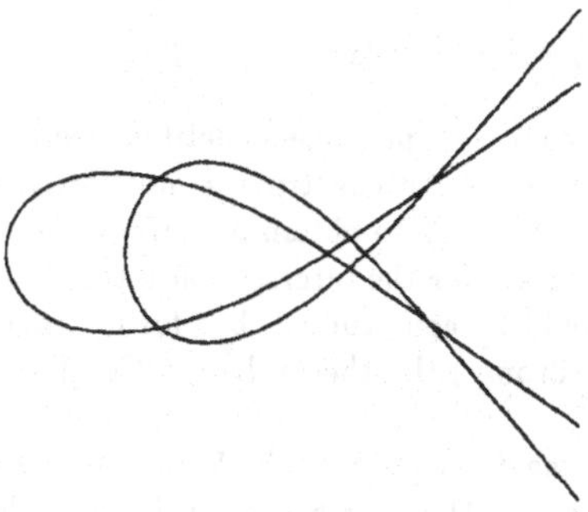

Figure 5: Doubling a morsification of $(y^2 - x^3)^2$

We can take small common neighbourhoods M of the m_i new branches in the morsification diagram of X, and N of the n_j new branches in the morsification diagram of Y. Because of the splice condition we may assume that the intersection of M with the other branches of X consists of n_j segments, and the intersection of N with the other branches of Y consists of m_i segments. The splice operation is just patching those segments on the multiplied branches of the other.

This procedure works by inspection of the proof of the A'Campo–Gusein-Sade method, bearing in mind that the splice components are related to partial resolutions of the singularity. This was shown to me by Jan Stevens, who communicated a proof of the case of "cabling" of morsifications.

5.4 Example (See figure 6.) We apply our method in the simple case of $J_{3,p}$ (equation: $f(x, y) = y^3 + x^3y^2 + x^{9+p}$, $p \geq 0$). It has two splice components. The first is $J_{k,\infty}$ (equation $y^3 + x^3y^2$). The second is isomorphic to $x^3(y^2 + x^{6+p})$. In figure 6 there are morsification diagrams for both, and we can identify the 2×3 lines along which the splicing takes place.

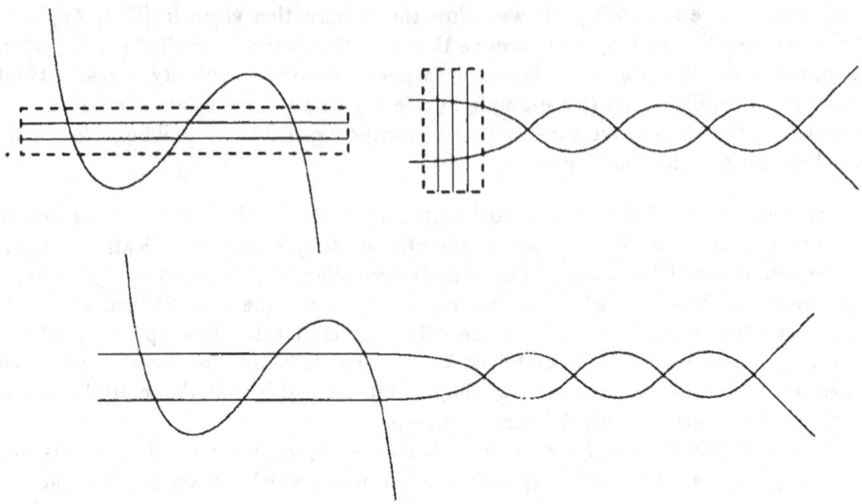

Figure 6: The splice decomposition of $J_{3,2}$

One can verify that $\mu(J_{3,\infty}) = 7$ and $\mu(J_{3,p}) = 16 + p$.

5.5 The multiplication of branches is in practice sometimes confusing, a notation as in [SR] is useful. As in [SR] we have now a way to construct a morsification for each topological type. For isolated singularities, we can use this to obtain a distinguished basis of vanishing cycles for the homology and a Dynkin diagram for the intersection form. For non-isolated singularities, a basis and a Dynkin diagram could be constructed, but for distinguishedness (that we want for the relationship with the monodromy) the theory lacks. Compare [Si4], where the curve case must be excluded.

More can be expected from the description of the Waldhausen decomposition in terms of the polar filtration and the Gabrielov-method for finding a distinguished base, see [Eb], chapter 2 or [AGV].

5.6 In [Sch], the author defined topological series of plane curve singularities using splice decompositions. It follows for instance, that in a topological series of a singularity with transversal type A_1, the Dynkin diagram of the intersection form contains r tails of the form

corresponding to each of the r branches of the singular locus.

References

[A'C] N. A'Campo, Le groupe de monodromie du déploiement des singularités isolées de courbes planes I, *Math. Annalen*, **213** (1975), pp. 1–32

[AGV] V.I. Arnol'd, S.M. Gusein-Sade, A.N. Varchenko: *Singularities of Differentiable Maps II*, Birkhäuser, 1988

[BG] R.-O. Buchweitz and G.-M. Greuel: The Milnor Number and Deformations of Complex Curve Singularities, *Invent. Math.*, **58** (1980), pp. 241–281

[Eb] W. Ebeling: *The Monodromy Groups of Isolated Singularities of Complete Intersections*, Springer LNM 1293, 1987

[EN] D. Eisenbud and W.D. Neumann: *Three-Dimensional Link Theory and Invariants of Plane Curve Singularities*, Annals of Mathematics Study 110, Princeton U.P., 1985

[Jo] Th. de Jong: Some Classes of Line Singularities: *Math. Zeit.* **198** (1988), pp. 493–517

[LMW] Lê D.T., F. Michel and C. Weber: *Courbes polaires et topologie des courbes planes*, preprint Genève 1988

[Pe1] R. Pellikaan: *Hypersurface Singularities and Resolutions of Jacobi Modules*, Thesis, Utrecht 1985

[Pe2] R. Pellikaan: Finite Determinacy of Functions with Non-Isolated Singularities, *Proc. London Math. Soc.* (3) **57** (1988) pp. 357–382

[Pe3] R. Pellikaan: Deformations of Hypersurfaces with a One-Dimensional Singular Locus, preprint Vrije Universiteit Amsterdam 1987, to appear in *Journal of Pure and Applied Algebra*

[Sch] R. Schrauwen: Topological Series of Isolated Plane Curve Singularities, preprint Utrecht January 1989, to appear in *L'Enseignement Mathématique*

[Si1] D. Siersma: Isolated Line Singularities *Proc. Symp. Pure Maths* **40** (1983) Part 2, pp. 485–496

[Si2] D. Siersma: Singularities with Critical Locus a One-Dimensional Complete Intersection and Transversal Type A_1, *Topology and its Applications* **27** (1987), pp. 51–73

[Si3] D. Siersma: Hypersurfaces with Singular Locus a Plane Curve and Transversal Type A_1, *Proc. Warsaw Semester on Singularities*, Banach Center Publ. **20**, PWN-Polish Scientific Publ., Warsaw 1988, pp. 397–410

[Si4] D. Siersma: Variation Mappings on 1-Isolated Singularities, Preprint Utrecht, to appear in *Topology*.

[Sp] E.H. Spanier: *Algebraic Topology*, New York etc., McGraw-Hill, 1966

[SR] Th. Schulze-Röbbecke: *Algorithmen zur Auflösung und Deformation von Singularitäten ebener Kurven*, Bonner Math. Schrift Nr. 96, Bonn 1977

Rob Schrauwen
Mathematisch Instituut
Rijksuniversiteit Utrecht
P.O. Box 80010
3508 TA Utrecht, The Netherlands

Vanishing Cycles and Special Fibres

Dirk Siersma

Abstract

We show that the homotopy type of certain special fibres in a perturbation of a holomorphic function is a wedge of spheres of middle dimension. We also define a basis of the homology of the special fibre.

1 Introduction

In the case of an isolated singularity $\mathbf{C}^{n+1} \to \mathbf{C}$ the vanishing cycles play an important role in the description of the general non-singular fibre. This Milnor fibre F is homotopy equivalent to a wedge of spheres:

$$F \overset{h}{\simeq} S^n \vee \cdots \vee S^n$$

where the wedge is taken over μ spheres in the middle dimension. Here μ is Milnor's number, which can be identified with the number of Morse-points (or A_1-points) in a generic perturbation. The vanishing cycles correspond to the spheres in the wedge. We show in this note, that the same conclusion is true for certain singular fibres which occur in the case of non-isolated singularities. The proof is similar to the case of the Milnor fibre of an isolated singularity ([AGV-II], [Lo]).

We also define a basis of the homology of the special fibre. Each basis element corresponds to an A_1-point. At the end we discuss monodromies of the special fibre, which occur in 1-dimensional families.

This note was written after discussions with David Mond and Duco van Straten about Mond's Theorem, that the homotopy type of the generic fibre of a disentanglement is a wedge of spheres.

2 Homotopy type of the special fibres

2.1 We consider non-zero holomorphic function germs $f : (\mathbf{C}^{n+1}, 0) \to (\mathbf{C}, 0)$ and allow arbitrary singularities (isolated or non-isolated). We recall the definition of the Milnor fibration. For $\epsilon > 0$ small enough there exist an ϵ-ball B_ϵ in $\mathbf{C}^{n+1}$ and an η-disc D_η in $\mathbf{C}$ such that the restriction:

$$f : f^{-1}(D_\eta) \cap B_\epsilon \longrightarrow D_\eta$$

is a locally trivial fibre bundle over $D_\eta \setminus \{0\}$. The fibres are called Milnor fibres. The boundary ∂B_ϵ is called the Milnor sphere.

2.2 We next consider a deformation F of f, i.e. a holomorphic mapgerm:

$$F : (\mathbf{C}^{n+1} \times \mathbf{C}^r, 0) \to (\mathbf{C} \times \mathbf{C}^r, 0)$$

of the form

$$F(x, a) = (f_a(x), a)$$

such that $f_0(x) = f(x)$.
The mapgerm f_a is called a *perturbation* of f.

We require that the deformation be *topologically trivial over the Milnor sphere* ∂B_ϵ.
This means: For η and ρ small enough

$$F : \partial B_\epsilon \times D_\rho \cap F^{-1}(D_\eta \times D_\rho) \to D_\eta \times D_\rho$$

must be a stratified submersion with strata $\{0\} \times D_\rho$ and $(D_\eta \setminus \{0\}) \times D_\rho$ on $D_\eta \times D_\rho$ and some stratification on $\partial B_\epsilon \times D_\rho \cap F^{-1}(D_\eta \times D_\rho)$.
This condition implies:

- $f_a^{-1}(t)$ is (stratified) transversal to ∂B_ϵ for all $|t| < \eta$ and for all $|a| < \rho$.
- $f_a^{-1}(D_\eta) \cap \partial B_\epsilon$ is homeomorphic to $f^{-1}(D_\eta) \cap \partial B_\epsilon$ and therefore contractible.
- the Milnor fibres of f and f_a are diffeomorphic.

2.3 Theorem *Let F be a deformation of f, which is topologically trivial over the Milnor sphere as defined in 2.2. Let $a \in D_\rho$ and suppose that all fibres of f_a are smooth or have isolated singularities except for one special fibre $X_t = f_a^{-1}(t) \cap \partial B_\epsilon$. Then X_t is homotopy equivalent to a wedge of spheres:*

$$X_t \overset{h}{\simeq} S^n \vee \cdots \vee S^n$$

The number of spheres is equal to the sum of the Milnor numbers in the fibres different from X_t.

Proof In the following we use the notation:

$$g : X \to D$$

for the perturbation:

$$f_a : f_a^{-1}(D_\eta) \cap \partial B_\epsilon \to D_\eta.$$

We denote: $X_Y = g^{-1}(Y)$.

Let $x_1, \dots, x_\sigma$ be the critical points outside X_t and $c_1, \dots, c_\tau$ be the critical values, different from t. Take small disjoint discs $D_0, D_1, \dots, D_\tau$ around $t, c_1, \dots, c_\tau$ and join them with a point s on ∂D_0 with the help of a system of non-intersecting paths Γ (in the usual way, cf. figure 1). Call the endpoints $s_1, \dots, s_\tau$.

We mention the homotopy equivalence:

$$X_t \overset{h}{\simeq} X_{D_0}.$$

This equivalence is well known in the local case (i.e. in a small neighborhood of a singular point),see proposition 2.A.3.(b) of [GM]. Since our map is proper one can patch together these local equivalences to a global homotopy equivalence. One can also apply directly lemma 2.A.2. of [GM].

Next we use homotopy lifting properties and have first:

$$X_{D_0} \overset{h}{\simeq} X_{D_0 \cup \Gamma}$$

and second:

$$X_D \overset{h}{\simeq} X_{D_0 \cup \Gamma} \cup X_{D_1} \cup \cdots \cup X_{D_\tau}$$

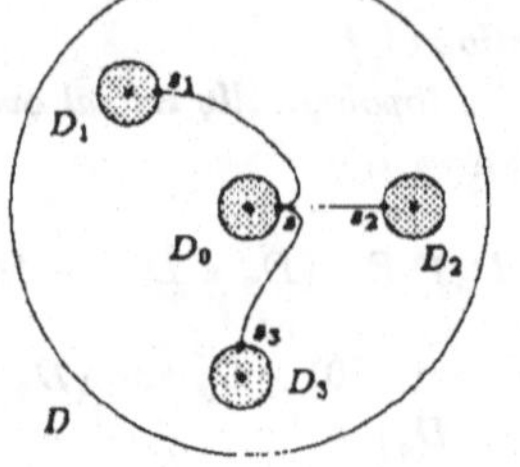

Figure 1: Perturbation g.

Similar homotopy equivalences occur in [Lo] and [Si-1].

All X_{c_i} contain only isolated singularities. Let μ_i be the sum of the Milnor numbers in the fibre X_{c_i}. Each X_{D_i} can be obtained (up to homotopy equivalence) from X_{s_i} by attaching μ_i cells of dimension $n+1$ in order to kill the vanishing cycles.

After retraction of Γ to the point s, it follows that

$$X_D \overset{h}{\simeq} X_{D_0} \cup \{\textstyle\sum \mu_i \text{ cells of dimension } n+1\ \}.$$

Since X_D is diffeomorphic to $f^{-1}(D_\eta) \cap \partial B_\epsilon$, which is contractible (as total space of the Milnor fibration), we have that:

$$\pi_k(X_t) = \pi_k(X_{D_0}) = 0 \text{ for all } k < n.$$

Since X_t has the homotopy type of a CW-complex of dimension n, see [GM](p.152), it follows that:

$$X_t \overset{h}{\simeq} S^n \vee \cdots \vee S^n,$$

the number of spheres being equal to $\nu = \sum \mu_i$. □

2.4 Remark In the case of an isolated singularity $f : (\mathbf{C}^{n+1}, 0) \to (\mathbf{C}, 0)$ the above theorem shows that for any perturbation $g : X \to D$ of f, *all* fibres X_s (including the singular fibres) have the homotopy type of a wedge of n-spheres. The same conclusion can be deduced from corollary (5.10) on page 75 of Looijenga's book [Lo], where the following is proved:

Let $f : X \to D$ be a proper good representative of a germ $f : (\mathbf{C}^{n+k}, x) \to (\mathbf{C}^k, 0)$ defining an isolated complete intersection singularity of dimension n. Then every fibre X_s has the homotopy type of a wedge of n-spheres.

2.5 Remark If X is not smooth but if $X \setminus X_t$ is smooth the same theorem as 2.3 applies since X_D is still contractible.

3 Homology of special fibres

3.1 We are especially interested in describing a basis for the homology of the special fibre in such a way that this basis is related to the A_1-points of the perturbation.

We have according to [Si-2] the following direct sum decomposition of the vanishing homology:

$$H_\bullet(E,X_s) \cong H_\bullet(\tilde{X}_t,X_s) \oplus \oplus_{i=1}^{q} H_\bullet(E_i,F_i)$$

where

$$\begin{array}{lcl} X_t & = & \text{special fibre} \\ \tilde{X}_t & = & X_{D_0}\ \text{, a neighbourhood of } X_t \\ X_s & = & \text{Milnor fibre} \\ E & = & \text{Milnor ball} \\ E_i & = & \text{Milnor ball of the isolated singularity at } x_i \\ F_i & = & \text{Milnor fibre of the isolated singularity at } x_i \end{array}$$

This direct sum decomposition can depend on the choice of the system of paths Γ.

3.2 Since $H_\bullet(E_i,F_i)$ is concentrated in dimension $n+1$ it follows from 3.1 that:

$$\begin{array}{rcll} H_{n+1}(E,X_s) & = & H_{n+1}(\tilde{X}_t,X_s) \oplus \mathbf{Z}^\nu & \text{where } \nu = \sum_{i=1}^{q} \mu_i \\ H_k(E,X_s) & = & H_k(\tilde{X}_t,X_s) & \text{if } k \neq n+1 \end{array}$$

All these isomorphisms are induced by inclusions. Therefore the exact sequence of the triple $(E,\tilde{X}_t,X_s)$ splits into short pieces:

$$0 \longrightarrow H_n(\tilde{X}_t,X_s) \longrightarrow H_n(E,X_s) \longrightarrow 0$$

and

$$0 \longrightarrow H_{n+1}(\tilde{X}_t,X_s) \xrightarrow{\alpha} H_{n+1}(E,X_s) \xrightarrow{\beta} H_{n+1}(E,\tilde{X}_t) \longrightarrow 0$$

Remark that independently from theorem 2.3 this gives again $\tilde{H}_k(X_t) = 0$ for $k \neq n$.

3.3 This above sequence extends to the diagram:

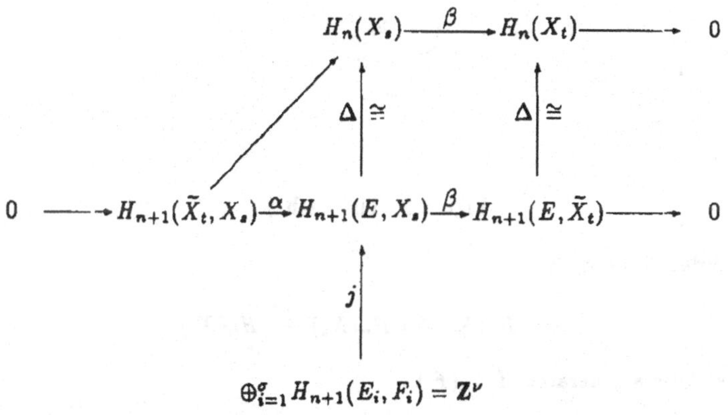

Because of the direct sum splitting 3.1 it follows that the composition

$$\oplus_{i=1}^{q} H_{n+1}(E_i,F_i) \xrightarrow{j} H_{n+1}(E,X_s) \xrightarrow{\beta} H_{n+1}(E,X_t)$$

is an isomorphism. Moreover also the induced map

$$\oplus_{i=1}^{q} H_n(F_i) \xrightarrow{j} H_n(X_s) \xrightarrow{\beta} H_n(X_t)$$

is an isomorphism. We summarize:

3.4 Proposition

$$H_n(X_t) \cong \oplus_{i=1}^{\sigma} H_n(F_i).$$

3.5 From now on we assume that all critical points outside X_t are of type A_1.

Definition

$$L = j(\oplus_{i=1}^{\sigma} H_n(F_i)) \subset H_n(X_s)$$

is called the *A_1-lattice in* $H_n(X_s)$. It has a natural basis (up to a sign), which depends via the isomorphism 3.4 on the choice of the paths, which constitute Γ. The A_1-lattice L is via β isomorphic to $H_n(X_t)$.

If necessary we write L_Γ and j_Γ to distinguish between different systems of paths.

3.6 We next adapt the definition of vanishing cycle to our situation. Let again $g : X \to D$ be a perturbation with the properties of 2.3. Let γ be a continuous path in $D \setminus \{t, c_1, \ldots, c_r\}$ from s to some s_i (cf. figure 2).

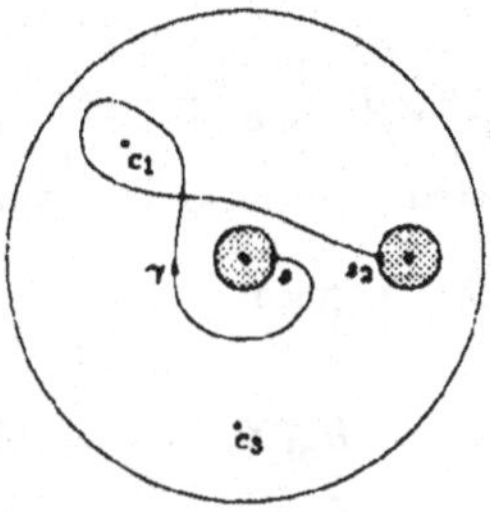

Figure 2: The path γ.

There are induced maps

$$\mathbb{Z} = H_n(F_i) \xrightarrow{j_\gamma} H_n(X_s) \xrightarrow{\beta} H_n(X_t)$$

Definition Let δ be a generator of $H_n(F_i)$.

$$\begin{aligned} \delta_\gamma &:= j_\gamma(\delta) \in H_n(X_s) \\ \Delta_\gamma &:= \beta(\delta_\gamma) \in H_n(X_t) \end{aligned}$$

$\Delta_\gamma \in H_n(X_t)$ is called the *cycle vanishing along* γ.

The fundamental group $\pi_1(D \setminus \{t, c_1, \ldots, c_r\}, s)$ acts on the homotopy classes from paths from s to s_i. The next lemma shows that loops around t do not affect a vanishing cycle.

3.7 Lemma *The definition of the vanishing cycle Δ_γ depends only on the homotopy class of γ in $D \setminus \{c_1, \ldots, c_r\}$.*

Proof A loop u_0 which goes from s around ∂D_0 induces a monodromy homomorphism:

$$h_0 : H_n(X_s) \longrightarrow H_n(X_s).$$

We claim that this monodromy is the identity modulo Ker ($\beta : H_n(X_s) \to H_n(X_t)$). This is intuitively clear for geometrical reasons, since one can choose a geometric monodromy, which is the identity on a sufficient big part of X_s. To be more precise:
Let E_0 be a tubular neighbourhood of the critical locus of X_t. We denote:

$$\begin{aligned} Y_s &= X_s \cap E_0 \\ A_s &= \overline{X_s \setminus Y_s} \\ \partial_1 E_0 &= \overline{\partial E_0 \setminus \partial X} \text{ , the inner boundary of } E_0 \\ \partial_1 Y_s &= X_s \cap \partial_1 E_0 \text{ , the inner boundary of } Y_s \end{aligned}$$

Since the restriction of g to $\overline{f^{-1}(D_0) \setminus E_0}$ has maximal rank over D_0 there exists a geometric monodromy

$$h_0 : (X_s, A_s) \longrightarrow (X_s, A_s)$$

which is the identity on A_s. We can also consider the restriction

$$h_0' : (Y_s, \partial_1 Y_s) \longrightarrow (Y_s, \partial_1 Y_s),$$

which is the identity on $\partial_1 Y_s$.
We can use now the theory of the variation mapping. We refer to Lamotke [La] or [Si-3] and especially to [Lo] p.35. The properties of the variation mapping, which we denote by VAR[1] imply that $h_0 - 1$ is the following composition

$$h_0 - 1 : H_n(X_s) \longrightarrow H_n(X_s, A_s) \xrightarrow{\text{exc}} H_n(Y_s, \partial_1 Y_s) \xrightarrow{\text{VAR}^1} H_n(Y_s) \xrightarrow{j_*} H_n(X_s),$$

We claim that

$$\text{Im } j_* \subset \text{Ker } \beta.$$

Consider the diagram:

$$\begin{array}{ccccccc} 0 & \longrightarrow & H_{n+1}(E_0, Y_s) & \longrightarrow & H_n(Y_s) & \longrightarrow & H_n(E_0) \\ & & \downarrow \alpha & & \downarrow j & & \\ 0 & \longrightarrow & H_{n+1}(E, X_s) & \xrightarrow{\cong} & H_n(X_s) & \longrightarrow & 0 \end{array}$$

Since $H_{n+1}(E_0, Y_s) = H_{n+1}(\tilde{X}_t, X_s)$ the vertical map is indeed α. So one has:

$$\text{Ker } \beta = \text{Im } \alpha \supset \text{Im } j$$

This completes the proof. □

3.8 Definition Let Γ be a system of non-intersecting paths. According to proposition 3.4 we get in $H_n(X_t)$ a basis of vanishing cycles $\Delta_1, \ldots, \Delta_\sigma$. This basis is called *distinguished* if we take into account a cyclic numbering condition as well. Compare ([AGV-II], p.14).

A distinguished basis only depends on the relative position of Γ with respect to $c_1, \ldots, c_\tau$ and not with respect to t. But the A_1-lattice $L_\Gamma \subset H_n(X_s)$ shall in general also depend on the relative position of t.

3.9 Theorem *A system of paths Γ gives a well-defined (distinguished) basis of vanishing cycles in $H_n(X_t)$ depending only on the isotopy class in $D \setminus \{c_1, \ldots, c_\tau\}$.*

4 Concluding remarks and questions

4.1 Remarks about disentanglements

Mond [Mo] considered in his work finitely determined map germs $F : \mathbf{C}^2 \to \mathbf{C}^3$. The image $F(\mathbf{C}^2)$ is a hypersurface $f = 0$ and has a 1-dimensional singular locus Σ with transversal singularity type A_1 on $\Sigma \setminus \{0\}$.

A stable perturbation G of F induces a *disentanglement* $G(\mathbf{C}^2)$ of $F(\mathbf{C}^2)$. A disentanglement has only singularities of type A_∞, D_∞ or $T_{\infty,\infty,\infty}$ (ordinary double curve, ordinary pinch point, ordinary triple point). Let $g = 0$ be the equation of $G(\mathbf{C}^2)$. According to [Mo] the function $g : \mathbf{C}^3 \to \mathbf{C}$ has outside $g^{-1}(0)$ only isolated singularities.

The notion of disentanglement was introduced in a more general context by De Jong en Van Straten [Jo-St]. Our theorem 2.3 implies that the surface $G(\mathbf{C}^2) = g^{-1}(0)$ is homotopy equivalent to a wedge of spheres:

$$g^{-1}(0) \overset{h}{\simeq} S^2 \vee \cdots \vee S^2$$

This was proved by Mond [Mo].

More general one can consider the versal unfolding:

$$\tilde{G} : \mathbf{C}^2 \times \mathbf{C}^d \longrightarrow \mathbf{C}^3 \times \mathbf{C}^d$$

Let $\tilde{G}(x, a) = (G_a(x), a)$ and let the image $G_a(\mathbf{C}^2)$ be the hypersurface with equation $g_a = 0$. According to Mond the map $g_a : \mathbf{C}^3 \to \mathbf{C}$ has for all $a \in \mathbf{C}^d$ only one fibre with non-isolated singularities. According to theorem 2.3 all the surfaces $G_a(\mathbf{C}^2) = g_a^{-1}(0)$ have the homotopy type of a wedge of spheres.

One can also consider the non-singular fibres of g_a. The general theory tells us, that they must be connected, but not necessarily simply connected. But in Mond's case one knows that $f(\mathbf{C}^2)$ is irreducible and so the Lê-Saito theorem implies that the fundamental group is trivial. So also the general fibre is a wedge of spheres. The number of these spheres is equal to:

$$2\#D_\infty - 1 + 2\#T_{\infty,\infty,\infty} - \chi(\tilde{\Sigma}) + \#A_1,$$

where $\tilde{\Sigma}$ is the normalisation of Σ. This formula can be shown in the same way as the formulas in [Si-2].

The only fibres we haven't discussed are those with isolated singularities. One can obtain such a fibre X_c from nearby smooth fibres by attaching 3-cells in order to kill the vanishing homology. Since the general smooth fibre is simply connected this implies that X_c is also simply connected.

Conclusion *In the case of disentanglements all fibres of g_a are wedges of 2-spheres.*

4.2 Remarks about monodromies of the special fibre

Let $G_a : \mathbf{C}^{n+1} \to \mathbf{C}$ be a 1-dimensional family of functions such that the map

$$\tilde{G} : \mathbf{C} \times \mathbf{C}^{n+1} \to \mathbf{C} \times \mathbf{C},$$

defined by $\tilde{G}(a, x) = (G_a(x), a)$ satisfies the conditions of theorem 2.3. We can suppose that for $a \neq 0$ small enough the singularity types of the isolated critical points $x_1, \ldots, x_\sigma$ are constant.

Consider in $\mathbf{C} \setminus \{0\}$ a small loop ω around $a = 0$. If we follow the loop, the critical values of the corresponding functions g_a will behave like braids (cf. figure 3).

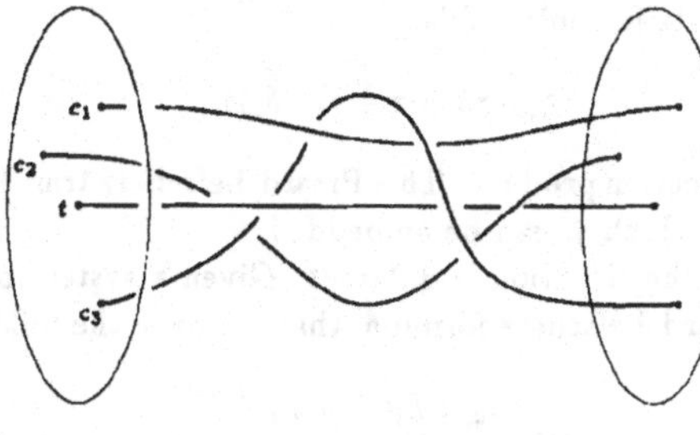

Figure 3: The moving critical values.

We can follow this by an isotopy of the disc D. At the end we get a permutation of the critical values. This permutation has to respect the "type" of the singularities in the singular fibre.

We now consider the case of one special fibre X_t with non-isolated singularities and/or several isolated singularities. Moreover we assume that all other singular fibres have only one singularity, which must be of type A_1. The above loop ω defines a monodromy map $X_t \to X_t$, which induces:

$$h_\omega : H_n(X_t) \longrightarrow H_n(X_t)$$

During this *special fibre monodromy* not only X_t is moving, but also the A_1-points move! Let a system of paths Γ be given and suppose that during the isotopy of the disc the system Γ move to a system Γ' (cf. figure 4).

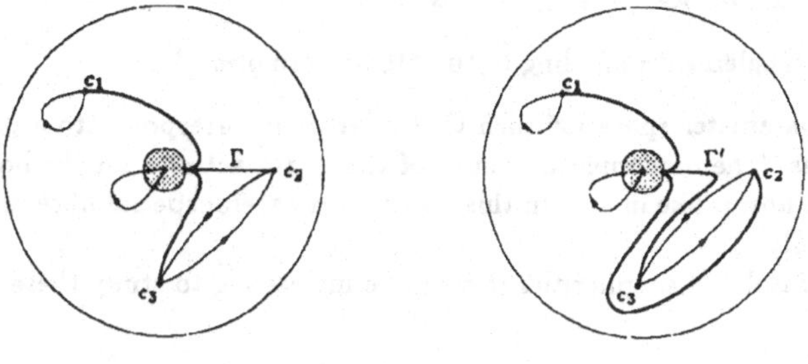

Figure 4: Γ and Γ'.

Using the isomorphisms 3.5 of $H_n(X_t)$ with the A_1-lattices L_Γ and $L_{\Gamma'}$,we see that the map $h_\omega : H_n(X_t) \to H_n(X_t)$ is just given by the base change from L_Γ to $L_{\Gamma'}$.

It could be interesting to study the special fibre monodromies in more detail.

4.3 Remarks about Picard-Lefschetz transformations

Let γ_i be a path joining s with s_i, defining vanishing cycles $\delta_i \in H_n(X_s)$ and $\Delta_i \in H_n(X_t)$. The simple loop u_i around the critical value c_i which corresponds to γ_i defines a Picard-Lefschetz transformation:

$$h_{u_i} : H_n(X_s) \longrightarrow H_n(X_s)$$

for which the Picard-Lefschetz formula holds:

$$h_{u_i}(x) = x + (x \cdot \delta_i)\delta_i,$$

where ($\cdot$) denotes the intersection product. This Picard-Lefschetz transformation is independent of the system of paths Γ in which γ_i can be embedded.

For the special fibre X_t the situation is different. Given a system of paths Γ, containing γ_i it follows e.g. from the Picard-Lefschetz formula, that we have the following restriction:

$$h_{u_i} : L_\Gamma \longrightarrow L_\Gamma.$$

Given an other system Γ', containing γ_i, we get the restriction:

$$h'_{u_i} : L_{\Gamma'} \longrightarrow L_{\Gamma'}.$$

Warning: The induced mappings h_{u_i} and $h'_{u_i} : H_n(X_t) \longrightarrow H_n(X_t)$ can in general be different.

This phenomenon already occurs in simple examples, such as isolated plane curve singularities with several A_1-points in the special fibre.

We conclude that there do not exist Picard-Lefschetz transformations of the above type on $H_n(X_t)$.

4.4 Remarks about special fibre monodromy groups

We consider next a d-parameter deformation:

$$\mathbf{C}^{n+1} \times \mathbf{C}^d \longrightarrow \mathbf{C} \times \mathbf{C}^d,$$

which satifies the conditions of theorem 2.3. We have two special examples in mind:

- The universal unfolding of isolated singularities.
- The universal disentanglement unfolding in the Mond examples.

One should stratify the parameter space $\mathbf{C}^d$ such that a stratum corresponds to a partition of the singular set into types. The fundamental groups of the strata act now on the homology of the corresponding special fibres, like in 4.2. In this way we get several special fibre monodromy groups related to our family.

Already in the case of isolated singularities it could be interesting to study these groups.

References

[AGV-II] V.I. Arnol'd, S.M. Gusein Zade, A.N. Varchenko: *Singularities of Differentiable Maps II*, Birkhäuser, 1988.

[GM] M. Goreski, R. MacPherson: *Stratified Morse Theory*, Ergebnisse der Mathematik und ihrer Grenzgebiete 3.Folge, Band 14, Springer Verlag 1988.

[Jo-St] T. de Jong, D. van Straten: *Disentanglements*, This Volume.

[La] K. Lamotke: *Die Homologie isolierter Singularitäten*, Math. Zeitschrift 143, 27-44 (1975).

[Lo] E.J.N. Looijenga: *Isolated Singular Points on Complete Intersections*, LMS-Lecture Note Series 77, Cambridge University Press, 1984.

[Mo] D.M.Q. Mond: *Vanishing cycles for analytic maps*, This Volume.

[Si-1] D. Siersma: *Isolated line singularities*, Proceedings of Symposia in Pure Mathematics Volume 40 (Part2) (1983), 485-496.

[Si-2] D. Siersma: *Singularities with critical locus a 1-dimensional complete intersection and transversal type* A_1, Topology and its applications 27 (1987) 51-73.

[Si-3] D. Siersma: *Variation Mappings on singularities with a 1-dimensional critical locus*, Topology (to appear).

Mathematisch Instituut
Rijksuniversiteit te Utrecht
P.O. Box 80.010
3508 TA UTRECHT
The Netherlands

On the versal deformation of cyclic quotient singularities

Jan Stevens

Abstract

In this paper we prove that the reduced base space of the versal deformation of a cyclic quotient singularity of embedding dimension e has at most $\frac{1}{e-2}\binom{2(e-3)}{e-3}$ irreducible components. All these components are smooth. The components are parametrised by certain continued fractions (introduced by Jan Christophersen), which represent zero. To prove our result, we determine all P-resolutions [K–S-B] by an inductive procedure parallel to the inductive generation of Christophersen's continued fractions. Smoothness of the components is deduced from an explicit description of the infinitesimal deformations of P-resolutions.

One of the goals of complex analytic singularity theory is to understand the versal deformations of singularities. In general the base space itself is a highly singular and complicated object. Cyclic quotients (in dimension two) are among the simplest and best understood singularities. Yet it took almost 15 years since Riemenschneider determined the infinitesimal deformations [R1], before Arndt gave in his thesis [A] an algorithm to find equations of the base space. Although the structure of the base is difficult to find from the equations, Arndt conjectured on the basis of computations for low embedding dimension e that the number of irreducible components should not exceed $\frac{1}{e-2}\binom{2(e-3)}{e-3}$ (the $(e-2)$-nd Catalan number C_{e-2}). In this paper I give a proof of this conjecture.

The components can be parametrised by the set K_{e-2} of continued fractions $[k_2, \ldots, k_{e-1}]$ with k_i positive integers, which represent zero. The importance of these continued fractions was first noticed by Jan Christophersen, who from his work on equations of the base formulated some conjectures on its structure. Christophersen has also obtained a proof of Arndt's conjecture [Ch].

Kollár and Shepherd-Barron gave an interpretation of the components of the base of quotient singularities in terms of certain partial resolutions: by taking a canonical model of a generic one parameter smoothing on a component one finds in the special fibre a modification $Y \to X$, whose properties lead to the definition of a P-resolution. On the other hand, the versal deformation of a P-resolution maps onto a component (deformations of the RDP-resolution give the Artin-component). The result is a (1–1)-correspondence between P-resolutions and components [K–S-B]. We give for general cyclic quotients an inductive construction of C_{e-2} P-resolutions, which parallels the construction of the set K_{e-2} of continued fractions by 'blowing up'. We then prove that no other P-resolutions are possible. This construction gives an abstract identification of P-resolutions and components. We obtain a description in coordinates on the base from an explicite basis of the vector space T_Y of infinitesimal deformations of Y, from which we compute the map $T_Y \to T_X$. Using a formula of Wahl for the dimension of smoothing components [W], we conclude from the calculation that all components are smooth, a fact no longer true for dihedral singularities.

A more precise formulation of our results is the following; for the notation used we refer to the relevant sections of this paper.

Theorem. *Let $X(a_2, \ldots, a_{e-1})$ be a cyclic quotient singularity, and let*

$$K_{e-2}(X) = \{[k_2, \ldots, k_{e-1}] \in K_{e-2} \mid k_i \leq a_i \text{ for all } i\}.$$

The base S_{red} of the versal deformation of X has exactly $\#K_{e-2}(X)$ irreducible components. Each component is smooth and its tangent space is spanned by:

$$\begin{array}{ll} \vartheta_\varepsilon^{(a)}, & 1 \leq a \leq a_\varepsilon - k_\varepsilon, \quad \varepsilon = 2, \ldots, e-1 \\ \sigma_\varepsilon + (\alpha_{\varepsilon-1} - 1)\vartheta_\varepsilon^{(1)}, & \text{if } \alpha_\varepsilon = 1, \quad \varepsilon = 3, \ldots, e-2. \end{array}$$

Jan Christophersen also obtained this theorem [Ch]. My approach does not use equations of the base, although their study has been vital in finding the statements.

The organisation of the paper is as follows: in the first section we introduce Christophersen's continued fractions, and prove that K_{e-2} has C_{e-2} elements by relating the continued fractions to subdivisions of polygons in triangles (in this context Catalan wrote down the number $\frac{1}{e-2}\binom{2(e-3)}{e-3}$ [Ca]). The next two sections review the infinitesimal deformations of cyclic quotients and P-resolutions. In Section 4 we prove the formula for the number of components, while the next section describes their tangent space. In Section 6 we explain how to write down explicitely the components from a given subdivision of a polygon. In the last section we give some corollaries and generalisations; in particular we show an example of a dihedral singularitiy with a non smooth component.

Notation. We will use some unusual notations for dual graphs of resolutions of rational singularities. A (-1)-curve will be denoted by a ○, a ■ denotes a (-3), a ■ with $-b$ above it is a $(-b)$ with $b \geq 3$, a ✱ is a curve with arbitrary self-intersection; as usual, a ● is a (-2) and a ● with $-b$ above it is a $(-b)$-curve.

Acknowledgements. The point of departure for this work was Arndt's beautiful result. I want to thank Jan Christophersen for sharing his ideas on cyclic quotients. For useful discussions thanks also go to the (quotient) singularity group in Hamburg.

1 Continued fractions

1.1. To fix notations, we recall some properties of continued fractions (cf. [O–W]). Let $[c_2, \ldots, c_{e-1}]$ be an (improper) continued fraction; the indices are numbered in this way to conform with the application to cyclic quotient singularities. We consider the symbol $[c_2, \ldots, c_{e-1}]$ as a rational function of its entries. One has the recursive definition: $[c_{e-1}] = c_{e-1}$, and

$$[c_i, c_{i+1}, \ldots, c_{e-1}] = c_i - 1/[c_{i+1}, \ldots, c_{e-1}].$$

One defines numbers $p_1, \ldots, p_e$ and $q_1, \ldots, q_e$ by

$$\begin{array}{lll} p_1 = 0, & p_2 = 1, & p_{i-1} + p_{i+1} = c_i p_i, \\ q_e = 0, & q_{e-1} = 1, & q_{i+1} + q_{i-1} = c_i q_i. \end{array}$$

With these numbers the partial continued fractions can be expressed as $[c_k, \ldots, c_{e-1}] = q_{k-1}/q_k$ and $[c_k, \ldots, c_2] = p_{k+1}/p_k$. Because a continued fraction $[c]$ of integers describes a chain of smooth rational curves on a surface, we use by analogy the term 'blowing up' for the following operations:

$$\begin{aligned}
[c_2, \ldots, c_{i-1}, c_{i+1}, \ldots, c_{e-1}] &\mapsto [c_2, \ldots, c_{i-1}+1, 1, c_{i+1}+1, \ldots, c_{e-1}] \\
[c_3, \ldots, c_{e-1}] &\mapsto [1, c_3+1, \ldots, c_{e-1}] \\
[c_2, \ldots, c_{e-2}] &\mapsto [c_2, \ldots, c_{e-2}+1, 1],
\end{aligned}$$

and for their inverses 'blowing down'.

1.2. A rational number r/s has a canonical continued fraction expansion of integers with $[c]$ defined by: $c_2 = \lceil \frac{r}{s} \rceil$ and $[c_3, \ldots, c_{e-1}]$ is the continued fraction expansion of $s/(sc_2 - r)$. Let n and q be positive integers with $n > q$ (as one has for a cyclic quotient $X_{n,q}$). One forms the continued fraction expansions

$$\begin{aligned}
\frac{n}{q} &= [b_1, \ldots, b_r] \\
\frac{n}{n-q} &= [a_2, \ldots, a_{e-1}].
\end{aligned}$$

The numbers p and q for $[a]$ are traditionally called i and j. The expansions of n/q and $n/(n-q)$ are dual; one finds the one from the other with Riemenschneider's point diagram [R1]: place in the i-th row $a_i - 1$ dots, the first one under the last one of the $(i-1)$-st row; the column j contains $b_j - 1$ dots.

Example. $[a] = [2, 5, 2, 3]$, $[b] = [3, 2, 2, 4, 2]$.

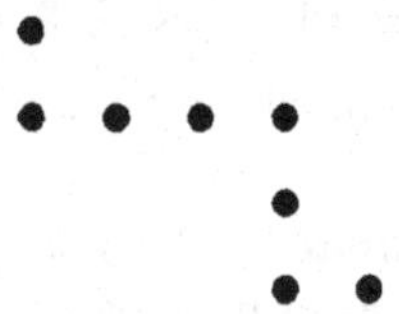

We remark that $[b_1, \ldots, b_r, 1, a_{e-1}, \ldots, a_2] = 0$.

1.3 Continued fractions which represent zero.

1.3.1 Definition (Christophersen). The set K_{e-2} is the set of continued fractions $[k] = [k_2, \ldots, k_{e-1}]$ such that $[k] = 0$.

1.3.2 Lemma. Every $[k] \in K_{e-2}$ can be obtained from $[0]$ by blowing up.

1.3.3 Lemma. Let $[k] \in K_{e-2}$. For the corresponding denominators of the partial continued fractions one has $p_i = q_i$. We write α_i for them.

1.3.4 Theorem. The number of elements of K_{e-2} is $\frac{1}{e-2}\binom{2(e-3)}{e-3}$.

Proof. The number on the right is the Catalan number C_{e-2} [B, N]. It describes, among other things, the number of ways to subdivide an $(e-1)$-gon in triangles [Ca]. We take an $(e-1)$-gon and mark a distinguished vertex; we number the remaining ones as $v_2, \ldots, v_{e-1}$. Given a subdivision we define numbers k_i as the number of triangles of which v_i is a vertex. This gives a bijection from K_{e-2} to the set of subdivisions. To blowing up corresponds the insertion of a new vertex, joined to two consecutive vertices.

1.3.5 Example.

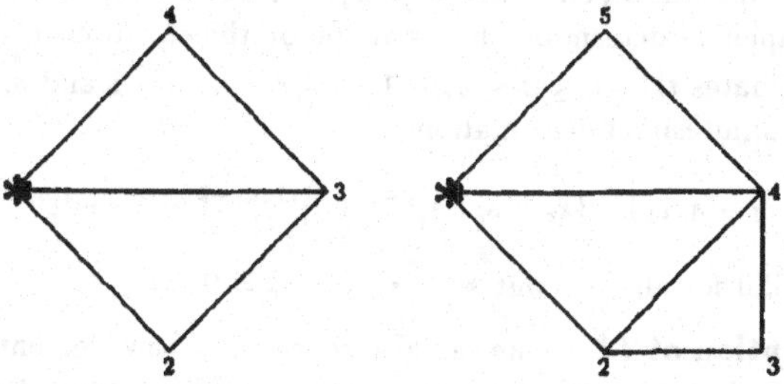

$[k] = [1,2,1]$ $\quad$ $[k] = [2,1,3,1]$
$\alpha = (1,1,1)$ $\quad$ $\alpha = (1,2,1,1)$

1.3.6. One obtains the α_i as follows: if the vertex v_i is joined to the distinguished vertex, then $\alpha_i = 1$; in a triangle (v_i, v_j, v_k) with $i < j < k$ one has $\alpha_i + \alpha_k = \alpha_j$.

1.3.7 Remark. One can also define the number k_*: it is the number of triangles coming together in the distinguished vertex. One has $k_* + \sum k_i = 3(e-3)$. Define $l = (e-3) - k_*$; the number l usually describes the codimension of smoothing components [W], see also Lemma 5.3. For fixed l the set $\{[k] \in K_{e-2} \mid k_* = (e-3) - l\}$ has $\frac{e-3-l}{e-3}\binom{e-4+l}{e-4}$ elements.

2 Cyclic quotient singularities

2.1. Let $X = \mathbf{C}^2/G_{n,q}$ be a cyclic quotient singularity ($G_{n,q}$ is the subgroup of $Gl(2,\mathbf{C})$, generated by $\begin{pmatrix} \zeta_n & 0 \\ 0 & \zeta_n^q \end{pmatrix}$, where ζ_n is a primitive n-th root of unity and $(n,q) = 1$). The continued fraction $n/q = [b_1, \ldots, b_r]$ gives the resolution: the exceptional divisor on the minimal resolution is a chain of smooth rational curves with self intersection $-b_i$.

The dual continued fraction $n/(n-q) = [a_2, \ldots, a_{e-1}]$ is important for the equations: first of all, one has the invariant monomials

$$z_\varepsilon = x^{i_\varepsilon} y^{j_\varepsilon}, \qquad \varepsilon = 1, \ldots, e.$$

The monomials z_ε generate the ring of invariants. The equations satisfied by these generators can be given by a quasi-determinantal [R2]:

$$\left|\begin{array}{ccccc} z_1 & z_2 & \ldots & z_{e-2} & z_{e-1} \\ z_2 & z_3 & \ldots & z_{e-1} & z_e \end{array}\right|$$
$$\qquad z_2^{a_2-2} \qquad \ldots \qquad z_{e-1}^{a_{e-1}-2}$$

The generalised minors of a quasideterminantal:

$$\left|\begin{array}{ccccc} f_1 & f_2 & \ldots & f_{k-1} & f_k \\ g_1 & g_2 & \ldots & g_{k-1} & g_k \end{array}\right|$$
$$\qquad h_{1,2} \qquad \ldots \qquad h_{k-1,k}$$

are $f_i g_j - g_i (\prod_{\varepsilon=i}^{j-1} h_{\varepsilon,\varepsilon+1}) f_j$.

Among the equations for X the following ones are of special importance:

$$z_{\varepsilon-1} z_{\varepsilon+1} - z_\varepsilon^{a_\varepsilon} \tag{1}$$

2.2 Infinitesimal deformations (the case $e > 3$). Arndt has proven that an infinitesimal deformation of X is completely determined by its action on the equations (1) [A]. On the vector space T^1_X we take coordinates $t^{(a)}_\varepsilon$, $1 \le a \le a_\varepsilon - 1$, $\varepsilon = 2, \dots, e-1$ and s_ε, $\varepsilon = 3, \dots, e-2$. Then we can write the infinitesimal deformations as:

$$z_{\varepsilon-1}(z_{\varepsilon+1} - s_{\varepsilon+1}) - (z_\varepsilon - s_\varepsilon)(z_\varepsilon^{a_\varepsilon-1} + t^{(1)}_\varepsilon z_\varepsilon^{a_\varepsilon-2} + \dots + t^{(a_\varepsilon-1)}_\varepsilon) \tag{2}$$

To make this formula valid for all ε we put $s_2 = s_{e-1} = s_e = 0$.

2.3 Pinkham's description of T^1. One can also represent T^1 by elements of $H^1(\Theta_{X-\{0\}}) = H^1(\Theta_{\mathbf{C}^2-\{0\}})^{G_{n,q}}$ [P]. The link with the description of Section 2.2 is given by the Čech-complex for the cohomology of $X' = X - \{0\}$:

$$\begin{array}{ccccccc} & & C^0(X',\Theta_{\mathbf{C}^e}|X) & \longrightarrow & C^0(X',N_X) & \longrightarrow & 0 \\ & & \downarrow\delta & & \downarrow\delta & & \\ 0 & \longrightarrow C^1(X',\Theta_X) & \longrightarrow C^1(X',\Theta_{\mathbf{C}^e}|X) & \longrightarrow & C^1(X',N_X) & \longrightarrow & 0 \end{array}$$

We take the covering $U_0 = \{z_1 \neq 0\}$, $U_{r+1} = \{z_e \neq 0\}$ of X', which corresponds to the covering $U_0 = \{x \neq 0\}$, $U_{r+1} = \{y \neq 0\}$ of $\mathbf{C}^2 - \{0\}$. Remember that $\delta(\varphi_1, \varphi_2) = \varphi_2 - \varphi_1$.

We define:

$$\vartheta^{(a)}_\varepsilon = -\frac{1}{n}\frac{1}{(x^{i_\varepsilon}y^{j_\varepsilon})^a}(j_\varepsilon x\frac{\partial}{\partial x} - i_\varepsilon y\frac{\partial}{\partial y}),\ 1 \le a \le a_\varepsilon - 1,\ 2 \le \varepsilon \le e-1$$

$$\sigma_\varepsilon = -\frac{1}{n}\frac{1}{x^{i_\varepsilon}y^{j_\varepsilon}}((j_\varepsilon - j_{\varepsilon-1})x\frac{\partial}{\partial x} - (i_\varepsilon - i_{\varepsilon-1})y\frac{\partial}{\partial y}),\ 3 \le \varepsilon \le e-2.$$

Because $x\frac{\partial}{\partial x} = \sum i_\rho z_\rho \partial_\rho$ and $y\frac{\partial}{\partial y} = \sum j_\rho z_\rho \partial_\rho$, one has:

$$\vartheta^{(a)}_\varepsilon = -\frac{1}{n}\frac{1}{z_\varepsilon^a}\sum(i_\rho j_\varepsilon - i_\varepsilon j_\rho)z_\rho\partial_\rho$$

$$\sigma_\varepsilon = -\frac{1}{n}\frac{1}{z_\varepsilon}\sum(i_\rho(j_\varepsilon - j_{\varepsilon-1}) - (i_\varepsilon - i_{\varepsilon-1})j_\rho)z_\rho\partial_\rho.$$

Because $a_\varepsilon i_\varepsilon = i_{\varepsilon-1} + i_{\varepsilon+1}$, one computes that $i_{\varepsilon-1}j_\varepsilon - i_\varepsilon j_{\varepsilon-1}$ is independent of ε, so equal to $i_0 j_1 - i_1 j_0 = n$. So the above formulae give:

$$\vartheta^{(a)}_\varepsilon = \frac{1}{z_\varepsilon^a}(\dots - z_{\varepsilon-1}\partial_{\varepsilon-1} + z_{\varepsilon+1}\partial_{\varepsilon+1} + \dots)$$

$$\sigma_\varepsilon = -\frac{1}{z_\varepsilon}(\dots + z_{\varepsilon-1}\partial_{\varepsilon-1} + z_\varepsilon\partial_\varepsilon + (a_\varepsilon - 1)z_{\varepsilon+1}\partial_{\varepsilon+1} + \dots).$$

The coefficients c_i of such vector fields $\sum c_i z_i \partial_i$ satisfy the equation $a_i c_i = c_{i-1} + c_{i+1}$. So two coefficients completely determine the field.

The vector fields are chosen in such a way that one can find a lift to $C^0(N_X)$, which gives exactly the infinitesimal deformations (2); one has to take $\rho \ge \varepsilon$ on U_{r+1}. Then:

$$\vartheta^{(a)}_\varepsilon = \frac{\partial}{\partial t^{(a)}_\varepsilon},\quad \sigma_\varepsilon = \frac{\partial}{\partial s_\varepsilon}.$$

2.4 A_k-singularities. For $e = 3$ one has to modify the previous formulae. The versal deformation of $A_{n-1}: z_1z_3 - z_2^n$ is

$$z_1z_3 - (z_2^n + t^{(2)}z_2^{(n-2)} + \ldots + t^{(n-1)}z_2 + t^{(n)})$$

and

$$\frac{\partial}{\partial t^{(a)}} = \vartheta^{(a)} = -\frac{1}{n}\frac{1}{(xy)^a}(x\frac{\partial}{\partial x} - y\frac{\partial}{\partial y}),\ 2 \leq a \leq n.$$

3 *P*-resolutions

3.1 Definition. A *P*-resolution of a quotient singularity X is a partial resolution $Y \to X$ (i.e. a modification), such that $K_Y \cdot E_i > 0$ for all exceptional divisors E_i and such that Y has only singularities Y_i with $K^2_{Y_i} \in \mathbb{Z}$.

3.1.1 Notation. The minimal resolution of a quotient singularity X will be denoted by $\tilde{X}$, and $\sigma: \tilde{Y} \to Y$ is the minimal resolution of the partial resolution Y of X; by $\tilde{Y}_i$ we denote the resolution of the germ Y_i.

3.1.2 Remark. One obtains $\tilde{Y}$ from $\tilde{X}$ by blowing up in intersection points of exceptional curves [K-S-B, 3.12-14].

3.1.3. The intersection number $K_Y \cdot E_i$ is defined as $\sigma^*K_Y \cdot \sigma^*E_i = \sigma^*K_Y \cdot \bar{E}_i$ (cf. [M]), where $\bar{E}_i$ is the strict transform of E_i on $\tilde{Y}$. Because $K_{\tilde{Y}} = \sigma^*K_Y + \sum K_{\tilde{Y}_j}$, one has $K_Y \cdot E_i = K_{\tilde{Y}} \cdot \bar{E}_i - \sum K_{\tilde{Y}_j} \cdot \bar{E}_i$.

3.2 Definition. A quotient singularity is of type T, when its resolution graph is of type T. Such a graph is of type T (a T-graph), when it is

i)  or , or

ii)  , if  is a T-graph.

Remark. Of course one can also add a (-2) on the side of b_r; a graph only obtains a direction by the choice of a numbering.

3.3 Lemma. A quotient singularity X has $K^2 \in \mathbb{Z}$ if and only if it is a rational double point or a singularity of type T [L-W, Prop. (5.9)].

3.4 Lemma. Let X be of type T. Let the rational cycle $K = -\sum k_iE_i$ represent the canonical divisor on the minimal resolution. In the case of 3.2 i) we have $K = -1/2E$, while in ii), if  has $k_i = p_i/q_i$ with $p_1 + p_r = q$, then 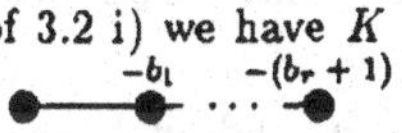 has $k_0 = p_1/(q+p_1), k_i = (p_i + p_1)/(q + p_1)$.

Proof. Let $(p_i/q)E_i^2 + (p_{i-1} + p_{i+1})/q = E_i^2 + 2$. Then $(p_i + p_1)/(q + p_1)E_i^2 + (p_{i-1} + 2p_1 + p_{i+1})/(q + p_1) = E_i^2 + 2$. An analogous computation applies for E_r.

3.4.1 Corollary. Let  and  have $k_r \leq k'_s$, then 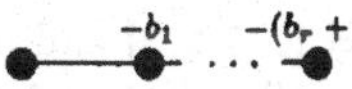 and 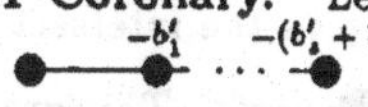 have $k_r \leq k'_s$.

3.5. A quotient singularity X of type T is also a quotient of a certain A_{m-1}-singularity [L-W]: we shall show that for some index δ one has $i_\delta = j_\delta$; so $z_\delta = (xy)^{i_\delta}$. We define δ to be the *central index* and z_δ the *central coordinate* of the singularity of type T. One can compute that

$n = (a_\delta - 1)i_\delta^2$; therefore $m = (a_\delta - 1)i_\delta$. In fact, A_{m-1} is the canonical cover of X, and the order of the covering group is usually called the *Index* of X; so the Index is in this case i_δ.

One can also describe the process of Definition 3.2 in terms of the numbers $\boldsymbol{a}$. In i) we have $[\boldsymbol{a}] = [2, a_3, 2]$, while in ii) we increase a_2 by 1 and add an $a_e = 2$. The final result will have the original value a_3 on position δ . Induction shows that $i_\delta = j_\delta$, but one can also argue as follows: if one replaces a_δ by 1, one obtains an element of K_{e-2}: $[a_2, \ldots, a_{\delta-1}, 1, a_{\delta+1}, \ldots, a_{e-1}] = 0$. Conversely, every element of K_{e-2} with only one "1" gives an $[\boldsymbol{a}]$ of a singularity of type T. Because up to δ the computation of i, resp j coincides with that of α for $[\boldsymbol{k}]$ with $k_i = a_i$, $i \neq \delta$ and $k_\delta = 1$, one has $i_\delta = j_\delta$.

The deformation space of X has a component S', which consists of the quotient of the equivariant deformations of the $A_{(a_\delta-1)i_\delta-1}$.

3.6 Proposition (Kollár–Shepherd-Barron). Let Y be a P-resolution of the quotient singularity X, with singularities Y_i. Let S'_{Y_i} be the component of the base S_{Y_i} of the versal deformation of Y_i, which consists of quotients of equivariant deformations and let S'_Y be the inverse image of $\prod S'_{Y_i}$ under the natural map $S_Y \to \prod S_{Y_i}$. The image of S'_Y in the base space S_X of the versal deformation of X is an irreducible component of S_X. In this way one obtains a (1–1)-correspondence between P-resolutions and components of $(S_X)_{red}$.

4 The number of components

In this section we will give an inductive construction of P-resolutions. As main result we obtain the following theorem.

4.1 Theorem. Let $X(a_2, \ldots, a_{e-1})$ be a cyclic quotient singularity, and let $K_{e-2}(X) = \{[k_2, \ldots, k_{e-1}] \in K_{e-2} \mid k_i \leq a_i \text{ for all } i\}$. The base S_{red} of the versal deformation of X has exactly $\#K_{e-2}(X)$ irreducible components.

4.2 Convention. I will use a rather sloppy terminology. A resolution graph will also serve as 'name' for a singularity. A curve on a resolution may be denoted by the same symbol as the corresponding vertex; also its strict transform on other (partial) resolutions has the same name. The sentence: 'the curve E_r does not belong to a singularity of type T' says that the strict transform of E_r on $\widetilde{Y}$ is not contained in $\sigma^{-1}(p_i)$ for some (Y, p_i) of type T, where (according to our notation convention 3.1.1) $\sigma: \widetilde{Y} \to Y$ is a resolution of a P-resolution $Y \to X$ with singularities (Y, p_i).

We will describe Y by showing resolution graphs of $\widetilde{Y}$, where the resolutions of the singularies of type T and rational double points are put between square brackets [ ]; sometimes we will delete the brackets.

4.3. Let $[\boldsymbol{k}] \in K_{e-3}$, $\boldsymbol{k} \leq \boldsymbol{a}$, and let $Y \to X$ be a P-resolution of $X(\boldsymbol{a})$. Let $[\boldsymbol{k}'] \in K_{e-2}$ be constructed from $[\boldsymbol{k}]$ by blowing up once. We shall construct for certain $[a'_2, \ldots, a'_{e-1}]$ a P-resolution Y' of $X(\boldsymbol{a}')$. There are no maps from the objects with primes to the ones without, but the graph of $\widetilde{Y}'$ is constructed from parts of that of $\widetilde{Y}$. As an example we consider $[\boldsymbol{a}] = [2,3,2,3]$ and $[\boldsymbol{k}] = [2,2,1,3]$, which gives the P-resolution: 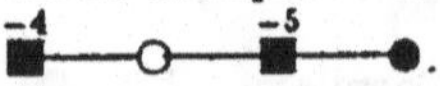

. Now take $[\boldsymbol{k}'] = [2,3,1,2,3]$ and $[\boldsymbol{a}'] = [2,4,2,3,3]$. On $\widetilde{X}'$ we can still blow up in the intersection point of the (−3) and the (−4)-curve to obtain 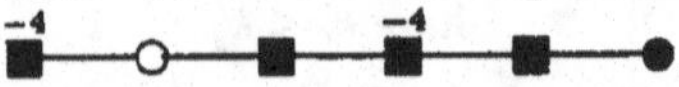

. This graph looks like that of $\widetilde{Y}$, but is somewhat changed on the right hand side. We want it to be as much the same as possible, so we want a $[b] = [5,2]$ singularity; this is achieved by blowing up in the intersection point of the (-4) and the (-3)-curve to get:

We observe that we obtain a new singularity of type T in the middle.

To descibe our construction we distinguish four cases:

I. Let $[k] =: [k_2, \ldots, k_{e-2}]$, $[k'] = [k_2, \ldots, k_{e-2} + 1, 1]$.

a) If $a_{e-2} = 2$, $k_{e-2} = 1$, let $[a'] = [a_2, \ldots, a_{e-2}, a_{e-1}]$.

b) For all a_{e-2}, k_{e-2} (also if $a_{e-2}.k_{e-2} = 2$): let $[a'] = [a_2, \ldots, a_{e-2} + 1, a_{e-1}]$.

II. Let $[k] =: [k_2, \ldots, k_{\varepsilon-1}, k_{\varepsilon+1}, \ldots, k_{e-1}]$, $[k'] = [k_2, \ldots, k_{\varepsilon-1} + 1, 1, k_{\varepsilon+1} + 1, \ldots, k_{e-1}]$.

a) Let $a_{\varepsilon-1}.k_{\varepsilon-1} = 2$ or $a_{\varepsilon+1}.k_{\varepsilon+1} = 2$; suppose $a_{\varepsilon-1}.k_{\varepsilon-1} = 2$. Let $[a'] = [a_2, \ldots, a_{\varepsilon-1}, a_\varepsilon, a_{\varepsilon+1} + 1, \ldots, a_{e-1}]$.

b) For all values of $[a]$, $[k]$: let $[a'] = [a_2, \ldots, a_{\varepsilon-1} + 1, a_\varepsilon, a_{\varepsilon+1} + 1, \ldots, a_{e-1}]$.

4.4 Claim. Let $E_1, \ldots, E_r$ be the exceptional curves on $\tilde{X}(a_2, \ldots, a_{e-2})$.If $a_{e-2}.k_{e-2} = 2$, then the curve E_r, which has $b_r = -E_r^2 > 2$ (as one sees from the point diagram), does not belong to a singularity of type T.

This follows from our inductive construction.

4.5 Construction Ia. By the above claim the P-resolution Y coincides in a neighbourhood of E_r with the minimal resolution. For Y' this will also be the case: let the right-hand side of the graph of $\tilde{Y}'$ be:

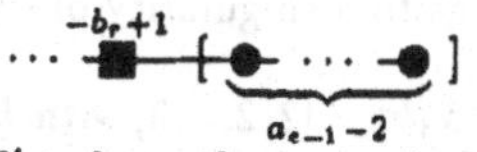

So we take the P-resolution Y' of X' to have the same singularities as Y plus an extra $A_{a_{e-1}-2}$.

4.6 Construction Ib. In this case the graph of $\tilde{Y}$ ends with an E_r with $E_r^2 = -b_r$, which may belong to a singularity of type T. The curve E_r occurs already on the minimal resolution; this follows again inductively or from the fact that exceptional curves of $\tilde{Y} \to \tilde{X}$ lie above intersection points of exceptional curves on $\tilde{X}$ (Remark 3.1.2). From the point diagram for $[a']$ we see that we can take as graph of $\tilde{Y}'$ a graph that ends like:

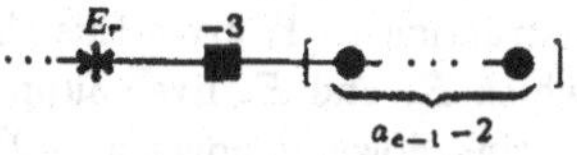

So again we take the P-resolution Y' of X' to have the same singularities as Y plus an extra $A_{a_{e-1}-2}$.

4.7 Construction IIa. Here we distinguish two subcases:

i) $\varepsilon = 3$: by the above claim we have on $\tilde{X}(a)$ and $\tilde{Y}(a)$ an E_1 with $b_1 \geq 3$. We get on $\tilde{Y}'$:

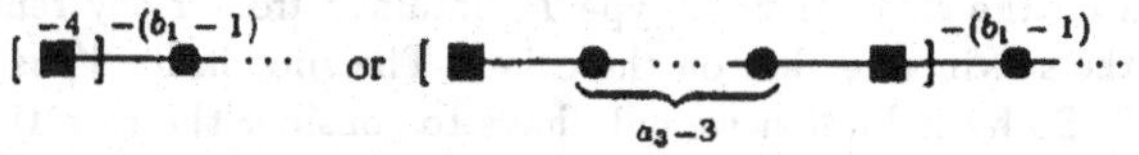

and we have one singularity of type T more.

ii) $\varepsilon > 3$: from the point diagram we have that the position ε determines a curve E_ρ with $b_\rho \geq 4$ on the minimal resolution. I claim that by construction E_ρ is part of a singularity of type T, obtained by applying the inductive process of Definition 3.2 on a (-4). We replace this singularity by one, obtained with the same process, but starting with:

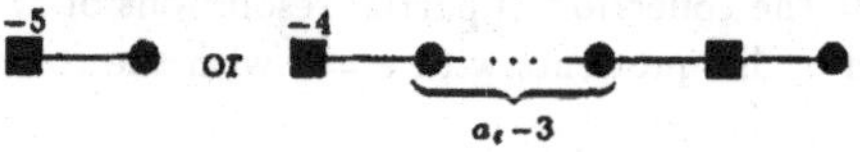

The resolution $\tilde{Y}'$ blows down to a minimal resolution, which has the required self-intersection numbers.

4.8 Construction IIb. We first look at the minimal resolution. From the point diagram we see that the insertion of a row a_ε leads to the replacement of a curve E_ρ with $b_\rho > 2$ by:

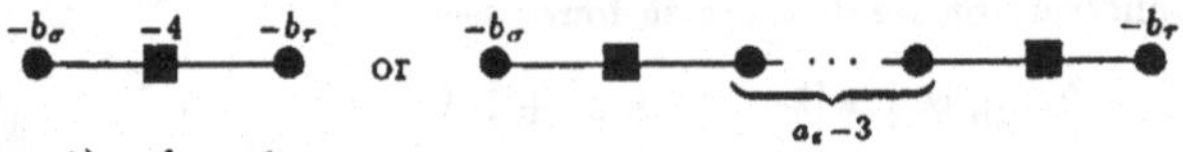

with $(b_\sigma - 1) + (b_\tau - 1) = b_\rho - 1$.

We have two subcases:

i) if the curve E_ρ does not belong to a singularity of type T, then 

is a new singularity of type T.

ii) let E_ρ belong to a singularity Y_i of type T. We want on Y' a singularity with the same resolution graph as Y_i. To obtain $\tilde{Y}'$ we first blow up $\tilde{X}'$ just as $\tilde{X}$: in stead of blowing up the intersection of E_ρ with other curves one blows up in the intersection of these curves with E_σ or E_τ. Then we 'restore' the graph of Y_i by repeatedly blowing up above the intersection point of E_σ or E_τ with the newly introduced (-3) (resp. (-4)); which possibility we take, depends on the position of the index ε with respect to the central index of Y_i: one obtains the point diagram of Y_i from those columns in the point diagram of X, which correspond to the curves belonging to Y_i, by adding some dots at the ends; we say that central index of Y_i is smaller than ε , if its row lies above the level where we insert the new row. In this case we 'restore' the singularity on the left hand side. This is our $\tilde{Y}'$.

4.9 Claim. Also the curve E_τ belongs to a singularity of type T.

Example. Consider $[a] = [2,3,2,3]$ and $[k] = [2,2,1,3]$ with P-resolution 

.
For $[a'] = [2,4,2,3,3]$ and $[k'] = [2,3,1,2,3]$ we blow up from 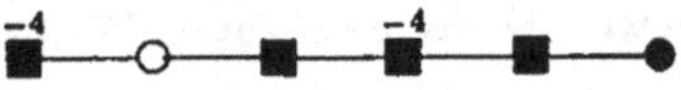

to obtain:

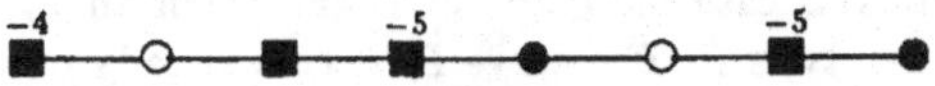

Proof of the claim. Let Y'_σ be the singularity of type T to which E_σ belongs, and let Y'_τ be the singularity, whose resolution consists of the new curves to the right of the (-1)-curve E_0, together with those that occurred in Y_i. Consider the singularity, consisting only of Y'_σ, the curve E_0 and Y'_τ (applying our construction to Y_i gives this singularity). Its resolution blows down to a minimal resolution, on which E_σ and E_τ live. Suppose that blowing down E_0 gives again a (-1)-curve. For $\tilde{Y}'_\sigma$ this blowing down removes a (-2) or alters a self-intersection; it becomes again a singularity Y''_σ of type T, if we also alter the self-intersection or remove a (-2) at the other end. Call this respectively a right hand side and a left hand side operation. We perform the same right-hand side operation on Y'_τ as on Y'_σ. The resulting surface Y'', consisting of one exceptional curve of self-intersection (-1) on $\tilde{Y}''$ with two singularities Y''_σ and Y''_τ, is the result of our operation on the singularity of type T, obtained from Y_i by removing a (-2) on one side and altering the self-intersection on the other. The singularity Y'_τ is of type T if and only if Y''_τ is of type T. So by induction we only have to consider the case that blowing down E_0 gives the minimal resolution. Then E_ρ is an end curve of Y_i and $b_\sigma = -E_\rho^2 - 1$, $b_\tau = 2$ and E_τ is an end curve of Y'_τ. Therefore Y'_τ is a singularity of type T.

4.10 Definition. Let $\mathcal{P}$ be the collection of partial resolutions of cyclic quotient singularities, which can be obtained from cyclic quotients with $e = 4$ with the constructions above.

4.11 Proposition. Every element of $\mathcal{P}$ is a P-resolution.

Proof. On a Y as above there are only singularities of type T or rational double points, so we have to check that the condition '$K \cdot E_i > 0$ for all (-1)-curves' is preserved under the operations **IIaii** and **IIbii**: if E_0 is a (-1)-curve, which intersects on $\tilde{Y}$ the curves E_1 and E_2, lying in singularities Y_1 and Y_2 of type T with $K_{Y_i} = -\sum k_j E_j$, then one needs $k_1 + k_2 > 1$. We first look at the (-1) between Y'_σ and Y'_τ in construction **IIbii**. By Corollary 3.4.1 and the construction of the previous proof we may assume that blowing down the (-1) gives the minimal resolution. This means that $-E_i^2 > 2$, so $k_i \geq \frac{1}{2}$ for $i = 1,2$ and $k_2 > \frac{1}{2}$, if E_2 belongs to Y'_τ. Therefore the condition is satisfied.

In the constructions **IIa** and **IIb** only one coefficient k_j in K_{Y_i} of a curve, which intersects the unaltered part of the graph, is made smaller (again an application of Corollary 3.4.1). Again we look only at the singularity, consisting of the (-1) and two singularities of type T; as before, we reduce by blowing down the (-1) (in **IIbii** both (-1)'s), altering a self-intersection and deleting a (-2), to the case that blowing down the (-1) (between the singularity of type T, which was already on Y, and the one resulting from the construction) gives the minimal resolution; because it is by construction not possible that the (-1) intersects two singularities of type T with $K = \frac{1}{2}E$, we find that also in this case the condition is satisfied.

4.12 Proposition. Every P-resolution of a cyclic quotient singularity belongs to the collection $\mathcal{P}$.

Proof (by double induction). We will use induction on the number of singularities of type T on Y. If Y contains no singularities of type T, then Y is the rational double point resolution, and one obtains Y with the constructions **I**. If Y has one singularity of type T, then Y is dominated by the minimal resolution; from the definition of singularities of type T one sees that (repeated) application of the inverse of **IIa** gives a P-resolution without singularities of type T.

Suppose Y has m singularities of type T, $m > 1$. If Y contains a $(-k)$-curve E_0, $k > 1$, which lies between singularities of type T, then the parts of Y on both sides of E_0 are P-resolutions of cyclic quotient singularities with less than m singularities of type T; as one can perform the constructions of type **II** always after those of type **I**, one can obtain Y from a P-resolution with less than m singularities of type T with the same constructions as for the parts of Y. So we may assume that the Y_i are separated by (-1)-curves. We now use induction on the number of blow-ups needed to obtain $\tilde{Y}$ from $\tilde{X}$. We blow down all (-1)-curves on $\tilde{Y}$; if the image point does not lie on a (-1), we insert at this point a curve with self-intersection (-2). At the ends of the singularities Y_1 and Y_m, which do not intersect a (-1), we lower the self-intersection or delete a (-2). We have a surface $\tilde{Y}'$ with a chain E of rational curves. Blowing down the maximal exceptional configurations gives a surface X' with possibly some complete curves on it; let $\tilde{X}'$ be its minimal resolution. A (-2)-curve E'_0, inserted after blowing down a (-1)-curve E_0, does not appear on $\tilde{X}'$, if at some stage of the blowing down from $\tilde{Y}'$ to $\tilde{X}'$ the curve E'_0 intersects a (-1)-curve E_1. Let $\tilde{Y}_1$ be the singularity consisting of E_0 and all curves at the same side of it as E_1; it is a P-resolution of a cyclic quotient singularity X_1. At some stage of the blowing down from $\tilde{Y}_1$ to $\tilde{X}'_1$ a (-1)-curve occurs at an end, contradicting Remark 3.1.2. The same argument shows that an end curve of Y_1 or Y_m does not become a (-1)-curve. Therefore the chain E is exceptional.

So we obtain a P-resolution Y', which is in $\mathcal{P}$ by the (second) induction hypothesis. If no inverse operation of type **II** can be applied on Y', then Y' is dominated by the minimal resolution; because on Y the (-1)-curves cannot separate two singularities with $K = -\frac{1}{2}E$, the partial resolution Y has only one singularity of type T, contradicting our assumption. So an inverse operation of type **II** can be applied on Y'. With the same arguments as in the proof of Proposition 4.11 one sees that a corresponding operation of exactly the same type can be applied to Y at the same spot, except when we can apply **IIai** or **IIbi** on $\widetilde{Y}'$. Suppose one has on $\widetilde{Y}'$ a  , where E_τ is an inserted (-2)-curve. So on $\widetilde{Y}$ one has 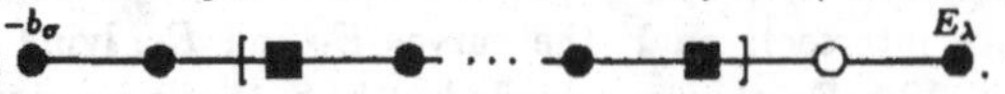 . This is the result of Construction **IIbii** on E_ρ with $b_\rho = b_\lambda + 1$. Similarly an operation of type **IIai** on Y' corresponds to Construction **IIbii** (delete E_σ). Therefore Y is the result of a construction on an element of $\mathcal{P}$ and so Y itself is in $\mathcal{P}$.

4.13. One can find the element of K_{e-2}, corresponding to a P-resolution directly from the singularities of Y. This is particularly easy in the case that Y has $(e-2)$ singularities Y_2, ..., Y_{e-1}: one takes $\alpha_i = \text{Index}(Y_i)$; this determines $[k]$. In general $[k]$ is determined if one knows for all i either α_i or k_i, because of the equations $k_i\alpha_i = \alpha_{i-1} + \alpha_{i+1}$, $\alpha_1 = \alpha_e = 0$, $\alpha_2 = \alpha_{e-1} = 1$. For every singularity Y_i of type T on Y the central index $\delta(i)$ corresponds to an index ε of X; we define $\alpha_\varepsilon = \alpha_{\delta(i)}$. If $a_\varepsilon = a_{\delta(i)}(Y_i)$, let $I = [\varepsilon - j_1, \varepsilon + j_2]$ be the largest interval on which $a_{\varepsilon+j} = a_{\delta(i)+j}(Y_i)$; we define $k_i = a_i$ for $i \in I, i \neq \varepsilon$. For the remaining indices we define $\alpha_i = 1$.

4.14 Remark. To finish the proof of Theorem 4.1 one has to show that the map from $\mathcal{P}(X)$, the P-resolutions of X, to $K_{e-2}(X)$ is injective. In the next section we shall give another description of this map, using the tangent space to a component; this description makes the injectivity evident.

A proof in the spirit of this Section consists in showing that for every "1" in $[k_2, \ldots, k_{e-1}]$ there exists a corresponding inverse construction. Here I give only some indications of the proof and an example. If Y has $e-2$ singularities, then only constructions of type **b** are possible; in **IIb** a singularity Y_i 'vanishes' into its neighbour, and the possibility to do this is clearly not affected by a construction, which leaves the "1" belonging to Y_i unchanged. In general the resolution of a singularity Y_i of type T may coincide with the minimal resolution in a neighbourhood of the curves, corresponding to the central index. This means that Y_i is obtained by constructions of type **IIa**; after an inverse construction, which makes the segment of coincidence of $\widetilde{Y}_i$ and the minimal resolution smaller, it is still possible to do an inverse of **IIa**, until Y_i can 'vanish' with **IIb**.

Example. Let $[k] = [2,2,1,3]$, $[a] = [2,2,a,3]$. Make with **IIb** and **IIa** the singularity

with $[k] = [3,2,1,4,1,3]$. This is also the result of Construction **IIb** on $[a] = [3,2,2,3,2]$.

5 The tangent space to the components

5.1 Theorem. Each irreducible component $S_{[k]}$ of the base of the versal deformation of a cyclic quotient singularity $X(a)$ is smooth and its tangent space is spanned by:

$$\begin{aligned} &\vartheta_\varepsilon^{(a)}, && 1 \le a \le a_\varepsilon - k_\varepsilon,\ \varepsilon = 2, \ldots, e-1 \\ &\sigma_\varepsilon + (\alpha_{\varepsilon-1} - 1)\vartheta_\varepsilon^{(1)}, && \text{if } \alpha_\varepsilon = 1,\ \varepsilon = 3, \ldots, e-2. \end{aligned}$$

5.2 Lemma. Consider in the point diagram the row ε ; it determines $a_\varepsilon - 1$ curves. If $\alpha_\varepsilon = 1$, then the number c_ε of these curves belonging to a singularity of type T with central index less than ε , is equal to $\alpha_{\varepsilon-1} - 1$.

Proof. By construction c_e equals the number of diagonals in the corresponding subdivision of the polygon, which connect the vertex v_e with vertices of lower index; but this is $\alpha_{e-1}-1$.

5.3 Lemma. The codimension of a component $S_{[k]}$ is $e-4+2l$, where $l=(e-3)-k_*$.

Proof. By [W] this statement is equivalent to the formula $b_2(F)=r-(2e-7)-l$ for the second Betti number of the fibre of a general smoothing on the component (if $e\geq 4$). We can compute this Milnor number from the corresponding P-resolution: it is equal to the number of exceptional curves plus the Milnor numbers of the singularities Y_i. So we have to show that for the constructions in Section 4 we have $b_2(F')=b_2(F)+a_e-2$ in the cases **a**, and $b_2(F')=b_2(F)+a_e-1$ (cases **b**). This is immediate from the construction, if one remembers that a singularity of type T, which is a $\mathbb{Z}/\alpha_\delta\mathbb{Z}$-quotient of $A_{\alpha_\delta(a_\delta-1)-1}$, has a Milnor fibre with $\chi(F)=\frac{1}{\alpha_\delta}\chi(F_{A_{\alpha_\delta(a_\delta-1)-1}})=a_\delta-2$.

5.4 Plan of the proof of the theorem. Let $Y\to X$ be a P-resolution. We shall describe $r-(e-4+2l)$ independent elements in T^1_Y, which are tangent to S'_Y, the space of deformations of Y which project on $S_{[k]}$ (see Proposition 3.6). We show that they project onto the space spanned by the elements $\vartheta^{(a)}_e$, $1\leq a\leq a_e-k_e$ and $\sigma_e+c_e\cdot\vartheta^{(1)}_e$, $\alpha_e=1$. This proves the theorem.

5.5 Description of the map $T^1_Y\to T^1_X$. Let $E_1,\ldots,E_r$ be the exceptional curves on Y; we remark that $r\leq e-3$. Let E_0 be the strict transform of the curve $\{z_e=0\}$ and E_{r+1} that of $\{z_1=0\}$. We cover Y with coordinate patches $(y^{(i)})$ with center in $E_i\cap E_{i+1}$, $i=0,\ldots,r$: if Y has in $E_i\cap E_{i+1}=p_i$ a singularity Y_i of embedding dimension e_i, one takes coordinates $y^{(i)}_1,\ldots,y^{(i)}_{e_i}$ with $E_i=Y_i\cap\{y^{(i)}_{e_i}=0\}$ and $E_{i+1}=Y_i\cap\{y^{(i)}_1=0\}$.

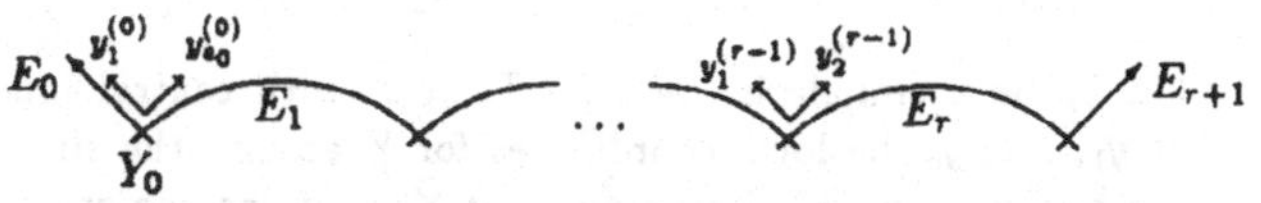

If Y is smooth, these coordinates reduce to the usual ones (cf. [L]). The transition functions are easily computed from the data on $\tilde{Y}$; the same holds for the map $Y\to X$. In particular, if Y_0 is not smooth, one has:

$$z_1=y^{(0)}_1,\quad z_2=y^{(0)}_2,\quad z_{e+1}=z_e^{a_e}z_{e-1}^{-1}$$

where the z_e are coordinates for X.

One computes T^1_Y with the exact sequence

$$0\longrightarrow H^1(Y,T^0_Y)\longrightarrow T^1_Y\longrightarrow\oplus_i T^1_{Y_i}\longrightarrow 0$$

and a Čech-covering $\{U_i\}$ with $U_i=Y-\bigcup_{j\neq i}E_j$, $i=0,\ldots,r+1$. Elements of $T^1_{Y_i}$ can be represented in the coordinates $y^{(i)}$ as in (2.3): as elements of $H^1(Y_i-\{p_i\},\Theta)$, so as vector fields on $U_i\cap U_{i+1}=\{z_1\neq 0\}\cap\{z_e\neq 0\}$. The elements of $H^1(Y,T^0_Y)$ also can be given as vector fields on $\{z_1\neq 0\}\cap\{z_e\neq 0\}$. The image of these vector fields in $H^1(X-\{p\},\Theta)$ determines their image in T^1_X.

5.6 S' for singularities of type T. Let X be a singularity of type T. The tangent space to the subspace S' of S, consisting of the quotient of equivariant deformations, is spanned by the vectors $\vartheta^{(1)}_\delta,\ldots,\vartheta^{(a_\delta-1)}_\delta$, where δ is the central index. Indeed, from Sections 2.3 and 3.5 one sees that $\vartheta^{(a)}_\delta=-\frac{i_\delta}{n}\frac{1}{(xy)^{i_\delta a}}(x\partial_x-y\partial_y)$; this are precisely the invariant elements of $T^1_{A_m}$, $m=i_\delta(a_\delta-1)-1$.

5.7 Definition. Let E_ρ be a curve with $b := b_\rho \geq 2$ (so it occurs also on the minimal resolution with the same self intersection); let it be covered there by two coordinate patches, (u_1, v_1) and (u_2, v_2), with $u_2 = v_1^{-1}, v_2 = v_1^b u_1$ and $E_\rho = \{u_1 = 0\} \cup \{v_2 = 0\}$. Define vector fields $\sigma_\rho^{(1)}$, $\ldots, \sigma_\rho^{(b-1)} \in H^1(Y, T_Y^0)$ by $\sigma_\rho^{(i)} = v_1^{-i}\frac{\partial}{\partial u_1} = u_2^{b-i}\frac{\partial}{\partial v_2}$.

5.8 Lemma (description of TS'_Y). The tangent space of S'_Y is spanned by the following elements:

- $\vartheta_{\delta(i)}^{(1)}, \ldots, \vartheta_{\delta(i)}^{(a_\delta - 1)} \in H^1(Y_i - \{p_i\}, \Theta)$ for each singularity Y_i of type T on Y with central index $\delta(i)$
- $\vartheta_i^{(2)}, \ldots, \vartheta_i^{(k+1)} \in H^1(Y_i - \{p_i\}, \Theta)$ for each A_k- singularity Y_i
- $\sigma_\rho^{(1)}, \ldots, \sigma_\rho^{(b_\rho - 1)} \in H^1(Y, T_Y^0)$ for each curve E_ρ with $b_\rho \geq 2$.

5.9 Lemma. Let the curve E_ρ with $b := b_\rho \geq 3$ be covered on the minimal resolution by two coordinate patches, (u_1, v_1) and (u_2, v_2), with $E_\rho = \{u_1 = 0\} \cup \{v_2 = 0\}$. Suppose that the column ρ in the point diagram contains points of the rows $\varepsilon + 1, \ldots, \varepsilon + b - 1$. The map $\pi: Y \to X$ is induced by the map $\widetilde{X} \to X$, partially given by:

$$\begin{cases} z_\varepsilon &= u_1^{a_{\varepsilon+1}-1} v_1^{a_{\varepsilon+1}-2} & \\ z_{\varepsilon+1} &= u_1 v_1 &= u_2^{b-1} v_2 \\ &\vdots & \\ z_{\varepsilon+b-1} &= u_1 v_1^{b-1} &= u_2 v_2 \\ z_{\varepsilon+b} &= & u_2^{a_{\varepsilon+b-1}-2} v_2^{a_{\varepsilon+b-1}-1}. \end{cases}$$

5.10 Lemma. Let Y_ε be a singularity of type T on Y with central index $\delta(\varepsilon)$. Suppose $\mathrm{Index}(Y_\varepsilon) = \alpha_\varepsilon$. Let $y_1, \ldots, y_{e'}$ be local coordinates for Y around the singular point p_ε. For the map $Y \to X$ one has $z_\varepsilon = y_{\delta(\varepsilon)}$. Furthermore, if $\widetilde{Y}_\varepsilon$ contains the strict transform of the non (-2)-curve E_ρ on $\widetilde{X}$, determined by the rows ε and $\varepsilon + 1$ in the point diagram, then also $z_{\varepsilon+1} = y_{\delta(\varepsilon)+1}$. In particular, if $k_\varepsilon = 1$, then $z_{\varepsilon-1} = y_{\delta(\varepsilon)-1}$ and $z_{\varepsilon+1} = y_{\delta(\varepsilon)+1}$.

5.11 Proofs. Lemma 5.9 is well-known. We shall prove the lemmata 5.8 and 5.10 together with the main theorem 5.1 by analyzing the effect of the constructions from Section 4.

5.11.1 Construction Ia. Suppose that $a_{e-1} > 2$. Let (y_1, y_2, y_3) be local coordinates on Y', such that the $A_{a_{e-1}-2}$-singularity is given by $y_2^{a_{e-1}-1} = y_1 y_3$. One has $z_e = y_3$, $z_{e-1} = y_2$, $z_{e-2} = y_1 y_2$. So:

$$\frac{1}{y_2^a}\left(y_1 \frac{\partial}{\partial y_1} - y_3 \frac{\partial}{\partial y_3}\right) = \frac{1}{z_{e-1}^a}\left(\ldots + z_{e-2}\frac{\partial}{\partial z_{e-2}} - z_e \frac{\partial}{\partial z_e}\right).$$

Therefore $\vartheta^{(2)}, \ldots, \vartheta^{(a_{e-1}-1)}$ map onto $\vartheta_{e-1}^{(2)}, \ldots, \vartheta_{e-1}^{(a_{e-1}-1)}$.

Let (u, v) be coordinates on $\widetilde{Y}'$ with center in the intersection of E_r with the exceptional locus of $A_{a_{e-1}-2}$ (so $y_1 = u, y_2 = uv$). Then $z_e = u^{a_{e-1}-2} v^{a_{e-1}-1}$, $z_{e-1} = uv$, $z_{e-2} = u^2 v$. The same formulae hold when $a_{e-1} = 2$; then (u, v) are the coordinates $(y_1^{(r+1)}, y_2^{(r+1)})$. One has:

$$u^{-i}\partial_v = \frac{1}{u^i v} v \frac{\partial}{\partial v} = \frac{1}{z_{e-i}}\left(\ldots + z_{e-2}\frac{\partial}{\partial z_{e-2}} + z_{e-1}\frac{\partial}{\partial z_{e-1}} + (a_{e-1} - 1) z_e \frac{\partial}{\partial z_e}\right).$$

The field $u^{-1}\partial_v$ has image $-(a_{e-1}-1)\vartheta_{e-1}^{(1)}$ in $T^1_{X'}$, because

$$\frac{1}{y^{j_{e-1}}}\frac{\partial}{\partial x} = \frac{1}{z_{e-1}}(\ldots + a_{e-2}z_{e-2}\frac{\partial}{\partial z_{e-2}} + z_{e-1}\frac{\partial}{\partial z_{e-1}}) \in \Gamma(U_{r+1},\Theta).$$

The field $u^{-2}\partial_v$ maps to $-\sigma_{e-2} = \frac{1}{z_{e-2}}(\ldots + z_{e-3}\partial_{e-3} + z_{e-2}\partial_{e-2} + (a_{e-2}-1)z_{e-1}\partial_{e-1} + \ldots)$, because we have $a_{e-2} = 2$.

For application of the construction on the left side of the singularity, introducing a new a_2, one finds similar results: the field $u^{-1}\partial_v$ has image $-\vartheta_2^{(1)}$ in $T^1_{X'}$, and $u^{-2}\partial_v$ maps to $-\sigma_3$.

Before the construction we had vector fields, projecting on the relevant $\vartheta_\varepsilon^{(a)}$, σ_ε, $\varepsilon \le e-3$. Their image is computed with local coordinates on Y and two functions z_{e-1}, z_e. One has the same local coordinates on Y' and the same formulae for z'_{e-1}, z'_e. So these vector fields also exist on Y'.

5.11.2 Construction Ib. In this case the vector fields on Y also exist on Y' and are computed with the same formulae. In particular, $\vartheta_{e-2}^{(1)}, \ldots, \vartheta_{e-2}^{(a_{e-2}-k_{e-2})}$ give $\vartheta_{e-2}^{(1)}, \ldots, \vartheta_{e-2}^{(a'_{e-2}-k'_{e-2})}$. The same computations as for **Ia** give vector fields, which map onto $\vartheta_{e-1}^{(2)}, \ldots, \vartheta_{e-1}^{(a_{e-1}-1)}$. Furthermore, one sees that $-u^{-1}\partial_v$ projects onto $(a_{e-1}-1)\vartheta_{e-1}^{(1)}$ and $-u^{-2}\partial_v$ onto $\sigma_{e-2}+(a'_{e-2}-2)\vartheta_{e-2}^{(1)}$; one has $c_{e-2} = k'_{e-2}-2$.

For an operation **Ib**, applied on the left, we have that the analogous $-u^{-2}\partial_v$ projects onto σ_3.

5.11.3 Construction IIa.
i) For the new coordinates $y^{(0)}$ around the singularity Y'_3 of type T one has $z_i = y_i^{(0)}$ for $1 \le i \le 4$, because a_2 and a_3 are equal for X' and Y'_3. So one finds a_3 vectors, projecting onto the $\vartheta_3^{(a)}$. The expression of z_ε, $\varepsilon \ge 4$, in the other local coordinates on Y' is not changed, so we find the same vectors, except for σ_4. We take local coordinates (u,v) with center in the intersection of E_1 and E_2, to compute the fields for $\varepsilon = 4$. If $a_4 = k_4$, we had before the construction exactly one field, projecting onto $\sigma_4 = (\ldots - \partial_4 - (a_4-1)\partial_5 - \ldots)$; this field still exists and projects onto an element with the same formula, but now $(\ldots - \partial_4 - (a_4-1)\partial_5 - \ldots) = (\ldots - \partial_4 - (a'_4-1)\partial_5 - \ldots) + (\ldots + \partial_5 - \ldots) = \sigma_4 + \vartheta_4^{(1)}$. Indeed we have that one curve, determined by a_4 belongs to the singularity of type T.
ii) Before the construction we had $z_{\varepsilon-2} = y_{\delta(\varepsilon)-2}$, $z_{\varepsilon-1} = y_{\delta(\varepsilon)-1}$ and $z_{\varepsilon+1} = y_{\delta(\varepsilon)+1}$. The formulae to compute z_{e-2}, z_{e-1}, and $y_{\delta(\varepsilon)-2}$, $y_{\delta(\varepsilon)-1}$ from a resolution are not changed. Because $a'_{\varepsilon-1} = a'_{\delta(\varepsilon)-1} = 2$ and $a'_\varepsilon = a'_{\delta(\varepsilon)}$, one has also $z'_\varepsilon = y'_{\delta(\varepsilon)}$ and $z'_{\varepsilon+1} = y'_{\delta(\varepsilon)+1}$. So one finds the required vector fields; the computation for $\sigma_{\varepsilon+1} + c_{\varepsilon+1}\cdot\vartheta_{\varepsilon+1}^{(a_{\varepsilon+1}-1)}$ is similar to the previous case.

5.11.4 Construction IIb. By Lemma 5.9 one has for coordinates on the minimal resolution with center in the intersection of E_σ with the (-3) (resp. (-4)) curve the equations $z_{\varepsilon-1} = uv$, $z_\varepsilon = uv^2$ and $z_{\varepsilon+1} = u^{a_\varepsilon-1}v^{2a_\varepsilon-1}$. On the minimal resolution of the newly introduced singularity of type T we can take the corresponding coordinate system; we find that $y_{\delta-1} = uv$, $y_\delta = uv^2$ and $y_{\delta+1} = u^{a_\varepsilon-1}v^{2a_\varepsilon-1}$. For the other singularity of type T involved one can compute coordinates from an unchanged chart at the right-hand side. This completes the proof of Lemma 5.10. Computations of vector fields as before yields Theorem 5.1.

6 Components and equations

6.1. As is well known, the two components of the cone over the rational normal curve of degree four are related to the two different ways of writing the equations:

$$\operatorname{rk}\begin{vmatrix} z_1 & z_2 & z_3 & z_4 \\ z_2 & z_3 & z_4 & z_5 \end{vmatrix} \leq 1 \quad \text{or} \quad \operatorname{rk}\begin{vmatrix} z_1 & z_2 & z_3 \\ z_2 & z_3 & z_4 \\ z_3 & z_4 & z_5 \end{vmatrix} \leq 1.$$

Riemenschneider has remarked that this generalises to all cyclic quadruple points [R1]. One can even give the equations as quasi-determinantals; contrary to the notation in (2.1) we write here the extra entries between those, with which they are to be multiplied. One has for the Artin component:

$$\begin{vmatrix} z_1 & z_2 & & z_3 & z_4^{a_4-1} \\ & & z_3^{a_3-2} & & \\ z_2^{a_2-1} & z_3 & & z_4 & z_5 \end{vmatrix}$$

and for the other component:

$$\begin{vmatrix} z_1 & & z_2 & & z_3^{a_3-1} \\ & z_2^{a_2-2} & & & \\ z_2 & & z_3 & & z_4 \\ & & & z_4^{a_4-2} & \\ z_3^{a_3-1} & & z_4 & & z_5 \end{vmatrix}.$$

In this case one multiplies the second term of the minor with every extra entry in the box determined by the minor. One obtains the equations of the deformation over the corresponding component by perturbing the entries of the matrices. For higher embedding dimension such a determinantal description is only known for the Artin component.

6.2. I will now describe a method to write down the equations of a cyclic quotient singularity from a subdivision of a polygon. We give each vertex v_i a weight $Z_i := z_i^{a_i-2}$; a line connecting the vertices v_i and v_j will get two weights: $w_{i,j}$ and $w_{j,i}$. We define these inductively:

$$w_{i,i+1} = \frac{z_i}{z_{i+1}}, \qquad w_{i+1,i} = \frac{z_{i+1}}{z_i};$$

if in the triangle (v_i, v_j, v_k) with $i < j < k$ the weights on the sides (v_i, v_j) and (v_j, v_k) are already defined, we set:

$$w_{i,k} = w_{i,j}(w_{j,i} Z_j w_{j,k}), \qquad w_{k,i} = (w_{j,i} Z_j w_{j,k}) w_{k,j}.$$

One gets the equations 1 of Section (2.1) by putting $w_{i,i-1} Z_i w_{i,i+1} = 1$; the equations $z_{i-1} z_{j+1} = \ldots$ are obtained by multiplying the weights from $w_{i,i-1}$ up to $w_{j,j+1}$ along the shortest path from v_i to v_j.

Example.

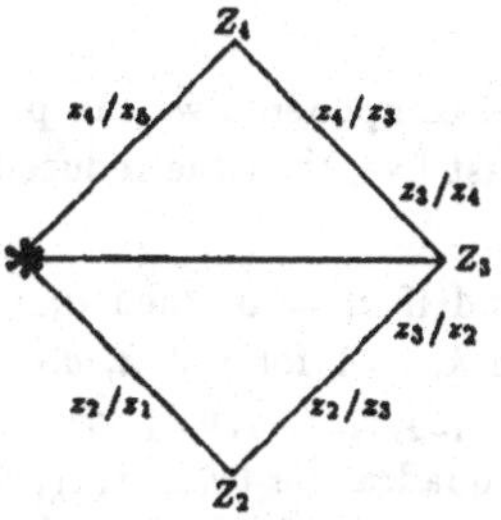

$z_1 z_5 = z_2 Z_2 Z_3 Z_4 z_4$

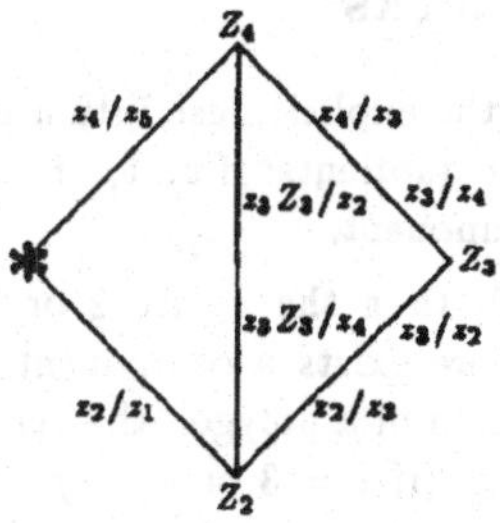

$z_1 z_5 = Z_2(z_3 Z_3)^2 Z_4$

To describe the equations it suffices to know the products $w_{i,j}w_{j,i}$. In the formula enters a continued fraction $[k'_i, \ldots, k'_j]$ with all $k'_\varepsilon \leq k_\varepsilon$. If one collapses all vertices v_ε with $\varepsilon < i$ or $\varepsilon > j$ into the distinguished vertex (so v_j, v_* and v_j are consecutive vertices in this $(j-i+1)$-gon), then the induced subdivision of this $(j-i+1)$-gon determines $[k'_i, \ldots, k'_j]$. It suffices to describe $w_{2,e-1}w_{e-1,2}$ for an arbitrary subdivision.

6.2.1 Lemma. One has:

$$w_{2,e-1}w_{e-1,2} = \frac{1}{z_2^{\alpha_3 - 1}} \left(\prod_{\varepsilon=3}^{e-2} (Z_\varepsilon z_\varepsilon^{2-k_\varepsilon})^{\alpha_\varepsilon} \right) \frac{1}{z_{e-1}^{\alpha_{e-2}-1}}.$$

Proof by induction. We only consider the introduction of a new vertex v_ε between $v_{\varepsilon-1}$ and $v_{\varepsilon+1}$ in the case $\varepsilon \neq 3, e-2$. Without v_ε the product $w_{2,e-1}w_{e-1,2}$ contains a factor

$$\left(\frac{z_{\varepsilon-1}}{z_{\varepsilon-2}} Z_{\varepsilon-1} \frac{z_{\varepsilon-1}}{z_{\varepsilon+1}}\right)^{\alpha_{\varepsilon-1}} \left(\frac{z_{\varepsilon+1}}{z_{\varepsilon-1}} Z_{\varepsilon+1} \frac{z_{\varepsilon+1}}{z_{\varepsilon+2}}\right)^{\alpha_{\varepsilon+1}}.$$

Because in the new situation $w_{\varepsilon-1,\varepsilon+1} = \dfrac{z_{\varepsilon-1}}{z_\varepsilon}\left(\dfrac{z_\varepsilon}{z_{\varepsilon-1}} Z_\varepsilon \dfrac{z_\varepsilon}{z_{\varepsilon+1}}\right)$, this factor has to be replaced by:

$$\left(\frac{z_{\varepsilon-1}}{z_{\varepsilon-2}} Z_{\varepsilon-1} \frac{z_{\varepsilon-1}}{z_\varepsilon}\right)^{\alpha_{\varepsilon-1}} \left(\frac{z_\varepsilon}{z_{\varepsilon-1}} Z_\varepsilon \frac{z_\varepsilon}{z_{\varepsilon+1}}\right)^{\alpha_{\varepsilon-1}+\alpha_{\varepsilon+1}} \left(\frac{z_{\varepsilon+1}}{z_\varepsilon} Z_{\varepsilon+1} \frac{z_{\varepsilon+1}}{z_{\varepsilon+2}}\right)^{\alpha_{\varepsilon+1}} =$$

$$\left(\frac{z_{\varepsilon-1}}{z_{\varepsilon-2}} Z_{\varepsilon-1} \frac{z_{\varepsilon-1}}{z_{\varepsilon+1}}\right)^{\alpha_{\varepsilon-1}} z_{\varepsilon-1}^{-\alpha_{\varepsilon-1}} (z_\varepsilon Z_\varepsilon)^{\alpha_\varepsilon} z_{\varepsilon+1}^{-\alpha_{\varepsilon+1}} \left(\frac{z_{\varepsilon+1}}{z_{\varepsilon-1}} Z_{\varepsilon+1} \frac{z_{\varepsilon+1}}{z_{\varepsilon+2}}\right)^{\alpha_{\varepsilon+1}}.$$

This proves the formula.

6.2.2. One can write down the deformation over the component by perturbing the weights: if $\alpha_\varepsilon = 1$, then $w_{\varepsilon-1,\varepsilon} = (w_{\varepsilon,\varepsilon-1})^{-1} = z_{\varepsilon-1}/(z_\varepsilon + s_\varepsilon)$ and

$$Z_\varepsilon = (z_\varepsilon + s_\varepsilon)^{\alpha_{\varepsilon-1}-1}(z_\varepsilon^{a_\varepsilon - \alpha_{\varepsilon-1}-1} + \ldots + \tilde{t}_\varepsilon^{(a_\varepsilon - k_\varepsilon)} z_\varepsilon^{k_\varepsilon - \alpha_{\varepsilon-1}-1}).$$

Expanding the right hand side of this formula gives the relation between $\tilde{t}_\varepsilon$ and t_ε (cf. [A]). If $\alpha_\varepsilon \neq 1$, one has the formula:

$$Z_\varepsilon = z_\varepsilon^{a_\varepsilon - 2} + \ldots + t_\varepsilon^{(a_\varepsilon - k_\varepsilon)} z_\varepsilon^{k_\varepsilon - 2}.$$

7 Further remarks

7.1. As corollary to the explicit description of the components we can prove Conjecture 7 of [K] in the case of cyclic quotients: if all b_i are at least five, then the reduced deformation space has only the Artin component.

Proof. If all $b_i \geq 5$, then the a_i are 2 or 3, and if $a_i = 3$, then $a_{i-2} = a_{i-1} = a_{i+1} = a_{i+2} = 2$. Suppose there exists a component with $k_i = 1$ for $i \neq 2, e-1$. Consider in the corresponding subdivision of a polygon the vertices $v_{i-2}, v_{i-1}, v_i, v_{i+1}, v_{i+2}$; one of them may be the distinguished vertex (if $i = 3$ or $e-2$). The quadrangle $(v_{i-2}, v_{i-1}, v_{i+1}, v_{i+2})$ has to be subdivided, so for $j = i-1$ or $j = i+1$ one has $k_j = a_j = 3$. Say $j = i-1$. Then $a_{i+1} = 2$, so v_{i-1} and v_{i+2} are connected, as are v_{i-2} and v_{i+2} (because $a_{i-1} = 3$). Since $i - 2 > 2$, the vertex v_{i-2} is connected to at least one other vertex; the same has to hold for v_{i+2}, so $a_{i+2} = 3$. Subdivision of the quadrangle $(v_{i-3}, v_{i-2}, v_{i+2}, v_{i+3})$ leads to a contradiction.

7.2 Dihedral singularities. One has the dual graph:

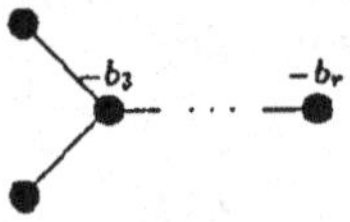

with $b_1 = b_2 = 2$.

7.2.1 Lemma. For a given P-resolution $Y \to X$ at most one of the curves E_1, E_2 belongs to a singularity of type T.

Proof. The minimal resolution $\tilde{Y}$ of Y has two arms which connect E_3 with E_1, respectively E_2. Suppose E_3 belongs to a singularity of type T, which contains also curves on the arm to E_1. Because singularities of type T are cyclic quotients, the curve E_3 intersects either E_2 or a (-1)-curve. In the last case we consider the resolution of the cyclic quotient, consisting of the curves on the arm to E_2 with E_3 replaced by a $(-N)$-curve E_N, followed by $(N-4)$ curves with self-intersection (-2). For N big enough the coefficient k_N of E_N in its singularity of type T is greater than that of E_3. So we have a P-resolution of a cyclic quotient, whose minimal resolution consists of the (-2)-curve E_2, a curve E_N with high self-intersection and $(N-4)$ (-2)-curves. For such a singularity the (-2)-curve E_2 does not belong to a singularity of type T.

7.2.2. So every P-resolution of a dihedral singularity gives a P-resolution of a cyclic quotient with $[b] = [2, b_3, \ldots, b_r]$. Some of these can be realized in two ways on the dihedral singularity.

Example. Let $e = 6$ and $X_a = X(a, 3, 3, 2)$. For $[k] = [3, 1, 2, 2]$ the singularity X_2 has two P-resolutions, whereas X_a, $a > 2$, has only one. The deformation to X_2 is realisable on the corresponding component, so this component of the base of X_a, $a > 2$, has to be singular, with a pinch point like singularity.

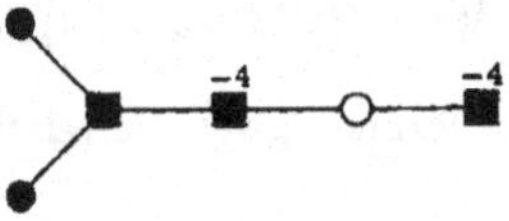

References

[A] Jürgen Arndt, *Verselle Deformationen zyklischer Quotientensingularitäten.* Diss. Hamburg 1988

[B] William G. Brown, *Historical note on a recurrent combinatorial problem.* Am. Math. Monthly **72** (1965), 973–977

[Ca] E. Catalan, *Note sur une équation aux différences finies.* J. Math. pures et appl. **3** (1838), 508–516

[Ch] Jan Christophersen, *Versal deformations of cyclic quotient singularities.* To appear.

[K] János Kollár, *Some conjectures about the deformation space of rational surface singularities.* Preprint 1988

[K-S-B] J. Kollár and N. I. Shepherd-Barron, *Threefolds and deformations of surface singularities.* Invent. math. **91** (1986), 299–338

[L] H.B. Laufer, *Normal two-dimensional singularities.* Princeton University Press: Princeton 1971 (Ann. of Math. studies 71)

[L–W] Eduard Looijenga and Jonathan Wahl, *Quadratic functions and smoothing surface singularities.* Topology **25** (1986), 261–291

[M] D. Mumford, *The topology of normal singularities of an algebraic surface and a criterion for simplicity.* Publ. Math. IHES **9** (1961), 5–22

[N] Eugen Netto, *Lehrbuch der Combinatorik.* Teubner: Leipzig, 1901

[O–W] P. Orlik and P. Wagreich, *Algebraic Surface with k^*-action.* Acta Math. **138** (1977), 43–81

[P] H. Pinkham, *Deformations of quotient surface singularities.* Proc. Symp. Pure Math. **30**, Part 1 (1977), 65–67

[R1] Oswald Riemenschneider, *Deformationen von Quotientensingularitäten (nach zyklischen Gruppen).* Math. Ann. **209** (1974), 211–248

[R2] Oswald Riemenschneider, *Zweidimensionale Quotientensingularitäten: Gleichungen und Syzygien.* Arch. Math. **37** (1981), 406–417

[W] Jonathan Wahl, *Smoothings of normal surface singularities.* Topology **20** (1981), 219–246

Address of the author:
Mathematisches Seminar der Universität Hamburg,
Bundesstraße 55,
2000 Hamburg 13,
Federal Republic of Germany

On Canny's roadmap algorithm: orienteering in semialgebraic sets
(an application of singularity theory to theoretical robotics).

by David Trotman (Aix-Marseille I)

I will describe an application of singularity theory, and in particular an application of Thom-Whitney stratifications of semialgebraic sets, to the generalised mover's problem - the abstract path planning problem in robotics. The geometric construction studied here, due to John Canny, is contained in his M.I.T. thesis, "The complexity of robot motion planning" and was announced at the 28th IEEE Symposium on the Foundations of Computer Science at Los Angeles in 1987 [3,4]. The general set-up is of a configuration space for a moveable object or a robot-arm, and a set of constraints on the possible motion. One wishes to find a continuous path between any two specified configurations , which is collision-free. Usually the constraints will have an algebraic description, so that the obstacle-free subset of the configuration space, known as free-space, will be a semialgebraic set.

Since the work of G.E.Collins [6] , cylindrical algebraic decomposition algorithms have been much studied : one decomposes the free-space into cells on each of which members of a given finite family of polynomials have fixed sign (negative, zero, or positive) and then constructs a path between any two positions by computing cell-adjacency (work of Collins, Ben-Or, Kozen, MacCullum, Prill, Reif, Schwartz, Sharir, Yap, etc.). The time required to carry out these algorithms is however doubly-exponential in the number of variables of the configuration space (its dimension), and this is best possible, as shown in a recent paper of J.H.Davenport and J.Heintz [7] .

Theorem (Davenport-Heintz,1988): *The time required to decompose* $\mathbb{R}^{6k+2}$ *cylindrically according to* $8k+2$ *polynomials of degree at most* 4 *is at least* $\exp(2^{k+1}\log 2)$.

The interest of Canny's method, which uses singularity theory, including Morse functions, 2-generic maps to the plane, Whitney stratifications of semialgebraic sets, and the Thom-Mather isotopy theorem, is that it provides a single exponential time algorithm. Also a recent paper of Canny [5] shows that the primitive element theorem of classical algebra can be used to perform the geometric algorithm described below in PSPACE, i.e. with polynomial-size storage space. Hence this work of Canny is particularly exciting to specialists in the complexity of geometric algorithms (see Sharir [15]), and those of us interested in the stratification theory of semialgebraic sets might well pay attention to these novel developments. The algebraic procedures described by Canny, used to perform the geometric construction of the semialgebraic paths (or roadmaps in Canny's terminology) will not be discussed here. My aim is to publicise Canny's application of singularity theory, to provide proofs that the roadmap constructions work (some of Canny's arguments are incorrect or incomplete) and to discuss the applicability of his method — see our objections at the end of this paper — from the viewpoint of a specialist in stratification theory.

I thank Jean-Jacques Risler, Bernard Teissier, Ricardo Benedetti, Jean-Pierre Françoise, Joel Robbin and Marie-Françoise Coste-Roy for enlightening discussions.

Aim. Suppose we are given a compact semialgebraic set S in $\mathbb{R}^r$. We want a procedure for defining a one-dimensional semialgebraic subset R of S, which is connected in each connected component of S, and can be determined quickly from the data defining S. Moreover one wants a canonical and quick procedure for reaching R from an arbitrary given point of S . Then one will have an algorithm for travel between any two points on S. The question of reducing the time of travel is not considered here, although Canny has also improved on previous results on this question. A general theory of geodesics on semialgebraic sets awaits its Maker, and seems very hard with present tools.

Silhouettes and roadmaps

Let S in $\mathbb{R}^r$ be a compact semialgebraic set, with a given Whitney stratification, denoted by $\underline{S}$. Let $\pi_{12} : \mathbb{R}^r \dashrightarrow \mathbb{R}^2$ be a linear map, where the notation is chosen to signify that there are coordinates $x_1,\ldots,x_r$ such that $\pi_{12}(x_1,\ldots,x_r) = (x_1,x_2)$.

Definition. The *silhouette* of $\underline{S}$ with respect to π_{12} is

$$\Sigma = \Sigma(\pi_{12}|_{\underline{S}}) = \bigcup \{\Sigma(\pi_{12}|_X) : X \in \underline{S}\}$$

the union of the critical sets of π_{12} restricted to the strata of $\underline{S}$. Then Σ is compact and semialgebraic . A theorem of Mather [11] (or a result of Teissier [16]) implies that for a generic set of linear maps $\pi_{12} : \mathbb{R}^r \dashrightarrow \mathbb{R}^2$, $\Sigma(\pi_{12}|_X)$ is a smooth one-dimensional submanifold of X , for each stratum X of $\underline{S}$. Note also that

$$\Sigma(\pi_{12}|_X) = \Sigma(\pi_{12}|_X) \cup \{\Sigma(\pi_2|_{X(v)} : v \in \mathbb{R}\}$$

where $X(v) = X \cap \pi^{-1}(v)$.

<u>Remarks</u>

1. The 1-skeleton of the stratification, i.e. the union of the strata of dimensions 0 and 1 , is always in Σ .

2. The silhouette Σ is not well-defined by $\underline{S}$, but depends on the choice of linear projection $\mathbb{R}^r \dashrightarrow \mathbb{R}^2$.

3. The silhouette is not always connected.

Definition. A one-dimensional semialgebraic subset R of a semialgebraic set S satisfies the roadmap condition if each connected component of S contains precisely one connected component of R (in particular R is connected if S is connected).

If, in addition, R contains the silhouette Σ of S with respect to some (generic) linear projection $\mathbb{R}^r \dashrightarrow \mathbb{R}^2$, R will be called a roadmap.

Thus Σ itself will be a roadmap if it is connected within each connected component of S. However because Σ need not be connected, we are thus led to give explicit constructions of roadmaps.

Note. The existence of some subset R satisfying the roadmap condition is trivial. Just take a small arc or circle on a stratum. However if we impose that the silhouette Σ is contained in R, any explicit construction is nontrivial. It is because Σ is contained in the two roadmaps R_0 and R_1 defined below that one can construct a path from an arbitrary point of S to the roadmap (in fact, to Σ) in linear time.

Construction of the basic roadmap $R_0(\underline{S})$

(i) If dim S = 1, set $R_0(\underline{S}) = S$.

(ii) If dim S ≥ 2, let $\underline{S}'$ denote the smallest Whitney stratification of S compatible with both $\underline{S}$ and the canonical stratification of Σ. Let $V = \{$ critical values of $\pi_1|\underline{S}'\}$. If π_1 is sufficiently generic (see below) then V will be a finite set, and critical values can be assumed distinct. Now set

$$R_0(\underline{S}) = \Sigma \cup \{ R_0(\underline{S}' \cap \pi_1^{-1}(v)) : v \in V \}.$$

Thus we have an inductive definition of $R_0(\underline{S})$, since

$$\dim (S \cap \pi_1^{-1}(v)) = \dim S - 1 .$$

However one needs to show that $R_0(\underline{S})$ is well-defined in the sense that $\underline{S}' \cap \pi_1{}^{-1}(v)$ inherits a canonical Whitney stratification from $\underline{S}'$. Of course $R_0(\underline{S})$ will depend topologically on the choices of linear projection π_{12} and π_{23} ,etc.(following the inductive definition down), so is not uniquely determined, even topologically. However standard stratification arguments should allow one to prove topological stability with respect to small changes in the linear maps used.

$R_0(\underline{S})$ makes sense.

By a theorem of Mather [11] , there exists a generic set of linear projections $\pi_{12} : \mathbb{R}^r \dashrightarrow \mathbb{R}^2$ such that $\pi_{12}|_X$ is locally infinitesimally stable for all strata X of $\underline{S}'$. This uses that the number of strata is finite, because S is compact and semialgebraic, and also that (r,2) belongs to the nice dimensions (n,p) for which Mather's theorem is true. Now another theorem of Mather (in Stability of C^∞ Mappings IV [10,12]), shows that if a smooth map f between smooth manifolds is infinitesimally stable, then $\Sigma(f)$ is smooth, and $f|_{\Sigma(f)}$ is a finite map. Thus $\pi_{12}|_\Sigma$ will have finite fibres. (Alternatively one might adapt Teissier's proof of the analogous result for the polar variety defined by generic projection of a complex analytic variety onto $\mathbb{C}^2$ — see [16, V.1.2.2(Lemme clé)].)

Now choose a linear map $p_1 : \mathbb{R}^2 \dashrightarrow \mathbb{R}$ which is a stratified Morse function with respect to the semialgebraic stratified set $\pi_{12}(\Sigma)$ in $\mathbb{R}^2$ (this uses the Tarski-Seidenberg principle, stratifiability of semialgebraic sets, and the density, hence existence, of stratified Morse functions among linear maps to $\mathbb{R}$ on a semialgebraic set — see Pignoni [13], and Goresky-MacPherson [8]). Finally set

$$\pi_1 = p_1 \circ \pi_{12} .$$

Then the fibres $\pi_1{}^{-1}(v)$ will be transverse to the strata of $\underline{S}'$ off Σ , and $\pi_1|_\Sigma$ is a finite map.

Now adding the finitely many points of $\Sigma \cap \pi_1^{-1}(v)$ to $\underline{S}'$, where v is a critical value of $\pi_1|_{\underline{S}'}$, we get a Whitney stratification of $S \cap \pi_1^{-1}(v)$ compatible with $\underline{S}' \cap \pi_1^{-1}(v)$ (and differing from it precisely by this finite set of point-strata). This uses the semialgebraicity of $\underline{S}'$. It presents only mild calculation difficulties. (See below for a discussion of the Whitney stratifications of Canny's algorithm and their calculation).

Theorem. *$R_0(S)$ is a roadmap.*

Proof. We must show, by definition,that $R_0(\underline{S})$ is connected in each component of S, since $\Sigma \subset R_0(\underline{S})$. First, order the critical values of $\pi_1|_{\underline{S}'}$: $v_1 < v_2 < v_3 < \ldots < v_m$. Set $u_0 = v_1$, $u_i = (v_i + v_{i+1})/2$ $(1 \leq i \leq m-1)$, and $u_m = v_m$. Now the Goresky-MacPherson technique of Moving the Wall (from their book, "Stratified Morse Theory" [8]) shows, using the Thom-Mather (stratified) isotopy theorem [9,17] for Whitney stratified sets, that $S \cap \pi_1^{-1}(v_i)$ is a deformation retract of $S \cap \pi_1^{-1}([u_{i-1}, u_i])$ by a stratum-preserving deformation Φ_t, in particular respecting the strata of the silhouette Σ. Note also that the silhouette of the fibre $\pi_1^{-1}(v_i)$ contains the fibre of the silhouette.

Lemma: *If for all critical values v of $\pi_1|_{S'}$, $R_0(S' \cap \pi_1^{-1}(v))$ is a roadmap, then $R_0(S)$ is also a roadmap.*

Proof. We first prove the lemma with S replaced by $S \cap \pi_1^{-1}([u_{i-1}, u_i])$, which we write as $S|^{[u_{i-1}, u_i]}$, following Canny. Let C^{ij} be a connected component of the slab $S|^{[u_{i-1}, u_i]}$. Because π_1

has only a single critical value v_i in the slab,

$$\Phi_1(C^{ij}) = C^{ij} \cap \pi_1{}^{-1}(v_i) = C^{ij}|^{v_i} .$$

A continuous image of a connected set is connected, hence $C^{ij}|^{v_i}$ is a connected component of $S|^{v_i}$, and so, by hypothesis of the lemma, $C^{ij}|^{v_i}$ contains a single (nonempty) connected component of $R_0(\underline{S}'\cap\pi_1{}^{-1}(v_i)) = R_0(\underline{S}'|^{v_i})$. But, by definition of $R_0(\underline{S})$,

$$R_0(\underline{S}) \cap C^{ij} = (\Sigma \cup R_0(\underline{S}'|^{v_i})) \cap C^{ij} ,$$

which is connected since points on $\Sigma \cap C^{ij}$ are deformed along arcs in Σ onto points on $\Sigma \cap C^{ij} \cap \pi_1{}^{-1}(v_i)$ (so in the fibre of the silhouette), which belong to $R_0(S'|^{v_i})$ (the silhouette of the fibre) , and this is connected, by induction on dimension.

It remains to extend to the general case. If C is a connected component of S , then C can be written as a finite union of C^{ij} 's , $C_1, \ldots, C_N$ say. Order the closed sets $C_1, \ldots, C_N$ such that if $j \geq 1$, then there exists $k < j$ such that $C_k \cap C_j \neq \emptyset$, and then , because $C_k \cap C_j$ is compact, $\pi_2|_{C_k \cap C_j}$ will admit an extremal point which will be in the silhouette $\Sigma(\pi_{12}|_{\underline{S}'})$ hence in $R_0(\underline{S}) \cap C_k \cap C_j$. Thus, applying the lemma, $R_0(\underline{S}) \cap C_1, \ldots, R_0(\underline{S}) \cap C_N$ is an ordered sequence of connected sets each of which meets an earlier element. It follows that $R_0(\underline{S}) \cap C$, which is the union of $\{ R_0(\underline{S}) \cap C_i : 1 \leq i \leq N \}$, is connected. This completes the proof that $R_0(\underline{S})$ is a roadmap.

Note. It is clear that for each stratum X of $\underline{S}$, $Cl(X) \cap R_0(\underline{S})$ is a roadmap for the semialgebraic set $Cl(X)$. Moreover, given any one-dimensional semialgebraic subset R of S with the property that $Cl(X) \cap R$ is a roadmap of $Cl(X)$ for every stratum X , then the previous connectivity argument implies that R is a roadmap for $\underline{S}$.

(Components of S are unions of finitely many closures of strata $\{Cl(X_i)\}$ of $\underline{S}$ which can be ordered so that each $Cl(X_j)$ intersects an earlier $Cl(X_k)$. The intersection $Cl(X_j) \cap Cl(X_k)$ contains a stratum of $\underline{S}$ and its connected roadmap, and hence the connected roadmaps of $Cl(X_j)$ and $Cl(X_k)$ intersect. It follows that R is a roadmap for $\underline{S}$.)

Thus to construct a roadmap for $\underline{S}$ it would be enough to do so for the closure $Cl(X)$ of a given stratum X. It turns out that constructing linking curves between components of the silhouette of a single $Cl(X)$ provides a more efficient algorithm. Canny has a particular way of doing this, to give what he denotes by $R_1(\underline{S})$, the full roadmap, defined below.

Remark. Another way to define a roadmap for S is to set

$$R(\underline{S}) = \bigcup \{ R_0(Cl(X)) : X \in \underline{S} \}.$$

This clearly is contained in $R_0(\underline{S})$, but will in general be smaller, because one will retain in $R_0(\underline{S}' \cap \pi_1^{-1}(v))$ only those pieces contained in the closures of strata adherent to the critical point p whose image is v. By the previous discussion $R(\underline{S})$ will be a roadmap. However, there may still be superfluous pieces of the roadmap $R(\underline{S})$ not needed to ensure connectivity. Canny gives a more detailed construction defined locally near each critical point of $\pi_1|_{\underline{S}'}$, adding just enough linking curves to connect the critical point with some distinct point both on the silhouette and in the same fibre of π_1, for each component of the silhouette in adjacent strata.

Linking Curves

These are defined successively by order of complexity of the situation, and later ones depend on previous constructions, so pay attention !

Step I.

First, let p be an arbitrary point in the closure of a stratum U, and let $v = \pi_1(p)$. Write $Cl(U)^v = Cl(U) \cap \pi_1^{-1}(v)$, and stratify $Cl(U)^v$ by $\underline{S} \cap \pi_1^{-1}(v)$ and $\{p\}$, using the genericity of π_1. Then $Cl(U)^v$ inherits a Whitney stratification from that of $\underline{S}$, written $Cl(\underline{U})^v$.

Define $L_0(p,U)$ to be the connected component of $R_0(Cl(\underline{U})^v)$ containing p, where $R_0(Cl(\underline{U})^v)$ is obtained by a generic choice of π_{23}, etc. Note as before in the construction of $R_0(\underline{S})$, that $R_0(Cl(\underline{U})^v)$ is not even topologically well-defined, but will depend on the choices of generic projections π_{23}, π_{23}, etc.

We wish to obtain a point $q \in L_0(p,U) \cap \Sigma(Cl(U))$. Canny's argument is to choose a point p_1 in $L_0(p,U)$ extremal with respect to π_2, so that either p_1 is in $\Sigma(U)$ or p_1 belongs to some frontier stratum U_1 in $Cl(U) - U$.

If p_1 belongs to a frontier stratum U_1, then $L_0(p_1,U_1)$ will contain a point p_2 which is extremal with respect to π_2, so that either p_2 is in $\Sigma(U_1)$ or p_2 belongs to a frontier stratum U_2 in $Cl(U_1) - U_1$. This process terminates after finitely many steps, and we define

$$L_1(p,U) = L_0(p,U) \cup \bigcup \{ L_0(p_i,U_i) : 1 \leq i \leq k-1 \}$$

where k is the smallest integer such that $p_i \in \Sigma(Cl(U))$ with p_i in $L_0(p_{i-1},U_{i-1})$. This is defined since $\dim U_i < \dim U_{i-1}$, and if $\dim U_i$ is 0 or 1, U_i is contained in $\Sigma(Cl(U))$. Then $L_1(p,U)$ is a connected one dimensional subset of $\pi_1^{-1}(v)$ containing p and some silhouette point q in $\Sigma(Cl(U))$.

Step II. Suppose now that p is a critical point of $\pi_1|_U$ of Morse index not equal to 0 or $\dim U$. For each connected component C_i of

$L_0(p,U) - \{p\}$, let q_i be an extremal point of C_i with respect to π_2 , and let V_i be the stratum containing q_i if $q_i \notin \Sigma(U)$. Then set

$$L_2(p,U) = L_0(p,U) \cup \bigcup \{ L_1(q_i,V_i)\}$$

where the union is over i such that $q_i \notin \Sigma(U)$. Again $L_2(p,U)$ is a connected one dimensional semialgebraic subset of $\pi_1^{-1}(v)$ containing p and a point of $\Sigma(U)$. Moreover $L_2(p,U)$ connects p , via each connected component C_i of $L_0(p,U) - \{p\}$, to the silhouette $\Sigma(Cl(U))$, and by curves within $\pi_1^{-1}(v)$.

<u>Step III.</u> Suppose finally that p is a critical point of $\pi_1|_U$ of index 0 or $\dim U$, so that $U \cap \pi_1^{-1}(v) = \{p\}$ in some neighbourhood of p . Let $\{W_j\}$ be the strata of $\underline{S}$ with $p \in Cl(W_j \cap \pi_1^{-1}(v))$, if there are any such strata. Then set

$$L_3(p,\underline{S}) = \bigcup \{ L_1(p,W_j)\}.$$

Thus p will be linked to the silhouette of each stratum adherent to p in $\pi_1^{-1}(v)$. There remain two cases, when no such W_j exist. In the case when p is a critical point of index 0 , a new connected component is created at p as one allows the values of π_1 to increase through $\pi_1(p)$. When p is a critical point of maximal index on U , an isolated connected component in previous slices disappears as values increase through $\pi_1(p)$. In both these situations p is on an arc of the silhouette of U connecting it to other critical slices. We set $L_3(p,\underline{S})$ equal to the empty set.

<u>The "Full" Roadmap $R_1(S)$</u>

Let P denote those critical points of $\pi_1|_{\underline{S}}$, <u>not</u> of the form ∂A where A is a one-dimensional arc of Σ . Then set

$$R_1(\underline{S}) = \Sigma \cup \bigcup \{ L(p,\underline{S}) : p \in P \}$$

where $L(p,S)$ is $L_1(p,U)$, $L_2(p,U)$, or $L_3(p,\underline{S})$ according as whether p is regular for $\pi_1|_U$, a critical point of index different from 0 or dim U, or a critical point of index 0 or dim U.

It remains to prove that $R_1(\underline{S})$ is a roadmap. We shall follow Canny's method of proof, adding details when necessary.

Let $v_1 < v_2 < \ldots < v_{i-1} < v_i < \ldots < v_m$ be the critical values of $\pi_1|_{\underline{S}'}$ and p_i the unique critical point such that $\pi_1(p_i) = v_i$. Suppose that we have proved that $R_1(\underline{S})|^{(-\infty,v_{i-1}]}$ is a roadmap for $S|^{(-\infty,v_{i-1}]}$, and that for each stratum U of $\underline{S}$ the restriction of $R_1(\underline{S})$ to $Cl(U|^{(-\infty,v_{i-1}]})$ is a roadmap. We will show that the restriction of $R_1(\underline{S})$ to each $Cl(U|^{(-\infty,v_i]})$ is a roadmap; this impies that $R_1(\underline{S})|^{(-\infty,v_i]}$ is a roadmap, as in the case of $R_0(\underline{S})$, and then the general result follows by induction on the number of critical values.

If U is a stratum of $\underline{S}$ such that $p_i \notin Cl(U)$, the (stratified) isotopy theorem allows one to deform $Cl(U|^{(-\infty,v_i]})$ to $Cl(U|^{(-\infty,v_{i-1}]})$ by a stratified isotopy compatible with Σ. In particular, connected components of $Cl(U|^{(-\infty,v_i]})$ will be preserved. So we may assume that $p_i \in Cl(U)$.

Now suppose that $p_i \in \partial A_1$ where A_1 is an arc of Σ. Then p_i belongs to a stratum V distinct from the stratum containing A_1 because $\Sigma|_U$ is a union of smooth one-dimensional submanifolds (arcs or circles), for every stratum U of $\underline{S}$. In V, $p_i \in \Sigma(\pi_{12}|_V)$ and so p_i belongs to (the interior of) a one-dimensional component A_2 of $\Sigma(\pi_{12}|_V)$. Now π_1 is a stratified Morse function on $\underline{S}'$, and hence $d(\pi_1|A_2)(p_i) \neq 0$. It follows that p_i is linked to $\underline{S}|^{(-\infty,v_{i-1}]}$ by the curve A_2 on the silhouette and in $Cl(U)$, and hence the roadmap property for $R_1(\underline{S}|^{(-\infty,v_i]})$ reduces to that for $R_1(\underline{S}|^{(-\infty,v_{i-1}]})$, which we assume holds by induction.

Next let p_i be a regular point of $\pi_1|_U$ (possible for cusp points of π_{12} for example). In $R_1(\underline{S}|^{(-\infty,v_i]})$ we have $L_1(p_i,U)$ which is a connected one-dimensional semialgebraic set linking p_i to a silhouette point q of U or some frontier stratum of U . The silhouette links q , and hence p_i in $R_1(\underline{S}|^{(-\infty,v_i]})$ to $R_1(\underline{S}|^{(-\infty,v_{i-1}]})$, and the roadmap condition follows.

Now we consider the case of a critical point p_i of $\pi_1|_U$ of index different from 0 or $\dim U$. The definition of $L_2(p_i,U)$ ensures that p_i is joined to a distinct point in the silhouette via <u>each</u> connected component of $L(p_i,U) - \{p_i\}$, ensuring connectivity.

One concludes similarly in the remaining case, when p_i is a critical point of $\pi_1|_U$ of index 0 or $\dim U$.

Thus $R_1(\underline{S})$ has been shown to be a roadmap.

<u>The problem of constructing a Whitney stratification of a given semialgebraic set.</u>

We have assumed above that our given compact semialgebraic set S already has a known given Whitney stratification , and it is with respect to this stratification that the silhouette and roadmap constructions have been made. Canny remarks that given any semialgebraic set in $\mathbb{R}^n$ defined by polynomials $f_1 ,\ldots, f_N$, the addition of arbitrarily small real numbers $\varepsilon_1,\ldots, \varepsilon_N$ (possibly all equal) to the f_i , will make the map $F + \varepsilon = (f_1+ \varepsilon_1 ,\ldots, f_N + \varepsilon_N) : \mathbb{R}^n \longrightarrow \mathbb{R}^N$ transverse to the canonical stratification of $\mathbb{R}^N$ by coordinate hyperplanes and their intersections, and one can then consider the pullback stratification into sign-invariant sets of the new semialgebraic set S_ε defined by $(F + \varepsilon)$. He then assumes that all the semialgebraic sets to which the algorithm is applied are already presented in this way, i.e. the defining polynomials define a map $F = (f_1, \ldots, f_N)$ transverse to the canonical coordinate stratification of $\mathbb{R}^N$, and that S is stratified by the sign-invariant sets defined by pulling back the strata in $\mathbb{R}^N$ .This is unsatisfactory in that one cannot treat arbitrary semialgebraic sets successfully like this. The topology of S_ε can be quite different from that of S . Connectivity properties will change in general, thus affecting the validity of the algorithm. As we have shown, the geometric construction works for any Whitney stratification , or any Bekka stratification ([1], [2]) for that matter, so that one would like an algorithm producing such a stratification for a given semialgebraic set. Recent work of Bekka is promising here. Bekka regularity is weaker than Whitney (b)-regularity, but does imply (a)-regularity, so one does at least need a canonical process giving an (a)-regular stratification . Specialists in geometric algorithms say that even an algorithm for the singular stratification of an algebraic set is not efficiently realisable at present, and this would be the first step in any process giving a regular stratification. One might be able to use Teissier's characterisation of Whitney regularity as equimultiplicity of polar varieties (generalised silhouettes!) in the complex analytic case. Certainly there is no need to find the <u>minimal</u> Whitney stratification (cf. my calculations in [17]) ; it is easier to complexify and to use the highly developed complex theory where there are algebraic criteria.

Note that it is not enough to consider the singular stratification of S for $R_0(\underline{S})$ to be a roadmap . Consider the union S of a sphere and a tangent circle of distinct radius in $\mathbb{R}^3$. Then $R_0(\underline{S})$ is not connected for generic projections.

Objections to Canny's algorithm.

1. The algorithm is not applicable to arbitrary semialgebraic sets : it applies only to semialgebraic sets in "general position" , hence with a prescribed Whitney stratification, or to those whose Whitney stratification is already known. Checking explicitly that one is in general position will take time, perhaps too much time. Presumably one could ignore this, applying the algorithm without knowing if one is in general position, and justifying the procedure by saying that in most cases (say "generically", i.e. off a nowhere dense subset of the semialgebraic sets of given diagram) one will be in general position.

2. Canny states that he uses the coarsest possible stratification . But this is so only when the semialgebraic set is defined by N polynomials in general position and one stratifies by pulling back the coordinate hyperplane stratification of $\mathbb{R}^N$. A stratification not satisfying Whitney regularity may be locally topologically trivial. In particular the recent (1988) Bekka regularity condition [1,2] is enough to ensure that the appropriate versions of the Thom-Mather isotopy theorems are valid, and Bekka regularity is much weaker than Whitney regularity. Thus the minimal Bekka stratification will have less strata, and be a coarser stratification.

3. Canny observes that transversality of the map F , defined by the defining polynomials associated to the semialgebraic set, to the coordinate hyperplanes and their intersections ensures Whitney regularity of the pullback stratification. In fact this pullback stratification will be much better than just Whitney regular : it will be locally smoothly, even algebraically, trivial along strata. This does not change the nature of the proof of the connectivity of the roadmaps, but it does allow one to justify the construction without using the difficult Thom-Mather isotopy lemmas.

4. The roadmap constructions, and even the notion of a one-dimensional silhouette, depend on being able to choose, or produce,

a generic linear map onto $\mathbb{R}^2$. Here the objection is similar to 2. Determining when a particular linear map is (sufficiently) generic may be expensive in time. And the calculation of the time required has not been carried out by Canny : one would have to determine explicitly once and for all the (equations of the) relevant Thom-Boardman strata and then provide an algorithm determining when one is transverse to these strata. Because Canny does not attempt to do this one must assume that he implicitly invokes the "generic" nature of the maps required : one can take any linear map at random, and most of the time (practically all of the time) it will be sufficiently generic.

5. Canny claims that it is enough that π_{12} be one-generic. But one needs two-genericity to ensure that $\pi_{12}|_{\Sigma}$ is a finite map, and this was used to prove the inductive step in the definition of $R_0(\underline{S})$. One is then led to use two deep theorems of Mather here. Alternatively one could perform an explicit calculation showing genericity of linear projections with the right properties, similar to Teissier's argument in the theory of polar varieties [15].

6. Canny claims to have constructed an algorithm for an arbitrary algebraic set, with the proviso that a small generic perturbation of the defining polynomials allows one to obtain easily a Whitney stratification. But the generic perturbation will change the topology, and in particular the number and the nature of the connected components will change. Points previously in the same connected component of S will cease to be so. A counter-argument might be that such changes are small, near boundary points, and so in practice they will make little difference to the realisable motions.

References

1. K. Bekka, *Propriétés métriques et topologiques des espaces stratifiés*, Thèse, Orsay, 1988.

2. K. Bekka, C-régularité et trivialité topologique, *Warwick Singularity Theory Symposium* 1989, to appear.

3. J. Canny, *The complexity of robot motion planning*, M.I.T. Thesis, 1986, ACM doctoral dissertation series, MIT Press, Cambridge, Mass. 1988.

4. J. Canny, A new algebraic method for robot-motion planning and real geometry, *Proceedings of the 28th IEEE Symposium on the Foundations of Computer Science, Los Angeles, 1987*, pp.39-48.

5. J. Canny, Some algebraic and geometric computations in PSPACE, *20th Annual ACM Symposium on Theory of Computing, 1988*, pp.460-467.

6. G. E. Collins, Quantifier elimination for real closed fields by cylindrical algebraic decomposition, *Lecture Notes in Computer Science, No.33*, Springer-Verlag, New York, 1975.

7. J. H. Davenport and J. Heintz, Real quantifier elimination is doubly exponential, *Journal of Symbolic Computation 5* (1988), pp.29-35.

8. M. Goresky and R. MacPherson, *Stratified Morse Theory*, Springer, Ergebnisse, 1988.

9. J. Mather, *Notes on topological stability*, Harvard University, 1970.

10. J. Mather, Stability of C^∞ Mappings : IV, Classification of stable germs by $\mathbb{R}$-algebras, *Publ. Math. I.H.E.S. 37* (1970), pp.223-248.

11. J. Mather, Generic projections, *Annals of Math. 98* (1973), pp.226-245.

12. J. Mather, How to stratify mappings and jet spaces, *Singularités d'Applications Différentiables, Plans-sur-Bex 1975, Lecture Notes in Math. 535,* Springer, Berlin-New York, 1976, pp.128-176.

13. R. Pignoni, Density and stability of Morse functions on a stratified space, *Ann. Scuo. Norm. Sup. Pisa (4) 4*, 1979, pp.592-608.

14. J. T. Schwartz and M. Sharir, Motion planning and related geometric algorithms in robotics, *Proceedings of the International Congress of Mathematicians, Berkeley 1986, Vol.2*, A. M. S. 1987, pp.1594-1611.

15. M. Sharir, Algorithmic motion planning in robotics, *Computer,* March 1989, pp. 9-20.

16. B. Teissier, Variétés Polaires II. Multiplicités polaires, sections planes et conditions de Whitney, *Algebraic Geometry Proceedings, La Rabida 1981, Lecture Notes in Mathematics 961,* Springer-Verlag, New York, 1982, pp.314-491.

17. R. Thom, Ensembles et morphismes stratifiés, *Bull. Amer . Math. Soc. 75,* 1969, pp.240-284.

18. D. Trotman, On the canonical Whitney stratification of algebraic hypersurfaces, *Séminaire sur la Géometrie Algébrique Réelle (sous la direction de J.-J. Risler), Publications Mathématiques de l'Université Paris VII , No. 24,* 1987, pp. 123-152.

Added in proof: I have added two references together with an extract from a letter from J. Canny dated February 8th 1990. The reference [20] is confirmation of a single exponential time "roadmap algorithm", by a different method based on techniques of Grigoriev and Vorobojov.

Here are the comments of John Canny about the various objections:

"1. I didn't assume the input polynomials were in general position. A random perturbation is always applied to the constant coefficients, which gives general position with probability one (this is a loose analytical argument which a priori might not seem to apply when the perturbations are rational numbers of a certain size, but see (ii) below).

2. It is coarsest possible after the perturbation.

3. OK. In an earlier proof [19], I didn't use them but I was trying to simplify!

4. There is another way to bound the time, which uses algebraic arguments, and gives numerical bounds on the size of random integers and probability of

success. The original roadmap algorithm does not attempt to verify the genericity of the map. It is a "Monte-Carlo" algorithm, which gives the right answer with some probability, but may also give the wrong answer. The running time of such an algorithm is proportional to the log of the failure probability. But notice that since good maps are dense (in the semi-algebraic context) it suffices to show that the algorithm gives the same answer on an open neighbourhood of maps about any given map. This is the basis of the algebraic approach.

5. Some very explicit proofs of genericity were given in a paper I wrote before I knew about stratifications or generic maps ([19], from an Oxford workshop on mathematical methods in robotics from the summer of 1986).

6. Hmmm. I never claimed that you got the right topology of a semi-algebraic set when you do the perturbation. A number of mathematicians have raised this objection, but it comes from thinking that I was trying to solve a different problem. In the context of motion planning, which is the problem I dealt with, it is completely valid, as I explained in the thesis, and in (i) below.

Note that the roadmap algorithm also assumes compactness. In fact in the extension to general SA sets, non-compactness is much harder to deal with than singular sets.

It seems that these objections spring from two problems, which are the following:

(i) The perturbations of the polynomials can change the topology. Yes, but if small changes DO change the topology, a robot has no business moving through the disputed territory. One must keep in mind where the semi-algebraic sets are coming from. They derive from the constraints between features of a robot and obstacles. Real robots have limited accuracy, and our knowledge of the placement and dimensions of obstacles is also crude. Even if these uncertainties are taken to be very small, we can also choose perturbations of the constraint polynomials much smaller than the uncertainties. At this scale, the question of whether the robot can move in a collision-free path does not have a unique answer. So no matter what the resolution of the problem is, we can make our perturbation smaller (at the cost of writing down the small number) and give an answer correct at that resolution. In fact we should always err on the side of safety, and grow the obstacles a little, so that any path found will be truly collision-free. This argument doesn't seem to satisfy everyone, but in the robotic context, it is really overkill. One can solve an abstract problem which is much more stringent than it needs to be given the physical uncertainties.

(ii) There was no analysis of how to verify good maps. This is true, but I was not trying to test this, the algorithm merely tries a map and gives an answer, which will be correct with high probability. However, it is not difficult to make this deterministic. Again, the trick is not to verify the map, merely to show that the algorithm gives the same answer on an open set. This makes use of the algebra.

Take the linear projection map(s) to be symbolic, i.e. a matrix whose entries are indeterminates A_{ij}. The basic roadmap algorithm is largely symbolic, i.e. symbolic means no branching is involved. The projection step and slice generation for the basic algorithm (without link curves) is purely symbolic. The projections are planar curves, whose coefficients will be polynomials in A_{ij}. The curve ordering algorithm is not symbolic, but the number of polynomials that might ever be generated down all branches is small. Finally, once can bound the number of polynomials generated by the BKR algorithm on polynomials with symbolic coefficients. Overall, one gets an admissible bound (single exponential) on the total number of polynomials that might ever be generated by the algorithm for any choice of the projection maps. One can compute all these polynomials in A_{ij}'s, and then the task is to find a value for the A_{ij}'s such that the algorithm gives the same result on a neighbourhood. But this means simply finding a set of values for which none of the polynomials is zero, since only the signs of these polys matters to BKR. Since the polynomials can be explicitly computed by "unravelling" the algorithm (following all branches), a non-root can be found in admissible time. This must be a good projecion map. The time required for the algorithm overall is $(kd)^{O(n^4)}$, where n=dimension, k=number of constraints, d=degree. This also gives a bound for the size of numbers required by the random version of the algorithm, which is $k^n d^{O(n^2)}$ as originally claimed."

Additional references:

[19] J. Canny, Constructing roadmaps of semi-algebraic sets I: Completeness, *Artificial Intelligence*, *37*, 1988, pp. 203-222

[20] J. Heintz, M.-F. Roy, P. Salerno, Single exponential path finding in semialgebraic sets, manuscript, 1990.

Unité de Recherche Associé 225 du C.N.R.S.,

Département de Mathématiques,

Université de Provence,

3 Place Victor Hugo,

F-13331 Marseille Cedex 3

Elliptic complete intersection singularities

C.T.C. Wall

The object of this paper is to study the class of singularities described in the title. The fundamental results here are due to Laufer [7] and Reid [10], and we use these freely. We first derive a list of the singularities, then study their defining equations. In the weighted homogeneous cases, we also list the weights.

The motivation for doing this is to understand this class of singularities better, to seek characterisations from other viewpoints, and to collect data for further study (e.g. topology of the Milnor fibre, monodromy, deformations, real forms). Two characterisations are in fact obtained.

Let us say that X is *prepared* if X is either a hypersurface singularity in $\mathbb{C}^3$ of multiplicity 3, or the intersection of two hypersurfaces of multiplicity 2 in $\mathbb{C}^4$ with distinct tangent cones. According to [7] and [9] any elliptic complete intersection singularity is either a hypersurface of multiplicity 2 or is prepared.

THEOREM 1 (see §4). *Suppose* X *is prepared; let* $p:\tilde{X} \to X$ *be defined by blowing up the maximal ideal. Then* X *is elliptic if and only if* $\tilde{X}$ *is normal, with rational singularities.*

The direct implication here follows from [7] and [10]. There are corresponding facts for double points, but these do not lead to such a neat formulation.

THEOREM 2 (see §6). *Let* X *be a normal surface singularity in* $\mathbb{C}^3$, *not belonging to Arnold's* E, Z^1 *or* Q *series. Then* X *is* 3-*modal if and only if it is elliptic, of type* α.

In the complete intersection case, $\mathscr{R}$-modality is not available and one must turn to $\mathscr{K}$-modality. We have not yet obtained any clear results in this context.

In addition to the above theorems, we feel the structure of the relation between equations and resolutions, which we explore in some detail, is of intrinsic interest. Also, the list of weights of the weighted homogeneous elliptic complete intersections seems difficult to obtain by any other method.

The paper consists of 6 sections, as follows.

§1. Enumerating the Graphs

We first recall the method of classification of elliptic Gorenstein singularities by weighted graphs: this is due to Laufer [7] and Reid [10]; we prefer the terminology of the latter.

The first class are those where the exceptional set in a resolution is a **Kodaira elliptic curve**: those curves which occur as singular elements in a family of elliptic curves [6]. We recall Kodaira's classification of these exceptional fibres.

Type	I_o	E is an elliptic curve
	$I_n(n > 0)$	E is a cycle of n rational curves, each meeting the next just once.
	II	Rational curve with a cusp
	III	Two rational curves touching.
	IV	Three rational curves with a single common point.
	$I_n^*(n \geq 0)$, II^*, III^*, IV^* (see below).	

Each of these cases gives rise to a series of elliptic Gorenstein singularities, classified by the selfintersections of those components of multiplicity 1, which can take any value up to and including those occuring in Kodaira's case.

An elliptic Gorenstein singularity not of type I_o has each component (of the exceptional set E in a minimal resolution) rational. Except for types I_1 and II, each such component is smooth. Except for type III, any intersection of two components is transverse. Except for type I_2 no two components meet more than once. Except for type IV, no three meet in a point. The intersections can thus be represented by a (connected) graph Γ with one vertex c for each component of E and two vertices joined by a point if the corresponding components meet. If we exclude the remaining types I_n, Γ is a tree.

We attach weights $a_c \geq 2$, $n_c \geq 1$ to the vertices of Γ: here a_c is minus the selfintersection number of c.

In general, a weighted graph is a connected graph G, with integers $a_c \geq 2$, $n_c \geq 1$ for each vertex c of G. We define an ordering on the 0th chain group H of G by defining the positive cone to be the semigroup generated by the vertices. We also define a symmetric bilinear form by setting c.c′ equal to the number of edges joining c to c′ (if $c \neq c'$), and $c.c = -a_c$. Write Z for the cycle $Z = \sum n_c c$. Then we say that the graph is *balanced* if for all c, $c.(Z-c) = 2$. It is *simple* if for any pair c,c′ we have $c.c' \leq 2$. The cycle Z is *irreducible* if there is no cycle d with $0 < d < Z$, $d = \sum r_c c$ such that for all c with $r_c \neq 0$ we have $c.(d-c) = 2$. In this situation, we also say that Z (or G) is *minimal*.

It follows at once from these conditions at $Z.c \leq 0$ for all c, so that $D = -Z.Z \geq 0$. It can also be deduced that the bilinear form is negative definite unless $D = 0$, when it is semidefinite with radical generated by Z. The number D is called the degree of the singularity.

Using the balance condition, we obtain $D = \sum n_c(a_c - 2)$.

Balanced minimal graphs correspond to elliptic Gorenstein singularities. Their structure is determined by D as follows:

$D \geq 5$:	not a complete intersection
$D = 4$:	the intersection of two hypersurfaces in $\mathbb{C}^4$
$D = 3$:	a hypersurface in C^3 (of multiplicity 3)
$D = 1,2$:	a hypersurface in C^3 (of multiplicity 2)

$D = 0$: these cases do not correspond to isolated singularities, but to Kodaira's exceptional fibres. In these cases, each $Z.c = 0$ and $c^2 = -2$. The corresponding graphs are shown below (n_c is indicated).

I_n^*

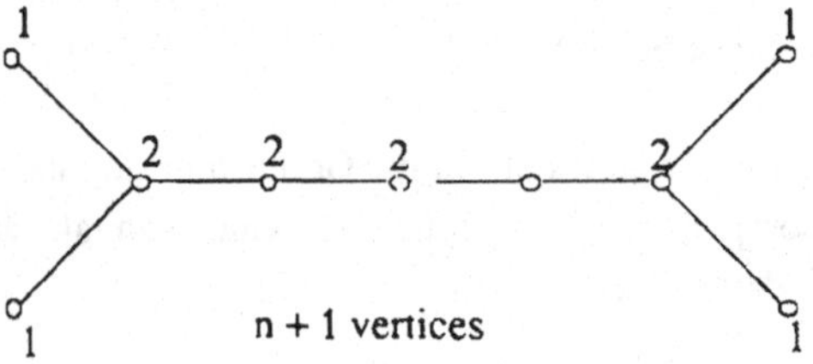

(type $\tilde{D}_{n+4}$)

II^*

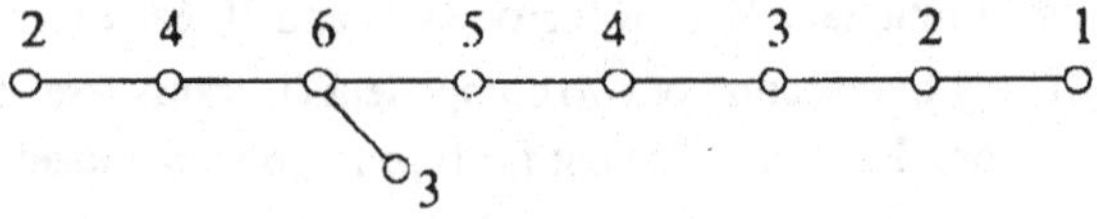

(type $\tilde{E}_8$)

III^*

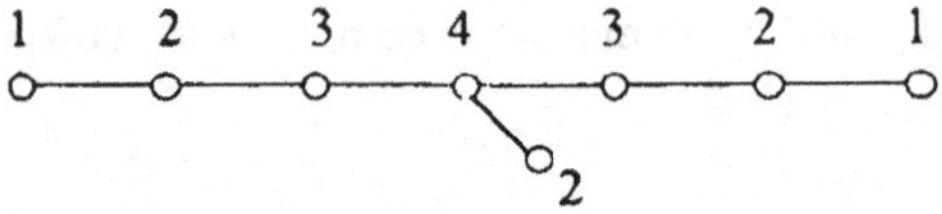

(type $\tilde{E}_7$)

IV^*

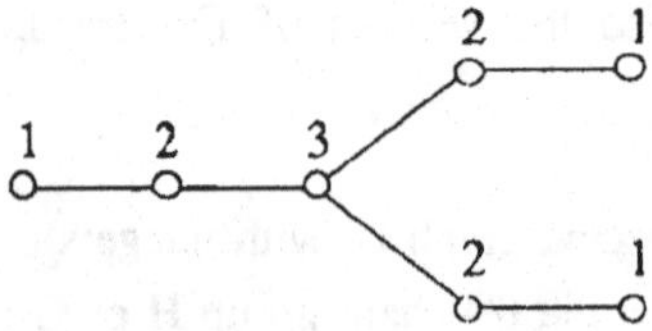

(type $\tilde{E}_6$)

The cases of interest for the present paper are those with $1 \leq D \leq 4$. In particular we have the Kodaira singularities, whose balanced graphs are obtained from those with $D = 0$ by increasing the value of a_c at one or more vertices with $n_c = 1$. The total amount by which these values are increased is D. The classification is thus easy. We refer to [4] for a full discussion of these cases.

For any case not of Kodaira type there is at least one vertex with $n_c \geq 2$, $a_c \geq 3$. Since we are assuming $D = \sum n_c(a_c - 2) \leq 4$, the possibilities for such a vertex are:

α	$n_c = 2$	$a_c = 3$	$n_c(a_c - 2) = 2,$
β	$n_c = 3$	$a_c = 3$	$n_c(a_c - 2) = 3,$
γ	$n_c = 4$	$a_c = 3$	$n_c(a_c - 2) = 4,$
δ	$n_c = 2$	$a_c = 4$	$n_c(a_c - 2) = 4.$

Thus if $D = 1$, the singularity is necessarily of Kodaira tpe. For $D = 2$, we may have one vertex of type α; for $D = 3$, we may have one of type α or β; and for $D = 4$, we may have one vertex (type α, β, γ or δ), or two (each of type α).

The balance condition may be written as

$$\sum\{n_d \mid c.d = 1\} = a_c(n_c - 1) + 2;$$

the right hand side takes the respective values 5, 8, 11, 6 for types α, β, γ and δ.

For enumeration, first remove the exceptional vertices from the graph. A component of the complement has at most one branch point, and that only of valence 3 (this follows from minimality of Z). If there is no branching, we can only have a chain

Here, kh may be a branch point or exceptional vertex. If the component is T-shaped, let k be the length of the leg to the exceptional vertex and k', k'' the lengths of the other two. By minimality, these cannot all be ≥ 3.

We give some more details on these arguments to illustrate the use of minimality. If there is a vertex of valence ≥ 4, resp. at least two branch points, we define a cycle d by the weighted subgraph

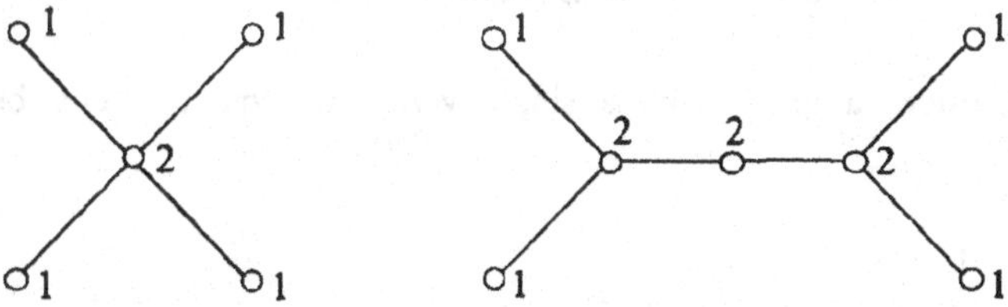

If there is a branch point of valence 3 and each leg has length ≥ 3, we contradict minimlity by picking a cycle d of the form

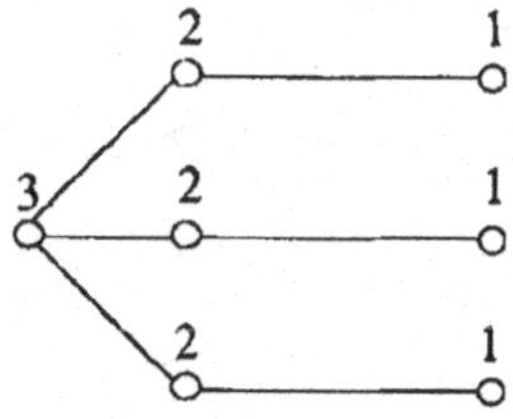

It is no accident that these are extended Coxeter–Dynkin diagrams!

If $k = 2$, the component is of type

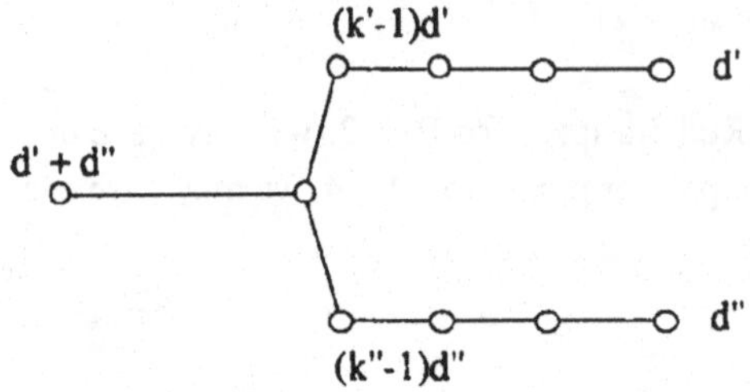

with $k'd' = k''d''$

If $k' = 2$, $k = 3$ the component is of type below (with the n_i all multiplied by some number a if k'' is odd, $^a/_2$ if k'' is even)

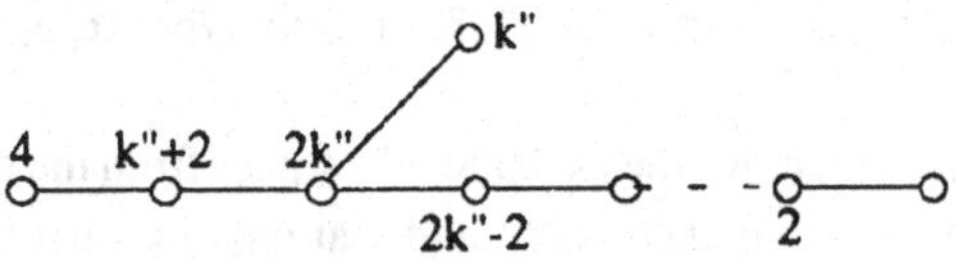

Cases $k > 3$ arise from this by extending to the left

$k'' = 1$ gives a chain

$k'' = 2$ is a tree

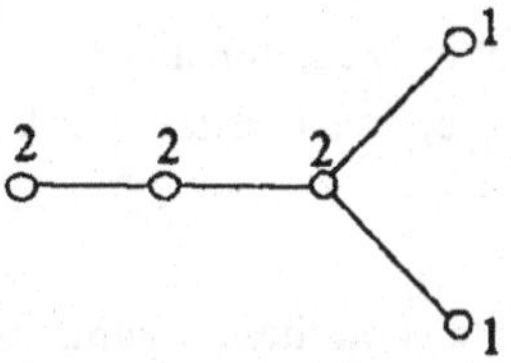

$k'' = 3$ gives part of a graph of type $\tilde{E}_8$ ($k = 3$ or 4 or 5)

$k'' = 4$ gives part of a graph of type $\tilde{E}_7$ ($k = 3$)

$k'' = 6$ also yields an $\tilde{E}_8$ graph ($k = 3$).

Now consider a graph with a single vertex of type α. Possibilities for the components are:

(C1)

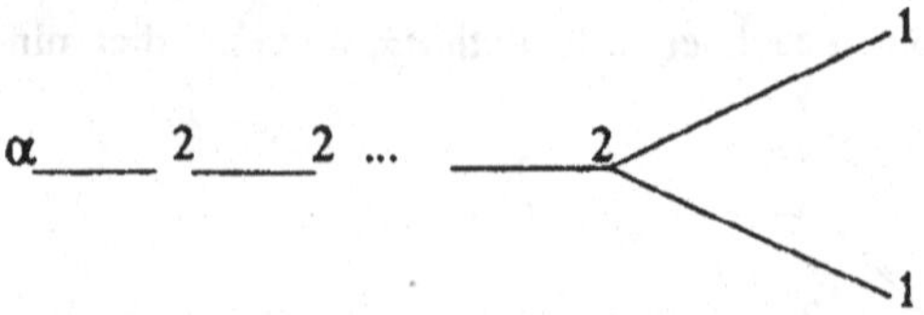

(k 2's) ($C2_k$)

(k = 2, d′ = d″ = 1)

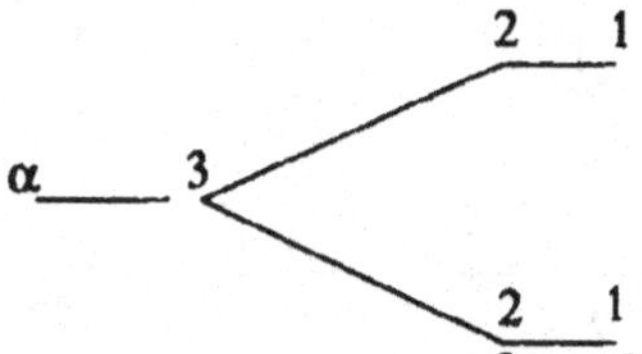

(C3)

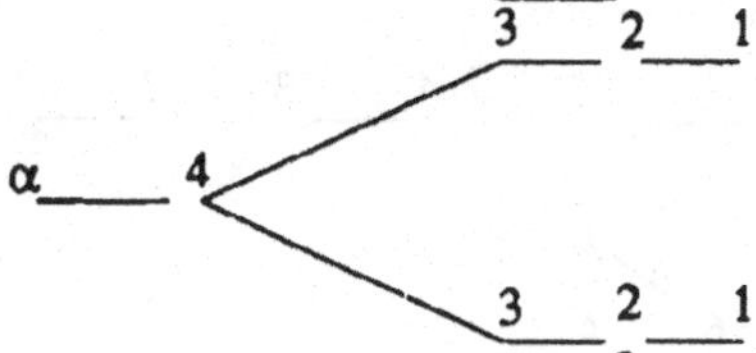

(C4)

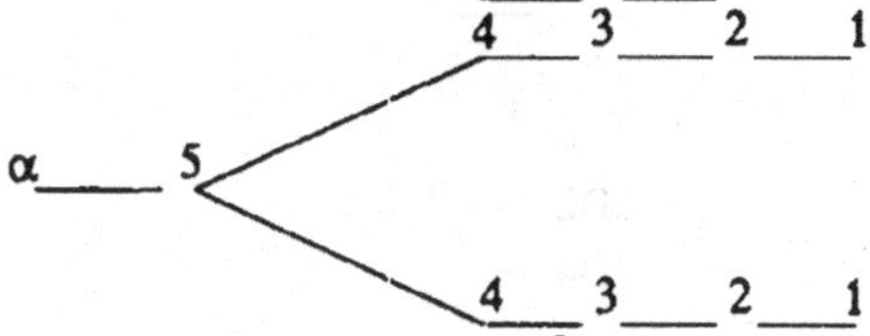

(C5)

(k = 3)

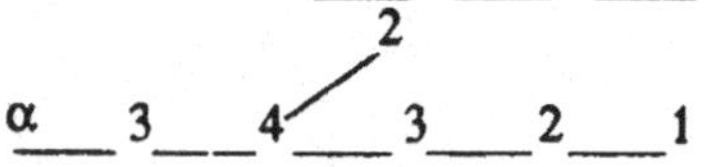

(C3′)

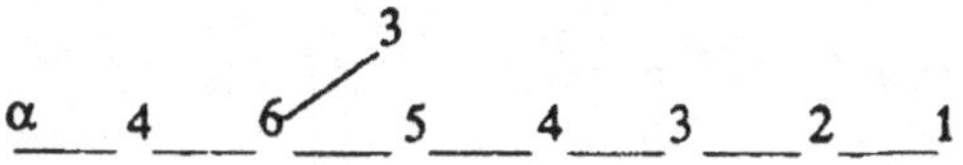

(C4′)

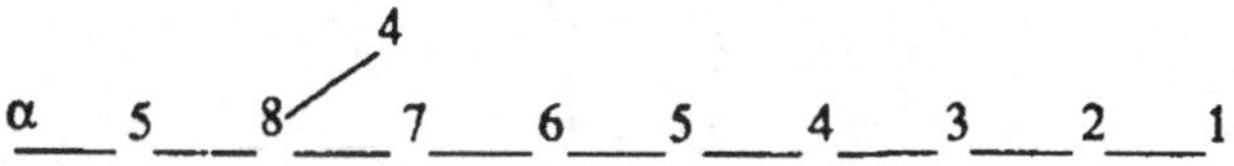

(C5′)

(k > 3)

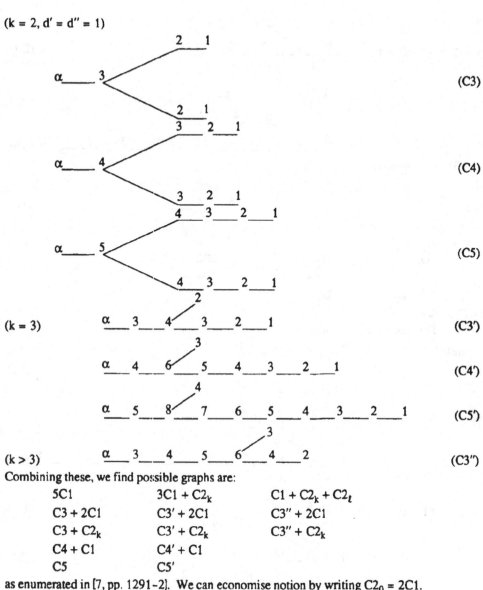

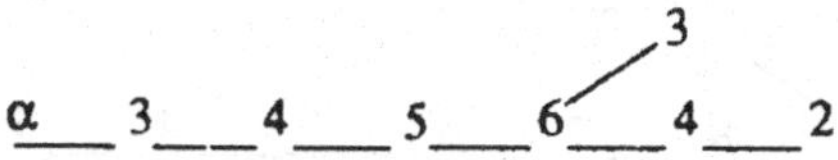

(C3″)

Combining these, we find possible graphs are:

5C1	3C1 + $C2_k$	C1 + $C2_k$ + $C2_\ell$
C3 + 2C1	C3′ + 2C1	C3″ + 2C1
C3 + $C2_k$	C3′ + $C2_k$	C3″ + $C2_k$
C4 + C1	C4′ + C1	
C5	C5′	

as enumerated in [7, pp. 1291–2]. We can economise notion by writing $C2_0$ = 2C1.

We proceed similarly for type β. Components are:

(chain) β____ 2____ 1 (D2)

$(k = 2, d' = 2, d'' = 1)$

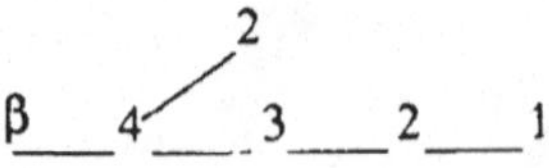

$$\beta \underline{\quad} 4 \nearrow^{2} \underline{\quad} 3 \underline{\quad} 2 \underline{\quad} 1 \tag{D4}$$

$$\beta \underline{\quad} 6 \nearrow^{4 \underline{\quad} 2} \underline{\quad} 5 \underline{\quad} 4 \underline{\quad} 3 \underline{\quad} 2 \underline{\quad} 1 \tag{D6}$$

$$\beta \underline{\quad} 8 \nearrow^{6 \underline{\quad} 4 \underline{\quad} 2} \underline{\quad} 7 \underline{\quad} 6 \underline{\quad} 5 \underline{\quad} 4 \underline{\quad} 3 \underline{\quad} 2 \underline{\quad} 1 \tag{D8}$$

$(k = 3)$ No solutions.

$(k > 3)$

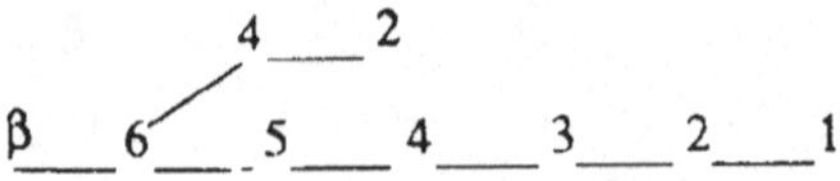

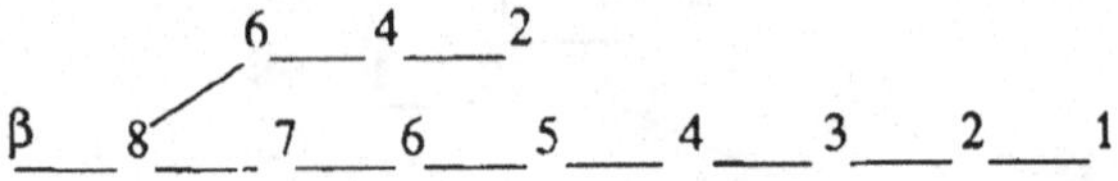

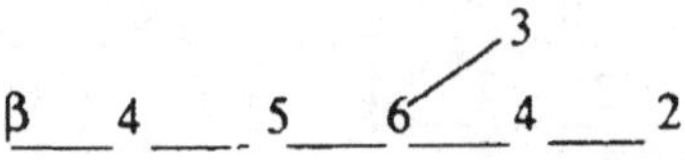

$$\beta \underline{\quad} 4 \underline{\quad} 5 \underline{\quad} 6 \nearrow^{3} \underline{\quad} 4 \underline{\quad} 2 \tag{D4'}$$

thus possible graphs are

4D2	2D2 + D4	2D2 + D4'
2D4	D4 + D4'	2D4'
D6 + D2	D8	

cf. [7, p. 1294].

For type γ we find components

(chain)

$$\gamma \underline{\quad} 3 \underline{\quad} 2 \underline{\quad} 1 \tag{E3}$$

$$\gamma \underline{\quad} 2 \tag{E2}$$

$(k = 2, \quad d' = 3, d'' = 1)$

$$\gamma \underline{\quad} 6 \nearrow^{3} \underline{\quad} 5 \underline{\quad} 4 \underline{\quad} 3 \underline{\quad} 2 \underline{\quad} 1 \tag{E6}$$

$$\gamma \underline{\quad} 9 \nearrow^{6 \underline{\quad} 3} \underline{\quad} 8 \underline{\quad} 7 \underline{\quad} 6 \underline{\quad} 5 \underline{\quad} 4 \underline{\quad} 3 \underline{\quad} 2 \underline{\quad} 1 \tag{E9}$$

$(d' = d'' = 2)$

$$\gamma \underline{\quad} 6 \nearrow^{4 \underline{\quad} 2} \underline{\quad} 4 \underline{\quad} 2 \tag{E6'}$$

$$\gamma \underline{\quad} 8 \nearrow^{6 \underline{\quad} 4 \underline{\quad} 2} \underline{\quad} 6 \underline{\quad} 4 \underline{\quad} 2 \tag{E8}$$

$$\gamma \underline{\quad} 10 \nearrow^{8 \underline{\quad} 6 \underline{\quad} 4 \underline{\quad} 2} \underline{\quad} 8 \underline{\quad} 6 \underline{\quad} 4 \underline{\quad} 2 \tag{E10}$$

$(k = 3)$

$$\gamma \underline{\quad} 5 \underline{\quad} 6 \nearrow^{3} \underline{\quad} 4 \underline{\quad} 2 \tag{E5}$$

$$\gamma \underline{\quad} 6 \underline{\quad} 8 \nearrow^{4} \underline{\quad} 6 \underline{\quad} 4 \underline{\quad} 2 \tag{E6''}$$

$$\gamma \underline{\quad} 7 \underline{\quad} 10 \nearrow^{5} \underline{\quad} 8 \underline{\quad} 6 \underline{\quad} 4 \underline{\quad} 2 \tag{E7}$$

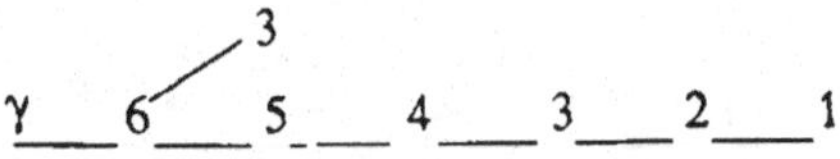

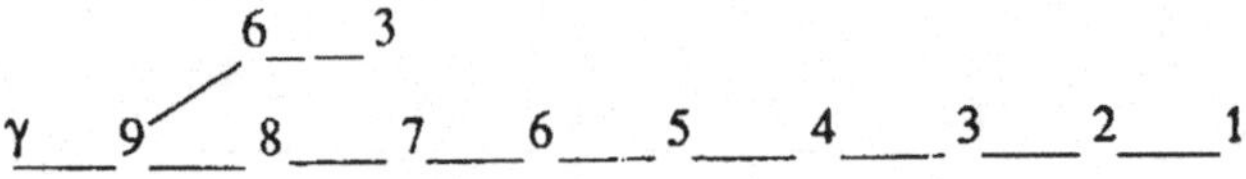

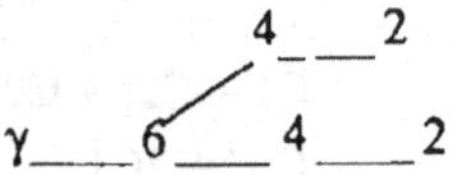

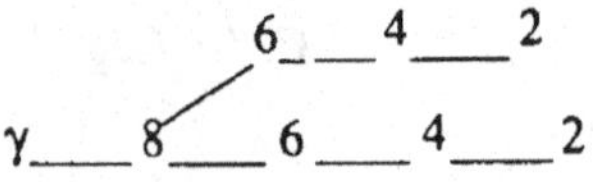

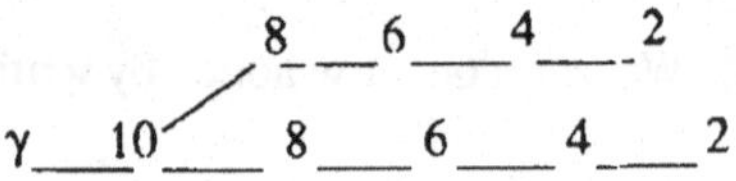

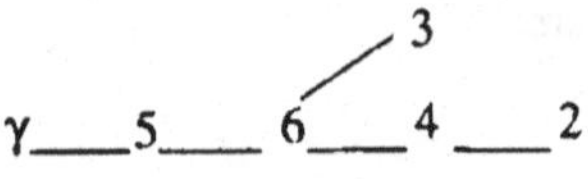

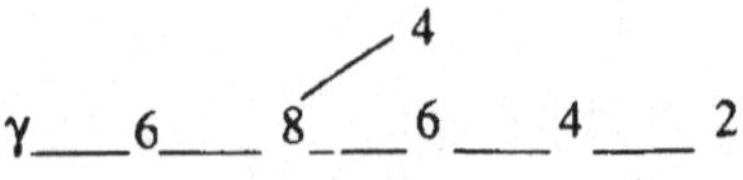

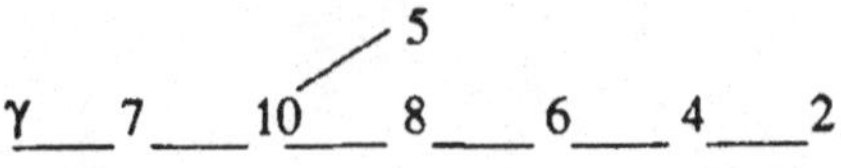

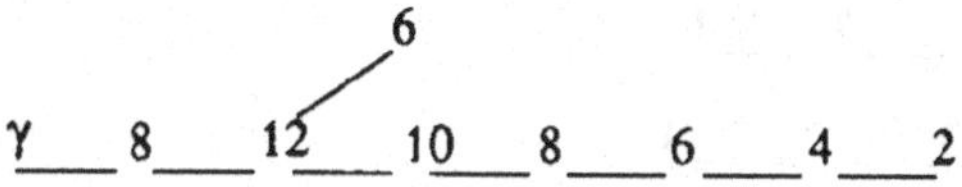

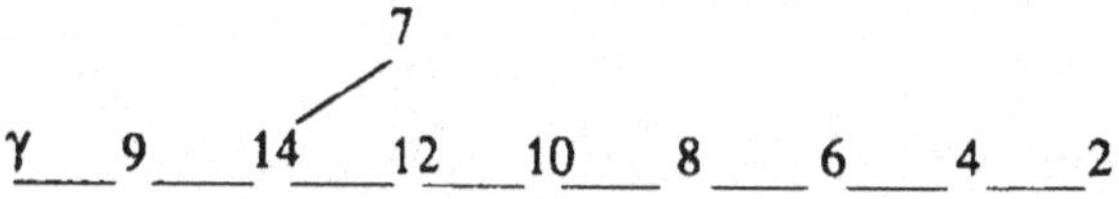

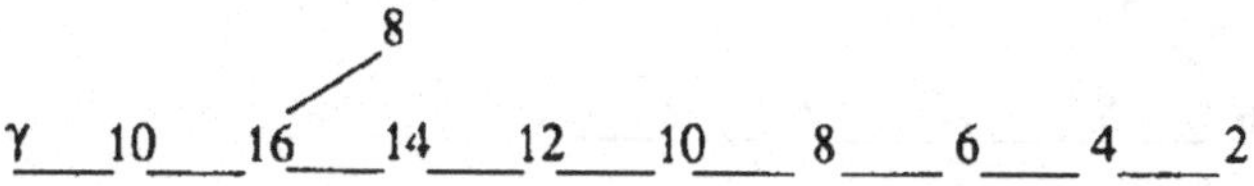

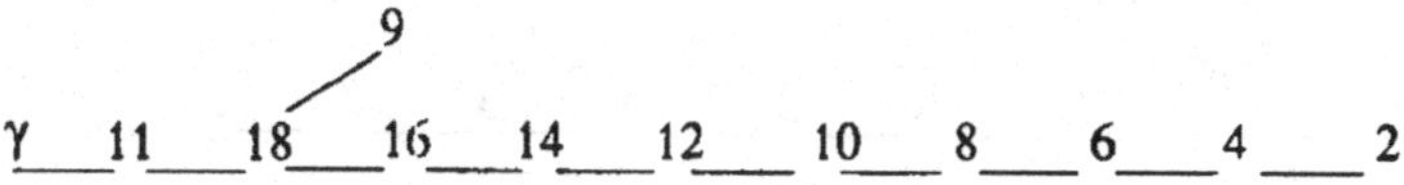

6

γ___8___12___10___8___6___4___2 (E 8′)

7

γ___9___14___12___10___8___6___4___2 (E 9′)

8

γ___10___16___14___12___10___8___6___4___2 (E 10′)

9

γ___11___18___16___14___12___10___8___6___4___2 (E 11)

2

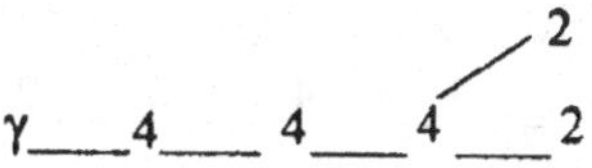

(k > 3) γ___4___4___4___2 (m 4′s) ($E4_m$)

6

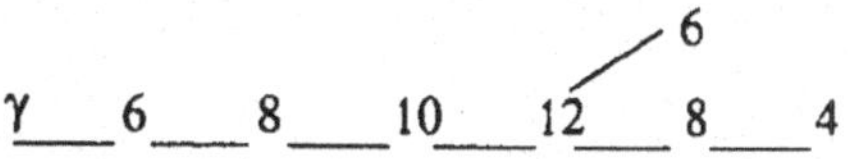

γ___6___8___10___12___8___4 (E 6‴)

There are many ways to fit these together, but most of these do not yield minimal graphs. Again write $E\,4_0$ for 2E2.

For	E11,	compare	C5′;
	E9′ + E2,	compare	C4′ + C1
	E8 + E3	compare	C4 + C1
	E8′ + E3	compare	C4′ + C1
	$E7 + E4_m$	compare	$C3' + C2_m$
	E6′ or E6″ or E6‴ with E5 or E3 + E2	compare	C3 or C3′ or C3″ with $C2_2$ or 2C1
	$E5 + E4_m + E2$	compare	$C2_2 + C2_m + C1$
	$E3 + E4_m + E4_n$	compare	$C1 + C2_m + C2_n$

The only cases to survive are thus:

E9 + E2, E6 + E5, E6 + E3 + E2, E5 + 2E3, 3E3 + E2.

For type δ we find components

(chain) δ___1 F 1

2___1

(k = 2, d′ = d″ = 1) δ___3___2___1 F 3

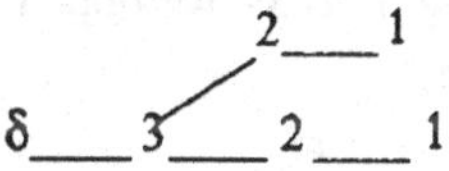

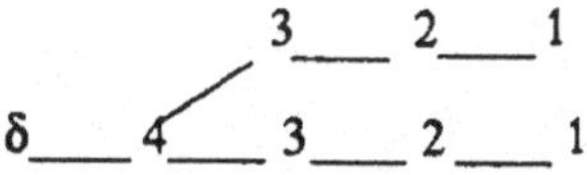

3___2___1

δ___4___3___2___1 F 4

4___3___2___1

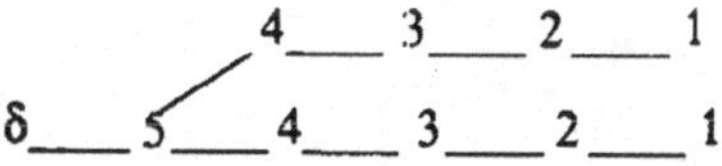

δ___5___4___3___2___1 F 5

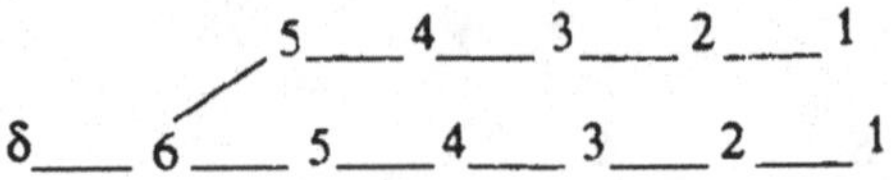

$$\delta \underline{\quad} 6 \underline{\quad} \overset{\displaystyle 5 \underline{\quad} 4 \underline{\quad} 3 \underline{\quad} 2 \underline{\quad} 1}{5 \underline{\quad} 4 \underline{\quad} 3 \underline{\quad} 2 \underline{\quad} 1} \qquad \text{F 6}$$

$$(k = 3) \qquad \delta \underline{\quad} 3 \underline{\quad} 4 \overset{\displaystyle 2}{\diagup} 3 \underline{\quad} 2 \underline{\quad} 1 \qquad \text{F 3}'$$

$$\delta \underline{\quad} 4 \underline{\quad} 6 \overset{\displaystyle 3}{\diagup} 5 \underline{\quad} 4 \underline{\quad} 3 \underline{\quad} 2 \underline{\quad} 1 \qquad \text{F 4}'$$

$$\delta \underline{\quad} 5 \underline{\quad} 8 \overset{\displaystyle 4}{\diagup} 7 \underline{\quad} 6 \underline{\quad} 3 \underline{\quad} 4 \underline{\quad} 3 \underline{\quad} 2 \underline{\quad} 1 \qquad \text{F5}'$$

$$\delta \underline{\quad} 6 \underline{\quad} 10 \overset{\displaystyle 5}{\diagup} 9 \underline{\quad} 8 \underline{\quad} 7 \underline{\quad} 6 \underline{\quad} 3 \underline{\quad} 4 \underline{\quad} 3 \underline{\quad} 2 \underline{\quad} 1 \qquad \text{F}$$

$$(k > 3) \qquad \delta \underline{\quad} 3 \underline{\quad} 4 \underline{\quad} 5 \underline{\quad} 6 \overset{\displaystyle 3}{\diagup} 4 \underline{\quad} 2 \qquad \text{F 3}''$$

$$\delta \underline{\quad} 2 \ \ldots \ \underline{\quad} 2 \underline{\quad} 1 \overset{\displaystyle 1}{\underset{\displaystyle 1}{\diagup \diagdown}} \qquad (k \text{ 2's}) \qquad \text{F2}_k$$

This time there is no problem with minimality : we obtain the list

$F2_\ell + F2_k + F2_m$		
$F3 + F2_k + F1$	$F3' + F2_k + F1$	$F'' + F2_k + F_1$
$2F3$	$F3 + F3'$	$F3 + F3''$
$2F3'$	$F3' + F3''$	$2F3''$
$F4 + F2_k$	$F4' + F2_k$	
$F5 + F1$	$F5' + F1$	
$F6$	$F6'$	

We turn to cases with two vertices of type α. There will be a single component containing both; we consider this. Now if the component starts $\alpha - 2$ it must be of type

$$\alpha - 2 - 2 \ldots -2 - \alpha \qquad (k \text{ 2's}) \qquad (G\ 22_k)$$

If it starts $\alpha - 3-$, then by replacing by $1-2-3-$ we obtain a balanced component with just one α. From the list of these we deduce

$$\overset{\displaystyle \alpha}{\underset{\displaystyle \alpha}{\diagdown \diagup}} 3 \underline{\quad} 2 \underline{\quad} 1 \qquad (G\ 33)$$

$$\alpha \underline{\quad} 3 \underline{\quad} 4 \overset{\displaystyle \alpha}{\diagup} 3 \underline{\quad} 2 \underline{\quad} 1 \qquad (G\ 34)$$

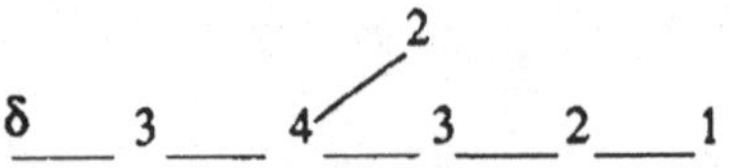

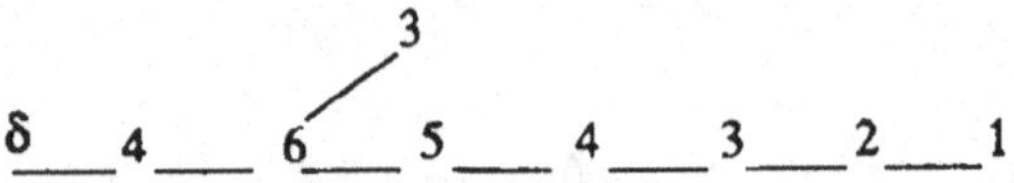

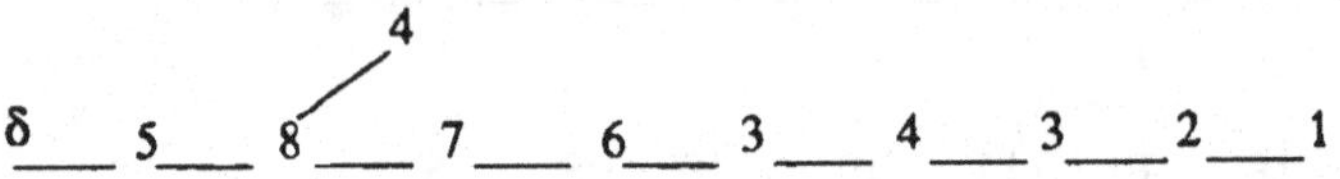

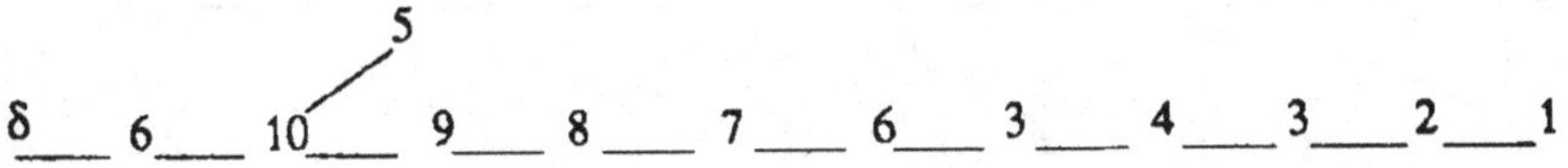

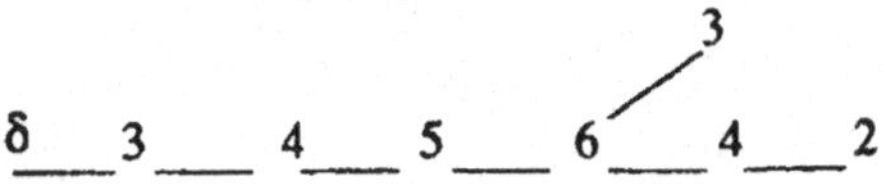

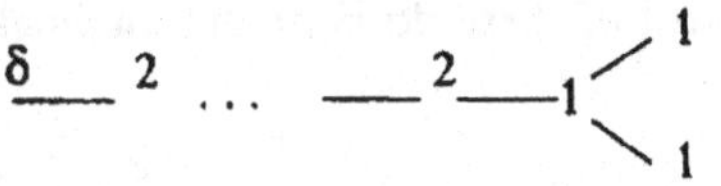

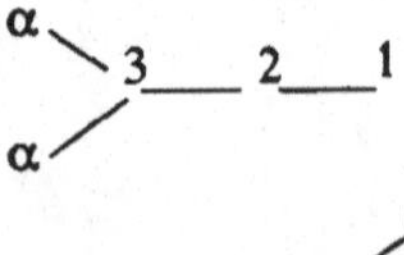

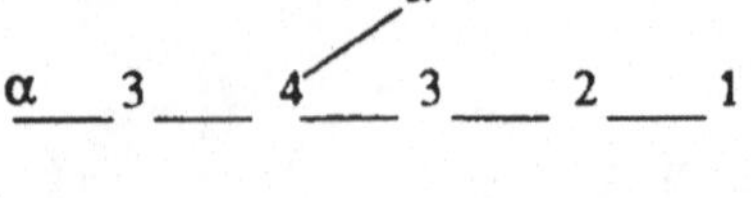

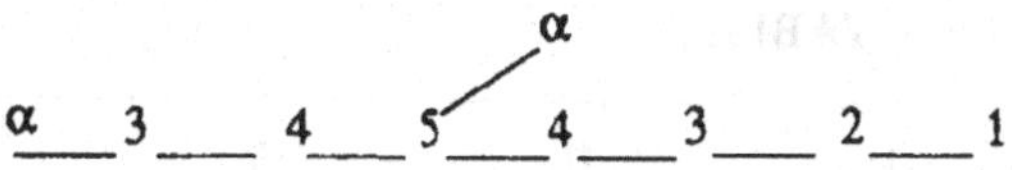

(G 35)

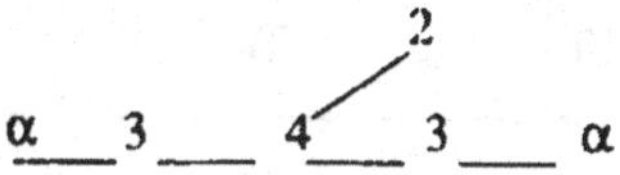

(G 33′)

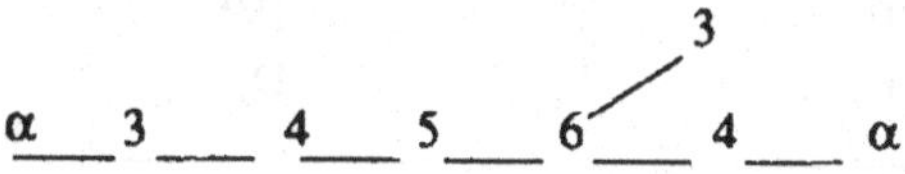

(G34′)

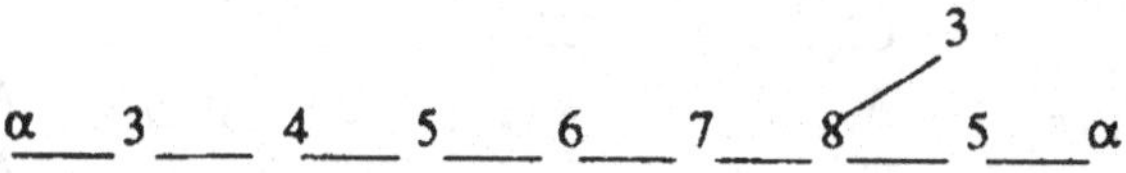

(G 35′)

Similarly α - 4 - can be replaced by 2 - 4 -; this yields no new cases. Finally a direct argument yields just one further case

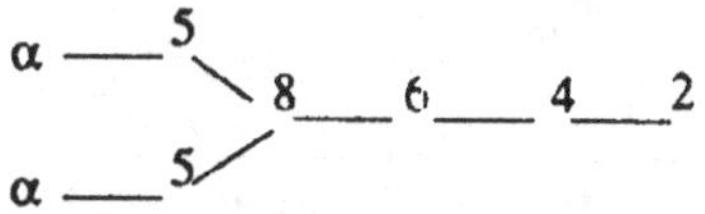

(G 55);

however (compare $\tilde{E}_6$) this is not minimal. Putting these components together we obtain graphs

$C1 + C_\ell + G22_k + C2_n + C1$

$C2_k + G33 + C2_\ell$	$C2_k + G33' + C2_\ell$
$C2_k + G34 + C1$	$C2_k + G34' + C1$
$C2_k + G35$	$C2_k + G35'$

$((C1 + C2_k)$ or $C3$ or $C3'$ or $C3'') + G22_\ell + (C1 + C2_m$ or $C3$ or $C3'$ or $C3'')$.

We now give a concordance of our own notation with that of [7]. The following table has several sections. For Kodaira singularities, we give the Kodaira type and the notation of [7]. For the remaining elliptic singularities we list the possible components of the complement of the vertices c with $n_c \geq 2$, $a_c \geq 3$ in the resolution graph giving first the notation above, then (for α and β types) that of [7], and then the notation for the graph formed by the complement.

The suffix in this last gives the number of vertices of the graph. Summing over all components and adding 1 (types α, β, γ and δ) or 2 (type α α) gives the total number N of components of the exceptional set. Its Euler characteristic is then (N + 1), and the Milnor number can be obtained from Laufer's formula [8]

$$\mu = \chi(E) + Z^2 + 12h - 1$$

which reduces here to $N - D + 12$ or, if $D = 4$, to $N + 8$.

TABLE

Notation	I_0	$I_n(n \geq 1)$	II	III	IV		
[7]	$E\ell$	No	Cu	Ta	Tr		
Notation	$I_n^*(n \geq 0)$		II^*	III^*	IV^*		
[7]	$A_{n+1,*,*,*,*}$		$E_{8,*}$	$E_{6,*,*}$	$D_{4,*,*,*}$		
Notation	C1	$C2_1$	$C2_{k+1}$	C3	C4	C5	
[7]	$A_{*,0}$	$A'_{1,*,*,0}$	$A_{k,*,*,0}$	$A'_{3,*,*,0}$	$A'_{5,*,*,0}$	$A'_{7,*,*,0}$	
Graph	A_1	A_3	D_{k+3}	A_5	A_7	A_9	
Notation	C3′	C4′	C5′	C3″			
[7]	$D_{5,*,0}$	$D_{7,*,0}$	$D_{9,*,0}$	$E_{7,0}$			
Graph	D_6	D_8	D_{10}	E_7			
Notation	D2	D4	D6	D8	D4′		
[7]	$A_{1,*,0}$	$A_{4,*,0}$	$A_{7,*,0}$	$A_{10,*,0}$	$E_{6,0}$		
Graph	A_2	A_5	A_8	A_{11}	E_6		
Notation	E2	E5	E3	E6	E9		
Graph	A_1	D_5	A_3	A_7	A_{11}		
Notation	Fm $(1 \leq m \leq 6)$		$F2_k(k \geq 2)$	Fm′$(3 \leq m \leq 6)$	F3″		
Graph	A_{2m-1}		D_{k+2}	D_{2m}	E_7		
Notation	$G2,2_k$	G3,3	G3,4	G3,5	G3,3′	G3,4′	G3,5′
Graph	A_k	A_3	A_5	A_7	D_4	D_6	D_8

§2. EQUATIONS: LOW ORDER TERMS.

Equations in the hypersurface case are given by Laufer [7], but we will review the case D = 3 (the case D = 1,2 are somewhat exceptional). Here the lowest order term f_0 of f define a cubic in 3 variables x,y, and z. The cases are as follows:

(i) Smooth cubic; f has a simple elliptic singularity.

(ii) Nodal cubic, conic and chord, triangle.
These all yield cusp singularities for f.

(iii) Cuspidal cubic, conic and tangent, 3 concurrent lines.

These include the remaining cases when f has a Kodaira singularity.

(iv) $f = x^2y$ resp. x^3. These include the remaining elliptic singularities of Laufer.

The above cubic is the tangent cone to X at P, which is the fibre of the blow-up of 0. We recall that for Kodaira singularities, the full resolution is obtained from a Kodaira fibre by increasing the value of a_c at the vertices c of the graph with $n_c = 1$ from 2 to $k_c + 2$, say. Those such components of the exceptional set with $k_c > 0$ are the ones obtained at the first stage above, and k_c is the degree in $P^2\mathbb{C}$ of the corresponding curve: note that $D = \Sigma k_c = 3$ in these cases.

In the remaining cases, the graph has just one vertex c with $a_c = 3$ and either (α) $n_c = 2$ or (β) $n_c = 3$. This component of the exceptional set appears as the repeated factor in (α) x^2y or (β) x^3 respectively.

The complete intersection case D = 4 is similar. The terms of degree 2 in $f:(\mathbb{C}^4,0) \to (\mathbb{C}^2,0)$ define a pencil of quadrics in $P^3(\mathbb{C})$, whose intersection is a curve B (by [7,3.13]).

We have the following cases (our notation for Segre symbols is explained on [12, p.477]):

(i) (Segre symbol (1,1,1,1)) B a smooth elliptic quartic. This is the simple elliptic case.

(ii) B is a cycle of rational components, meeting transversely (degrees 4, 3 + 1, 2 + 2, 2 + 1 + 1, 1 + 1 + 1 + 1 correspond to Segre symbols ((2,1,1), (2,2), ((1,1,),1,1), (2,(1,1), ((1,1), (1,1))). This is the case of cusp singularities.

(iii) B as in (ii), but with all the intersection points coincident (Segre symbols (3,1), (4), ((2,1),1), ((3,1)), (1,1,1;1) respectively; "series" J', L, K', M and I in notation of [11]). This includes the remaining Kodaira singularities.

(iv) The cases where B has dimension 1 but is not reduced are as follows.

Symbol	(1;2)	((2,2))	((1,1,1),1)	(2,1;1)	((2,1,1))	(3;1)	((1,1),1;1)	((2,1);1)	(1,1;1,1)
B	$\ell^2 s$	$\ell^2 mn$	s^2	$\ell^2 mn$	$\ell^2 m^2$	$\ell^3 m$	$\ell^2 m^2$	ℓ^4	ℓ^4
Codim	5	5	5	6	6	7	7	8	10
Type	$\alpha^{(1)}$	$\alpha^{(2)}$	δ	$\alpha^{(3)}$	$\alpha\alpha^{(1)}$	β	$\alpha\alpha^{(2)}$	γ	-

In the table ℓ,m and n denote lines, s an irreducible conic. The codimension is that of the type of pencil. The type is as in §1: but one of the cases yields no elliptic singularities.

As in the case D = 3, the curve B is the tangent cone to X at P. Its description in the cases of Kodaira type is as before. In general B consists of the exceptional curves E_c corresponding to those vertices c of the graph for which $a_c > 2$. As a curve in projective space, this has multiplicity n_c and degree (a_c-2). This holds for all elliptic singularities except the simple elliptic. Observe that $D = \Sigma n_c(a_c-2)$, thus D = 4 is compatible with B having degree 4.

We observe for later reference that the cases where there is a component of multiplicity 1 in B are $\alpha^{(1)}$, $\alpha^{(2)}$, $\alpha^{(3)}$ and β.

§3. LINEAR REDUCTION.

Although we shall discuss equations in the next section directly from the resolution, we first describe here a direct relation between equations in the D = 3 case and those of type α and β in the D = 4 case which gives a simpler approach to these cases. We then discuss equations using this relation. For these cases, this gives an alternative approach to that in §4, which we develop independently. However, knowing this relation helps explain the repetition of certain patterns of results in §4.

As we saw in [12], a complete intersection singularity whose equation has the form $f(w,x,y,z) = (wx + a(x,y,z), wy + b(x,y,z))$ defines, by eliminating w, a 'linear reduction'

$$L_w f(x,y,z) = xb(x,y,z) - ya(x,y,z)$$

which has an isolated singularity if f has, and satisfies $\mu(L_w f) = 1 + \mu(f)$. Geometrically, of course, eliminating w means projecting the quartic curve in P^3 to a cubic curve in P^2. If we resolve the singularities of $f^{-1}(0)$ and $(L_w f)^{-1}(0)$ we obtain exceptional sets E and E' with identical resolution graphs except that one vertex c of E with multiplicity $n_c = 1$ has $a'_c = 1 + a_c$.

The equation of f can be put in the above form if and only if the above quartic curve B has a simple point (or equivalently, a component of multiplicity 1). Thus we can recover equations for $\alpha^{(1)}$, $\alpha^{(2)}$, $\alpha^{(3)}$ and β by reversing the process. Moreover if $\varphi(x,y,z)$ vanishes along the line x = y = 0 we can write it in the form $\varphi = L_x f$. The different maps f so arising are equivalent by linear coordinate changes. (However, the choice of a line - or of a smooth curve that can be transformed into a line - in $\varphi^{-1}(0)$ is less trivial, and we only claim to obtain examples, not normal forms, for each type.)

A similar 'quadratic reduction' is available for D = 3 cases whose equation can be put in the form

$$f(x,y,z) = yz^2 + 2z\,a(x,y) + b(x,y): \quad \text{then} \quad \Delta_z f = a(x,y)^2 - yb(x,y) + w^2.$$

Corresponding remarks apply to this case.

We apply these remarks to strata of type α. There are 3 possible forms for the 2-jet, as noted above; unfolding and reducing leads to the following:

$\alpha^{(1)}$ (symbol 1;2) 2-jet $\langle w^2 + xz, xy \rangle$
$f = \langle w^2 + xz + a(y), xy + wb(z) + c(y,z) \rangle$
$L_x f = w^2 y - wzb(z) + ya(y) - zc(y,z)$
$\Delta_w L_x f = z^2(b(z))^2 - 4y^2 a(y) + 4yz\,c(y,z)$, with 5-jet
$4y(-a_3 y^4 + z\,c_3(y,z))$.

$\alpha^{(2)}$ (symbol (2,2)) 2-jet $\langle wx, wy + xz \rangle$
$f = \langle wx + wa(z) + xb(y) + c(y,z), wy + xz \rangle$
$L_x f = (w + b(y))\,wy - z\,(wa(z) + c(y,z))$
$\Delta_w L_x f = (y\,b(y) - za(z))^2 + 4yz\,c(y,z)$, with 5-jet
$4yz\,c_3\,(y,z)$

$\alpha^{(3)}$ (symbol (2.1;1)) 2-jet $\langle xz, y(x+z)\rangle$

$f = \langle xz + a(w,y), y(x+z) + b(w) + yc(w) + zd(w)\rangle$

$L_x f = yz^2 + z(b(w) + yc(w) + zd(w)) - ya(w,y)$

$\Delta_z L_x f = (b(w) + yc(w))^2 + 4(y + d(w))\, ya(w,y)$, with 5-jet $y^2 a_3(w,y)$.

We can interpret these calculations as follows. For any f of type α with D = 4, Lf (with D=3) and ΔLf (with D=2) can be defined, and have type α.

In the case D = 2, we can write the equation as $w^2 = a(x,y)$ with $j^4 a = 0$ and the factorisation of the 5-jet of a into distinct linear factors gives a first indication of the stratum: indeed, the multiplicities are precisely the numbers in our notation (e.g. C1, C3 etc.). When D = 2 we have $a_c = 2$ for all the vertices with $n_c = 1$.

In the case D = 3 we can write the equation as $f = yz^2 + 2za(x,y) + b(x,y)$, and $\Delta_z f$ leads to the quintic $yb(x,y)$. One of the roots of this is y, and for one of the vertices with $n_c = 1$ corresponding to this factor we must increase a_c from 2 to 3.

The case D = 4 splits as above into three subcases. For $\alpha^{(1)}$ we have a preferred root y of the quintic, and for some corresponding c we have $a_c = 4$. For $\alpha^{(2)}$ we have two preferred roots y and z: each has a corresponding c with $a_c = 3$. Finally for $\alpha^{(3)}$ we have a preferred repeated root y^2, and for two of the vertices c with $n_c = 1$ corresponding to this factor will have $a_c = 3$. This rule is of course as in §2.

Thus, in particular, no preferred root may have type C3″ and for $\alpha^{(3)}$ the preferred root must have one of the types $C2_m$, C3, C4 and C5.

We now show how explicit equations can be obtained for the various α strata. For D = 2, we have $w^2 = a(x,y)$. The factorisation of $j^5 a$ extends (by Hensel's lemma) to a factorisation of a, and the various types appear as follows:

C1	y	(smooth)
$C2_m$	$y^2 - x^{m+2}$ $(m \geq 0)$	(A_{m+1})
C3, C3′, C3″	$y^3 - x^4$, $y^3 - yx^3$, $y^3 - x^5$	(E_6, E_7, E_8)
C4, C4′	$y^4 - x^5$, $y^4 - yx^4$	(W_{12}, W_{13})
C5, C5′	$y^5 - x^6$, $y^5 - yx^5$.	

For D = 3 we can write

$$f = yz^2 + 2za(x) + b(x,y) \qquad (*)$$

$$\Delta_z f = a(x)^2 - yb(x,y) + w^2;$$

Conversely, a function of x and y can be put in the form $a(x)^2 - yb(x,y)$ if and only if **either** y = 0 is a factor **or** y = 0 has even intersection multiplicity with it. Now for C1, C3′, C4′ and C5′ y is a factor; as it is for $C2_{2k}$ if we rewrite this as $y(y - x^{k+1})$. For $C2_{2k-1}$, C3, C4 and

C5, the intersection multiplicity of $y = 0$ with this component is $2k + 1$, 4, 5 and 6 respectively; adding the contributions of the other components yields $2k + 4$, 6, 6 and 6; all of which are even. The argument thus falls only for C3″ which, as noted above, must be excluded.

For $D = 4$, we have linear reduction $\alpha^{(1)}\ w^2y - wzb(z) + (ya(y) - zc(y,z))$ and (∗) can always be put in this form.
For $\alpha^{(2)}$, $\Delta_w L_x f = (yb(y) - za(z))^2 = 4yz\, c(y,z)$ and $\varphi(y,z)$ can be put in this form provided it is a square or (zero) modulo each of y and z: a condition which, as above, excludes only C3″.
For $\alpha^{(3)}$, we seek to put φ in the above form with $d = 0$.

We may write $\varphi(x,y) \equiv (b(x))^2 + y\gamma(x) \pmod{y^2}$ (provided we exclude C3″) and then need ord $\gamma \geq$ ord $b + 2$.
For $C2_{2k}$ take $y(y - x^{k+1})$ as $y(y + d(w))$ and then take $b(w)$ and $c(w)$ as 0.
For $C2_{2k-1}$ (and other cases below) take $d(w) = 0$.

Modulo y^2, $\varphi \equiv x^{2k+1}\,\psi$ with ord $\psi = 3$, so $\varphi \equiv x^{2k+4} + y\gamma(x)$ with ord $\gamma \geq 2k + 3 \geq (k+2) + 2$ provided $k \geq 1$.
For C3, C4 and C5 similarly, $\varphi \equiv x^{4,5,6}\,\psi$ with ord $\psi = 2,1,0$. Then $\varphi \equiv x^6 + y\gamma(x)$ with ord $\gamma \geq 5 = 3 + 2$ as required. We have already observed that cases C1, C3′, C4′, C5, C3″ do not occur. We can thus find appropriate equations in all cases. We abstain from exhibiting them.

We turn to strata of type β. We again borrow Laufer's [7] list of equations: in our notation, this runs

4D2	$z^3 = x^4 + y^4$
2D2 + D4	$z^3 = y^4 + x^2y^2 + x^3z$
2D2 + D4′	$z^3 = y^4 + x^2y^2 + x^5$
2D4	$z^3 = x^2y^2 + x^3z + y^3z$
D4 + D4′	$z^3 = x^2y^2 + x^3z + y^5$
2D4′	$z^3 = x^2y^2 + x^5 + y^5$
D2 + D6	$z^3 + xy^3 + x^3z$
D8	$z^3 = y^4 + x^3z$

Here the curves from the first blow-up are seen on setting $z = 0$ and factorising the quartic terms in x and y; the multiplicities of the factors must then be doubled.

Our pencil of quadrics has symbol (3;1) and normal form $\langle xz + y^2, xy\rangle$: here the 3-fold line ℓ is $x = y = 0$ and the other line $y = z = 0$. We unfold to

$$f = (xz + y^2 + a(w),\ xy + b(z,w) + yc(z,w) + xd(w)), \text{ with}$$

$$L_x f = zb(z,w) + yzc(z,w) - (y^2 + a(w))\,(y + d(w)).$$

To put any function of the form $y^3 + y\varphi(w,z) + \psi(w,z)$ in this shape (with $d(w) = 0$) is (uniquely) possible, provided z divides ψ. We observe that in the examples above we can extract linear factors corresponding to all roots of the quartic other than those of type D4′. Clearly the preferred factor contains the exceptional curve for which a_c is altered from 3 to 2 by the reduction: observe that there is a unique curve c with $n_c = 1$ in each case except D4′ where there are none. It is thus relatively easy to find an appropriate equation in each case.

§4. EQUATIONS VIA RESOLUTION.

Since we have classified our singularities by means of resolutions, to find appropriate equations it is necessary to calculate the resolution from the equation. This is a routine process, and we will describe it in summary form: it does not seem of interest to exhibit explicit equations in all cases. We treat the possible 2-jets in turn, beginning with the cases $D = 2, 3$ in the order given in §1.

In detail, we first give a 'prenormal form' [12] for functions with the given 2-jet, and then blow up the origin once and describe the singular points in the result. We verify in each case (after a preliminary 'normalisation' when $D = 2$) that if the result has rational double points only, it is one of the cases listed in §1, and the remaining cases are either singular along a component of the exceptional curve or deform to cases where there is a simple elliptic singularity at some point. This proves the result asserted in the introduction, and also shows that there is an increase in modality associated with passing from elliptic to non-elliptic singularities: this is further considered in §6 below.

Our notation for blowing up is as follows. If we have 4 variables (w,x,y,z) then, for example, the "y-blow-up", or "blow-up using y" is obtained from the substitution

$$w = WY, \quad x = XY, \quad y = Y, \quad z = YZ$$

and then dividing through by Y^2 (or Y^3 in the 3 variable case).

We recognise rational double points, as in [3], using the results of Arnold [1] on semiquasihomogeneous functions. In the 3 variable cases, we assign weights to the variables and then check terms of degree ≤ 1 with respect to these weights (in fact there will be no terms of degree <1). In the 4 variable cases, we sometimes make a substitution to reduce to the cases above and sometimes assign weights directly: here there are two equations f_1, f_2 and we check only terms of degree $\leq d_i$ in f_i. We specify d_1 and d_2, but these are in fact the least degrees arising.

In the majority of cases arising, these terms will determine the singularity up to isomorphism. However, if we are able to show that the 2-jet of $\varphi(w,x,y)$ is $w^2 + x^2$, or that the 3-jet is $w^2 + x^2y$, then $\varphi = 0$ has a singularity of type A_{k+1} or D_{k+2} (for some $k \geq 1$ or 3) respectively, and it is not always possible to determine k by assigning weights. Thus in a number of cases we content outselves with leaving our conclusion in this form. Also, for the occurrences of the simple elliptic singularities $\tilde{E}_8, \tilde{E}_7, \tilde{E}_6, \tilde{D}_5$ we do not need to verify the nondegeneracy conditions showing that these (and no higher) singularities are present, though the reader may check that for appropriate choices of the coefficients this is in fact the case.

With these explanations, we are ready to begin.

α: D = 2 Although there is one slight difference in this case, we choose it to discuss fully as an introduction to the rest, since all the arguments will be entirely analogous.

We begin with the equation

$$0 = w^2 + a(x,y),$$

where a has order ≥5. The first blow-up produces the exceptional curve α, corresponding to w = 0, but this is at least a double line. For the blow-up we write w = WX, y = YX, x = X and divide by X^2. We now further set W = VX and divide by a further X^2. This is a normalisation (if a has order 5); or we may regard the two operations together as a (2,1,1) blow-up.

If we write $a(x,y) = \sum a_{ij}x^iy^j$, we obtain

$$0 = V^2 + X \sum_{i+j=5} a_{i,j}Y^j + X^2 \sum_{i+j=6} a_{i,j}Y^j + \dots$$

This has a singular point at the origin if and only if $a_{5,0} = 0$. The singular points thus correspond to the directions given by the roots of quintic $\sum_{i+j=5} a_{i,j}\, x^i y^j = 0$. If this quintic vanishes identically, the surface we have constructed still has a non-isolated singularity. Such a surface cannot be elliptic. We can thus exclude this case, and let the equation

$$\sum_{i+j=5} a_{i,j}\, Y^j = 0$$

have a root of multiplicity m ($1 \le m \le 5$) at Y = 0.

If m = 1, we have an A_1 singularity.

If m ≥ 2, and $a_{6,0} \neq 0$, we have an A_{2m-1} singularity, as we see by taking weights $V \to \frac{1}{2}$ $X \to \frac{1}{2}$ $Y \to \frac{1}{2m}$.

If m = 2 and $a_{6,0} = 0$ we have a D_{k+2} singularity for some k ≥ 2, since the 3-jet is $a_{3,2}XY^2 + a_{5,1}X^2Y + a_{7,0}X^3$ ($a_{3,2} \neq 0$). An algorithm to decide the value of k in all cases is complicated and we do not pursue it.

If m ≥ 3, $a_{6,0} = 0$ and $a_{5,1} \neq 0$, we have a D_{2m} singularity as we see by taking weights

$$V \to \frac{1}{2} \quad Y \to \frac{1}{2m-1} \quad X \to \frac{m-1}{2m-1} \quad .$$

Otherwise m ≥ 3, $a_{6,0} = a_{5,1} = 0$. Now if $a_{7,0} = 0$ we have a singularity that deforms to $\tilde{E}_7$ $\left(\text{take weights } \left(\frac{1}{2}, \frac{1}{4}, \frac{1}{4}\right)\right)$. If $a_{7,0} \neq 0$ and m ≥ 4, our singularity deforms to $\tilde{E}_8$ $\left(\text{take weights } \left(\frac{1}{2}, \frac{1}{3}, \frac{1}{6}\right)\right)$. Finally if $a_{7,0} \neq 0$ and m = 3, we have an E_7 singularity $\left(\text{take weights } \left(\frac{1}{2}, \frac{1}{3}, \frac{2}{9}\right)\right)$.

The above paragraphs enumerate all the cases when the first blow-up has rational singularities; and give the precise condtions under which they occur. The cases arising are:

A_{2m-1} ($1 \le m \le 5$), D_{k+2} (m=2), D_{2m} ($3 \le m \le 5$) and E_7 (m = 3).

This coincides precisely with the list of cases obtained in §1: we see that corresponding cases can be identified. Moreover, the fact that the values of m at the different critical points add up to 5 also concides with the condition $\sum n_d = 5$ (for α singularities) of §1.

To introduce the notion to be used in the other cases, we now summarise the calculations above.

$a_{6,0} \neq 0$ or $m = 1$	A_{2m-1}	$\left(\frac{1}{2}, \frac{1}{2}, \frac{1}{2m}\right)$	Cm
$a_{6,0} = 0, m = 2$	D_{k+2}	(3-jet)	$C2_k$
$a_{6,0} = 0, a_{5,1} \neq 0, m \geq 3$	D_{2m}	$\left(\frac{1}{2}, \frac{m-1}{2m-1}, \frac{1}{2m-1}\right)$	Cm'
$a_{6,0} = a_{5,1} = 0, a_{7,0} \neq 0, m = 3$	E_7	$\left(\frac{1}{2}, \frac{1}{3}, \frac{2}{9}\right)$	C3"
$a_{6,0} = a_{5,1} = a_{7,0} = 0, m \geq 3$	$\tilde{E}_7$	$\left(\frac{1}{2}, \frac{1}{4}, \frac{1}{4}\right)$	
$a_{6,0} = a_{5,1} = 0, m \geq 4$	$(\tilde{E}_8)$	$\left(\frac{1}{2}, \frac{1}{3}, \frac{1}{6}\right)$	

With the same notation, the remaining hypersurface cases are as follows.

α: D = 3 We see from the equation

$$0 = x^2y + xa(z) + b(y,z)$$

that the first blow-up produces the exceptional curve α (x = 0) and one other curve C (y = 0) with $n_C = 1$. The singular points on α correspond to the roots of $b_4(y,z) = 0$: again we use m to denote the multiplicity of the root. These sum to 4: the additional 1 (to make 5) is the point y = 0.

At a point on α other than y = 0 we may suppose z = 0. Use the y-blow-up.

$b_{5,0} \neq 0$ or $m = 1$	A_{2m-1}	$\left(\frac{1}{2}, \frac{1}{2}, \frac{1}{2m}\right)$	Cm
$b_{5,0} = 0, m = 2$	D_{k+2}	(3-jet)	$C2_k$
$b_{5,0} = 0, b_{4,1} \neq 0, m \geq 3$	D_{2m}	$\left(\frac{1}{2}, \frac{m-1}{2m-1}, \frac{1}{2m-1}\right)$	Cm'
$b_{5,0} = b_{4,1} = 0, b_{6,0} \neq 0, m = 3$	E_7	$\left(\frac{1}{2}, \frac{1}{3}, \frac{2}{9}\right)$	C3"
$b_{5,0} = b_{4,1} = b_{6,0} = 0, m \geq 3$	$(\tilde{E}_6)$	$\left(\frac{1}{3}, \frac{1}{3}, \frac{1}{3}\right)$	
$b_{5,0} = b_{4,1} = 0, m \geq 4$	$(\tilde{E}_8)$	$\left(\frac{1}{2}, \frac{1}{3}, \frac{1}{6}\right)$	

At the point x = y = 0 we set x = XZ., y = YZ, z = Z. If m = 0, the point is nonsingular. Otherwise, we have

$a_3 \neq 0$	A_{2m}	$\left(\frac{m}{2m-1}, \frac{1}{2m-1}, \frac{m+1}{2m-1}\right)$	C(m+1)
$a_3 = 0, m = 1$	A_{k+1}	(2-jet)	$C2_k$
$a_3 = 0, b_{0,5} \neq 0, m \geq 2$	D_{2m+1}	$\left(\frac{2m-1}{4m}, \frac{1}{2m}, \frac{1}{2}\right)$	C(m+1)′
$a_3 = b_{0,5} = 0, m \geq 2$	$(\tilde{E}_6)$	$\left(\frac{1}{3}, \frac{1}{3}, \frac{1}{3}\right)$	

Again these correspond precisely to the cases in §1, once we note that the curve C (with $n_C = 1$) has already been blown up, and that this singular point counts as (m+1) of the 5 neighbours of α.

β: D = 3. We have the prenormal form

$$x^3 + xa(y,z) + b(y,z)$$

and the first blow-up yields the curve β (x = 0). Singular points of β correspond to roots of $b_4(y,z) = 0$: a root of multiplicity m counts 2m towards the 8 neighbours of β. We analyse a typical root y = 0 using the z-blow-up

$a_{0,3} \neq 0$ or $m = 1$	A_{3m-1}	$\left(\frac{1}{3}, \frac{1}{3m}, \frac{2}{3}\right)$	D2m
$a_{0,3} = 0,\ b_{0,5} \neq 0, m = 2$	E_6	$\left(\frac{1}{3}, \frac{1}{4}, \frac{1}{2}\right)$	D4′
$a_{0,3} = b_{0,5} = 0, m \geq 2$	$(\tilde{E}_6)$	$\left(\frac{1}{3}, \frac{1}{3}, \frac{1}{3}\right)$	
$a_{0,3} = 0, m \geq 3$	$(\tilde{E}_8)$	$\left(\frac{1}{3}, \frac{1}{6}, \frac{1}{2}\right)$.	

We now turn to the complete intersection cases. The 2-jets were listed in §2: we discuss them in turn using an unfolding of each.

$\alpha^{(1)}$: D = 4. $\langle w^2 + xz + a(y),\ xy + wb(z) + c(y,z) \rangle$. The first blow-up produces the curve α (x = w = 0) and the conic C ($y = w^2 + xz = 0$), which is thus one of the curves in the resolution with $n_c = 1$. These intersect in a single point (w = x = y = 0). The singular points on α are given by the roots of $ya_3(y) - zc_3(y,z) = 0$ together with the intersection (y = 0) with C.

A typical point other than this intersection is z = 0: we analyse using the y-blow-up. The second component reduces to $X + \varphi(W,Y,Z)$: substituting in the first gives the same singularity as we find on the exceptional curve of the blow-up of the reduction

$$y(w^2 + a(y)) - z(wb(z) + c(y,z)) = 0.$$

The discussion is thus reduced to the previous case.

For α ∩ C we blow up using z. Here we find the same situation as above with the roles of the two components reversed.

$\alpha^{(2)}$: **D = 4.** $\langle wy + xz,\ wx + wa(z) + xb(y) + c(y,z)\rangle$. The first blow-up produces 3 exceptional curves: α ($w = x = 0$), C_1 ($w = z = 0$) and C_2 ($x = y = 0$), thus C_1 and C_2 meet α in distinct points P_1, P_2. Apart from these, the singular points on α correspond to roots of $c_3(y,z) = 0$: as usual write m for the multiplicity of these roots.

At P_2 we blow-up using Z, and substitute $x = -WY$ from the first equation. Then:

$m = 0$	smooth	(Z)	C1
$m = 1$	A_{k+2}	(2-jet)	$C2_k$
$a_2 \neq 0,\ m \geq 2$	A_{2m}	$\left(\frac{m}{2m+1}, \frac{1}{2m+1}, \frac{m+1}{2m+1}\right)$	C(m+1)
$a_2 = 0,\ c_{0,4} \neq 0,\ m \geq 2$	D_{2m+1}	$\left(\frac{2m-1}{4m}, \frac{1}{2m}, \frac{1}{2}\right)$	C(m+1)'
$a_2 = c_{0,4} = 0,\ m \geq 2$	$(\tilde{E}_6)$	$\left(\frac{1}{3}, \frac{1}{3}, \frac{1}{3}\right)$.	

We shall not give calculations for the other points on α, as the coordinates are less convenient here: as for $\alpha^{(1)}$, they can be reduced to the case D = 3 by linear reduction.

$\alpha^{(3)}$: **D = 4.** $\langle xz + a(w,y),\ y(x + z) + b(w) + yc(w) + zd(w)\rangle$. Here the first blow-up yields exceptional curves α ($x = z = 0$), C_1 ($x = y = 0$) and C_2 ($y = z = 0$) which are concurrent in a point P. Apart from this, the singular points on α are given by roots of $a_3(w,y) = 0$. Taking $w = 0$ as a typical such point, we blow up using y and then substitute for X from the second equation.

$m = 1$	A_1	$\left(\frac{1}{2}, \frac{1}{2}, \frac{1}{2}\right)$	C1
$a_{0,4} \neq 0,\ m \geq 2$	A_{2m-1}	$\left(\frac{1}{2m}, \frac{1}{2}, \frac{1}{2}\right)$	Cm
$a_{0,4} = 0,\ m = 2$	D_{k+2}	$\left(\frac{1}{3}, \frac{1}{3}, \frac{1}{2}\right)$	$C2_k$
$a_{0,4} = 0,\ a_{1,3} \neq 0,\ m = 3$	D_6	$\left(\frac{1}{5}, \frac{2}{5}, \frac{1}{2}\right)$	C3'
$a_{0,4} = a_{1,3} = 0,\ a_{0,5} \neq 0,\ m = 3$	E_7	$\left(\frac{2}{9}, \frac{1}{3}, \frac{1}{2}\right)$	C3''
$a_{0,4} = a_{1,3} = a_{0,5} = 0,\ m = 3$	$\tilde{E}_7$	$\left(\frac{1}{4}, \frac{1}{4}, \frac{1}{2}\right)$	

For the point P we blow up using w. Here if $b_3 \neq 0$ we can substitute for W from the second equation into the first. Taking weights

$$W, X, Y, Z \to \frac{m+2}{2m+2}, \frac{1}{2}, \frac{1}{2m+2}, \frac{1}{2}$$

and ignoring terms of weight $> \frac{m+2}{2m+2}$ in f_2 allows us to determine the result modulo terms of weight >1: it has type A_{2m+1}. If $b_3 = 0$ but $m = 0$, we can substitute for W from the first equation into the second: we see from the 2-jet that the result has type A_k for some k. If $b_3 = 0$ and $m \geq 1$ we do not have a hypersurface singularity. We summarise these arguments as follows.

$b_3 \neq 0$	A_2	$\left(\frac{m+2}{2m+2}, \frac{1}{2}, \frac{1}{2m+2}, \frac{1}{2} \middle/ 1, \frac{m+2}{2m+2}\right)$	C(m+1)
$b_3 = 0, m = 0$	A_{k+1}	(2-jet)	$C2_k$
$b_3 = 0, m \geq 0$	$(\tilde{D}_5)$	(1,1,1,1/2,2).	

β: D = 4. $\langle xz + y^2 + a(w), xy + b(z,w) + yc(z,w) + xd(w)\rangle$. The first blow-up yields the exceptional curves β (x = y = 0) and C_1 (y = z = 0): the remaining singularities on β correspond to roots of $b_3(z,w) = 0$.

At the typical point w = 0 (other than $\beta \cap C_1$) of β we blow up using z and substitute for X from the first equation into the second.

$c_{2,0} \neq 0$ or $m = 1$	A_{3m-1}	$\left(\frac{1}{3m}, \frac{1}{3}, \frac{2}{3}\right)$	D2m
$c_{2,0} = 0, b_{4,0} \neq 0, m = 2$	E_6	$\left(\frac{1}{4}, \frac{1}{3}, \frac{1}{2}\right)$	D4′
$c_{2,0} = b_{4,0} = 0, m \geq 2$	$(\tilde{E}_6)$	$\left(\frac{1}{3}, \frac{1}{3}, \frac{1}{3}\right)$	
$c_{2,0} = 0, m \geq 3$	$(\tilde{E}_8)$	$\left(\frac{1}{6}, \frac{1}{3}, \frac{1}{2}\right)$	

At $\beta \cap C_1$ we blow up using w

$a_3 \neq 0$ or $m = 0$	A_{3m+1}	$\left(\frac{2m+2}{3m+2}, \frac{2m+1}{3m+2}, \frac{m+1}{3m+2}, \frac{1}{3m+2} \middle/ \frac{2m+2}{3m+2}, 1\right)$	D(2m+2)
$a_3 = 0, m \geq 0$	$(\tilde{D}_5)$	(1,1,1,1/2,2).	

γ: D = 4 $\langle wz + y^2 + d(x), z^2 + y^2a(x,w) + 2yb(x,w) + c(x,w)\rangle$. The first blow-up yields just the exceptional curve γ (y = z = 0). The singular points on this are w = 0 and the roots of $c_3 = 0$: to obtain the correct multiplicity of 11 (eleven) we must count the former as 2 and the roots of $c_3 = 0$ as 3 each.

For a point other than w = 0 we take the root as x = 0 and blow-up using w: then use the first equation to substitute for Z into the second.

$b_{0,2} \neq 0$ or $m = 1$	A_{4m-1}	$\left(\frac{3}{4}, \frac{1}{4m}, \frac{1}{4}\right)$	E(3m)

$b_{0,2} = 0, m \geq 2$	$(\tilde{E}_6)$	$\left(\frac{1}{3}, \frac{1}{3}, \frac{1}{3}\right)$	

At the point $w = 0$ we blow up using x.

$m = 0$	A_1	$\left(\frac{1}{2}, 1, \frac{1}{2}, \frac{1}{2} \middle/ 1, 1\right)$	E2
$d_3 \neq 0, m = 1$	D_5	$\left(\frac{1}{4}, \frac{3}{4}, \frac{3}{8}, \frac{1}{2} \middle/ \frac{3}{4}, 1\right)$	E5
$d_3 = 0, m \geq 1$	$(\tilde{D}_5)$	$\left(\frac{1}{2}, \frac{1}{2}, \frac{1}{2}, \frac{1}{2} \middle/ 1, 1\right)$	
$m \geq 2$	$(\tilde{E}_8)$	$\left(\frac{1}{6}, \frac{2}{3}, \frac{1}{3}, \frac{1}{2} \middle/ \frac{2}{3}, 1\right)$	

δ: D = 4 $\langle xy - z^2, w^2 + \varphi(x,y,z)\rangle$. The first blow-up yields the curve $\delta{:}0 = w = xy - z^2$, and the singular points on the conic are those where φ vanishes. The total of their multiplicities is 6. We investigate in particular the point $y = z = 0$ of δ, noting that we may reduce φ to the form $a(x,y) + zb(x,y)$. Apply the y-blow-up, and substitute $x = z^2$ in the second equation.

$a_{0,4} \neq 0$ or $m = 1$	A_{2m-1}	$\left(\frac{1}{2}, \frac{1}{2}, \frac{1}{2m}\right)$	F(m)
$a_{0,4} = 0, m = 2$	D_{k+2}	(3-jet) $\left(\frac{1}{2}, \frac{1}{3}, \frac{1}{3}\right)$	$F2_k$
$a_{0,4} = 0, b_{0,3} \neq 0, m \geq 3$	D_{2m}	$\left(\frac{1}{2}, \frac{m-1}{2m-1}, \frac{1}{2m-1}\right)$	Fm′
$a_{0,4} = b_{0,3} = 0, a_{0,5} \neq 0, m = 3$	E_7	$\left(\frac{1}{2}, \frac{1}{3}, \frac{2}{9}\right)$	F3″
$a_{0,4} = b_{0,3} = a_{0,5} = 0, m \geq 3$	$\tilde{E}_7$	$\left(\frac{1}{2}, \frac{1}{4}, \frac{1}{4}\right)$	
$a_{0,4} = b_{0,3} = 0, m \geq 0$	$(\tilde{E}_8)$	$\left(\frac{1}{2}, \frac{1}{3}, \frac{1}{6}\right)$.	

$\alpha\alpha^{(1)}$: D = 4 $\langle wx - yz, w^2 + a(x,y) + b(x,z) + yzc(x)\rangle$. Here the first blow-up yields the curves α_1: $w = y = 0$ and α_2: $w = z = 0$ meeting in a point P. We may assume $b_{r,0} = 0$ for any r, as terms x^r can be absorbed in $a(x,y)$. The singular points on α_1 are P (which counts twice) with the roots of $a_3x^3 + b_3(x,z) = 0$, counting once each.

For a point other than P we suppose $x = 0$ and form the z-blow-up: then substitute $Y = WX$ in the second equation.

$b_{0,4} \neq 0$ or $m = 1$	A_{2m-1}	$\left(\frac{1}{2}, \frac{1}{2m}, \frac{1}{2}\right)$	Cm
$b_{0,4} = 0, m = 2$	D_{k+2}	$\left(\frac{1}{2}, \frac{1}{3}, \frac{1}{3}\right)$	$C2_k$

$b_{0,4} = 0, b_{1,3} \neq 0, m = 3$	D_6	$\left(\frac{1}{2}, \frac{1}{5}, \frac{2}{5}\right)$	C3′
$b_{0,4} = b_{1,3} = 0, b_{0,5} \neq 0, m = 3$	E_7	$\left(\frac{1}{2}, \frac{2}{9}, \frac{1}{3}\right)$	C3″
$b_{0,4} = b_{1,3} = b_{0,5} = 0, m = 3$	$(\tilde{E}_7)$	$\left(\frac{1}{2}, \frac{1}{4}, \frac{1}{4}\right).$	

For the point P we use the x-blow-up and substitute $W = YZ$ in the second equation.

$m = 0$	smooth	$(1,1,1)$	$G2,2_0$
$b_{2,1} \neq 0$ or $m = 1$	A_{2m+1}	$\left(\frac{m+2}{2m+2}, \frac{1}{2m+2}, \frac{m}{2m+2}\right)$	G3, (m+2)
$b_{2,1} = 0, m \geq 2$	$(\tilde{E}_6)$	$\left(\frac{1}{3}, \frac{1}{3}, \frac{1}{3}\right)$	

$\alpha\alpha^{(2)}$: D = 4. $\langle yz + a(w) + xb(w), x^2 + yc(y,w) + zd(z,w) + e(w)\rangle$. We have curves α_1 $(x = y = 0)$ and α_2 $(x = z = 0)$ meeting in a point P. Other singular points on α_1 are given by $zd_2(z,w) + e_3w^3 = 0$.

For a point other than P we suppose $w = 0$ and use the z blow-up. We can substitute for Y from the first equation into the second: in fact in each case below, this has too high weight to make a contribution.

$d_{3,0} \neq 0$ or $m = 1$	A_{2m-1}	$\left(\frac{1}{2m}, \frac{1}{2}, \frac{1}{2}\right)$	Cm
$d_{3,0} = 0, m = 2$	D_{k+2}	$\left(\frac{1}{3}, \frac{1}{2}, \frac{1}{3}\right)$	$C2_k$
$d_{3,0} = 0, d_{2,1} \neq 0, m = 3$	D_6	$\left(\frac{1}{5}, \frac{1}{2}, \frac{2}{5}\right)$	C3′
$d_{3,0} = d_{2,1} = 0, d_{4,0} \neq 0, m = 3$	E_7	$\left(\frac{2}{9}, \frac{1}{2}, \frac{1}{3}\right)$	C3″
$d_{3,0} = d_{2,1} = d_{4,0} = 0, m = 3$	$(\tilde{E}_7)$	$\left(\frac{1}{4}, \frac{1}{2}, \frac{1}{4}\right)$	

For the point P we use the w blow-up. Observe that there are two multiplicities: one as singular point on α_1 and one as singular point on α_2: they are independent save that $m_1 \geq 1 \Leftrightarrow e_3 = 0 \Leftrightarrow m_2 \geq 1$.

$a_3 \neq 0, m = (0,0)$	A_1	$\left(1, \frac{1}{2}, \frac{1}{2}, \frac{1}{2} \middle/ 1, 1\right)$	$G22_1$
$a_3 = 0, m = (0,0)$	A_k	$\left(1, \frac{1}{2}, \frac{1}{2}, \frac{1}{2} \middle/ 1, 1\right)$	$G22_k$

$a_3 \neq 0$, m = (1,n)	D_{2n+2}	$\left(\frac{n+1}{2n+1}, \frac{1}{2}, \frac{1}{2n+1}, \frac{n}{2n+1} \Big/ \frac{n+1}{2n+1}, 1\right)$	G(3,n+2)′
$a_3 = 0$, m ≥ (1,1)	$(\tilde{D}_5)$	$\left(\frac{1}{2}, \frac{1}{2}, \frac{1}{2}, \frac{1}{2} \Big/ 1, 1\right)$	
m ≥ (2,2)	$(\tilde{E}_7)$	$\left(\frac{1}{2}, \frac{1}{2}, \frac{1}{4}, \frac{1}{4} \Big/ \frac{1}{2}, 1\right)$	

Taking cases $\alpha\alpha^{(1)}$ and $\alpha\alpha^{(2)}$ together, we observe that in either of them we can have any combination of the Cm, $C2_k$, C3′ and C3″ with the appropriate total multiplicity at points other than P; at P itself we have the possibilities

$\alpha\alpha^{(1)}$:	$G22_0$	G33,	G34,	G35,
$\alpha\alpha^{(2)}$:	$G22_k (k \geq 1)$	G33′	G34′	G35′ .

Thus each case from the previous enumeration occurs precisely once. It is not entirely clear why the list of cases splits into two in this particular way.

The remaining case. $\langle x^2 + a(w,z) + yb(w,z), y^2 + c(w,z) + xd(w,z)\rangle$. Blowing up once here produces a whole singular line corresponding to x = 0: we never have just isolated singularities.

Indeed, the line x = y = 0 defined by the intersection of the two quadrics has, in this case, multiplicity 4 and its generic hyperplane section gives the ideal $\langle x^2,y^2\rangle$ in $\mathcal{O}_{x,y}$ (in all previous cases, the ideal was of the form $\langle x^m,y\rangle$).

This concludes our proof of Theorem 1 via enumeration, except for the cases of Kodaira type, the analogous calculations for which we leave to the reader. However, a direct proof can also be given, and I am grateful to the referee for supplying the following argument.

Write ω, ω_i for the canonical classes of X, X_i. Then if $f_1: X_1 \to X$ is a resolution, X is elliptic Gorenstein if and only if $f_{1*}\omega_1 = \mathfrak{m}_x.\omega$. Now if we suppose X elliptic Gorenstein, and X_2 a partial resolution defined by blowing up $\mathfrak{m}_x$, so f_1 factors as

$$X_1 \xrightarrow{f_3} X_2 \xrightarrow{f_2} X,$$

then either $f_{2*}\omega_2 = \omega$ (if X is a hypersurface double point) or $f_{2*}\omega_2 = \mathfrak{m}_x.\omega = f_{1*}\omega_1$, so $f_{3*}\omega_1 = \omega_2$ and X_2 then has rational singularities.

Conversely, if X is prepared and X_2 obtained by blowing up $\mathfrak{m}_x$, then $f_{2*}\omega_2 = \mathfrak{m}_x.\omega$, so if X_2 has rational singularities so that $f_{3*}\omega_1 = \omega_2$, it follows that $f_{1*}\omega_1 = \omega$, so that X is indeed elliptic.

§5. The Weighted Homogeneous Case.

We now investigate the weighted homogeneous singularities included in the above. For such a singularity, the resolution graph must be star-shaped. One sees at once (e.g. consider whether or not the exceptional vertex α,β ... can be the centre of the star) that these cases are as follows. In our notation below we indicate the integer $r = n_{c_0} - 1$ where c_0 is the vertex of the star.

α1 = 5C1 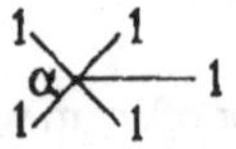

α3 = C4 + C1 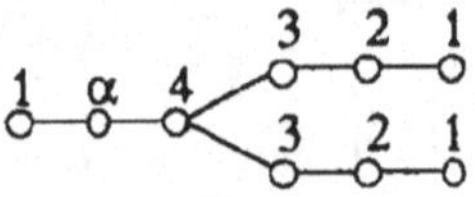

α4 = C5 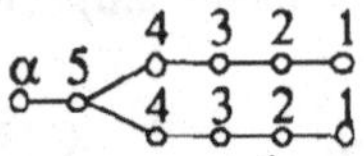

α5 = C4′ + 1 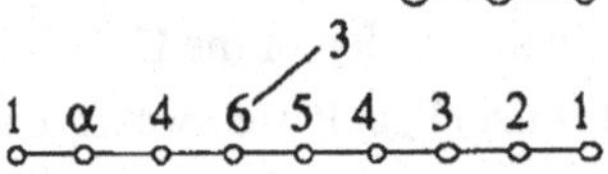

α7 = C5′ 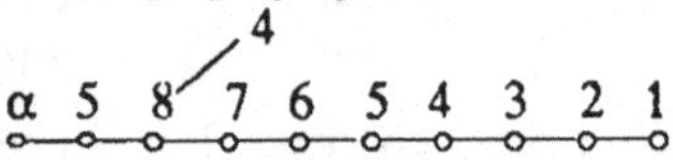

β2 = 4D2 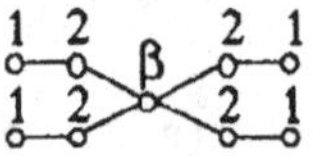

β5 = D6 + D2 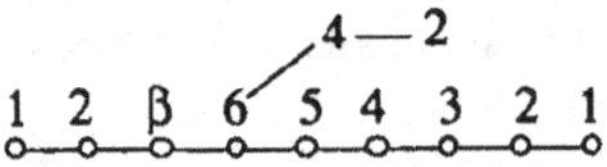

β7 = D8 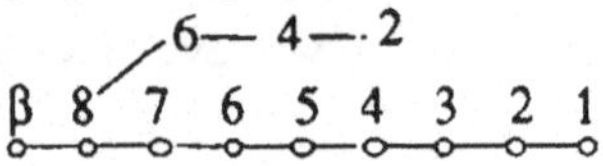

γ3 = 3E3 + E2 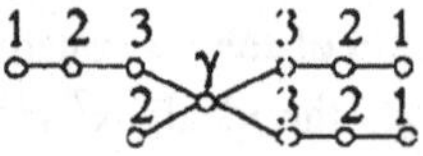

γ8 = E9 + E2 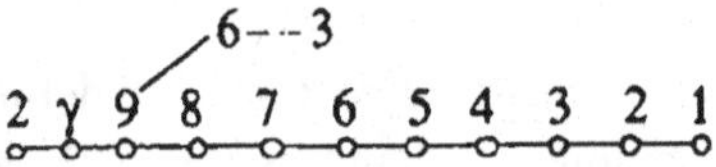

δ1 = 6F1 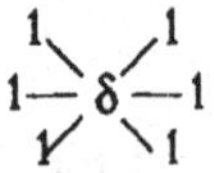

δ4 = F5 + F1 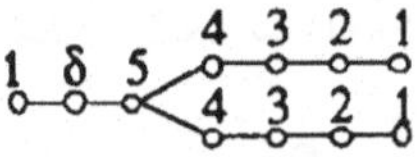

δ5 = F6 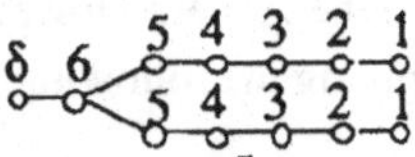

δ7 = F5′ + F1 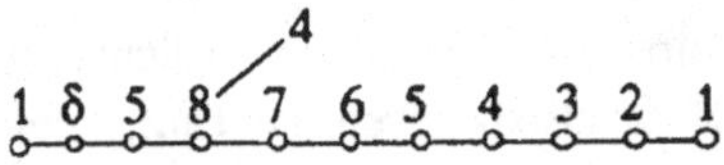

δ9 = F6′ 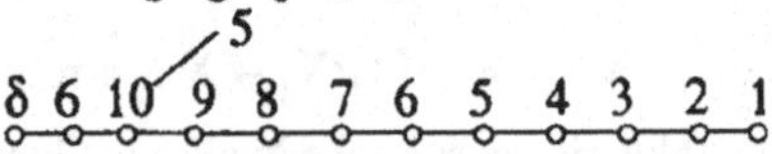

In each case there do exist corresponding weighted homogeneous singularities: we next determine the corresponding weights and degrees. For example, a hypersurface singularity with weights u,v,w and degree d is denoted by u,v,w/d (some authors write $\frac{1}{d}(u,v,w)$). We treat cases in the same order as before.

α: D = 2. Here we have Laufer's equations for curves, and add w^2:

α1:	$x^5 + y^5$	2,2,5/10
α3:	$x(y^4 - x^5)$	4,5,12/24
α4:	$y^5 - x^6$	5,6,15/30
α5:	$x(y^4 - yx^4)$	6,8,19/38
α7:	$y^5 - yx^5$	8,10,25/50

α: D = 3. When we have vertices with $n_c = 1$ and $a_c > 2$, we list the corresponding values of $a_c - 2$. We make the convention that these numbers are separated by commas if the vertices play a symmetrical role; otherwise by semicolons. When there may be ambiguity, we list vertices from left to right in the above diagram.

We can obtain weights using quadratic reduction. If $f = yx^2 + 2za(x) + b(x)$ then if x,y,z have weights w_1, w_2 and w_3, f has degree $w_2 + 2w_3$. The reduction $\Delta_z f = w^2 + a(x)^2 - yb(x)$ then has weight $(x,y,z) = (w_1, w_2, w_2 + w_3)$ and degree $2w_2 + 2w_3$. We can reverse this once we have decided which of w_1 and w_2 is which. Observe that w_2 is the weight of y, which is tangent to the factor where a multiplicity is to be modified. We thus deduce

α1	(0,0,0,0,1)	2,2,3/8		
α3	(1;0,0)	4,5,8/20	(0;0,1)	4,5,7/19
α5	(1;0)	6,8,13/32	(0;1)	6,8,11/30
α4	(0,1)	5,6,9/24		
α7	(1)	8,10, 15/40		

α: D = 4. Here we use linear reduction. If we have

$$f(x,y,z,w) = (xy + \varphi(y,z,w), xz + \psi(y,z,w))$$

so that $\Delta_x f = y\psi - x\varphi$ then if $wt(x,y,z,w) = (w_1,w_2,w_3,w_4)$ we have deg $f = (w_1 + w_2, w_1 + w_3)$, while for $\Delta_x f$ we have $w_2,w_3,w_4/w_1 + w_2 + w_3$. Conversely given this, to recover the weights and degrees of f we must know which of w_2,w_3 and w_4 is the weight of w. Now for $\alpha^{(1)}$ strata, this "w" is the "z" produced by reversing the construction Δ. For $\alpha^{(2)}$ strata, it is the tangent to the already preferred branch of the curve. And for $\alpha^{(3)}$ it is the coordinate not tangent to the preferred branch. We thus find

α1	$(0,0,0,0,2)^{(1)}$	2,2,3,4/6,6	$(0,0,0,1,1)^{(2)}$	2,2,3,3/5,6
α3	$(2;0,0)^{(1)}$	4,5,8,11/15,16	$(1;0,1)^{(2)}$	4,5,7,8/12,15
	$(0;0,2)^{(1)}$	4,5,7,10/14,15	$(0;1,1)^{(3)}$	4,5,7,7/12,14
α5	$(2;0)^{(1)}$	6,8,13,18/24,26	$(1;1)^{(2)}$	6,8,11,13/19,24
	$(0;2)^{(1)}$	6,8,	11,16/22,24	
α4	$(0,2)^{(1)}$	5,6,9,13/18,19	$(1,1)^{(3)}$	5,6,9,9/15,18
α7	$(2)^{(1)}$	8,10,15,22/30,32		

β: D = 3.

β2	can take $\varphi = 0$, ψ of degree 4 general.		3,3,4/12
β5	Take $\psi = xy^3$.	For D6 need x^3z term.	6,7,9/27
β7	Take $\psi = y^4$.	For D8 need x^3z term.	8,9,12/36

β: D = 4.

β2	Take b_3 general, rest 0.	3,3,4,5/8,9
β5	(1,0) We take $b = w^3$, $c = z^2$ (rest 0)	6,7,9,12/18,21
	(0,1) Take $b = wz^2$, $a = w^3$	6,7,9,11/18,20
β7	Take $b = z^3$: need $a = w^3$	8,9,12,15/24,27

These could also be obtained by linear reduction methods.

γ.

γ3 Roots of c_3 must be distinct and general: need no more

$\langle wz + y^2, z^2 + x^3 + w^3\rangle$ 4,4,5,6/10,12

γ8 c_3 has 3-fold root not at w: take as x^3. Need $b_{0,2} \neq 0$

$\langle wz + y^2, z^2 + yw^2 + x^3\rangle$ 9,10,12,15/24,30

δ.

δ1 φ_3 meets the conic in 6 distinct points: we need no more

$\langle xy - z^2, w^2 + \varphi_3(x,y,z)\rangle$ 2,2,2,3/4,6

δ4 Take φ_3 as x^2z. We need $a_{0,4} \neq 0$, hence y^4 term

$\langle xy - z^2, w^2 + x^2z + y^4\rangle$ 5,6,7,10/12,20

δ5 Take φ_3 as x^3: again we need a y^4 term

$\langle xy - z^2, w^2 + x^3 + y^4\rangle$ 6,7,8,12/14,24

δ7 Here $\varphi_3 = x^3z$, $a_{0,4} = 0$, we need $b_{0,3} \neq 0$ i.e. y^3z term

$\langle xy - z^2, w^2 + x^2z + y^3z\rangle$ 8,10,12,17/20,34

δ9 In this case $\varphi_3 = x^3$ and we need y^3z

$\langle xy - z^2, w^2 + x^3 + y^3z\rangle$ 10,12,14,21/24,42

This - together with the list of weights of weighted homogeneous Kodaira singularities - completes the list of weights of elliptic isolated complete intersection singularities of surfaces. They are characterised among icis by the inequality $\Sigma w_j - \Sigma d_i < \min d_i$. In fact, in each of the above cases, $\Sigma w_j - \Sigma d_i = r$ and $\min d_i = r + 1$, where $r = n_c - 1$ as above.

The author has tried - and failed - to enumerate directly those sextuples $(w_1,w_2,w_3,w_4,d_1,d_2)$ which satisfy this inequality and define an icis.

For each weighted homogeneous singularity we have an induced action of S^1 on the link, which we can thus regard as a Seifert manifold. We can determine the Seifert invariants in two ways. If we start from the resolution, there will be an exceptional fibre corresponding to each branch of the graph from the bifurcation point. The $\mathbb{C}^\times$ action in a neighbourhood of a curve E of the resolution can be described as follows.

Suppose E has selfintersection -d. Then a neighbourhood of E is obtained by attaching two copies of $\mathbb{C} \times \mathbb{C}$ by the identification

$$(x,y) \sim (x^{-1}, x^d y),$$

where E itself is given by $y = 0$. The group $\mathbb{C}^\times$ acts by

$$t(x,y) = (t^u x, t^v y),$$

say. Thus $\mathbb{C}^\times$ acts on E by multiplication by t^u, on the adjacent fibre at $x = 0$ we have multiplication by t^v and on that at $x = \infty$ by t^{ud-v}. If E_0 is the branch curve of the resolution, $\mathbb{C}^\times$ acts trivially on E_0 and by multiplication by t on its normal bundle, hence on each adjacent curve. The invariant we seek corresponds to the action at the fixed point at the end curve of the branch other than its intersection point with the next curve.

We thus arrive at the following rule. List the values of a_c in order along the branch starting from (but excluding) E_0. Form the negative continued fraction

$$\frac{p}{q} = a_1 - \cfrac{1}{a_2 - \cfrac{1}{a_3 \ldots - \cfrac{1}{a_k}}}$$

Then p is the order of the corresponding isotropy group and (p,q) its Seifert invariant.

Alternatively we can determine the p_i directly from the equation: each coordinate axis lying on S corresponding to a variable x_i with weight w_i contributes one $p = w_i$; for each coordinate (x_i,x_j) plane, any remaining component of its intersection with S contributes a p equal to the h.c.f. (w_i,w_j) (if this exceeds one); similarly for any coordinate hyperplane. Now for each p, the action of $t = \exp(2\pi i/p)$ has eigenvalues t^{d_i} on the source and t^{w_j} on the target. The former subset must include the latter: one of the remaining eigenvlaues is 1 (we have a fixed curve), the other is t^{-r} (for $r = \Sigma w_j - \Sigma d_i$). We leave the reader to verify that this gives the same result in each case.

The invariants of the circle action on the link form a special case of the general result mentioned in [4, 1.2]. We have $q_i r \equiv 1 \pmod{p_i}$ for each i. The quotient orbifold is a sphere with cone points of orders p_i, thus has $\chi = 2 + \Sigma\left(p_i - 1\right)$, and the link is a circle bundle over it with characteristic class $e = r^{-1}\chi$. The first homology group of the link is a finite group of order $\Delta = |e| \Pi p_i$.

We now tabulate the values of the p_i, χ and Δ in the above cases (for the Kodaira cases see [4, p8]).

Name		Weight	Degrees	p	-e	Δ	μ
α1	0	2, 2, 5	10	2, 2, 2, 2, 2,	$\frac{1}{2}$	16	16
	1	2, 2, 3	8	2, 2, 2, 2, 3	$\frac{2}{3}$	32	15
	1, 1	2, 2, 3, 3	5, 6	2, 2, 2, 3, 3	$\frac{5}{6}$	60	14
	2	2, 2, 3, 4	6, 6	2, 2, 2, 2, 4	$\frac{3}{4}$	48	14

	Name	Weight	Degrees	p	-e	Δ	μ
α3	0; 0 0	4, 5, 12	24	5, 4, 4	$\frac{1}{10}$	8	19
	0; 0 1	4, 5, 7	19	5, 4, 7	$\frac{19}{140}$	19	18
	1; 0 0	4, 5, 8	20	8, 4, 4	$\frac{1}{8}$	16	18
	0; 0 1	4, 5, 7, 7	12, 14	5, 7, 7	$\frac{6}{35}$	42	17
	1; 0 1	4, 5, 7, 8	12, 15	8, 4, 7	$\frac{9}{86}$	36	17
	0; 0 2	4, 5, 7, 10	14, 15	5, 4, 10	$\frac{3}{20}$	30	17
	2; 0 0	4, 5, 8, 11	15, 16	11, 4, 4	$\frac{3}{22}$	24	17
α4	0, 0	5, 6, 15	30	3, 5, 5,	$\frac{1}{15}$	5	20
	0, 1	5, 6, 9	24	3, 5, 9	$\frac{4}{45}$	12	19
	1, 1	5, 6, 9, 9	15, 18	3, 9, 9	$\frac{1}{9}$	27	18
	0, 2	5, 6, 9, 13	18, 19	3, 5, 13	$\frac{19}{195}$	19	18
α5	0, 0	6, 8, 19	38	8, 2, 6	$\frac{1}{24}$	4	20
	0, 1	6, 8, 11	30	8, 2, 11	$\frac{5}{88}$	10	19
	1, 0	6, 8, 13	32	13, 2, 6	$\frac{2}{39}$	8	19
	0, 2	6, 8, 11, 16	22, 24	8, 2, 16	$\frac{1}{16}$	16	18
	1, 1	6, 8, 11, 13	19, 24	13, 2, 11	$\frac{19}{286}$	19	18
	2, 0	6, 8, 13, 18	14, 26	18, 2, 6	$\frac{1}{18}$	12	18

Name		Weight	Degrees	p	-e	Δ	μ
α7	0	8, 10, 25	50	5, 2, 8	$\frac{1}{40}$	2	21
	1	8, 10, 15	40	5, 2, 15	$\frac{1}{30}$	5	20
	2	8, 10, 15, 22	30, 32	5, 2, 22	$\frac{2}{55}$	8	19
β2	0	3, 3, 4	12	3, 3, 3, 3	$\frac{1}{3}$	27	18
	1	3, 3, 4, 5	8, 9	3, 3, 3, 5	$\frac{2}{5}$	54	17
β5	0, 0	6, 7, 9	27	7, 3, 6	$\frac{1}{14}$	9	20
	0, 1	6, 7, 9, 11	18, 20	7, 3, 11	$\frac{20}{231}$	20	19
	1, 0	6, 7, 9, 12	18, 21	12, 3, 6	$\frac{1}{12}$	18	19
β7	0	8, 9, 12	36	3, 4, 8	$\frac{1}{24}$	4	21
	1	8, 9, 12, 15	24, 27	3, 4, 15	$\frac{1}{20}$	9	20
γ3		4, 4, 5, 6	10, 12	2, 4, 4, 4	$\frac{1}{4}$	32	19
γ8		9, 10, 12, 15	24, 30	3, 5, 9	$\frac{2}{45}$	6	21
δ1		2, 2, 2, 3	4, 6	2, 2, 2, 2, 2, 2	1	64	15
δ4		5, 6, 7, 10	12, 20	5, 5, 7	$\frac{4}{35}$	20	19
δ5		6, 7, 8, 12	14, 24	4, 6, 6	$\frac{1}{12}$	12	20
δ7		8, 10, 12, 17	20, 34	2, 8, 12	$\frac{1}{24}$	8	20
δ9		10, 12, 14, 21	24, 42	2, 7, 10	$\frac{1}{35}$	4	21

In the table we have also listed the values of the Milnor number μ. This can be obtained from the genral formula of [5], in terms of weights but it is simpler to note that the follows from Laufer's formula [8] as in §1.

§6. MODALITY.

In the previous paragraphs we have discussed isolated surface singularities and given defining equations. We may also study these equations from the viewpoint of the theory of singularities of mappings; it is well known that the map germs $f_i:(\mathbb{C}^n,0) \to (\mathbb{C}^p,0)$ $(n > p)$ are $\mathcal{K}$-(contact-) equivalent if and only if the variety germs $f_i^{-1}(\underset{\sim}{0})$ are analytically equivalent.

In the hypersurface case $(n = 3, p = 1)$ it has been found (notably by Arnold [2]) that the modality $m(f)$ of f with respect to $\mathcal{R}$-(right-) equivalence is related to the structure of the singularity. He enumerated the cases $m(f) \leq 2$, and identified them. In the terminology of this paper, $m(f) = 0$ gives the rational double points, and $m(f) = 1$ or 2 the Kodaira singularities, where $m(f) = 1$ for those of types I_n, II, III and IV and $m(f) = 2$ for those of type I_n^*, II^*, III^* and IV^*.

Further classifications of functions f are known, and the modality determined for them (see e.g. [12]), but to the author's best knowledge these have not yet been given alternative characterisations. In the case of weighted homogeneous singularities, those of modality 3 or 4 were lised in [13]. For modality 3, in addition to the members of Arnold's series E, Z^1 and Q ($E_{4,0}$, $E_{4,(\varepsilon)}$, $Z_{2,0}$, $Z_{2,(\varepsilon)}$, $Q_{3,0}$, $Q_{3,(\varepsilon)}$, $\varepsilon = 0$ or ± 1, in the notation of [12]), the list consists precisely of the quasihomogeneous elliptic hypersurface singularities of type α, as listed in §5 above. In fact we have

THEOREM 2. *Let* f *define an isolated hypersurface singularity, with* $m(f) = 3$, *not belonging to the series* E, Z^1 *and* Q. *Then* $f = 0$ *is elliptic of type* α; *conversely, all these have modality* 3.

PROOF. For let f have corank 2. By the splitting theorem, we may suppose f a function of two variables. As it does not belong to the E series, it has order ≥ 4. If it has order 4 and is not a cusp singularity and does not belong to the Z^1 series, the 4-jet is of the form y^4. As it is not a Kodaira singularity in the W-series, the Newton polygon lies above that of $y^4 + z^8$. But then the modality is ≥ 4.

If f has order ≥ 6, its modality is ≥ 6. Hence f has order 5. Now for each elliptic singularity we can easily verify that the modality is 3. Otherwise, we have the $\widetilde{E}_7$ or $\widetilde{E}_8$ case in our enumeration. Checking the conditions for these, we see that the Newton polygon for f lies above (or coincides with) that of

$$y^5 + x^2y^3 + x^8 \quad \text{or} \quad x(y^4 + x^6)$$

respectively. But for each of these we have $m(f_0) = 4$ (and $\mu(f_0) = 22$) for the function given, hence for any nondegenerate function with the same Newton polygon, and $m(f) \geq m(f_0) = 4$.

If f has corank ≥ 4, then $m(f) \geq 5$. Thus f has corank 3. As it is not a cusp singularity and does not belong to the Q-series, the 3-jet is of type $S(x^2y + xz^2)$, $U(x^2y + y^3)$, $\alpha(x^2y)$ or $\beta(x^3)$. Now singularities in the S- and U- series are listed, [12], and there are none of modality 3. The generic singularity in the β-series is $\beta 2$, of modality 4. Hence again f belongs to the α-series.

But here again we can verify that for each elliptic case the modality is 3. Otherwise f has Newton polyhedron above that spanned by one of the following sets:

$$\langle x^2y, xz^3, yz^3, y^7 \rangle$$
$$\langle x^2y, xz^3, z^4, y^6 \rangle$$
$$\langle x^2y, xz^4, y^4, y^2z^2, z^6 \rangle$$

In each case a short computation shows that the function has modality ≥ 4.

This completes the proof. We have omitted the computations since, in fact, the claims follow from our results above. For when we impose successive conditions on the coefficients of f, at the point where these imply that the singularity is elliptic rather than rational the Milnor number increases by 2 rather than one. In our case we start with f elliptic and successively impose condtions. At the point where f ceases to be elliptic, one singularity in the blow-up becomes elliptic; its Milnor number increases by 2, hence so does that of f (e.g. using Laufer's formula for μ), and m(f) increases by 1.

The same argument shows that in the β-series the only functions with m(f) = 4 are the elliptic ones. However there are several functions with modality 4 in the other series and we do not get so simple a result in this case.

We observe that just 2 of the functions found above as giving boundaries are weighted homogeneous:

$w^2 + x(y^4 + x^6)$	2, 3, 7/14
$x^2y + y^6 + z^4$	5, 2, 3/12 .

These weights also are of some interest.

It seems that there should be analogous characterisations in the complete intersection case. However $\mathcal{R}$-modality is no longer available here. The notion of $\mathcal{K}$-modality was explored in [11], but seems too crude to yield the results we would like. The refinement by strict-$\mathcal{K}$-modality [11] depends on constructing bifurcations: this also is too imprecise as it seems that, for example, the 'boundaries of the elliptic singularities' as obtained above, like the elliptic singularities themselves, still have strict -$\mathcal{K}$-modality 2. We leave this question open for possible further work.

REFERENCES.

1. V.I. Arnol'd, Normal forms for functions near degenerate critical points, *Russian Math. Surveys* 29ii (1974) 11-49.

2. V.I. Arnol'd, Critical points of smooth functions and their normal forms, *Russian Math. Surveys* 30i (1975) 1-75.

3. J.W. Bruce & C.T.C. Wall, On the classification of cubic surfaces, *Journal London Math. Soc.* 19 (1979) 245-256.

4. W. Ebeling & C.T.C. Wall, Kodaira singularities and an extension of Arnol'd's strange duality, *Comp. Math.* 56 (1985) 3-77.

5. G-M. Greuel & H.A. Hamm, Invarianten quasihomogener vollständiger Durchschnitte, *Invent. Math.* 49 (1978) 67-86.

6. K. Kodaira, On compact complex analytic surfaces II, *Ann. of Math.* 77 (1963) 563-626.

7. H. Laufer, On minimally elliptic singularities, *Amer. J. Math.* 99 (1977) 1257-1295.

8. H. Laufer, On μ for surface singularities, pp. 45-49 in Proc. Symp. in Pure Math. 30, *Amer. Math. Soc.* 1977.

9. D.T. Lê, Calcul du nombre de Milnor d'une singularité isolée d'intersection complète, *Funct. Anal. Appl.* 8 (1974) 45-52.

10. M.A. Reid, Elliptic Gorenstein singularities, unpublished manuscript, 1975.

11. C.T.C. Wall, Classification of unimodal isolated singularities of complete intersections, pp. 625-640 in Proc. Symp. Pure Math. 40, vol. 2, *Amer. Math. Soc.* 1983.

12. C.T.C. Wall, Notes on the classification of singularities, *Proc. London Math Soc.* 48 (1984) 461-513.

13. E. Yoshinaga & H. Suzuki, Normal forms of nondegenerate quasihomogeneous functions with inner modality ≤4, *Invent. Math.* 55 (1979) 185-206.

Department of Pure Mathematics,
University of Liverpool,
P.O. Box 147,
Liverpool L69 3BX.

PENCILS OF CUBIC CURVES AND RATIONAL ELLIPTIC SURFACES

C.T.C.Wall

A pencil of cubic curves in the plane can be studied from several viewpoints. One may be interested in the geometry presented by the family of curves themselves. Or one may blow up the plane at the base points of the pencil, obtaining (if the general cubic is nonsingular) a rational elliptic surface whose own geometry is of interest per se, and as a limiting form of del Pezzo surfaces.

However, if one studies the family of all pencils, it is natural to apply the techniques of geometric invariant theory with the aim of understanding moduli, and of selecting "good" classes of stable or semi-stable pencils. Indeed in Miranda's paper [2], just this was done, and the result related to the geometry of the rational elliptic surface.

The object of the present paper is to go over this ground in more detail, seeking a fuller understanding of the interrelation between geometry in the plane, geometry of elliptic surfaces, and geometric invariant theory. The full results are not easy to summarise, but points we believe to be of particular interest are the description in §4 of exceptional fibres explicitly from the invariant theory, and the analysis in §5 of the different ways a rational elliptic surface can be blown down to a plane.

The sections of this paper are as follows:

Notation

We write our pencil as $\lambda f(x,y,z) + \mu g(x,y,z)$, where f and g are homogeneous cubics in x, y and z. We refer to the cubic curves f = 0 and g = 0 as Γ and Γ' respectively. Denote by $x^\vee$, $y^\vee$ and $z^\vee$ the dual coordinates in the dual vector space; this induces a notion of orthogonality between cubics in x, y and z and those in $x^\vee$, $y^\vee$ and $z^\vee$ which is termed *apolarity*.

§1. Rational elliptic surfaces

We call a pencil *regular* if it contains a smooth (i.e. elliptic) cubic; otherwise *singular*. For a regular pencil, if we blow up the ambient plane $P^2(\mathbf{C})$ at the 9 points of intersection of any two distinct members of the pencil, we obtain a (rational) elliptic surface. If the blowing-up is performed sequentially, the last exceptional curve yields a section of the surface. Conversely, any rational elliptic surface with a section arises in this way. This fact seems to be known, but I include a proof since I cannot find a reference.

More generally, consider an arbitrary elliptic surface, S, with base curve B. Now for any surface S, S is ruled if and only if the Kodaira dimension $\kappa(S) = -\infty$; the base curve of the ruling then has genus $1 - \chi(\mathcal{O}_S)$; so S is rational if and only if $\kappa(S) = -\infty$ and $\chi(\mathcal{O}_S) = 1$. For an elliptic surface S we have the formula $\kappa(S) = \mathrm{sgn}\,(\chi(\mathcal{O}_S) - \chi^{orb}(B))$ in the notation of [12], so our conditions reduce to $\chi(\mathcal{O}_S) = 1$ and $\chi^{orb}(B) > 1$.

Now the "orbifold Euler characteristic" is given by the formula

$$\chi^{orb}(B) = 2 - 2g(B) - \sum(1 - m_i^{-1}),$$

where the m_i are the multiplicities of the multiple fibres. The inequality $\chi^{orb}(B) > 1$ thus implies that B has genus 0 and that there is at most one multiple fibre. Write m for this multiplicity, or $m = 1$ if there is no multiple fibre: then $\chi^{orb}(B) = 1 + m^{-1}$. The canonical class K is rationally equivalent to $(\chi(\mathcal{O}_S) - \chi^{orb}(B)) = -m^{-1}$ times the class F of a fibre.

For any irreducible curve C on S not contained in a fibre, $C.K < 0$, so by the adjunction formula, $C.C \geq -1$. Thus any irreducible curve C on S with $C.C \leq -2$ is contained in a fibre; by Kodaira's enumeration [1] of exceptional fibres, we see that the only possibility is $C.C = -2$.

Now S is obtained by iterated blowing up from a minimal rational surface, which can only be $P^2(\mathbb{C})$, $P^1(\mathbb{C}) \times P^1(\mathbb{C})$ or the Hirzebruch surface F_2 (since there are no curves with self-intersection < -2). But we cannot blow up F_2 at a point of the (-2)-curve, since this would produce a (-3)-curve; blowing it up at any other point, or $P^1(\mathbb{C}) \times P^1(\mathbb{C})$ at any point, gives a surface obtainable from $P^2(\mathbb{C})$ by blowing-up twice. Thus in all cases, since $\chi(S) = 12$, S is obtained from $P^2(\mathbb{C})$ by blowing up 9 times.

Thus S contains exceptional curves. Such a curve has $E.K = -1$, so $E.F = m$: we have an m-section. If $m = 1$, a general fibre meets the exceptional curves in distinct points, so blows down to a smooth elliptic curve, which must thus be a cubic. Since two such fibres are disjoint, the 9 intersection points of the two cubics must be the points to be blown up. Thus the images of the fibres form a pencil of cubics.

Arguing similarly in the general case, we find that we have a pencil of plane curves of degree 3m with 9 assigned m^{tuple} points (which cannot be arbitrarily chosen). The cubic through these points, counted m times, then gives the multiple fibre.

For the remainder of this paper, we will tacitly restrict to the case $m = 1$ above. Here it follows from the adjunction formula that a curve is a section of the fibration if and only if it is an exceptional curve. Indeed, if $m > 1$, as there is a multiple fibre there can be no sections: thus sections exist if and only if $m = 1$.

We now recall from [1] Kodaira's classification of types of fibre of elliptic surfaces: we have the table

Name	$I_n (n\geq0)$	$I_n^* (n\geq0)$	II	III	IV	II^*	III^*	IV^*
χ	n	n+6	2	3	4	10	9	8
Δ	A_{n-1}	D_{n+4}	-	A_1	A_2	E_8	E_7	E_6

In each case, we have given the Euler characteristic χ of the fibre: for a rational surface, these add up to 12.

The components F_{ij} of a singular fibre F_i span a subgroup X_i of the Picard group P of S. The intersection form restricted to X_i is negative semidefinite, with radical generated by the class K.

If s_0 is a section of S, it meets just one component F_{i0} of F_i; the rest span a subgroup mapping isomorphically to $\bar{X}_i = X_i/\mathbb{Z}K$, and form a basis for a root system in this: the type of this system is indicated by Δ in the above table. All F_{ij} are rational: if we exclude cases II, III, IV they have normal crossings, and the dual graph represents the extended root system of the corresponding type. Cases II, III, IV have fibres isomorphic to plane cubic curves: cuspidal, conic & tangent, or concurrent lines respectively.

If $K^{\perp}$ is the orthogonal complement of K in P, then $Q = K^{\perp}/\mathbb{Z}K$ is isomorphic to a root lattice of type E_8. The root lattices $\bar{X}_i$ form an orthogonal direct sum $\bar{X} \subset Q$. Addition in the group structures of the fibres (taking s_0 as zero) defines a group structure on the set $\mathcal{S}$ of sections and [8] the map $\mathcal{S} \longrightarrow Q/\bar{X}$ induced by $s \longmapsto [s-s_0]$ is an isomorphism. This description of the situation is copied from [5].

We can also verify the following. If $X_i^{\#}$ denotes the dual lattice to $\bar{X}_i$, taking intersections of $s-s_0$ with the F_{ij} defines a natural map ϕ_i: $\mathcal{S} \longrightarrow X_i^{\#}/\bar{X}_i = D_i$, say. There is a bijection of D_i on the set of components of F_i of multiplicity 1 such that if $\phi_i(s)$ corresponds to F_{ij} then F_{ij} is the component of F_i which intersects s.

§2. Geometric invariant theory

We recall briefly the essential notions of geometric invariant theory. We consider only the case of a reductive group G (usually SL_n) acting linearly on a vector space V, or the restriction of such an action to some invariant subvariety W of V. A point P of W is said to be *stable* for the action if (i) the isotropy group G_P is finite and (ii) the orbit G.P is closed. The point P is called *semi-stable* if 0 does *not* belong to the closure of the orbit G.P.

These properties are detected by the Hilbert-Mumford criterion, which we now describe. Consider a 1-parameter subgroup $\gamma(t)$ of G. The image in GL(V) can be identified with a diagonal subgroup $t \rightarrowtail \mathrm{diag}(t^{a1},\dots,t^{an})$ by choosing suitable coordinates in V. Then P is not stable (resp. not semistable) if for all r with $x_r(P) \neq 0$, $a_r \geq 0$ (resp. $a_r > 0$). Moreover, the set of non-stable (resp. non-semistable) points is the union of the (linear) subspaces defined by this condition as γ varies over the set of 1-parameter subgroups of G.

We now discuss the application of this criterion to the family of pencils of plane cubics. SL_3 acts on the space C (of dimension 10) of plane cubics, hence on its second exterior power (of dimension 45), the decomposable points in which parametrise the pencils of plane cubics, using Plücker coordinates. In his paper [2] Miranda applies the criterion to the question of stability for this problem, and derives the following results.

Proposition 2.1 ([2,5.1]) The pencil is not semistable if and only if we can choose coordinates x, y and z and (distinct) members f and g of the pencil such that one of the following occurs:

(U1) $f \in \langle x^3 \rangle$.

The pencil contains a triple line.

(U2) $f \in \langle x^3, x^2y \rangle$, g apolar to $\langle y^\vee z^{\vee 2}, z^{\vee 3} \rangle$.

The cubic Γ has the form l^2m; Γ' touches l at $l \cap m$.

(U3) $f \in \langle x^3, x^2y, x^2z \rangle$, g apolar to $\langle y^{\vee 2}z^\vee, y^\vee z^{\vee 2}, z^{\vee 3} \rangle$.

Γ has the form l^2m; l is an inflexional tangent to Γ'.

(U4) $f \in \langle x^3, x^2y, xy^2, y^3 \rangle$, g apolar to $\langle x^\vee z^{\vee 2}, .y^\vee z^{\vee 2}, z^{\vee 3} \rangle$.

The point (0,0,1) is double on Γ' and triple on Γ.

(U5) $f, g \in \langle x^3, x^2y, x^2z, xy^2, xyz \rangle$.

Γ and Γ' have a common double point (0,0,1) and a common tangent $y = 0$ there.

Proposition 2.2 ([2,5.2]) The pencil is not stable if and only if we can choose f, g, x, y and z such that one (at least) of the following holds:

(U6) $f \in \langle x^3, x^2y, x^2z \rangle$.

$\Gamma = l^2m$ contains a repeated line.

(U7) $f, g \in \langle x^3, x^2y, x^2z, xy^2, xyz, xz^2 \rangle$.

Γ and Γ' have a common line component $x = 0$.

(U8) $f \in \langle x^3, x^2y, x^2z, xy^2 \rangle$, g apolar to $\langle y^\vee z^{\vee 2}, z^{\vee 3} \rangle$.

Γ consists of a conic S and a line $x = 0$ tangent to it; Γ' touches $x = 0$ at its point of contact with S.

(U9) $f \in \langle x^3, x^2y, xy^2, y^3 \rangle$, g apolar to $\langle z^{\vee 3} \rangle$.

Γ has a triple point (0,0,1) which lies on Γ'.

(U10) f, g apolar to $\langle x^{\vee}z^{\vee 2}, .y^{\vee}z^{\vee 2}, z^{\vee 3}\rangle$.

Γ and Γ' have a common double point (0,0,1).

In each of the cases above, Γ' may be a general element of the pencil; we will retain the convention that Γ is the special element. We will explore the geometric consequences of this result in more detail below. We observe that an alternative calculation leading to the same result was sketched in [11]. Instead of considering the action of SL_3 on the second exterior power (of dimension 45) of C, we consider the action of $SL_3 \times SL_2$ on $C \oplus C = C \otimes \mathbb{C}^2$ (of dimension 20). It follows from the main result of [11] that the notions of stability and semi-stability in the two cases correspond.

The main results of [2] can now be summarised as follows.

Proposition (2.3) [2,7.1] A pencil is stable if and only if it is regular, and every fibre is reduced.

Thus fibres of types I_n, II, III, IV are permitted; I_n^*, II^*, III^* and IV^* are not. As to singular pencils, we see [2,§7] by Bertini's theorem that either

(α) all members of the pencil have a common singular point,

(β) the curves of the pencil consist of a fixed line plus a pencil of conics, or

(γ) we have a fixed conic and a pencil of lines.

Now (α) coincides with case (U10) of (1.2) and (β) with case (U7); (γ) is a subcase of (U8), so indeed none of these can be stable.

PROPOSITION (2.4) [2,6.1] A regular pencil is semi-stable if and only if no component of a fibre has multiplicity ≥3.

This permits fibres of type I_n^*, but still excludes II^*, III^* and IV^*. Also certain singular pencils are semistable: we will study semistable singular pencils in §7 below. The results (2.3) and (2.4) follow from the more detailed assertions in §3 below, which are obtained by elaborating Miranda's arguments.

Geometric invariant theory also affords a characterisation of those semistable pencils whose orbits are closed (only the regular case is described in [2]: the regular case of the next result is taken from [2,8.3,8.5]).

PROPOSITION 2.5 A semistable pencil defines a closed orbit if and only if it is equivalent to one of

(CS) $\langle x^2z, \phi_3(y,z)\rangle$, where ϕ_3 has distinct linear factors,

(CT) $\langle x(y^2+ axz), z(y^2+ bxz)\rangle$, where $ab \neq 0$,

(CU) $\langle x^2z, y^2z\rangle$.

Case (CS) is singular if and only if $z|\phi_3$; (CT) only if $a = b$; (CU) is, of course, singular.

A regular pencil defines a closed orbit if and only if it has two fibres of type I_0^*.

§3. Identification of exceptional fibres

In this section we consider only regular pencils. It is routine to perform explicitly the blowings-up required to obtain an elliptic surface, and hence to identify the nature of the fibre in each case. We now list all the cases that arise in terms of the cubic Γ giving rise to the fibre F in question. Observe that since the pencil is regular, any further member Γ' of it is smooth at singular points of Γ.

(1) Γ is smooth. $F \cong \Gamma$ has type I_0.

(2) Γ is irreducible nodal, or conic and chord, or triangle. Each node of Γ not lying on Γ' is unaltered. If Γ' has intersection numbers (k,1) with the branches of Γ at the node, we acquire k new curves on blowing-up. If n_Γ and $n = n_F$ are the numbers of components of Γ and F respectively, we thus have

$$n = n_\Gamma + \sum_{nodes} k_i,$$

and F has type I_n.

Since this case includes many subcases, we may denote it by $(2)(n_\Gamma; k_1, k_2, \ldots)$.

Next suppose Γ cuspidal, with cusp P.

(3) If Γ' does not pass through P, $F \cong \Gamma$ has type II.

(4) If $P \in \Gamma'$ but Γ' is not tangent to Γ at P, F has type III.

(5) If Γ' touches Γ at P, F has type IV.

If Γ consists of a conic σ and a line l touching σ at P, the cases are determined by the intersection multiplicities (k,k') of Γ' with l and with σ at P.

(6) (0,0) type III.

(7) (1,1) type IV.

(8) (2,2) type I_0^* (general case of U8).

(9) (3,2) type I_1^*.

(10) (2,3) type I_1^*.

(11) (2,4) type I_2^*.

(12) (2,5) type I_3^*.

(13) (2,6) type I_4^*.

If Γ consists of three lines meeting in a point P, we need the intersection multiplicities of Γ· with the lines at P.

(14) (0,0,0) type IV.

(15) (1,1,1) type I_0^* (general case of U9).

(16) (1,1,2) type I_1^*.

(17) (1,1,3) type I_2^*.

If $\Gamma = x^2y$ contains a repeated line, we list cases in terms of the symbol $(a;b_0;b_1,..)$, where a is the intersection multiplicity of Γ· with y = 0 at x = y = 0, b_0 the intersection multiplicity of Γ· with x = 0 at x = y = 0, and $b_1, ...$ those with x = 0 at the remaining points of intersection.

(18) (0;0;1,1,1) type I_0^* (general case of U6).

(19) (0;0;2,1) type I_1^*.

(20) (0;0;3) type IV^* (general case of U3).

(21) (1;1;1,1) type I_1^*.

(22) (2;1;1,1) type I_2^*.

(23) (3;1;1,1) type I_3^*.

(24) (1;1;2) type I_2^*.

(25) (2;1;2) type I_3^*.

(26) (3;1;2) type I_4^*.

(27) (1;2;1) type IV^* (general case of U2).

(28) (1;3) type III^*.

Finally if Γ is a threefold line $x^3 = 0$ we list the intersection multiplicities of $\Gamma\cdot$ with $x = 0$.

(29) (1,1,1) type IV^* (general case of U1).

(30) (2,1) type III^*.

(31) (3) type II^*.

§4. Semistable regular pencils

It follows from general results in geometric invariant theory that the closure of any semistable orbit contains a unique closed orbit. From §2 we know that there are three types of closed orbit, and types CS and CT appear in the unfolding of type CU which we will describe below. Thus this unfolding contains all semistable orbits, and we will proceed to locate them all explicitly.

We can take our versal unfolding of $\langle xy^2, xz^2\rangle$ to be

$$\langle xy^2 + a_4y^2z + a_5yz^2 + a_6z^3 + a_7x^2z + a_8z^3,$$
$$xz^2 + a_1y^3 + a_2y^2z + a_3yz^2 + a_9x^3 + a_{10}x^2y\rangle.$$

The 2-dimensional torus acts on these pencils via

$$x\cdot = (uv)^{-1}x,\ y\cdot = uy,\ z\cdot = vz$$

and on $\lambda f + \mu g$ by $\lambda\cdot = u^{-1}v\lambda$, $\mu\cdot = uv^{-1}\mu$. However, it is simplest to exhibit the weights by forming the quintic $z^2f - y^2g$ and using trilinear coordinates to plot the coefficient of $x^iy^jz^k$ $(i+j+k=5)$.

This yields the diagram:

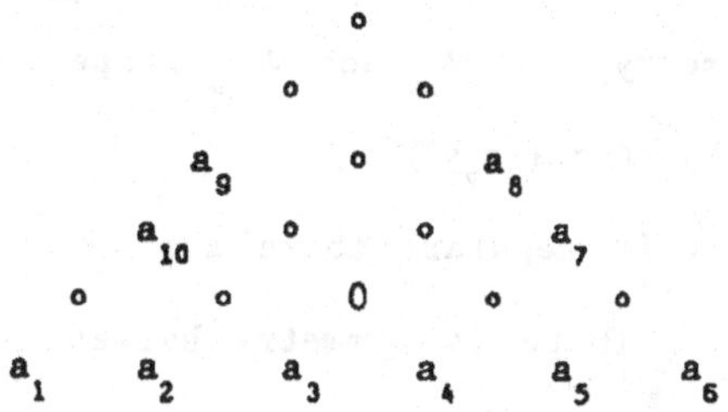

Denote by V_i the vertex in this diagram where the coefficient a_i appears ($1 \le i \le 10$). For each pencil of the above form, consider the subset $\{V_i | a_i \ne 0\}$ and the convex cover C of the (infinite) rays OV_i in the above plane. We will see that the nature of the pencil is largely determined by the sector C.

If C is contained in the sector defined by $V_1, \ldots, V_6$, the pencil consists of cubics with the common double point $y = z = 0$. It is thus singular of type α. If it is contained in the sector defined by $V_7, \ldots, V_{10}$ the cubics all contain the line $x = 0$, so the pencil is singular of type β.

For the rest of this section we exclude these cases, so C contains V_6 and V_7 (or V_1 and V_{10}: but this case is reduced to the other by interchanging y and z). First suppose C strictly contained in a half-space. We can list cases in terms of the bounding rays of C. We find that $g = 0$ is an exceptional fibre of type given by the following table:

V_6V_7 I_4^* (26)

V_5V_7 I_3^* (25) V_6V_8 I_3^* (23)

V_4V_7 I_2^* (24) V_5V_8 I_2^* (22) V_6V_9 I_2^* (17)

V_3V_7 I_1^* (19) V_4V_8 I_1^* (21) V_5V_9 I_1^* (16) V_6V_{10} I_1^* (9)

Observe that all the I_n^* cases from §2 appear here except (8), (15), (18) (all I_0^*) and (10)-(13) (I_1^*- I_4^*).

Next suppose C is a line. This must be V_2V_7 or V_3V_8 (equivalent by symmetry to V_5V_{10} and V_4V_9 respectively). For V_2V_7 we have $\langle x(y^2+ a_7xz),\ z(xz + a_2y^2)\rangle$.

If $a_2a_7 \neq 1$ this is regular; there are two singular fibres, each of type I_0^* (8). There is symmetry between these, which is different from the symmetry in our main diagram. However if $a_2a_7 = 1$ this pencil is singular of type γ. All these (with $a_2 \neq 0$, $a_7 \neq 0$) define closed orbits.

For the line V_3V_8 we have $\langle x(y^2+ a_8x^2),\ z^2(x + a_3y)\rangle$. If $a_3{}^2a_8 \neq -1$ this is regular; f=0 is a fibre of type I_0^* (15) and g=0 of type I_0^* (18). We have a closed orbit whenever $a_3 \neq 0$ and $a_8 \neq 0$; but if $a_3{}^2a_8 = -1$ the cubics have a common line $(x + a_3y = 0)$ so this reduces to a subcase of β.

If C is the whole plane, the pencil is stable. It thus remains to consider the case when C is a half-plane. Here the coefficients on the line bounding the half-plane play a rôle, as above. As before, we may assume C contains the sector OV_6V_7. We have 4 cases: V_2V_7, V_3V_8, V_4V_9 and V_5V_{10} in the above notation, though due to the additional symmetry mentioned above, we will find that the V_2V_7 and V_5V_{10} cases are essentially equivalent.

V_2V_7	V_5V_{10}	
$a_2a_7 \neq 1$	$a_5a_{10} \neq 1$	I_0^* (8)
$a_2a_7 = 1,\ a_3 \neq 0$	$a_5a_{10} = 1,\ a_6 \neq 0$	I_1^* (10)
$a_3 = 0,\ a_4 \neq 0$	$a_6 = 0,\ a_9 \neq 0$	I_2^* (11)
$a_4 = 0,\ a_5 \neq 0$	$a_9 = 0,\ a_7 \neq 0$	I_3^* (12)
$a_5 = 0,\ a_6 \neq 0$	$a_7 = 0,\ a_8 \neq 0$	I_4^* (13)

V_3V_8

$a_3^2a_8 \neq -1$	I_0^* (18)
$a_3^2a_8 = -1,\ a_3^2a_7 \neq -a_4$	I_1^* (21)
$a_3^2a_7 = -a_4,\ a_5 \neq 0$	I_2^* (22)
$a_5 = 0,\ a_6 \neq 0$	I_3^* (23)
$a_6 = 0$	β

V_4V_9

$a_4^2a_9 \neq -1$	I_0^* (15)
$a_4^2a_9 = -1,\ a_5 \neq 0$	I_1^* (16)
$a_5 = 0,\ a_6 + a_7a_4^2 \neq a_8a_4^3$	I_2^* (17)
$a_6 + a_7a_4^2 = a_8a_4^3$	β

We thus see each I_0^* type arising in an essentially unique way, and recover the relation (§2) between regular semistable pencils with closed orbits and elliptic surfaces with two exceptional fibres of type I_0^*. The remaining types (10-13) from §2 appear in the V_2V_7 list, unfolding the closed orbit of type γ. That the V_3V_8 and V_4V_9 lists produce nothing new was to be expected, since the case $a_3^2a_8 = -1$ (when C was the line V_3V_8) was a subcase of β and could itself be put in a form where the corresponding sector was contained in OV_6V_{10}.

§5. Unstable regular pencils

It follows from (2.4) that a regular pencil is unstable if and only if it has a fibre of type II^*, III^* or IV^*, and from the further details in §3 that these divide into subcases as follows:

(20) $\Gamma = x^2y$; $\Gamma\cdot$ meets $x=0$ thrice at a point other than $x=y=0$ IV^*

$\Gamma = x^2y$; $\Gamma\cdot$ touches $x=0$ at $x=y=0$;

(27) $\Gamma\cdot$ meets $x=0$ in one further point IV^*

(28) $\Gamma\cdot$ meets $x=0$ thrice at $x=y=0$ III^*

$\Gamma = x^3$; $\Gamma\cdot$ meets Γ in:

(29) 3 distinct points IV^*

(30) A repeated point and another III^*

(31) A threefold point II^*

In this paragraph we investigate the relation of this classification to that of elliptic surfaces.

To achieve this, we perform the indicated sequence of 9 blowings-up, and give the dual graphs of the set of exceptional curves that arise (including the components of Γ). For this we need to subdivide cases (20), (27) and (28) further as follows:

(a) $\Gamma\cdot$ meets $y=0$ in 3 distinct points,

(b) $\Gamma\cdot$ touches $y=0$ at a point other than $x=y=0$,

(c) $\Gamma\cdot$ has inflexional contact with $y=0$.

In the diagrams below, the vertices denoted o, ⊙ represent rational curves with selfintersection -2, -1 respectively.

(20) 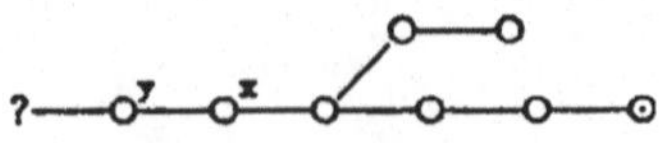

(27) 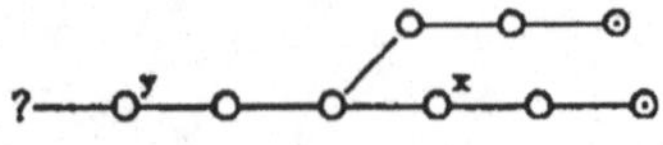

(28) 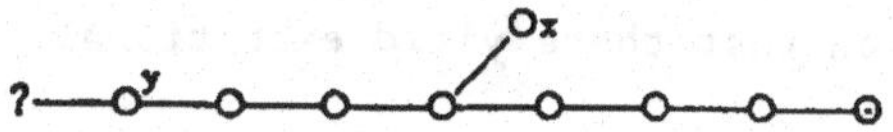

(29) 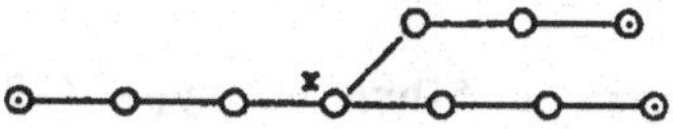

(30) 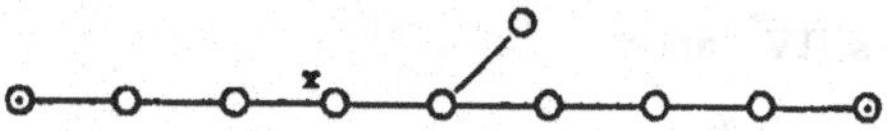

(31) 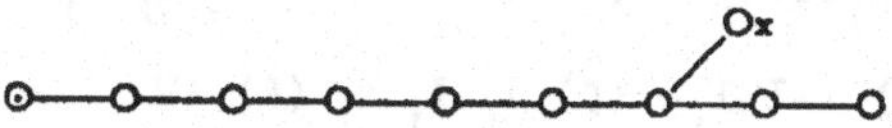,

where for (20) the ? represents

(a) 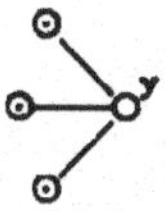

(b) 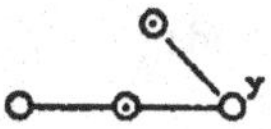

(c) ,

and for (27) and (28) the ? represents

(a) 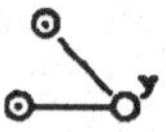

(b) .

We observe that the further (-2)-curves in cases (b) and (c) must belong to further exceptional fibres.

Now by §1, the root systems defined by the fibres together give a root subsystem of E_8. The cases arising here are thus

E_6, E_6A_1, E_6A_2, E_7, E_7A_1 and E_8.

The possible configurations of exceptional fibres are then as follows:

E_6	$IV^*.I_1^4$, $IV^*.II.I_1^2$, $IV^*.II^2$
E_6A_1	$IV^*.I_2.I_1^2$, $IV^*.I_2.II$, $IV^*.III.I_1$
E_6A_2	$IV^*.I_3.I_1$, $IV^*.IV$
E_7	$III^*.I_1^3$, $III^*.II.I_1$
E_7A_1	$III^*.I_2.I_1$, $III^*.III$
E_8	$II^*.I_1^2$, $II^*.II$.

In fact, all of these arise. To see this, we refer to our classification [10] of cubic functions on $\mathbf{C}^2$. We can consider these as defining pencils of cubic curves where one of the members is z^3. We see by inspection that these yield exceptional fibres as follows:

For family I of [10], z^3 gives a fibre of type (29) above. Singular fibres have types IV^* and:

(α) I_1^4 (β) $II.I_1^2$ (η) II^2 (γ) $I_2.I_1^2$

(δ) $II.I_2$ (ϵ) $III.I_1$ (ζ) $I_3.I_1$ (θ) IV.

For family II of [10], z^3 is a member of type (30). The singular fibres have types III^* and:

(α) I_1^3 (β) $II.I_1$ (γ) $I_2.I_1$ (δ) III.

For family III of [10], z^3 has type (31). We have singular fibres of types II^* and:

(α) I_1^2 (β) II.

We thus obtain all the possible configurations of exceptional fibres listed above.

For each such surface S, if we can find on S a configuration of exceptional curves whose dual graph is one of the above - and note that this includes the hypothesis that there are no intersections other than those shown - then by blowing down 9 curves in the inverse order to that of blowing-up we obtain a projective plane on which the fibres map as cubic curves. The exceptional curves were described in §1c: we now apply this information. Observe in each case that the first section s_0 chosen is unique up to an automorphism of S (translate in each fibre): choice of presentation of S by pencils of cubics thus depends on how much choice we have of the remaining sections that are needed. However, as the sections must be mutually disjoint, little choice is left. Recall also from §1c the discriminant groups Δ_i of the singular fibres F_i.

E_8 $\mathcal{S}$ and Δ are trivial. We have case (31) only.

E_7A_1 $\mathcal{S} \cong \mathbb{Z}/2\mathbb{Z} \cong \Delta_1 \cong \Delta_2$. The two sections are disjoint. We obtain cases (30), (28b) but not (28a).

E_7 $\mathcal{S} \cong \mathbb{Z}$ maps onto $\Delta_1 \cong \mathbb{Z}/2\mathbb{Z}$. The intersection number of s_n and s_{n+d} is $(d/2)^2 - 1$ if d is even and $(d/2)^2 - 1/4$ if d is odd: thus s_n is disjoint from $s_{n\pm1}$ and $s_{n\pm2}$ but from no others: a maximal mutually disjoint set is of the form (s_{n-1}, s_n, s_{n+1}). We have configurations on S of types (30) (using sections s_n, s_{n+1}) and (28a) (using s_{n-1} at the left hand end and s_n at the right), but not (28b) as no other fibre is reducible.

E_6A_2 $\mathcal{S} \cong \mathbb{Z}/3\mathbb{Z} \cong \Delta_1 \cong \Delta_2$. The three sections are pairwise disjoint. We thus have configurations of types (29), (27b) and (20c) on S, but not (27a), (20a) or (20b), which require at least 4 sections.

E_6A_1 $\mathcal{S} \cong \mathbb{Z}$ maps onto $\Delta_1 \times \Delta_2 \cong \mathbb{Z}/3\mathbb{Z} \times \mathbb{Z}/2\mathbb{Z}$. Two sections are disjoint if and only if their difference maps to ±1, ±2 or ±3 in $\mathbb{Z}$. We thus have configurations (29) (use s_{n-1}, s_n, s_{n+1}), (27a) (use s_n, s_{n+3} at y and s_{n+1}, s_{n+2} at the other end), (27b) (use s_n at y and s_{n-1}, s_{n+1} at the other end) and (20b) (use s_n, s_{n+3} at y and s_{n+1} at the other end: the component meeting s_n is disjoint from s_{n+1} and s_{n+3}); but not (20a) (we cannot find three disjoint sections meeting the same component of the IV^* fibre) or (20c) (which implies a second fibre of type A_2 at least).

E_6 $\mathcal{S} \cong \mathbb{Z} \oplus \mathbb{Z}$ maps onto $\Delta \cong \mathbb{Z}/3$. We visualise $\mathcal{S}$ as a hexagonal lattice of side 1: the kernel is a hexagonal sublattice of side $\sqrt{3}$. Sections are disjoint if their images are at distance 1 or $\sqrt{3}$. We cannot obtain subcases b or c of (20), (27) or (28), which require further reducible fibres, but get (29) (sections forming a triangle of side 1), (27a) (sections as in (D1) below) and (20a) (sections as in (D2) below).

To complete the above arguments, we now describe how we calculate intersections of sections. First note that if S is obtained by blowing $P^2(\mathbf{C})$ up at 9 distinct points there is a natural base of the Picard group Pic(S) consisting of:

ϵ_0, the proper transform of a general line in $P^2(\mathbf{C})$,

ϵ_i $(1\le i\le 9)$, the exceptional curves.

We have $\epsilon_0^2 = 1$, $\epsilon_i^2 = -1$ for $1\le i\le 9$, and $\epsilon_i\epsilon_j = 0$ for $i\ne j$. The class $\kappa = 3\epsilon_0 - \sum_1^9\epsilon_i$ is also the class of a fibre when we have an elliptic surface; the canonical class is $-\kappa$.

If the 9 points are not all distinct we have a deformation of the above surface and can describe its Picard group in the same way. In general we define fundamental roots

$$\rho_1 = \epsilon_0 - \epsilon_1 - \epsilon_2 - \epsilon_3, \qquad \rho_i = \epsilon_{i-1} - \epsilon_i \ (2\le i\le 9).$$

If S has a fibre of type II^*, we have a root system of type E_8, which may be taken to have basis $(\rho_i | 1\le i\le 8)$. The remaining component of the fibre then has class ρ_9 (since the sum of all components, with multiplicities, is κ). Any section is orthogonal to ρ_i for $1\le i\le 8$ and has intersection number 1 with ρ_9. Its class is thus of the form $\epsilon_9 + c\kappa$: since it has self-intersection -1, we have $c = 0$.

For a fibre of type III^* we start with the root system $(\rho_i | 1\le i\le 7)$. The remaining component C of the fibre has class $\epsilon_0 - \epsilon_1 - \epsilon_8 - \epsilon_9$. Calculating as above, we find that a section meeting C has class

$$\sigma_b = \epsilon_9 + b(\rho_9 + \kappa) + b^2\kappa$$

for some $b \in \mathbf{Z}$, while one meeting the ρ_7 component has class

$$\tau_b = \epsilon_7 + b\rho_9 + b^2\kappa \qquad (b \in \mathbf{Z}).$$

Consider the group structure on the sections, with τ_0 as zero. Reducing $\kappa^\perp$ modulo the components of the fibre gives a group spanned by ρ_8 and ρ_9 with $2\rho_8 + \rho_9 = 0$. We have

$$[\tau_b - \tau_0] = b\rho_9 = -2b\rho_8,$$

$$[\sigma_b - \tau_0] = b\rho_9 + \epsilon_9 - \epsilon_7 = b\rho_9 - \rho_8 - \rho_9 = (1-2b)\rho_8,$$

exhibiting the group structure. The intersection numbers are

$$\sigma_b.\sigma_c = \tau_b.\tau_c = (b-c)^2 - 1,$$

$$\sigma_b.\tau_c = (b-c)^2 - (b-c).$$

Thus for two sections σ and τ with $[\sigma-\tau] = \delta\rho_8$, we have

$$\sigma.\tau = \delta^2/4 - 1 \quad (\delta \text{ even}); \qquad (\delta^2-1)/4 \quad (\delta \text{ odd}).$$

In particular, $\sigma.\tau = 0$ if and only if $\delta = \pm 1$ or ± 2.

If there is a further exceptional fibre, the additional root system must have type A_1, and we can take the root as ρ_9, with the other component having class $\kappa - \rho_9$. Sections must have intersection 0 or 1 with ρ_9. This allows only τ_0 (product 0) and σ_0 (product 1). The group structure is trivially determined, and these two sections are disjoint.

The case of a fibre of type IV^* is more complicated. We start with the root system $\{\rho_i | 1 \le i \le 6\}$, and calculate the remaining component C of the fibre to have class $\gamma = \epsilon_0 - \epsilon_7 - \epsilon_8 - \epsilon_9$: denote the classes of the other components of multiplicity 1 by $\alpha = \rho_2$, $\beta = \rho_6$. We find that the sections meeting these respective components have classes

$$\alpha_{a,b} = (\epsilon_0 - \epsilon_1 - \epsilon_9) + a\rho_8 + b(\rho_9 - \kappa) + (a^2 - ab + b^2)\kappa,$$

$$\beta_{a,b} = \epsilon_6 + a\rho_8 + b\,\rho_9 + (a^2 - ab + b^2)\kappa,$$

$$\gamma_{a,b} = \epsilon_9 + a\rho_8 + b(\rho_9 + \kappa) + (a^2 - ab + b^2)\kappa.$$

As to the group structure, reducing $\kappa^{\perp}$ modulo the components of the fibre gives a group spanned by ρ_7, ρ_8 and ρ_9 with the relation $3\rho_7 + 2\rho_8 + \rho_9 = 0$. Rather than substitute for ρ_9 here - which destroys the symmetry - we substitute for ρ_7. We find

$$[\alpha_{a,b} - \beta_{0,0}] = (a + \tfrac{1}{3})\ [\rho_8] + (b + \tfrac{2}{3})\ [\rho_9],$$

$$[\beta_{a,b} - \beta_{0,0}] = a\ [\rho_8] + b\ [\rho_9],$$

$$[\gamma_{a,b} - \beta_{0,0}] = (a - \tfrac{1}{3})\ [\rho_8] + (b - \tfrac{2}{3})\ [\rho_9].$$

As before, the intersection number of two sections σ and τ depends only on the class of their difference

$$[\sigma - \tau] = a\rho_8 + b\rho_9:$$

the square of the right hand side here is $-2L = -2(a^2 - ab - b^2)$. We have $\sigma.\tau = L-1$ if $a, b \in \mathbf{Z}$ and $\sigma.\tau = L - \frac{1}{3}$ otherwise. We note that ρ_8 and ρ_9 generate a hexagonal lattice, and that adjoining $\frac{1}{3}(\rho_8 - \rho_9)$ gives another such lattice.

The sections disjoint from a given one thus correspond to its first and second neighbours in the lattice, as in the figure.

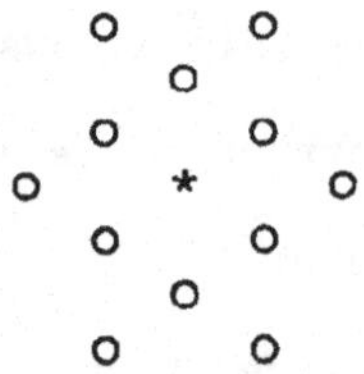

There are two types of maximal set of mutually disjoint sections, represented by the diagrams

(D1) 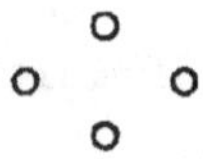 (D2) 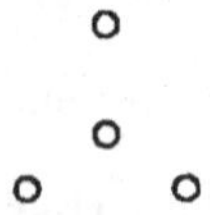

If there is an additional fibre of A_1 type, we can take its components to have classes ρ_8 and $\kappa - \rho_8$. We must then restrict the above sections to have product with ρ_8 equal to 0 or 1. Observe that each of $\alpha_{a,b}$, $\beta_{a,b}$ and $\gamma_{a,b}$ has product $(b - 2a)$ with

ρ_8. We have

$$[\alpha_{a,2a} - \beta_{0,0}] = (-2-6a)\,[\rho_7],$$
$$[\alpha_{a,2a+1} - \beta_{0,0}] = (-5-6a)\,[\rho_7],$$
$$[\beta_{a,2a} - \beta_{0,0}] = (-6a)\,[\rho_7],$$
$$[\beta_{a,2a+1} - \beta_{0,0}] = (-3-6a)\,[\rho_7],$$
$$[\gamma_{a,2a} - \beta_{0,0}] = (2-6a)\,[\rho_7],$$
$$[\gamma_{a,2a+1} - \beta_{0,0}] = (-1-6a)\,[\rho_7],$$

exhibiting the group structure and the projections on the discriminant groups. If $[\sigma-\tau] = m[\rho_7]$, then $\sigma.\tau$ is the integer part of $(m^2-1)/12$; thus $\sigma.\tau = 0$ if and only if $0 < |m| < 3$. A maximal set of mutually disjoint sections corresponds to a set of 4 consecutive elements of $\mathbb{Z}$.

Finally suppose there is, as well as the fibre of type IV^*, one of type A_2 (i.e. I_3 or IV). We can then take its components to have classes ρ_8, ρ_9 and $\kappa - \rho_8 - \rho_9$. A section in this case must have intersection number 1 with one of these, 0 with the other two. Since

$$\rho_9.\alpha_{a,b} = -1 + (a - 2b),$$
$$\rho_9.\beta_{a,b} = (a - 2b), \text{ and}$$
$$\rho_9.\gamma_{a,b} = 1 + (a - 2b),$$

we have just 3 solutions: $\alpha_{-1,-1}$ meeting ρ_8, $\gamma_{0,0}$ meeting ρ_9 and $\beta_{0,0}$ meeting $(\kappa - \rho_8 - \rho_9)$. These three are mutually disjoint.

§6 Direct methods

As an alternative to the approach via root systems in §5, we can proceed by examining the equation of each cubic in the pencil and determining its type and whether or not it is singular at a base point of the pencil. Such calculations rapidly become cumbersome, but it is not too hard to decide whether a cubic is reducible (so we can decide all fibres except those of type II). If the equation is in Weierstrass form, it is also easy to decide whether there is a cusp.

We now summarise the results we have obtained by such calculations. We begin with the case of unstable (regular) pencils. The first cases were worked out in [10].

(29) $f = x^3 + y^3 + 2axyz + (bx + cy)z^2, \quad g = z^3.$

Write $\Delta = a^8 - 6a^4bc - 4a^2(b^3 + c^3) - 3b^2c^2$. We have one fibre of type IV^*: the other exceptional fibres are:

I_1^4	$(\Delta \neq 0,\ b^3 \neq c^3)$	$I_2I_1^2$	$(\Delta \neq 0,\ b^3 = c^3 \neq 0)$
$II.I_1^2$	$(\Delta = 0,\ b^3 \neq c^3,\ a \neq 0)$	II^2	$(a = b = 0,\ c \neq 0)$
$II.I_2$	$(b^3 = c^3 = a^6/27 \neq 0)$	$III.I_1$	$(b^3 = c^3 = -a^6 \neq 0)$
I_3I_1	$(a \neq 0,\ b = c = 0)$	IV	$(a = b = c = 0)$.

(30) $f = x^2y + y^2z + (bx + cy)z^2, \quad g = z^3.$

The exceptional fibres have types III^* and:

I_1^3	$(b \neq 0,\ 4c^3 + 27b^2 \neq 0)$	$II.I_1$	$(b \neq 0,\ 4c^3 + 27b^2 = 0)$
$I_2.I_1$	$(b = 0,\ c \neq 0)$	III	$(b = c = 0)$.

(31) $f = x^3 + y^2z + axz^2, \quad g = z^3.$

We have exceptional fibres of types II^* and

I_1^2 (a ≠ 0) II (a = 0).

For the remaining cases we indicate the methods used in calculations.

(27) $f = xy^2 + x^2z + bxyz + cxz^2 + dz^3, \quad g = yz^2.$

The fibre λ=0 has type IV^*. For any cubic to be singular at a base point of the pencil we need $c^2 = 4d$ (defining case (27b)). For a cubic $\lambda f + \mu g$ to fctorise we need $d = 0$ and $\mu = 0$. Thus cases are:

(27a) ($c^2 \neq 4d$)

I_1^4 or $II.I_1^2$ or II^2 ($d \neq 0$, $b^2 \neq 4c$)

$I_2.I_1^2$ or $I_2.II$ ($d = 0$, $b^2 \neq 4c$) $III.I_1$ ($b^2 = 4c$)

(27b) ($c^2 = 4d$)

$I_2.I_1^2$ or $I_2.II$ ($d \neq 0$, $b^2 \neq -2c$) $III.I_1$ ($d \neq 0$, $b^2 = -2c$)

$I_3.I_1$ ($c = d = 0$, $b \neq 0$) IV ($b = c = d = 0$)

(28) $f = x^2z + y^3 + ay^2z + bz^3, \quad g = yz^2.$

Here the cubics are in Weierstrass form, so never reducible. We have case (28b) (singular point at a base point) when $b = 0$. The fibre λ=0 has type III^*: the rest are

I_1^3 ($b \neq 0$, $a^3 \neq 27b$) $II.I_1$ ($b \neq 0$, $a^3 = 27b$)

$I_2.I_1$ ($b = 0$, $a \neq 0$) III ($b = a = 0$).

(20) $f = x^3 + axyz + bxz^2 + y^2z + cz^3, \quad g = yz^2.$

We can put $f + \mu g$ in Weierstrass form by substituting $y' = y + (ax + \mu z)/2$, so it is never reducible. We have a singular point at a base point when $4b^3 + 27c^2 = 0$. Write

$$c_0 = \frac{b^2}{a^2} - \frac{a^2 b}{2^3 3} + \frac{7a^6}{2^8 3^3}$$

(20a) $(4b^3 + 27c^2 \neq 0)$

I_1^4 $(a \neq 0,\ c \neq c_0$ or $a = 0,\ b \neq 0)$ $\quad$ $II.I_1^2$ $(a \neq 0,\ c = c_0)$

II^2 $(a = b = 0,\ c \neq 0)$

(20b) $(4b^3 + 27c^2 = 0,\ c \neq 0)$. Write $b = -3t^2$, $c = 2t^3$.

$I_2.I_1^2$ $(a^2 \neq 12t,\ c \neq c_0)$ $\qquad$ $II.I_2$ $(c = c_0)$

$III.I_1$ $(a^2 = 12t)$. Note that if $a^2 = 12t$, c_0 reduces to $4t^3$.

(20c) $(b = c = 0)$

$I_3.I_1$ $(a \neq 0)$ $\qquad$ IV $(a = 0)$.

We proceed to the first few semistable cases.

I_4^* (V_6V_7) $\qquad$ $f = xy^2 + ax^2z + bz^3$, $\quad g = xz^2$ $\quad (a \neq 0,\ b \neq 0)$.

In Weierstrass form. $f + \mu g$ singular when $\mu^2 = 4ab$.

I_1^2 (only case).

I_3^* (V_6V_8) $\quad$ $f = xy^2 + ax^3 + bx^2z + cz^3$, $g = xz^2$, $(a \neq 0,\ c \neq 0)$.

Again in Weierstrass form; $f + \mu g$ never has a node on $z = 0$.

I_1^3 $(b^3 \neq 27a^2c)$ $\qquad$ $II.I_1$ $(b^3 = 27a^2c)$.

I_3^* (V_5V_7) $\quad$ $f = xy^2 + ax^2z + bz^3 + cyz^2$, $g = xz^2$, $(a \neq 0,\ c \neq 0)$.

Here $f + \mu g$ never has a node on $x = 0$.

I_1^3 $(16b^3 \neq 27ac^4)$ $\qquad$ $II.I_1$ $(16b^3 = 27ac^4)$.

I_2^* (V_6V_9) $\quad$ $f = -xy^2 + dx^3 + 3cx^2z + bz^3$, $\quad g = 3(xz^2 + ex^3)$,

with $b \neq 0$, $e \neq 0$. We have chosen this in Weierstrass form. There is a singular point of some $f + \mu g$ on $g = 0$, necessarily on $x = 0$, when $B = e(3c - be)^2 + d^2 = 0$. Write also $A = bd^2 - c(3be - c)^2$.

If $A = 0$, set $c = bt^2$, $d = bt^3 - 3bet$.

I_1^4 ($A \neq 0$, $B \neq 0$). $\quad$ $II.I_1^2$ ($A = 0$, $e \neq -t^2, t^2/3$)

$III.I_1$ ($A = 0$, $e = -t^2$ (so $B = 0$)) $\quad$ II^2 ($A = 0$, $e = t^2/3$)

$I_2.I_1^2$ ($B = 0$, $d \neq 0$, $c + be \neq 0$) $\quad$ I_2^2 ($B = 0$, $d = 0$).

I_2^* (V_5V_8) $\quad f = xy^2 + ax^3 + bx^2z + cz^3 + dyz^2, \quad g = xz^2,$

with $a \neq 0$, $d \neq 0$. There is no singular point of any $f + \mu g$ on $x = 0$. $f + \mu g$ is reducible if ($c = 0$ case) $b = 0$, $\mu^2 = -ad^2$, ($c \neq 0$ case) $4ac^2 + b^2d^2 = 0$, $-\mu = (bd^4 + 2c^3)/2cd^2$.

I_1^4 or $I_1^2.II$ or II^2 ($4ac^2 + b^2d^2 \neq 0$)

$I_2.I_1^2$ ($4ac^2 + b^2d^2 = 0$, $c \neq 0$, $c^3 \neq bd^4$)

$III.I_1$ ($4ac^2 + b^2d^2 = 0$, $c \neq 0$, $c^3 = bd^4$)

I_2^2 ($b = c = 0$).

I_2^* (V_4V_7) $\quad f = xy^2 + ax^2z + bz^3 + cyz^2 + dy^2z, \quad g = xz^2,$

with $a \neq 0$, $d \neq 0$. There is a singular point of $f + \mu g$ on $x = 0$ if $c^2 = 4bd$ ($c = 2dt$, $b = dt^2$, $\mu = -t^2$). $f + \mu g$ is reducible if $c = 0$ ($\mu = ad + bd^{-1}$).

I_1^4 or $I_1^2.II$ or II^2 ($c^2 \neq 4bd$, $c \neq 0$)

$I_2.I_1^2$ ($c^2 = 4bd \neq 0$, $b \neq -ad^2$ or $c = 0$, $b \neq 0, ad^2$)

$III.I_1$ ($c^2 = 4bd$, $b = -ad^2$ or $c = 0$, $b = ad^2$)

I_2^2 ($b = c = 0$).

REMARK We have seen that all combinations of fibres compatible with root systems containing E_6, E_7, E_8, D_6, D_7 or D_8 occur, except $I_4^*.II$, $I_2^*.I_2.II$ which do not.

Obtaining an explicit list of configurations of singular fibres was one of the original motivations for our work. However, this has now been achieved more efficiently by Persson [7]. See also Miranda [3] for an approach closer to that of the present article.

§7. Semistable singular pencils

As we pointed out in §4, the family exhibited there contains all semistable pencils. We also saw which members of this family were singular, viz.:

(α) C is contained in the sector OV_1V_6,

(β) C is contained in the sector OV_7V_{10},

(γ) $C = V_2V_7$, $a_2a_7 = 1$,

(β·) $C = V_3V_8$, $a_3^2a_8 = -1$, or $C = OV_3V_6V_8$, $a_3^2a_8 = -1$, $a_3^2a_7 = -a_4$, $a_5 = a_6 = 0$, or $C = OV_4V_5V_9$, $a_4^2a_9 = -1$, $a_5 = 0$, $a_4^2a_7 + a_6 = a_8a_4^3$.

However these cases (β·) were "accidental", and all the same isomorphism classes appear under (β): we will not consider (β·) further.

Recall that pencils of type (γ) consist of a fixed conic and a pencil of lines. If the conic is singular, this is a subcase of (β). If the vertex of the pencil of lines lies on the conic, we have a subcase of (α). Otherwise we have a unique isomorphism class, and this is the one which appears above.

We turn to case (β), where we have

$$\Gamma = xS = x(y^2 + a_7xz + a_8x^2),$$

$$\Gamma\cdot = xS\cdot = x(z^2 + a_{10}xy + a_9x^2).$$

The common line x does not pass through any point of $S \cap S'$. Conversely, if we take any pencil of conics, and add a fixed line not passing through a base point of the pencil, we have one of the above cases. For the pencil on the fixed line can be reduced to the normal form $\langle y^2, z^2 \rangle$, and substitutions $y \longrightarrow y + ax$, $z \longrightarrow z + bx$ will then eliminate terms in xy, xz in S, S' respectively.

We have thus characterised the semistable subcases of (β), and observe that for the remaining (unstable) cases there is a common double point of Γ and Γ' at which they have a common tangent (even a common component) - a subcase of (U5) of (2.1).

We can recognise the type of the pencil $\langle S, S' \rangle$ by forming its discriminant, which reduces to

$$-a_7^2\lambda^3 + 2a_8\lambda^2\mu + 2a_9\lambda\mu^2 - a_{10}^2\mu^3.$$

Repeated lines in the pencil can only occur at $\lambda = 0$ (or $\mu = 0$), and then only if $a_9 = a_{10} = 0$ (or $a_7 = a_8 = 0$). If neither λ nor μ is a repeated factor of the discriminant, the Segre symbol of the pencil is (1,1,1), (2,1) or (3) depending on coincidences of the roots; if λ or μ has multiplicity 2 resp. 3 as root, we have Segre symbol ((1,1),1) resp. ((2,1)); if the discriminant vanishes identically, we have symbol (;2). Notice that the appearance of λ or μ as repeated root corresponds to restriction of the sector C, so there is some analogy with the pattern of results in §4.

We now consider case (α):

$$\Gamma = xy^2 + a_4y^2z + a_5yz^2 + a_6z^3,$$
$$\Gamma' = xz^2 + a_1y^3 + a_2y^2z + a_3yz^2.$$

We first observe that the leading terms at the common double point $y = z = 0$ are xy^2, xz^2. Given any Γ, Γ' with a common double point, the leading terms define a pencil of quadratics which can be reduced to one of:

$$\text{(i) } \langle y^2, z^2\rangle \quad \text{(ii) } \langle y^2, yz\rangle \quad \text{(iii) } \langle yz, 0\rangle$$

$$\text{(iv) } \langle y^2, 0\rangle \quad \text{(v) } \langle 0,0\rangle,$$

each of these being a specialisation of the preceding one. Note that (ii) is the same as (U5) of (2.1), so this and later cases are not semistable. In fact, (iii) is the same as (U4) of (2.1), which is thus superfluous. On the other hand, any two cubics in case (i) can be written in the above form, using substitutions $x' = x + ay + bz$ to eliminate the terms y^3 in Γ, z^3 in Γ'. We can summarise these conclusions as follows.

PROPOSITION A singular pencil is semistable if and only if it is either

(α) with a common double point $y = z = 0$ and leading terms there $\langle xy^2, xz^2\rangle$,

(β) a pencil of conics with a fixed line not passing through a base point of the pencil, or

(γ) a fixed nonsingular conic, with a pencil of lines whise vertex is not on the conic.

A singular pencil which is not semistable has a common double point $y = z = 0$, where the leading terms are one of (ii) - (v) above.

It remains to analyse the cases arising under (α). Set

$$\begin{aligned}\Delta(y,z) &= y^2g - z^2f \\ &= a_1y^5 + a_2y^4z + a_3y^3z^2 - a_4y^2z^3 - a_5yz^4 - a_6z^5,\end{aligned}$$

a generic quintic in y and z. The cubic $f = 0$ is

cuspidal if $a_6 \neq 0$,

conic + tangent if $a_6 = 0$, $a_5 \neq 0$,

repeated line + line if $a_6 = a_5 = 0$,

and similarly for $g = 0$. For a general member $g = t^2 f$ of the pencil ($t \neq 0$) we have a node at $y = z = 0$. It is irreducible unless $\Delta(1,t)$ or $\Delta(1,-t)$ vanishes; a conic + secant if just one of these is zero; a triangle if both vanish. The types of all cubics in the pencil can thus be read off from the roots of Δ, together with their behaviour under $t \longrightarrow -t$.

We also observe that there are 4 moduli in this classification, arising from the roots of Δ with the preferred points $y = 0$, $z = 0$ in $P^1(\mathbf{C})$. Note finally that the above analysis shows that the types of $f = 0$ and $g = 0$ are determined by the cone C, again in analogy to the results of §4.

§8. Unstable singular pencils

There are numerous, increasingly degenerate cases here. It is possible to make lists of cases, and to apply theory, but no coherent pattern emerges, so we confine ourselves to a few remarks.

We have already given a provisional grouping into families: those with a common singular point $x = y = 0$ at which the terms linear in z are

(ii) $\langle y^2, yz\rangle$ (iii) $\langle yz, 0\rangle$ (iv) $\langle y^2, 0\rangle$ (v) $\langle 0,0\rangle$.

A much finer grouping can be obtained from invariant theory: one picks for each pencil P that 1-parameter group which not only selects coefficients showing P to be unstable, but also so that the plane π separating nonzero coefficients from 0 is as far from 0 as possible. This is related to, but different from the other finer groupings below (though for unstable regular pencils it does recover the same list of 6 main types as §5).

We can also attempt directly to list in each case the type (nodal cubic, for example) of the generic member of P, and the types of all non-generic members. One quickly finds that cases need to be subdivided according as the cubics contain a common line or not. For those where there is, we have a pencil of conics plus a line in the plane of the pencil: such pairs are easily listed (see e.g. [9]).

Finally, case (v) is that of pencils of binary cubics, where the classification is well known (see e.g. [6]).

References

1 K. Kodaira, On compact analytic surfaces II, Ann. of Math. 77 (1963) 563-626.

2 R. Miranda, On the stability of pencils of cubic curves, Amer. J. Math. 102 (1980) 1177-1202.

3 R. Miranda, Some remarks on Persson's list, preprint, Colorado State University, 1988.

4 I. Naruki, Configurations related to maximal rational elliptic surfaces, pp 314-340 in "Complex analytic singularities", Adv. Study in Pure Math. 8, Kinokuniya, 1986.

5 I. Naruki, K3 surfaces attached to root systems in E_8, Preprint, RIMS, 1987.

6 P.E. Newstead, Invariants of pencils of binary cubics, Math.Proc. Camb. Phil. Soc. 89 (1981) 201-209.

7 U. Persson, Configurations of Kodaira fibres on rational elliptic surfaces, preprint, Royal Institute of Technology, Stockholm, 1988.

8 T. Shioda, On elliptic modular surfaces, J. Math. Soc. Japan 25 (1972) 20-59.

9 C.T.C. Wall, Nets of quadrics and theta characteristics of singular curves, Phil. Trans. Roy. Soc. 289 (1978) 229-269.

10 C.T.C. Wall, Affine cubic functions I: the complex plane, Math.Proc. Camb. Phil. Soc. 85 (1979) 387-401.

11 C.T.C. Wall, Geometric invariant theory of linear systems, Math.Proc. Camb. Phil. Soc. 93 (1983) 57-62.

12 C.T.C. Wall, Geometric structures on compact complex analytic surfaces, Topology 25 (1986) 119-153.

ADDRESSES OF CONTRIBUTORS

D.Armbruster	Department of Applied Mathematics, Arizona State University, Tempe, AZ 85287-1804, USA
P.J.Aston	Department of Mathematics, University of Surrey, Guildford GU2 5XH, United Kingdom
David Chillingworth	Department of Mathematics, University of Southampton, Southampton SO9 5NH, United Kingdom
P.Chossat	Université de Nice, Laboratoire de Mathématiques (UA CNRS 168), Parc Valrose, F-06034 Nice Cedex, France
J.D.Crawford	Department of Physics and Astronomy, University of Pittsburgh, Pittsburgh, PA 15260, USA
James Damon	Department of Mathematics, University of North Carolina, Chapel Hill, NC 27599, USA
G.Dangelmayr	Institüt für Informationsverarbeitung, Universität Tübingen, Köstlinstrasse 6, D-7400 Tübingen 1, F R Germany
Odo Diekmann	Centre for Mathematics and Computer Science, PO BOX 4079, 1009 AB Amsterdam, The Netherlands
Mike Field	Department of Pure Mathematics, University of Sydney, Sydney NSW 2006, Australia
J.E.Furter	Mathematics Institute, University of Warwick, Coventry CV4 7AL, United Kingdom
Stephan A. van Gils	Department of Applied Mathematics, University of Twente, PO BOX 217, 7500 AE Enschede, The Netherlands
M.Golubitsky	Department of Mathematics, University of Houston, Houston, TX 77204-3476, USA
M.G.M.Gomes	Mathematics Institute, University of Warwick, Coventry CV4 7AL, United Kingdom
S. Janeczko	Instityut Matematyki PW, Pl. Jednošci Robotniczej 1, 00661 Warszawa, Poland
E.Knobloch	Department of Physics, University of California at Berkeley, Berkeley, CA 94720, USA
Reiner Lauterbach	Insitut für Mathematik, Universität Augsburg, Universitätsstrasse 8, D-8900 Augsburg, F R Germany

James Montaldi	Mathematics Institute, University of Warwick, Coventry CV4 7AL, United Kingdom
Irene M. Moroz	School of Mathematics, University of East Anglia, Norwich NR4 7TJ, United Kingdom
Martin Peters	Universität Karlsruhe, Fakultät für Informatik, Institut für Algorithmen and kognitive Systeme, Postfach 6980, D-7500 Karlsruhe 1, F R Germany
Mark Roberts	Mathematics Institute, University of Warwick, Coventry CV4 7AL, United Kingdom
Ian Stewart	Mathematics Institute, University of Warwick, Coventry CV4 7AL, United Kingdom
Yieh-Hei Wan	Department of Mathematics, State University of New York at Buffalo, 106 Diefendorf Hall, Buffalo, NY 14214-3093, USA
M.Wegelin	Institüt für Informationsverarbeitung, Universität Tübingen, Köstlinstrasse 6, D-7400 Tübingen 1, F R Germany

Vol. 1369: B. Jiang, Ch.-K. Peng, Z. Hou (Eds.), Differential Geometry and Topology. Proceedings, 1986–87. VI, 366 pages. 1989.

Vol. 1370: G. Carlsson, R.L. Cohen, H.R. Miller, D.C. Ravenel (Eds.), Algebraic Topology. Proceedings, 1986. IX, 456 pages. 1989.

Vol. 1371: S. Glaz, Commutative Coherent Rings. XI, 347 pages. 1989.

Vol. 1372: J. Azéma, P.A. Meyer, M. Yor (Eds.), Séminaire de Probabilités XXIII. Proceedings. IV, 583 pages. 1989.

Vol. 1373: G. Benkart, J.M. Osborn (Eds.), Lie Algebras. Madison 1987. Proceedings. V, 145 pages. 1989.

Vol. 1374: R.C. Kirby, The Topology of 4-Manifolds. VI, 108 pages. 1989.

Vol. 1375: K. Kawakubo (Ed.), Transformation Groups. Proceedings, 1987. VIII, 394 pages, 1989.

Vol. 1376: J. Lindenstrauss, V.D. Milman (Eds.), Geometric Aspects of Functional Analysis. Seminar (GAFA) 1987–88. VII, 288 pages. 1989.

Vol. 1377: J.F. Pierce, Singularity Theory, Rod Theory, and Symmetry-Breaking Loads. IV, 177 pages. 1989.

Vol. 1378: R.S. Rumely, Capacity Theory on Algebraic Curves. III, 437 pages. 1989.

Vol. 1379: H. Heyer (Ed.), Probability Measures on Groups IX. Proceedings, 1988. VIII, 437 pages. 1989.

Vol. 1380: H.P. Schlickewei, E. Wirsing (Eds.), Number Theory, Ulm 1987. Proceedings. V, 266 pages. 1989.

Vol. 1381: J.-O. Strömberg, A. Torchinsky, Weighted Hardy Spaces. V, 193 pages. 1989.

Vol. 1382: H. Reiter, Metaplectic Groups and Segal Algebras. XI, 128 pages. 1989.

Vol. 1383: D.V. Chudnovsky, G.V. Chudnovsky, H. Cohn, M.B. Nathanson (Eds.), Number Theory, New York 1985–88. Seminar. V, 256 pages. 1989.

Vol. 1384: J. Garcia-Cuerva (Ed.), Harmonic Analysis and Partial Differential Equations. Proceedings, 1987. VII, 213 pages. 1989.

Vol. 1385: A.M. Anile, Y. Choquet-Bruhat (Eds.), Relativistic Fluid Dynamics. Seminar, 1987. V, 308 pages. 1989.

Vol. 1386: A. Bellen, C.W. Gear, E. Russo (Eds.), Numerical Methods for Ordinary Differential Equations. Proceedings, 1987. VII, 136 pages. 1989.

Vol. 1387: M. Petkovi´c, Iterative Methods for Simultaneous Inclusion of Polynomial Zeros. X, 263 pages. 1989.

Vol. 1388: J. Shinoda, T.A. Slaman, T. Tugué (Eds.), Mathematical Logic and Applications. Proceedings, 1987. V, 223 pages. 1989.

Vol. 1000: Second Edition. H. Hopf, Differential Geometry in the Large. VII, 184 pages. 1989.

Vol. 1389: E. Ballico, C. Ciliberto (Eds.), Algebraic Curves and Projective Geometry. Proceedings, 1988. V, 288 pages. 1989.

Vol. 1390: G. Da Prato, L. Tubaro (Eds.), Stochastic Partial Differential Equations and Applications II. Proceedings, 1988. VI, 258 pages. 1989.

Vol. 1391: S. Cambanis, A. Weron (Eds.), Probability Theory on Vector Spaces IV. Proceedings, 1987. VIII, 424 pages. 1989.

Vol. 1392: R. Silhol, Real Algebraic Surfaces. X, 215 pages. 1989.

Vol. 1393: N. Bouleau, D. Feyel, F. Hirsch, G. Mokobodzki (Eds.), Séminaire de Théorie du Potentiel Paris, No. 9. Proceedings. VI, 265 pages. 1989.

Vol. 1394: T.L. Gill, W.W. Zachary (Eds.), Nonlinear Semigroups, Partial Differential Equations and Attractors. Proceedings, 1987. IX, 233 pages. 1989.

Vol. 1395: K. Alladi (Ed.), Number Theory, Madras 1987. Proceedings. VII, 234 pages. 1989.

Vol. 1396: L. Accardi, W. von Waldenfels (Eds.), Quantum Probability and Applications IV. Proceedings, 1987. VI, 355 pages. 1989.

Vol. 1397: P.R. Turner (Ed.), Numerical Analysis and Parallel Processing. Seminar, 1987. VI, 264 pages. 1989.

Vol. 1398: A.C. Kim, B.H. Neumann (Eds.), Groups – Korea 1988. Proceedings. V, 189 pages. 1989.

Vol. 1399: W.-P. Barth, H. Lange (Eds.), Arithmetic of Complex Manifolds. Proceedings, 1988. V, 171 pages. 1989.

Vol. 1400: U. Jannsen. Mixed Motives and Algebraic K-Theory. XIII, 246 pages. 1990.

Vol. 1401: J. Steprans, S. Watson (Eds.), Set Theory and its Applications. Proceedings, 1987. V, 227 pages. 1989.

Vol. 1402: C. Carasso, P. Charrier, B. Hanouzet, J.-L. Joly (Eds.), Nonlinear Hyperbolic Problems. Proceedings, 1988. V, 249 pages. 1989.

Vol. 1403: B. Simeone (Ed.), Combinatorial Optimization. Seminar, 1986. V, 314 pages. 1989.

Vol. 1404: M.-P. Malliavin (Ed.), Séminaire d´Algèbre Paul Dubreil et Marie-Paul Malliavin. Proceedings, 1987–1988. IV, 410 pages. 1989.

Vol. 1405: S. Dolecki (Ed.), Optimization. Proceedings, 1988. V, 223 pages. 1989. Vol. 1406: L. Jacobsen (Ed.), Analytic Theory of Continued Fractions III. Proceedings, 1988. VI, 142 pages. 1989.

Vol. 1407: W. Pohlers, Proof Theory. VI, 213 pages. 1989.

Vol. 1408: W. Lück, Transformation Groups and Algebraic K-Theory. XII, 443 pages. 1989.

Vol. 1409: E. Hairer, Ch. Lubich, M. Roche. The Numerical Solution of Differential-Algebraic Systems by Runge-Kutta Methods. VII, 139 pages. 1989.

Vol. 1410: F.J. Carreras, O. Gil-Medrano, A.M. Naveira (Eds.), Differential Geometry. Proceedings, 1988. V, 308 pages. 1989.

Vol. 1411: B. Jiang (Ed.), Topological Fixed Point Theory and Applications. Proceedings. 1988. VI, 203 pages. 1989.

Vol. 1412: V.V. Kalashnikov, V.M. Zolotarev (Eds.), Stability Problems for Stochastic Models. Proceedings, 1987. X, 380 pages. 1989.

Vol. 1413: S. Wright, Uniqueness of the Injective III_1 Factor. III, 108 pages. 1989.

Vol. 1414: E. Ramirez de Arellano (Ed.), Algebraic Geometry and Complex Analysis. Proceedings, 1987. VI, 180 pages. 1989.

Vol. 1415: M. Langevin, M. Waldschmidt (Eds.), Cinquante Ans de Polynômes. Fifty Years of Polynomials. Proceedings, 1988. IX, 235 pages. 1990.

Vol. 1416: C. Albert (Ed.), Géométrie Symplectique et Mécanique. Proceedings, 1988. V, 289 pages. 1990.

Vol. 1417: A.J. Sommese, A. Biancofiore, E.L. Livorni (Eds.), Algebraic Geometry. Proceedings, 1988. V, 320 pages. 1990.

Vol. 1418: M. Mimura (Ed.), Homotopy Theory and Related Topics. Proceedings, 1988. V, 241 pages. 1990.

Vol. 1419: P.S. Bullen, P.Y. Lee, J.L. Mawhin, P. Muldowney, W.F. Pfeffer (Eds.), New Integrals. Proceedings, 1988. V, 202 pages. 1990.

Vol. 1420: M. Galbiati, A. Tognoli (Eds.), Real Analytic Geometry. Proceedings, 1988. IV, 366 pages. 1990.

Vol. 1421: H.A. Biagioni, A Nonlinear Theory of Generalized Functions, XII, 214 pages. 1990.

Vol. 1422: V. Villani (Ed.), Complex Geometry and Analysis. Proceedings, 1988. V, 109 pages. 1990.

Vol. 1423: S.O. Kochman, Stable Homotopy Groups of Spheres: A Computer-Assisted Approach. VIII, 330 pages. 1990.

Vol. 1424: F.E. Burstall, J.H. Rawnsley, Twistor Theory for Riemannian Symmetric Spaces. III, 112 pages. 1990.

Vol. 1425: R.A. Piccinini (Ed.), Groups of Self-Equivalences and Related Topics. Proceedings, 1988. V, 214 pages. 1990.

Vol. 1426: J. Azéma, P.A. Meyer, M. Yor (Eds.), Séminaire de Probabilités XXIV, 1988/89. V, 490 pages. 1990.

Vol. 1427: A. Ancona, D. Geman, N. Ikeda, École d'Eté de Probabilités de Saint Flour

XVIII, 1988. Ed.: P.L. Hennequin. VII, 330 pages. 1990.

Vol. 1428: K. Erdmann, Blocks of Tame Representation Type and Related Algebras. XV. 312 pages. 1990.

Vol. 1429: S. Homer, A. Nerode, R.A. Platek, G.E. Sacks, A. Scedrov, Logic and Computer Science. Seminar, 1988. Editor: P. Odifreddi. V, 162 pages. 1990.

Vol. 1430: W. Bruns, A. Simis (Eds.), Commutative Algebra. Proceedings. 1988. V, 160 pages. 1990.

Vol. 1431: J.G. Heywood, K. Masuda, R. Rautmann, V.A. Solonnikov (Eds.), The Navier-Stokes Equations – Theory and Numerical Methods. Proceedings, 1988. VII, 238 pages. 1990.

Vol. 1432: K. Ambos-Spies, G.H. Müller, G.E. Sacks (Eds.), Recursion Theory Week. Proceedings, 1989. VI, 393 pages. 1990.

Vol. 1433: S. Lang, W. Cherry, Topics in Nevanlinna Theory. II, 174 pages.1990.

Vol. 1434: K. Nagasaka, E. Fouvry (Eds.), Analytic Number Theory. Proceedings, 1988. VI, 218 pages. 1990.

Vol. 1435: St. Ruscheweyh, E.B. Saff, L.C. Salinas, R.S. Varga (Eds.), Computational Methods and Function Theory. Proceedings, 1989. VI, 211 pages. 1990.

Vol. 1436: S. Xambó-Descamps (Ed.), Enumerative Geometry. Proceedings, 1987. V, 303 pages. 1990.

Vol. 1437: H. Inassaridze (Ed.), K-theory and Homological Algebra. Seminar, 1987–88. V, 313 pages. 1990.

Vol. 1438: P.G. Lemarié (Ed.) Les Ondelettes en 1989. Seminar. IV, 212 pages. 1990.

Vol. 1439: E. Bujalance, J.J. Etayo, J.M. Gamboa, G. Gromadzki. Automorphism Groups of Compact Bordered Klein Surfaces: A Combinatorial Approach. XIII, 201 pages. 1990.

Vol. 1440: P. Latiolais (Ed.), Topology and Combinatorial Groups Theory. Seminar, 1985–1988. VI, 207 pages. 1990.

Vol. 1441: M. Coornaert, T. Delzant, A. Papadopoulos. Géométrie et théorie des groupes. X, 165 pages. 1990.

Vol. 1442: L. Accardi, M. von Waldenfels (Eds.), Quantum Probability and Applications V. Proceedings, 1988. VI, 413 pages. 1990.

Vol. 1443: K.H. Dovermann, R. Schultz, Equivariant Surgery Theories and Their Periodicity Properties. VI, 227 pages. 1990.

Vol. 1444: H. Korezlioglu, A.S. Ustunel (Eds.), Stochastic Analysis and Related Topics VI. Proceedings, 1988. V, 268 pages. 1990.

Vol. 1445: F. Schulz, Regularity Theory for Quasilinear Elliptic Systems and – Monge Ampère Equations in Two Dimensions. XV, 123 pages. 1990.

Vol. 1446: Methods of Nonconvex Analysis. Seminar, 1989. Editor: A. Cellina. V, 206 pages. 1990.

Vol. 1447: J.-G. Labesse, J. Schwermer (Eds), Cohomology of Arithmetic Groups and Automorphic Forms. Proceedings, 1989. V, 358 pages. 1990.

Vol. 1448: S.K. Jain, S.R. López-Permouth (Eds.), Non-Commutative Ring Theory. Proceedings, 1989. V, 166 pages. 1990.

Vol. 1449: W. Odyniec, G. Lewicki, Minimal Projections in Banach Spaces. VIII, 168 pages. 1990.

Vol. 1450: H. Fujita, T. Ikebe, S.T. Kuroda (Eds.), Functional-Analytic Methods for Partial Differential Equations. Proceedings, 1989. VII, 252 pages. 1990.

Vol. 1451: L. Alvarez-Gaumé, E. Arbarello, C. De Concini, N.J. Hitchin, Global Geometry and Mathematical Physics. Montecatini Terme 1988. Seminar. Editors: M. Francaviglia, F. Gherardelli. IX, 197 pages. 1990.

Vol. 1452: E. Hlawka, R.F. Tichy (Eds.), Number-Theoretic Analysis. Seminar, 1988–89. V, 220 pages. 1990.

Vol. 1453: Yu.G. Borisovich, Yu.E. Gliklikh (Eds.), Global Analysis – Studies and Applications IV. V, 320 pages. 1990.

Vol. 1454: F. Baldassari, S. Bosch, B. Dwork (Eds.), p-adic Analysis. Proceedings, 1989. V, 382 pages. 1990.

Vol. 1455: J.-P. Françoise, R. Roussarie (Eds.), Bifurcations of Planar Vector Fields. Proceedings, 1989. VI, 396 pages. 1990.

Vol. 1456: L.G. Kovács (Ed.), Groups – Canberra 1989. Proceedings. XII, 198 pages. 1990.

Vol. 1457: O. Axelsson, L.Yu. Kolotilina (Eds.), Preconditioned Conjugate Gradient Methods. Proceedings, 1989. V, 196 pages. 1990.

Vol. 1458: R. Schaaf, Global Solution Branches of Two Point Boundary Value Problems. XIX, 141 pages. 1990.

Vol. 1459: D. Tiba, Optimal Control of Nonsmooth Distributed Parameter Systems. VII, 159 pages. 1990.

Vol. 1460: G. Toscani, V. Boffi, S. Rionero (Eds.), Mathematical Aspects of Fluid Plasma Dynamics. Proceedings, 1988. V, 221 pages. 1991.

Vol. 1461: R. Gorenflo, S. Vessella, Abel Integral Equations. VII, 215 pages. 1991.

Vol. 1462: D. Mond, J. Montaldi (Eds.), Singularity Theory and its Applications. Warwick 1989, Part I. VIII, 405 pages. 1991.